TRIGONOMETRIC FUNCT

	sin	cos	tan	cot	
0°	.000	1.000	.000	—	90°
5°	.087	.996	.087	11.430	85°
10°	.174	.985	.176	5.671	80°
15°	.259	.966	.268	3.732	75°
20°	.342	.940	.364	2.747	70°
25°	.423	.906	.466	2.144	65°
30°	.500	.866	.577	1.732	60°
35°	.574	.819	.700	1.428	55°
40°	.643	.766	.839	1.192	50°
45°	.707	.707	1.000	1.000	45°
	cos	sin	cot	tan	

Degrees	Radians		Radians	Degrees
5°	.09		0.1	5.73°
10°	.17		0.2	11.46°
15°	.22		0.3	17.19°
20°	.35		0.4	22.92°
25°	.44		0.5	28.65°
30°	.52		0.6	34.38°
35°	.61		0.7	40.11°
40°	.70		0.8	45.84°
45°	.78		0.9	51.57°
50°	.87		1.0	57.23°
55°	.96		1.1	63.03°
60°	1.05		1.2	68.76°
65°	1.13		1.3	74.48°
70°	1.22		1.4	80.41°
75°	1.31		1.5	85.94°
80°	1.40		1.6	91.67°
85°	1.48		2.0	114.59°
90°	1.57		3.0	171.89°

Calculus

Calculus

LEONARD GILLMAN
University of Texas at Austin

ROBERT H. McDOWELL
Washington University

W·W·NORTON & COMPANY·INC·
NEW YORK

COPYRIGHT © 1973 BY W. W. NORTON & COMPANY, INC.

First Edition

Library of Congress Cataloging in Publication Data

Gillman, Leonard.
  Calculus.

  1. Calculus. I. McDowell, Robert H., joint
author. II. Title.
QA303.G48 1973   515   72–13762
ISBN 0–393–09350–6

PRINTED IN THE UNITED STATES OF AMERICA

1 2 3 4 5 6 7 8 9 0

Contents

Contents

Contents

Preface

This book is designed to teach the student to solve calculus problems with facility. It is easy to study from. It avoids special symbols, long computations, and cluttered displays. Every new concept is illustrated by an example, worked out in detail. There are more than 3500 problems (most with answers). The bulk are routine. [If $f(x) = x^3$, what is $f'(x)$?] Others carry a little bite. [If $f(x) = 3x^2$, what is $f'(x^3)$?]

The exposition is relaxed but careful. Proofs that smack of advanced calculus have been set in small type, as material likely to be skipped. In brief, we have tried to make the theory accessible without implying that it must be mastered in order to solve routine problems.

We have taken a fresh look at every concept and every proof. The experienced teacher will recognize many niceties of detail in organization and exposition.

The integral

Our major innovation lies in the development of the integral, which becomes both simpler and more rigorous. We *define* the integral by two of its properties (called "Additivity" and "Betweenness"). This is a distinct departure from custom.

Advantages appear quickly.

The definition is easy to understand. There are no least upper bounds, no Riemann sums, no ϵ's.

The definition is easy to apply. Each integral is set up with a minimum of analytic fuss, and with no vague remarks about "negligible errors when Δx is small." A definition in terms of an integral is never something that "seems reasonable," but in every instance follows as a mathematical necessity from explicitly stated assumptions.

Coverage

The book is designed for a three-semester course, averaging three sections per week. It is organized into five major blocks.

Chapters 1–8 constitute a thorough treatment of the differential and integral calculus of a single variable. They begin with some preliminaries about numbers, functions, and analytic geometry (lines, circles, and parabolas), and carry through to a careful discussion of the exponential, logarithmic, trigonometric, and inverse trigonometric functions.

Chapters 9–11 study the analytic geometry and calculus of vectors in two and three dimensions. Topics in analytic geometry include polar coordinates in the plane and a detailed treatment of lines and planes in space (vector and parametric equations and coordinate equations).

Chapters 12–14 treat functions of several variables, partial derivatives, and multiple integrals.

Chapter 15 is devoted to infinite series.

Chapter 16 is a brief introduction to ordinary differential equations.

Flexibility

After Chapters 1–8, the various blocks may be taken up in almost any order. Hence the book is also well adapted for a one-year course, as recommended, for example, by CUPM: one can easily include particular topics and omit others. Likewise, there are many ways of arranging the material over the entire three semesters.

Every chapter begins with a detailed outline of its contents, followed by introductory comments. Together, they provide a great deal of information helpful in planning.

It is possible to proceed directly to multivariable calculus (Chapters 12–14) after first picking up vector notation and space coordinates in §§9A–B and §§11A–B. (In order to include tangent lines and planes (§§13F, J), one would also have to look first at parts of §§9C–E and §§11C–F.) Incidentally, Chapter 14 (multiple integrals) is independent of Chapter 13 (partial derivatives).

Chapter 15 on infinite series can be inserted at any time. Also, §15D, on L'Hôpital's Rule, is independent of the rest of the chapter.

Chapter 16 on differential equations can be taken up any time (with minor adjustments as mentioned in the chapter introduction).

Chapter 10 on polar coordinates is virtually independent of the rest of the book. Of course, it should be gone over before tackling the double integral in polar coordinates (§14G).

ACKNOWLEDGMENTS

It has been a pleasure to work with Joseph B. Janson II and Mary Pell of Norton.

L.G.
R.H.M.

January, 1973
Austin, Texas
St. Louis, Mo.

Calculus

CHAPTER 1

Numbers and Functions

This chapter consists of background material.

The first four sections are largely review. They should not require much time.

§1A is a brief review of algebra. The main emphasis is on inequalities, which are important in calculus.

In §§1B–1D, we look more closely at linear and quadratic equations and develop some basic analytic geometry. We assume the elementary facts about parallel and perpendicular lines, similarity and congruence of triangles, and the Theorem of Pythagoras.

§§1E–1G are devoted to intervals and functions. The ideas in these sections are used throughout the book. They must be mastered thoroughly.

In §1H we discuss the "completeness" of the set of real numbers. This property underlies many of the theoretical portions of calculus.

§§1E–1G are devoted to intervals and functions. The ideas in these sections are used throughout the book. They must be mastered thoroughly.

§1A. The real numbers

1A1. Elementary algebra. Calculus works with the system of real numbers, which we denote by $\mathscr{R}$. The real numbers are the rational and the irrational numbers, positive, negative, and zero. We recall that a rational number is a quotient of two integers. All integers are rational numbers. Each real number is either rational or irrational.

The reader is assumed to be familiar with elementary algebraic facts, such as the rules of signs or the rules about parentheses, and we will use them freely. We will not hesitate to write

$$(a - b)(c - d) = ac - ad - bc + bd$$

without further explanation.

Multiplications involving the number 0 often cause trouble. The following important facts are stated here for emphasis.

(i) Zero times any number is zero; that is, if a is any number, then $0 \cdot a = 0$.

(ii) If a product is zero, at least one of the factors must be zero; that is, if $ab = 0$ then either $a = 0$, or $b = 0$, or both.

(iii) There is no such thing as dividing by zero. If a is any number, the inkmark $a/0$ has no mathematical meaning. The reason is that no meaning can be assigned to it in any way that is useful, but only in ambiguous or inconsistent ways. So we don't assign any meaning. Division by zero is not defined.

On the other hand, 0 can be divided *by* any number not zero. The result is 0. For, if $b \neq 0$, then

$$\frac{0}{b} = 0 \cdot \frac{1}{b} = 0 \qquad\qquad (b \neq 0),$$

by (i).

1A2. Order. The notion of *order* in $\mathscr{R}$ is also familiar. Every number is either positive (i.e., greater than 0), negative (less than 0), or zero. To say that

$$a > b \qquad (a \text{ is greater than } b)$$

is to say that $a - b$ is positive: $a - b > 0$. The same fact is expressed by

$$b < a \qquad (b \text{ is less than } a),$$

i.e., $b - a$ is negative: $b - a < 0$. The expression

$$a \geq b \qquad (a \text{ is greater than or equal to } b)$$

states that either $a > b$ or $a = b$. The expression

$$b \leq a \qquad (b \text{ is less than or equal to } a)$$

asserts the same fact.

For example,

$$a > 0$$

states that a is positive;

$$a \geq 0$$

states that a is nonnegative, i.e., either positive or zero. Also,

$$3 > 2, \quad 2 < 3; \qquad 3 \geq 2, \quad 2 \leq 3; \qquad 3 \geq 3, \quad 3 \leq 3.$$

1A3. The real line. The connection between numbers and geometry is achieved in the usual way by the introduction of a coordinate system on a line. We pick an *origin* or point of reference and mark it 0, and we pick a positive direction on the line. Each point is then assigned a real number, representing its *directed* distance from the origin—that is, the distance itself if the point lies in the positive direction from the origin, and the negative of the distance otherwise (Figure 1). The assigned number is called the *coordinate* of the point. Each point has a coordinate, and each number is the coordinate of some point.

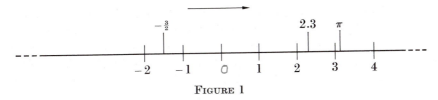

FIGURE 1

Because of this correspondence between points and numbers we speak of them synonymously. The coordinatized line is called the *real line* (in grade school, the "number line"), and is also denoted by $\mathscr{R}$. The point whose coordinate is a is called, for short, the point a.

1A4. Absolute value. The *absolute value* of a number a, denoted $|a|$, is defined to be a or $-a$, whichever is nonnegative. Thus, whatever a may be,

$$|a| \geq 0.$$

For example,

$$|4| = 4, \qquad |-4| = -(-4) = 4, \qquad |0| = 0.$$

In other words, $|a|$ is the distance between a and 0. Clearly, the distance between points a and b is $|a - b|$ (Figure 2).

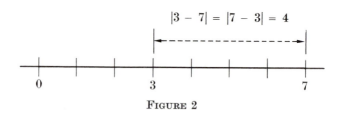

FIGURE 2

1A5. Inequalities. It is important to be able to combine assertions about order by means of the operations of algebra. This is called working with *inequalities*. We present the basic facts here, and provide enough exercises to develop or test the requisite skills.

We know that for each number a, exactly one of the following is true: $a > 0, a = 0, a < 0$. The basic relations with algebraic operations are:

(1) $\qquad a > b \quad means \quad a - b > 0.$

(2) $\qquad If \quad a > 0 \quad and \quad b > 0 \quad then \quad a + b > 0.$

(3) $\qquad If \quad a > 0 \quad and \quad b > 0 \quad then \quad ab > 0.$

We will take these for granted. All other relations follow from them and the rules of algebra. Some are natural enough and cause no difficulty. For instance, if $a > b$ and $b > c$, then $a > c$. (*Proof.* By (1), $a - b > 0$ and $b - c > 0$; by (2), $a - c = (a - b) + (b - c) > 0$; by (1), $a > c$.) Next, if $a > b$ then $-a < -b$. (*Proof.* (1).) In particular, if $a > 0$ then $-a < 0$; if $b < 0$ then $-b > 0$.

There are many others. If $a < 0$ and $b < 0$ then $a + b < 0$. Also, ab and a/b are both positive or both negative according as a and b have the same sign or opposite signs. In particular, squares are positive:

(4) $\qquad If \quad a \neq 0, \quad then \quad a^2 > 0.$

(Hence $1 = 1 \cdot 1 > 0$.) Also,

(5)   *If $a > b > 0$ then $0 < \dfrac{1}{a} < \dfrac{1}{b}$.*

(*Proof.* $\dfrac{1}{b} - \dfrac{1}{a} = \dfrac{a - b}{ab}$, the quotient of two positive numbers.)

A simple but important consequence of (1) is that adding the same number to both sides of an inequality preserves the inequality; that is, if $a > b$ then $a + c > b + c$. (*Proof.* Both statements assert that $a - b > 0$.) But there is no such simple rule about multiplying, and this situation is a major source of error. If $a > b$, is it true that $ca > cb$? *When in doubt, fall back on the basic facts,* (1), (2), *and* (3). The answer here depends on c. Multiplying by a *positive* number *preserves* an inequality:

(6)   *If $a > b$ and $c > 0$ then $ca > cb$.*

(*Proof.* $a - b > 0$ and $c > 0$. By (3), $ca - cb = c(a - b) > 0$.) Multiplying by a *negative* number *reverses* an inequality:

(7)   *If $a > b$ and $c < 0$ then $ca < cb$.*

(The proof is similar to the other.) And if $c = 0$ we get equality: $ca = cb$, because both sides are 0.

1A6. Remark on notation. By notational convention, a succession of equalities or inequalities written in tandem are understood to hold *simultaneously*. For instance,

$$a < b = c \le d$$

means that *$a < b$ and $b = c$ and $c \le d$*;

$$2 < x \le 3$$

means that x is a number between 2 and 3, excluding 2 but not 3. See Figure 3.

FIGURE 3

Figure 4

Inequalities presented as a *choice* have to be stated separately. To show that x is *either* less than 2 *or* greater than 3 (Figure 4), we write

$$x < 2 \quad \text{or} \quad x > 3.$$

We do *not* write $2 > x > 3$. That would mean $x < 2$ *and* $x > 3$, simultaneously, which is not only impossible but different from what we are trying to express.

1A7. Rational exponents; square root. The reader is assumed to be familiar with rational exponents and their properties, which we summarize here for convenience.

Let m and n be positive integers.

If a is any real number, then a^n (the nth power of a) is defined as $a \cdot a \cdot \cdots \cdot a$ (n factors). For $a \neq 0$ we also define $a^0 = 1$ and $a^{-n} = 1/a^n$.

For $a \geq 0$, $a^{1/n}$ (the nth root of a) is that nonnegative real number whose nth power is a. That is, by definition of $a^{1/n}$,

(8) $$a^{1/n} \geq 0 \quad \text{and} \quad (a^{1/n})^n = a.$$

(There is always exactly one such number.) Next, $a^{m/n}$ is defined to be $(a^{1/n})^m$. In case $a > 0$ we also define $a^{-m/n} = 1/a^{m/n}$.

For n *odd*, we define $(-a)^{m/n} = [-(a^{1/n})]^m$.

Square root. For $b \geq 0$ we define $\sqrt{b}$ (the square root of b) as $b^{1/2}$—that is, the nonnegative number whose 2nd power (square) is b. Hence for any number a,

(9) $$\sqrt{a^2} = |a|.$$

Laws of exponents. The following laws of exponents follow from the definitions. For $a \geq 0$ and rational numbers r and s,

(10) $$a^r a^s = a^{r+s} \quad \text{and} \quad (a^r)^s = a^{rs}$$

(where, it is understood, $r > 0$ and $s > 0$ in case $a = 0$).

If, in addition, $b \geq 0$,

(11) $$a^r b^r = (ab)^r$$

(where $r > 0$ in case $ab = 0$).

No familiarity with *irrational* exponents is assumed. They will be introduced in Chapter 7.

Problems (§1A)

1. Verify the following formulas.
   (a) $(a^3 - b^3) = (a - b)(a^2 + ab + b^2)$
   (b) $ab = \frac{1}{4}[(a + b)^2 - (a - b)^2]$
   (c) $(a^2 + b^2)(c^2 + d^2) = (ad - bc)^2 + (ac + bd)^2$

2. True or false?
   (a) $5 < 6$
   (b) $5 \leq 6$
   (c) $5 < -6$
   (d) $-5 < 6$
   (e) $-5 < -6$
   (f) $-5 \leq -6$
   (g) $6 < 5$
   (h) $6 < -5$
   (i) $6 \leq -5$
   (j) $-6 < 5$
   (k) $-6 < -5$
   (l) $-6 \leq -5$

3. True or false? [*True* means that the statement holds for all values of x. *False* means that there is at least one value of x for which the statement does not hold.]
   (a) If $x < 2$ then $x \leq 2$
   (b) If $x \leq 2$ then $x < 2$
   (c) If $x \leq 2$ then $x < 3$
   (d) If $x > 2$ then $-x < 2$
   (e) If $x > 2$ then $-x \leq -2$
   (f) If $x < 2$ then $-x < 2$
   (g) If $x \leq 2$ then $-x > -2$

4. True or false?
   (a) If $x < 2$ and $y < 3$ then $x + y < 5$
   (b) If $x < -4$ and $y < 11$ then $x + y < 7$
   (c) If $x < 2$ and $y < 3$ then $2 + y < 5$
   (d) If $x < 2$ and $y < 3$ then $x + y < 2 + y$
   (e) If $x < 2$ and $y < 3$ then $x < y$
   (f) If $x > 2$ and $y > 3$ then $2y > 6$
   (g) If $x > 2$ and $y > 3$ then $xy > 2y$
   (h) If $x > 2$ and $y < 3$ then $xy > 6$
   (i) If $x < 2$ and $y < 3$ then $xy < 6$
   (j) If $x > 4$ and $y > 5$ then $xy > 20$
   (k) If $x > -4$ and $y > -5$ then $xy > 20$
   (l) If $0 < x < 4$ and $0 < y < 5$ then $xy < 20$
   (m) If $x < 4$ and $1 < y < 5$ then $xy < 20$

5. True or false? If $1 \leq x \leq 2$ then:
   (a) $-2 < x \leq 2$
   (b) $0 < x < 3$
   (c) $-1 \leq x \leq -2$
   (d) $-2 \leq -x \leq -1$
   (e) $0 < x < 2$
   (f) $-1 \leq -x \leq -2$

6. True or false? If $-1 < x < 1$ then:
   (a) $-1 < -x < 1$
   (b) $|x| \leq 1$
   (c) $-1 \leq -x \leq 1$
   (d) $|x| \geq 1$
   (e) $-1 \leq x \leq 1$
   (f) $|x| < 1$
   (g) $0 < x < 1$

7. True or false? If $0 < x < y < 1$ then:
   (a) $xy < 1$
   (b) $\dfrac{1}{x} < 1$
   (c) $1 < \dfrac{1}{y} < \dfrac{1}{x}$
   (d) $0 < \dfrac{1}{y} < \dfrac{1}{x}$
   (e) $x + y < 1$
   (f) $0 < \dfrac{1}{x} < \dfrac{1}{y}$
   (g) $\dfrac{1}{y} > 1$
   (h) $x^2 < 1$
   (i) $x^2 < y^2$
   (j) $x^2 + y^2 < 1$
   (k) $y - x < 1$
   (l) $x - y < 1$
   (m) $xy < x + y$

8. (a) If $ab = ac$, does it follow that $b = c$?
   (b) If $a > b$, does it follow that $a^2 > b^2$?
   (c) If $x^2 > y^2$, does it follow that $x > y$?
   (d) Solve for x:
$$(x - 1)(x + 2) = (x - 1)(x + 1).$$
   (e) Solve for x: $x^3 = x$.
   (f) If $a^2 + b^2 = 0$, what can be said about the possible values of a and b?

9. Let $a < b$. Write down an expression for a number that lies between a and b.

10. (a) Give an example of a number a such that $4a < 3a$.
    (b) Give an example of a number b such that $4b = 3b$.
    (c) Give an example of a number c that is greater than its square.
    (d) Give an example of a number a such that
$$|3 + a| < 3 + |a|.$$
    (e) If $a < 0$ and $a + x > 0$, what can be said about x?
    (f) Give an example of numbers a and b such that $a < b$ *and* $a^2 > b^3$.
    (g) If $3x < 4x$, what can be said about x?
    (h) If $a^{m/2} = b$ and $b^{n/3} = a$, what can be said about m and n?
    (i) If $5^x = 25 \times (5^x)^3$, what is x?
    j) Which numbers have their square equal to their absolute value?
    (k) If $ab = a$, what are the possibilities for a and b?

11. Give an example of numbers a and b such that $a < b$ and
   (a) $a^b < b^a$ (b) $a^b = b^a$ (c) $a^b > b^a$

12. (a) Give an example of two numbers each less than the square of the other.
   (b) Give an example of two numbers each greater than the square of the other.
   (c) Are there two numbers each equal to the square of the other?

13. (a) Can two numbers each be equal to the absolute value of the other?
   (b) Can two numbers each be less than the absolute value of the other?
   (c) Can two numbers each be greater than the absolute value of the other?

14. Prove the following formulas.
   (a) $a^2 + b^2 \geq 2ab$
   (b) $x^2 \geq 2x - 1$
   (c) If $a > 0$ then $a + \dfrac{1}{a} \geq 2$
   (d) If $x > 0$ and $y > 0$ then $\dfrac{x + y}{2} \geq \sqrt{xy}$.

15. (a) Give an example of a number x for which $|x| = -x$.
   (b) Establish the formula $|ab| = |a| \cdot |b|$.

16. (a) Prove that $y \leq |y|$ for every number y.
   (b) Prove that if $x \leq c$ and $-x \leq c$ then $|x| \leq c$.
   (c) Give an example of two numbers x and y such that $|x + y| = |x| + |y|$.
   (d) Give an example of two numbers x and y such that $|x + y| < |x| + |y|$.
   (e) Prove the formula

$$|a + b| \leq |a| + |b|.$$

17. Prove the following:
   (a) If $a > b > 0$, then $a^2 > b^2$
   (b) If $a > b > 0$, then $\sqrt{a} > \sqrt{b}$
   (c) If $x > y$, then $x^3 > y^3$.

18. Prove that if $x > 0$ and $x^2 > 2$, then

$$x^2 > \left(\frac{x}{2} + \frac{1}{x}\right)^2 > 2.$$

[*Hint.* $x/2 > 1/x$. Also, $(x/2 - 1/x)^2 > 0$.]

19. Prove that if $x \geq 0$, then $x^3 \geq 3x - 2$. [*Hint.* Write $x^3 = x \cdot x^2$ and apply Problem 14(b).]

Answers to problems (§1A)

2. (a) *T*   (b) *T*   (c) *F*   (d) *T*   (e) *F*   (f) *F*   (g) *F*   (h) *F*
   (i) *F*   (j) *T*   (k) *T*   (l) *T*

3. (a) *T*   (b) *F*   (c) *T*   (d) *T*   (e) *T*   (f) *F*   (g) *F*

4. (a) *T*   (b) *T*   (c) *T*   (d) *T*   (e) *F*   (f) *T*   (g) *T*   (h) *F*
    (i) *F*   (j) *T*   (k) *F*   (l) *T*   (m) *T*
5. (a) *T*   (b) *T*   (c) *F*   (d) *T*   (e) *F*   (f) *F*
6. (a) *T*   (b) *T*   (c) *T*   (d) *F*   (e) *T*   (f) *T*   (g) *F*
7. (a) *T*   (b) *F*   (c) *T*   (d) *T*   (e) *F*   (f) *F*   (g) *T*   (h) *T*
    (i) *T*   (j) *F*   (k) *T*   (l) *T*   (m) *T*
10. (i) -1

§1B. Distance and slope

1B1. Graphs. We assume the reader has been introduced to graphs. He should also know the elementary facts about parallel and perpendicular lines, similarity and congruence of triangles, and the Theorem of Pythagoras.

A coordinate system is introduced in the plane in the usual way (Figure 1). Each point in the plane is assigned a pair of numbers (its coordinates), and vice versa. The point whose coordinates are (3, 2) is called, for short, the point (3, 2).

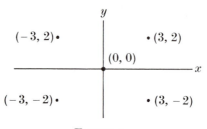

FIGURE 1

The graph of an equation in x and y is the set of all (x, y) that satisfy the equation. If (x, y) satisfies the equation, then (x, y) is on the graph; conversely, if (x, y) is on the graph, then (x, y) satisfies the equation.

The equation of a curve is an equation whose graph is the curve. (There is always more than one equation, but by custom, any one of them is called "the" equation.)

1B2. Distance formula. Let d denote the distance between two points (x_1, y_1) and (x_2, y_2). (See Figure 2.) By the Theorem of Pythagoras,

(1) $$d^2 = |x_1 - x_2|^2 + |y_1 - y_2|^2 = (x_1 - x_2)^2 + (y_1 - y_2)^2.$$

Therefore,

(2) $$d = \sqrt{(x_1 - x_2)^2 + (y_1 - y_2)^2}.$$

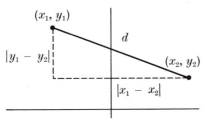

FIGURE 2

Example 1. The distance between $(3, 4)$ and $(2, -1)$ is

$$d = \sqrt{(3 - 2)^2 + [4 - (-1)]^2} = \sqrt{1^2 + 5^2} = \sqrt{26}.$$

1B3. Equation of the circle. Let us find the equation of the circle with center (h, k) and radius r (Figure 3).

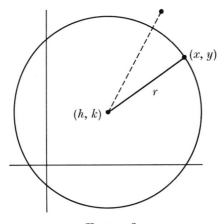

FIGURE 3

If (x, y) lies on the circle then $(x - h)^2 + (y - k)^2 = r^2$ and if (x, y) doesn't lie on the circle then $(x - h)^2 + (y - k)^2 \neq r^2$. Therefore the equation of the circle is

(3) $$(x - h)^2 + (y - k)^2 = r^2.$$

Example 2.

(a) The equation of the circle with center $(-3, 2)$ and radius 4 is

$$(x + 3)^2 + (y - 2)^2 = 16.$$

(b) The graph of the equation

$$x^2 + (y + 1)^2 = 7$$

is the circle with center $(0, -1)$ and radius $\sqrt{7}$.

1B4. Slope of a line. The *slope* of a nonvertical line l is the number

(4)
$$m = \frac{y_1 - y_2}{x_1 - x_2} = \frac{y_2 - y_1}{x_2 - x_1},$$

where (x_1, y_1) and (x_2, y_2) are any two distinct points on the line. Since l is not vertical, $x_1 \neq x_2$ and the above fractions make sense. It does not matter which two points are used, because the results are the same, as is clear from the similar triangles (Figure 4).

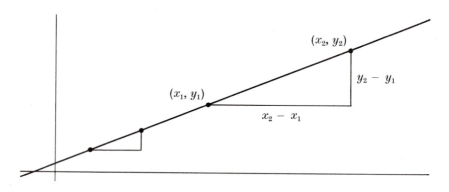

FIGURE 4

The slope of a line is positive, zero, or negative according as the line is rising, level, or falling (Figure 5).

We do not define slope for a vertical line.

(5)   *If two nonvertical lines are parallel their slopes are equal, and if their slopes are equal the lines are parallel.*

This is clear from Figure 6. A nonvertical line is therefore determined by its slope and a point through which it passes. Hence a line of given slope is determined by its y-intercept, b (Figure 7).

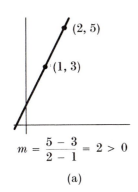
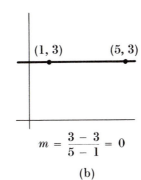
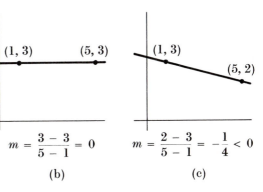

$$m = \frac{5-3}{2-1} = 2 > 0 \qquad m = \frac{3-3}{5-1} = 0 \qquad m = \frac{2-3}{5-1} = -\frac{1}{4} < 0$$

(a)   (b)   (c)

FIGURE 5

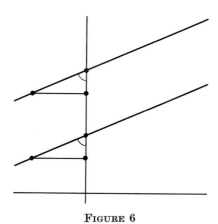

FIGURE 6

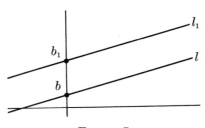

FIGURE 7

1B5. Slope-intercept equation of a line. *The equation of the line with slope m and y-intercept b is*

(6) $$y = mx + b.$$

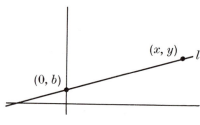

FIGURE 8

Proof. Let (x, y) be any point of the line other than $(0, b)$ (Figure 8). By definition of slope,

(7) $$\frac{y - b}{x - 0} = m.$$

Therefore $y - b = mx$, or

(8) $$y - mx = b.$$

Equation (8) is also satisfied by $(0, b)$. It therefore holds for all points of the line. These are the only points for which (8) holds, because if (x_1, y_1) is not on the line then $y_1 - mx_1 = b_1 \neq b$ (Figure 9). Hence (8), or (6), is the equation of the line.

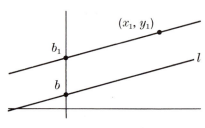

FIGURE 9

Example 3. Show that

$$3x + 4y - 8 = 0$$

is the equation of a nonvertical line and find its slope.

Solution. Solving for y we get

$$y = -\tfrac{3}{4}x + 2,$$

the equation of a line with slope $-\tfrac{3}{4}$ and y-intercept 2.

1B6. Point-slope equation of a line. *The equation of the line with slope m through the point (x_0, y_0) is*

(9)
$$y - y_0 = m(x - x_0).$$

Proof. Let b denote the y-intercept. By (6), the equation of the line is

(10)
$$y = mx + b.$$

Since (x_0, y_0) is on the line,

(11)
$$y_0 = mx_0 + b.$$

Subtracting, we get (9).

Example 4.

(a) The equation of the line with slope 4 and passing through the point $(2, 3)$ is

$$y - 3 = 4(x - 2).$$

(b) The graph of the equation

$$y - 6 = -3(x + 2)$$

is the line with slope -3 through the point $(-2, 6)$.

Example 5. Find the equation of the line through $(1, -4)$ and $(2, 1)$.

Solution. The slope is

(12)
$$m = \frac{1 - (-4)}{2 - 1} = 5.$$

Using the point $(1, -4)$, we write the equation as

$$y + 4 = 5(x - 1).$$

Using the point $(2, 1)$, we would get

$$y - 1 = 5(x - 2).$$

Both equations reduce to $y = 5x - 9$.

1B7. Lines parallel to an axis. Horizontal lines are the lines with slope 0. On a horizontal line, y is constant. The equation of the horizontal line through $(0, b)$ is $y = b$ (Figure 10).

Vertical lines have no slope. x is constant. The equation of the vertical line through $(a, 0)$ is $x = a$ (Figure 10).

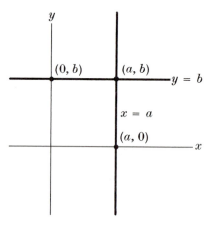

FIGURE 10

1B8. The linear equation. The equations of the line,

(13) $\qquad\qquad y = mx + b \qquad$ (nonvertical)

and

(14) $\qquad\qquad\qquad x = a \qquad$ (vertical),

may be written as $mx - y + b = 0$ and $x - a = 0$. As such, they are instances of the general form

(15) $\qquad\qquad\qquad Ax + By + C = 0 \qquad$ (A, B not both 0).

It is noteworthy that, conversely, every equation of the form (15) can be put into one or the other of the forms (13) or (14) and hence is the equation of a line. In fact, if $B \neq 0$ we can solve for y in (15), obtaining

(16) $\qquad\qquad\qquad y = -\dfrac{A}{B}x - \dfrac{C}{B}.$

This is in the form (13), with $m = -A/B$ and $b = -C/B$. On the other hand, if $B = 0$, then, by assumption, $A \neq 0$, and we can solve for x in (15). We get

$$(17) \qquad\qquad x = -\frac{C}{A},$$

which is in the form (14).

Because of these facts, equations of the form (15) are called *linear* equations.

Problems (§1B)

1. Find the distance between the two given points.
   (a) $(1, 3)$ and $(-1, 6)$ (d) $(-2, 5)$ and $(4, 4)$
   (b) $(1, 3)$ and $(5, 0)$ (e) $(4, -3)$ and $(4, 7)$
   (c) $(1, 3)$ and $(-4, 3)$ (f) $(0, -2)$ and $(3, 4)$

2. Specify four points equidistant
   (a) from $(1, 3)$ and $(6, 3)$
   (b) from $(1, 3)$ and $(5, 4)$

3. Write the equation of the circle with the given center and radius.
   (a) center $(3, -4)$,   radius 6
   (b) center $(2, 0)$,   radius $\sqrt{5}$
   (c) center $(0, 3)$,   radius 4

4. By inspection, find the center and radius of each of the following circles.
   (a) $x^2 + (y + 1)^2 = 6$
   (b) $(x + 3)^2 + (y - 4)^2 = 9$
   (c) $(x - 4)^2 + y^2 = 15$

5. Name a point on the given circle such that the diameter through the point is neither horizontal nor vertical.
   (a) $x^2 + y^2 = 1$ (b) $(x - 1)^2 + (y - 2)^2 = 6$

6. Give an equation of a circle that is tangent to
   (a) the x-axis but not the y-axis.
   (b) the y-axis but not the x-axis.
   (c) both axes.

7. Give equations of two circles with different centers, one circle lying wholly inside the other.

8. Find the slope of the line through the two given points.
   (a) $(1, 2)$ and $(3, 4)$ (d) $(-2, -3)$ and $(-3, -3)$
   (b) $(-1, 3)$ and $(1, -2)$ (e) $(4, 0)$ and $(5, 2)$
   (c) $(-2, -3)$ and $(-4, -1)$

9. Find the equation of the line through the given point and having the given slope m.
   (a) $(1, 3)$, $m = 2$ (d) $(0, -3)$, $m = 1$
   (b) $(-2, -5)$, $m = -3$ (e) $(6, 227/5)$, $m = 0$
   (c) $(-2, 1)$, $m = -2$

10. Name three points on the given line.
    (a) $y = 2x + 1$ (b) $x - y = 4$ (c) $2x + 3y - 5 = 0$

11. Find the equation of the line through the two given points.
    (a) $(1, 2)$ and $(4, 3)$
    (b) $(-1, 3)$ and $(-2, 1)$
    (c) $(1, 6)$ and $(1, 0)$

12. Find the slope of each of the following lines.
    (a) $3x - 4y + 6 = 0$ (c) $3y - 4x - 6 = 0$
    (b) $3y - 4x + 6 = 0$ (d) $3x + 4y - 6 = 0$

13. Write the equation of the line through $(1, 2)$ whose x-intercept is equal to its y-intercept.

14. Give an example of a point whose distance from the origin is equal to the slope of the line joining it to the origin.

15. (a) The line with slope 4 and y-intercept $b > 0$ cuts off a triangle with the axes of area 4. Find b.
    (b) Make up an example of a line that cuts off a triangle with the axes of area 5.

16. Decide whether the three given points lie on a line.
    (a) $(-1, 19)$, $(6, 5)$, and $(10, -3)$
    (b) $(1, -4)$, $(-1, -11)$, and $(3, 2)$
    (c) $(-2, 8)$, $(2, -4)$, and $(0, 2)$

17. Find the value of the unknown if the three points lie on a line.
    (a) $(2, -2)$, $(5, 13)$, and $(x, -12)$
    (b) $(-1, 6)$, $(6, -29)$, and $(2, y)$

18. Do the lines $y = 3x + \frac{13}{41}$ and $y = 5x - \frac{151}{19}$ intersect?

19. Draw a rough sketch to compare the graphs of
    (i) $y = 2x + 4$
    (ii) $y = 2x + 3$
    (iii) $y = 2x$
    (iv) $y = 2x - 1$

20. Draw a rough sketch to compare the graphs of
    (i) $y = 3$
    (ii) $y = x + 3$
    (iii) $y = 2x + 3$
    (iv) $y = 3 - x$
    (v) $y = 3 - 2x$

21. Decide whether the given point lies above, on, or below the given line.
    (a) $(1, 9)$, $y - 2x = 7$ (d) $(2, 3)$, $3y - x = 7$
    (b) $(4, 3)$, $2y = 3x - 4$ (e) $(3, 22)$, $y = 7x + 11$
    (c) $(2, 3)$, $y + 3x - 3 = 0$

22. Locate D if the points $A = (-1, -2)$, $B = (2, 4)$, $C = (3, -6)$, and D are the vertices of a parallelogram. [*Hint.* Either AB, AC, or BC can be a diagonal.]

23. Verify that the points (a_1, b_1), (a_2, b_2), and $((a_1 + a_2)/2, (b_1 + b_2)/2)$ lie on a line and that the third one is equidistant from the other two. Then state a formula for the midpoint of a line segment.

24. Find the equation of the circle passing through $(0, 0)$, $(4, 4)$, and $(2, 6)$.

25. Find the equations of the circles with center $(6, 7)$ and tangent to the circle $(x + 2)^2 + (y - 1)^2 = 9$.

26. Let l_1 and l_2 be lines with positive slopes m_1 and m_2. Suppose that the acute angle that l_1 makes with the x-axis is equal to the acute angle that l_2 makes with the y-axis. What is the relation between m_1 and m_2?

Answers to problems (§1B)

1. (a) $\sqrt{13}$ (c) 5   (e) 10
3. (a) $(x - 3)^2 + (y + 4)^2 = 36$
4. (a) center $(0, 1)$, radius $\sqrt{6}$
8. (a) 1   (c) -1
9. (a) $y = 2x + 1$ (c) $2x + y + 3 = 0$
11. (a) $3y = x + 5$
12. (a) $\frac{3}{4}$
13. $x + y = 3$ $y = 2x$
15. (a) $4\sqrt{2}$
16. (a) yes   (b) no
17. (a) 0
21. (a) on   (c) above   (e) below
22. $(-2, 8)$, $(0, -12)$, or $(6, 0)$
24. $(x - 1)^2 + (y - 3)^2 = 10$
25. $(x - 6)^2 + (y - 7)^2 = 49$, $(x - 6)^2 + (y - 7)^2 = 169$

§1C. The parabola

1C1. The parabola. A *parabola* is defined as the set of all points equidistant from a fixed point and a fixed line. The fixed point, F, is called the *focus* and the fixed line, l, the *directrix*.

To obtain a simple equation, we choose the coordinate axes so that $F = (0, p)$, where $p > 0$, and l is the line $y = -p$ (Figure 1).

Even before deriving the equation, we can make some observations directly from the definition and the choice of axes. First, the curve will pass through the origin. Next, consider any other point (x, y) on the curve. Evidently, $y > 0$, since otherwise the point would be closer to l than to F. Also, $x \neq 0$, as otherwise the point would be closer to F than to l. Next, $(-x, y)$ is also a point of the parabola; thus, the curve is its own reflection

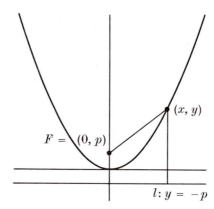

FIGURE 1

in the y-axis. Finally, it is clear that there are points on the curve with arbitrarily large x and y.

The line of symmetry is called the *axis* of the parabola. Its intersection with the parabola is the *vertex*. In the present case, the axis is the y-axis and the vertex is the origin. The distance between focus and vertex—the number p—is known as the *focal distance*.

To derive the equation, consider any point (x, y) in the plane. Its distance to l is

(1) $$|y + p|;$$

and its distance to F is

(2) $$\sqrt{x^2 + (y - p)^2}.$$

Therefore, (x, y) is on the parabola if

(3) $$|y + p| = \sqrt{x^2 + (y - p)^2},$$

and not otherwise. Thus, (3) is an equation of the parabola.

If we square both sides in (3) we get

(4) $$(y + p)^2 = x^2 + (y - p)^2.$$

Conversely, if we take the (positive) square root in (4), we get (3). Therefore (4) is also an equation of the parabola.

When we expand in (4), both y^2 and p^2 cancel out and the equation simplifies to

(5) $$4py = x^2.$$

Finally, putting

(6)
$$a = \frac{1}{4p},$$

we get, as the equation of the parabola,

(7)
$$y = ax^2.$$

The simplicity of this equation results from our choice of axes.

With $F = (0, -p)$ and l the line $y = p$, where $p > 0$, we again get equation (7), but this time with

(8)
$$a = -\frac{1}{4p} < 0.$$

We see that $a > 0$ means the parabola is opening upward and $a < 0$ means it is opening downward (Figure 2).

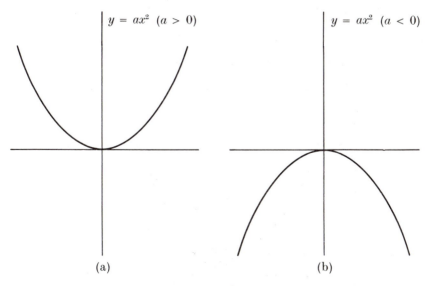

$$y = ax^2 \ (a > 0) \qquad\qquad y = ax^2 \ (a < 0)$$

(a)   (b)

FIGURE 2

Example 1. What is the equation of the parabola with focus $(0, 6)$ and directrix $y = -6$?

Solution. Here $p = 6$ and $a = 1/4p = \frac{1}{24}$. Hence the equation is

$$y = \tfrac{1}{24}x^2.$$

Example 2. What is the equation of the parabola with focus $(0, -\frac{1}{8})$ and directrix $y = \frac{1}{8}$?

Solution. Here $p = \frac{1}{8}$ and $a = -1/4p = -2$. The equation is

$$y = -2x^2.$$

Example 3. Describe the graph of the equation

$$y = \tfrac{1}{2}x^2.$$

Solution. Here $a = \frac{1}{2}$ and $p = 1/4a = \frac{1}{2}$. The graph is a parabola opening upward, with vertex the origin. The focus is the point $(0, \frac{1}{2})$. The directrix is the line $y = -\frac{1}{2}$.

1C2. *Vertex not the origin.* When the vertex is at a point (h, k) not the origin, the equation is still simple, so long as the directrix l is horizontal. Let p denote the focal distance and choose the focus as $F = (h, k + p)$. Then l is the line $y = k - p$ (Figure 3). The distance from a point (x, y) to l is

(9) $|y - k + p|$;

and its distance to F is

(10) $\sqrt{(x - h)^2 + (y - k - p)^2}.$

Reasoning as before, we arrive at the equation

(11) $[(y - k) + p]^2 = (x - h)^2 + [(y - k) - p]^2.$

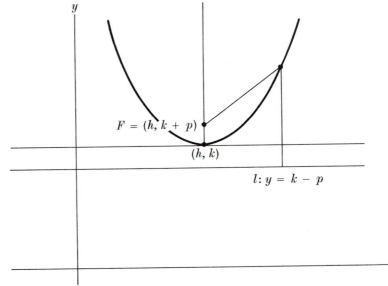

FIGURE 3

This time $(y - k)^2$ and p^2 cancel out and we get

(12) $$4p(y - k) = (x - h)^2,$$

or, as before, with $a = 1/4p$,

(13) $$y - k = a(x - h)^2.$$

The constants of the parabola can be read off from this equation: vertex (h, k), focal distance $p = 1/4a$ (since $a > 0$), axis the line $x = h$. The parabola opens upward.

If $F = (h, k - p)$ and l is the line $y = k + p$, we get the same equation (13), this time with $a = -1/4p$. This parabola opens downward.

When $a > 0$ in equation (13), the parabola opens upward and $p = 1/4a$. When $a < 0$ the parabola opens downward and $p = -1/4a > 0$.

Example 4. What is the equation of the parabola with vertex $(-3, 5)$ and directrix the line $y = 3$ (Figure 4)?

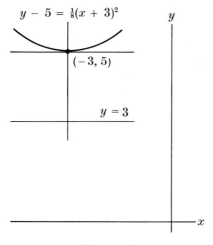

$$y - 5 = \tfrac{1}{8}(x + 3)^2$$

$(-3, 5)$

$y = 3$

F<small>IGURE</small> 4

Solution. Here $h = -3$ and $k = 5$. The distance from the vertex to the directrix is $p = 2$. Hence the focus is $F = (-3, 7)$. Next, $a = 1/4p = \tfrac{1}{8}$. The equation is

$$y - 5 = \tfrac{1}{8}(x + 3)^2.$$

Example 5. Describe the graph of the equation

$$y + 1 = -\tfrac{1}{4}(x - 2)^2.$$

Solution. In this equation, $a = -\frac{1}{4}$. The graph is a parabola opening downward, with vertex $(h, k) = (2, -1)$, focal distance $p = -1/4a = 1$, and axis the line $x = 2$ (Figure 5). Hence $F = (2, -2)$ and l is the x-axis.

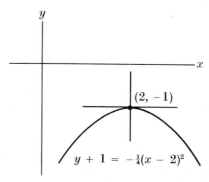

$$(2, -1)$$

$$y + 1 = -\tfrac{1}{4}(x - 2)^2$$

FIGURE 5

1C3. The quadratic equation. The equation of a parabola with vertical axis, (13), presents y as a quadratic expression in x. If we multiply out and collect terms we can put it in the general form

(14) $$y = Ax^2 + Bx + C \qquad (A \neq 0).$$

Explicitly, we get

(15) $$y = ax^2 + (-2ah)x + (ah^2 + k).$$

It is noteworthy that, conversely, every quadratic equation (14) is the equation of a parabola with vertical axis. That is, (14) can be put into the form (15) (or (13)). To see this, simply match coefficients between (14) and (15): first define $a = A$, then $h = -B/2a$, then $k = C - ah^2$.

In applications, we convert from (14) to (13) by "completing the square," i.e., by invoking the identity

(16) $$x^2 + px = \left(x + \frac{p}{2}\right)^2 - \left(\frac{p}{2}\right)^2.$$

Examples 6 and 7 show how this works.

Example 6. Describe the parabola

$$y = 3x^2 + 4x + 2.$$

Solution. We have

$$\frac{y}{3} = x^2 + \tfrac{4}{3}x + \tfrac{2}{3} = (x + \tfrac{2}{3})^2 + (\tfrac{2}{3} - \tfrac{4}{9}),$$

$$y = 3(x + \tfrac{2}{3})^2 + \tfrac{2}{3},$$

$$y - \tfrac{2}{3} = 3(x + \tfrac{2}{3})^2.$$

This is the parabola with vertical axis whose vertex is $(-\frac{2}{3}, \frac{2}{3})$, focal distance $1/(4 \times 3) = \frac{1}{12}$, and that opens upward (Figure 6).

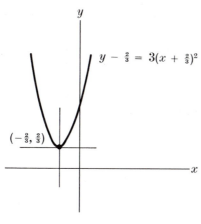

$$y - \tfrac{2}{3} = 3(x + \tfrac{2}{3})^2$$

$$(-\tfrac{2}{3}, \tfrac{2}{3})$$

FIGURE 6

Example 7. Describe the parabola

$$y = -2x^2 + 3x - 1.$$

Solution. We have

$$\frac{y}{-2} = x^2 - \tfrac{3}{2}x + \tfrac{1}{2} = (x - \tfrac{3}{4})^2 + (\tfrac{1}{2} - \tfrac{9}{16}),$$

$$y = -2(x - \tfrac{3}{4})^2 + \tfrac{1}{8},$$

$$y - \tfrac{1}{8} = -2(x - \tfrac{3}{4})^2.$$

This is the parabola with vertical axis whose vertex is $(\frac{3}{4}, \frac{1}{8})$, focal distance $1/(4 \times 2) = \frac{1}{8}$, and that opens downward (Figure 7).

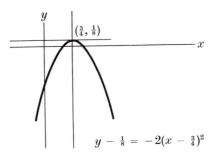

$$(\tfrac{3}{4}, \tfrac{1}{8})$$

$$y - \tfrac{1}{8} = -2(x - \tfrac{3}{4})^2$$

FIGURE 7

Problems (§1C)

1. Write the equation of the parabola with the given focus and directrix.
   - (a) focus $(0, 3)$, directrix $y = -3$
   - (b) focus $(0, -2)$, directrix $y = 2$
   - (c) focus $(0, 3)$, directrix $y = -2$
   - (d) focus $(4, 1)$, directrix $y = 0$
   - (e) focus $(4, 0)$, directrix $y = 2$

2. Find the focus and directrix of each of the following parabolas.
   - (a) $y = 7x^2$
   - (b) $y = -3x^2$
   - (c) $y = (x - 1)^2$
   - (d) $y - 2 = -2(x + 3)^2$
   - (e) $y = 3x^2 - 2x + 1$
   - (f) $y = -x^2 + 4x - 6$

3. Draw a rough sketch to compare the graphs of
   - (i) $y = x^2$
   - (ii) $y = 2x^2$
   - (iii) $y + 1 = x^2$

4. Draw a rough sketch to compare the graphs of
   - (i) $y = x^2$
   - (ii) $y = (x + 1)^2$
   - (iii) $y = (x - 2)^2$

5. Name four points on the parabola $y - 3 = 4(x - 2)^2$.

6. Find the center and radius of each of the following circles.
   - (a) $x^2 + y^2 - 2x + 4y + 2 = 0$
   - (b) $x^2 + y^2 + y = 5\frac{3}{4}$
   - (c) $2x^2 + 2y^2 + 4x = 4y + 1$

7. If the radius of the circle

$$x^2 + y^2 + Hx - 6y + 6 = 0$$

is $\sqrt{7}$, what is the center?

8. Can two parabolas both opening upward meet at exactly one point if it is the vertex
   - (a) of both?
   - (b) of one but not the other?
   - (c) of neither?

9. Can two parabolas, one opening upward and one opening downward, meet at exactly one point if it is the vertex
   - (a) of both?
   - (b) of one but not the other?
   - (c) of neither?

10. Consider those parabolas that open upward and whose axis is the y-axis.
    - (a) Are there two that never intersect?
    - (b) Are there two that intersect in exactly one point?
    - (c) Are there two that intersect in exactly two points?

11. The parabola $y = ax^2$ passes through the point $(1, 2)$. What is its equation?

12. Find all lines through $(1, 3)$ that intersect the parabola $y = 3x^2$ in that point only.

13. Find the equation of the parabola with vertical axis and passing through the points $(3, 0)$, $(-1, 0)$, and $(0, 3)$.

14. A parabola with vertical axis has its vertex on the x-axis and passes through the points $(2, 3)$ and $(3, 12)$. What is its equation?

15. Find the equation of the line through the points of intersection of the parabolas $y = (x - 6)^2$ and $5y = 54 - x^2$.

16. (a) For what values of a do the parabolas $y = ax^2 + 3$ and $y = x^2$ intersect?
   (b) For what values of k do the parabolas $y = x^2 + k$ and $y = 3x^2$ intersect?

17. Find the points of intersection of the line $y = 4x + 2$ with the parabola $y - 3 = 3(x - 2)^2$.

18. The line $y = 9$ cuts the parabola $y = 2kx^2 + k$ in a chord of length 4. What is k?

19. An equilateral triangle is inscribed in the parabola $y = 4x^2$ with one vertex at the origin. Where are the other two?

Answers to problems (§1C)

1. (a) $12y = x^2$ (d) $2y = (x - 4)^2 + 1$

2. (a) focus $(0, \frac{1}{28})$,   directrix $y = -\frac{1}{28}$
   (d) focus $(-3, \frac{15}{8})$,   directrix $y = \frac{17}{8}$
   (e) focus $(\frac{1}{3}, \frac{3}{4})$,   directrix $y = \frac{7}{12}$

6. (a) center $(1, -2)$,   radius $\sqrt{3}$

7. $(2, 3)$ or $(-2, 3)$

11. $y = 2x^2$

12. $x = 1$, $y = 6x - 3$

13. $y = 4 - (x - 1)^2$

14. $y = 3(x - 1)^2$, $y = 3(3x - 7)^2$

15. $2x + y = 15$

16. (a) $a < 1$ (b) $k \geq 0$

17. $(1, 6)$ and $(\frac{13}{3}, \frac{58}{3})$

18. 1

19. $(\sqrt{3}/4, 3/4)$ and $(-\sqrt{3}/4, 3/4)$

§1D. Perpendicularity

1D1. Right triangles. In deriving the distance formula we invoked the Theorem of Pythagoras, which states that in triangle ABC, if $AC \perp BC$, then $a^2 + b^2 = c^2$ (Figure 1). All students know this famous theorem. Not so many people have seen or remember the converse, or realize there is one, or know what a converse is.

Converse to the Theorem of Pythagoras. *In triangle ABC, if $a^2 + b^2 = c^2$, then $AC \perp BC$.*

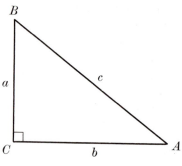

FIGURE 1

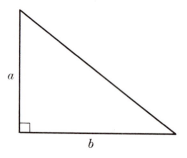

FIGURE 2

Proof. Construct a triangle with two perpendicular sides of lengths a and b (Figure 2). By the Theorem of Pythagoras the third side is c. The triangle is therefore congruent to triangle ABC and it follows that $AC \perp BC$.

The following proof uses both the Pythagorean Theorem and its converse.

1D2. Perpendicular lines. *For any two nonvertical lines:*

(a) *If the lines are perpendicular the product of their slopes is* -1.

(b) *Conversely, if the product of their slopes is* -1, *the lines are perpendicular.*

Proof. Call the slopes m_1 and m_2. Clearly, it is sufficient to work with the lines

$$y = m_1 x \qquad \text{and} \qquad y = m_2 x,$$

which have these slopes and pass through the origin.

The point $(1, m_1)$ lies on the first line and $(1, m_2)$ on the second. They are distinct points, as we shall now verify. In (a) it is given that the lines are perpendicular; hence they are different lines and so $m_1 \neq m_2$. In (b) it is given that $m_1 m_2 = -1$; since no square is equal to -1, we again have $m_1 \neq m_2$.

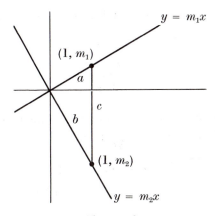

FIGURE 3

Let a, b, c denote the sides of the triangle as shown in Figure 3. Then $a^2 = 1 + m_1^2$, $b^2 = 1 + m_2^2$, and $c^2 = (m_1 - m_2)^2$. Thus,

$$a^2 + b^2 = m_1^2 + m_2^2 + 2, \qquad \text{and}$$

(1)

$$c^2 = m_1^2 + m_2^2 - 2m_1 m_2.$$

If the lines are perpendicular then by the Pythagorean Theorem, $a^2 + b^2 = c^2$. Then by (1), $2 = -2m_1 m_2$, and so $m_1 m_2 = -1$. This proves (a). Conversely, if $m_1 m_2 = -1$ then by (1), $a^2 + b^2 = c^2$. It then follows from the converse to the Pythagorean Theorem that the lines are perpendicular. This proves (b).

Example 1. The lines

$$y + 6x = 3 \qquad \text{and} \qquad y - \tfrac{1}{6}x = 2$$

are perpendicular, because the first has slope -6 and the second has slope $+\tfrac{1}{6}$.

Example 2. Find the equation of the line l_2 passing through the point $(-1, 3)$ and perpendicular to the line

$$l_1 : y = \tfrac{1}{2}x + 1.$$

Solution. Since the slope of l_1 is $\tfrac{1}{2}$, the slope of l_2 is -2. Hence the equation of l_2 is

$$y - 3 = -2(x + 1).$$

1D3. Distance between two parallel lines. We now derive formulas for the distance between two parallel lines and the distance from a point to a line. Let

$$y = mx + b \qquad \text{and} \qquad y = mx + b_1$$

be two lines with the same slope m, and let d be the distance between them. Then

(2)
$$d = \frac{|b_1 - b|}{\sqrt{1 + m^2}},$$

as we shall now show. We note that in Figure 4, triangle A is congruent to B, so that the legs of A are as marked. Consequently,

(3)
$$d^2 + m^2 d^2 = (b_1 - b)^2.$$

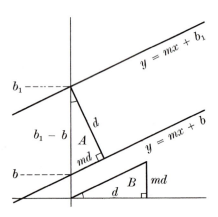

FIGURE 4

The figure is drawn for $m > 0$ and $b_1 > b$. If $m < 0$ or $b_1 < b$ the diagram or its labeling are modified, but equation (3) comes out the same because of the squaring. Solving in (3) for d we get (2). If $m = 0$ the lines are horizontal and $d = |b_1 - b|$. Hence (2) holds in that case too.

The distance between two vertical lines, $x = a$ and $x = a_1$, is simply $|a_1 - a|$.

Example 3. The distance between the parallel lines

$$y = 3x + 2 \qquad \text{and} \qquad y = 3x + 5$$

is

$$d = \frac{|2 - 5|}{\sqrt{1 + 3^2}} = \frac{3}{\sqrt{10}}.$$

1D4. Distance from a point to a line. We now prove that the distance from a point (x_1, y_1) to a line l: $Ax + By + C = 0$ is

(4)
$$d = \frac{|Ax_1 + By_1 + C|}{\sqrt{A^2 + B^2}}.$$

The distance means the distance along the perpendicular. Hence it is the same as the distance from l to the line l_1 through (x_1, y_1) parallel to l (Figure 5).

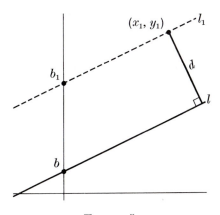

FIGURE 5

Assume first that $B \neq 0$. Then l is the line $y = mx + b$, where $m = -A/B$ and $b = -C/B$; and hence l_1 is the line $y = mx + b_1$, where $b_1 = y_1 - mx_1$. Therefore the distance between them is

(5) $\quad d = \dfrac{|b_1 - b|}{\sqrt{1 + m^2}} = \dfrac{|y_1 - mx_1 - b|}{\sqrt{1 + m^2}} = \dfrac{|y_1 + Ax_1/B + C/B|}{\sqrt{1 + A^2/B^2}}.$

Multiplying top and bottom by $|B|$, we get (4).

In case $B = 0$, l is the vertical line $x = -C/A$. The distance from (x_1, y_1) to l is then $|x_1 + C/A|$. Multiplying and dividing by $|A|$, we get (4).

Example 4. Find the distance from the point $(3, 5)$ to the line $y = 4x + 6$.

Solution. Writing the line as $4x - y + 6 = 0$, we have $A = 4$, $B = -1$, and $C = 6$. Hence the distance is

$$d = \frac{|(4)(3) + (-1)(5) + 6|}{\sqrt{4^2 + (-1)^2}} = \frac{13}{\sqrt{17}}.$$

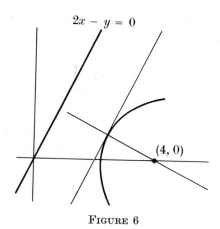

FIGURE 6

Example 5. Find the equation of the parabola with focus (4, 0) and directrix $2x - y = 0$ (Figure 6).

Solution. The distance from a point (x_1, y_1) to the directrix is

$$\frac{|2x_1 - y_1|}{\sqrt{5}}.$$

The distance to the focus is

$$\sqrt{(x_1 - 4)^2 + y_1^2}.$$

At this stage we no longer need the subscripts. Equating the distances and squaring, we have

$$\frac{(2x - y)^2}{5} = (x - 4)^2 + y^2.$$

This simplifies to

$$x^2 + 4xy + 4y^2 - 40x + 80 = 0.$$

Problems (§1D)

1. Determine whether the three points are the vertices of a right triangle.
   (a) (1, 2),   (4, 11),   and   (4, 1)
   (b) (0, 1),   (10, −1),   and   (1, 4)
   (c) (4, 11),   (−1, 1),   and   (9, −5)
   (d) (4, 11),   (−1, 1),   and   (1, 0)
   (e) (4, 11),   (−1, 1),   and   (8, 9)

2. Find the equation of the line through the given point and perpendicular to the given line.

   (a) (1, −3), $y - 3x = 4$ (d) (1, 3), $y = x$
   (b) (−1, 1), $2y - x = 3$ (e) (−3, 2), $y = 1$
   (c) (−2, −1), $3y + 2x = 7$

3. Let P be any point on the graph of $x^2 + y^2 = 6y$ not on the y-axis. Show that P, $(0, 6)$, and the origin are the vertices of a right triangle.

4. Find the areas of the triangles with the given vertices.
   (a) $(1, 3)$, $(1, 6)$, and $(-4, 3)$
   (b) $(-2, 4)$, $(-5, -5)$, and $(1, 3)$
   (c) $(-3, 0)$, $(0, -5)$, and $(4, 0)$
   (d) $(1, 2)$, $(-1, 4)$, and $(0, 5)$
   (e) $(1, 2)$, $(-1, 4)$, and $(6, 3)$

5. Two of the vertices of a right triangle are $(1, -2)$ and $(-1, 0)$. Find x if the third vertex is:
   (a) $(x, 5)$ (b) $(x, -1)$

6. Find the distance between the given parallel lines.
   (a) $2y - x = 7$ and $x - 2y = 3$
   (b) $3y + 2x = 1$ and $3y + 2x = 2$
   (c) $y = 4x - 1$ and $y = 4x + 1$

7. Find the distance from the given point to the given line.
   (a) $(3, 4)$, $4x + 3y + 6 = 0$ (c) $(-1, 3)$, $x - 1 = 0$
   (b) $(1, 5)$, $4x - 3y + 6 = 0$ (d) $(-2, -1)$, $3x - 2y = 5$

8. Show that the distance between $(1, 2)$ and any point on the line $y = 3x + 4$ is at least $\frac{1}{2}\sqrt{10}$.

9. Find the equations of the bisectors of the angles formed by the x-axis and the line $y = x$. [*Hint.* A point on the bisector is equidistant from the sides of the angle.]

10. A point moves so that its distance from $(3, 0)$ is always twice its distance from the origin. What is its path?

11. Exhibit a pair of perpendicular lines, one through the point $(1, 4)$ and the other through $(6, 2)$, neither of which has slope 0 or slope $\frac{5}{2}$.

12. Specify three points that are the vertices of a right triangle whose hypotenuse is 6, and no side of which is parallel to either of the coordinate axes.

13. Make up an example of two lines whose intersections with the coordinate axes are the vertices of a rectangle of area 10.

14. Find a so that the points $(0, 0)$, (x_1, y_1), and $(-x_1, y_1)$ of the parabola $y = ax^2$ are the vertices of a right triangle.

15. A right triangle of area 20 is inscribed in the parabola $y = \frac{1}{4}x^2$, with the vertex of the right angle at the origin. What are the other two vertices of the triangle?

16. Lines are drawn from a point of intersection of the parabolas $y = ax^2$ and $y = 4 - x^2$ to the vertex of each parabola. For what value of a are these lines perpendicular?

17. Find the equation of the parabola with the given focus and directrix.
   (a) focus $(1, 3)$, directrix $y = x$
   (b) focus $(0, 7)$, directrix $y = 2x + 3$

Answers to problems (§1D)

1. (a) yes   (b) no
2. (a) $x + 3y + 8 = 0$
4. (a) $\frac{15}{2}$ (c) $\frac{35}{2}$
5. (a) 4 or 8   (b) $2, -2, \sqrt{2}$, or $-\sqrt{2}$
6. (a) $4\sqrt{5}$
7. (a) 6
9. $y + (1 - \sqrt{2})x = 0$ and $y + (1 + \sqrt{2})x = 0$
10. circle with center $(-1, 0)$, radius 2
14. $1/y_1$
15. $(8, 16)$ and $(-2, 1)$, or $(-8, 16)$ and $(2, 1)$
16. $\frac{1}{3}$
17. (a) $x^2 + 2xy + y^2 - 4x - 12y + 20 = 0$
    (b) $x^2 + 4xy + 4y^2 - 12x - 64y + 236 = 0$

§1E. Intervals

1E1. Set notation. We shall occasionally make use of the following well-known symbols:

membership:	$x \in A$	x belongs to A, x is an element of A
inclusion:	$A \subset B$ or $B \supset A$	A is contained in B
set builder:	$\{x: \cdots\}$	the set of all x such that $\cdots$
union:	$A \cup B$	$\{x: x \in A$ or $x \in B$ or both$\}$
intersection:	$A \cap B$	$\{x: x \in A$ and $x \in B\}$

Example 1.

$\{x: x > 0\}$ is the set of all positive real numbers;

$1 \in \{x: x > 0\}$;

$\{x: x > 0\} \supset \{x: x \geq 5\}$;

$\{x: x > 0\} \cup \{x: x < 4\} = \mathscr{R}$;

$\{x: x > 0\} \cap \{x: x \leq 3\} = \{x: 0 < x \leq 3\}$.

1E2. Intervals. The following sets in $\mathscr{R}$ are called *intervals*: for $a < b$,

$$\{x: a \leq x \leq b\},$$
$$\{x: a < x < b\},$$
$$\{x: a \leq x < b\},$$
$$\{x: a < x \leq b\};$$

also,

$$\{x: x \geq a\}, \quad \{x: x > a\}, \quad \{x: x \leq a\}, \quad \{x: x < a\};$$

finally, at one extreme, $\mathscr{R}$ itself, and at the other, any single point. We sometimes say for short, "the interval $x > a$," etc.

The first two in the list are the most important and have special names and symbols. The interval

(1)
$$[a, b] = \{x : a \le x \le b\}$$

is called a *closed* interval, while

(2)
$$(a, b) = \{x : a < x < b\}$$

is an *open* interval. (The context will always make clear whether (a, b) denotes an open interval on the line or a point in the plane.) We also speak of the closed interval $[a, a]$, consisting of the point a alone. The notations $[a, b) = \{x : a \le x < b\}$ and $(a, b] = \{x : a < x \le b\}$ for the "half-open" intervals are also used.

A closed interval $[a, b]$ has two endpoints and contains both. The open interval (a, b) has the same two endpoints but excludes both. Some intervals, like $\{x : x > a\}$, have only one endpoint; $\mathscr{R}$ itself has none at all.

1E3. Interior. The *interior* of an interval is the interval without its endpoints (if any). The interior of the closed interval $[0, 1]$ is the open interval $(0, 1)$; the interior of $(0, 1)$ is also $(0, 1)$. The interior of $\mathscr{R}$ is $\mathscr{R}$.

Problems (§1E)

1. Which intervals are their own interiors?

2. Give an example of two open intervals
    (a) without any points in common.
    (b) of which one contains the other.
    (c) with points in common, though neither contains the other.

3. Let A and B be any two sets about which we know nothing except that $A \supset B$.
    (a) What is $A \cup B$?   (b) What is $A \cap B$?

4. State the relations of equality or inclusion among the following sets.
    $A = (0, 1)$ $D = (0, 1]$
    $B = \{t : 0 < t < 1\}$ $E = \{v : |v| \le 1\}$
    $C = \{y : 0 \le y \le 1\}$ $F = \{z : z^2 \le 1\}$

5. State the relations of equality or inclusion among the following sets.
    $A = \{x : |x| < 2\}$ $D = [0, 2)$
    $B = \{x : x^2 < 4\}$ $E = \{x : x < 1\} \cap \{x : x > 0\}$
    $C = (-2, 2)$ $F = [-2, 2]$

6. State the relations of equality or inclusion among the following sets.
$A = \{x: x^2 \geq 9\}$ $D = \{x: x < -3\}$
$B = \{x: x^2 > 9\}$ $E = \{x: x < -3\} \cup \{x: x > 3\}$
$C = \{x: |x| > 3\}$ $F = \{x: x < -3\} \cup \{x: x \geq 3\}$

7. State the relations of equality or inclusion among the following sets.
$A = \{x: |x - 1| < 2\}$ $D = [-1, 3)$
$B = \{y: y - 1 < 2\}$ $E = \{t: t - 1 < 2\}$
$C = (-1, 3)$ $F = (-1, 2) \cup (0, 3)$

8. Each of the following sets is a closed, open, or half-open interval; express it in the appropriate form $[a, b]$, (a, b), $[a, b)$, or $(a, b]$. Also, in each case, specify the interior.

(a) $(2, 4] \cap [3, 7)$
(b) $[2, 4) \cap (3, 7]$
(c) $(2, 4] \cup [3, 7)$
(d) $[2, 4) \cup (3, 7]$
(e) $\{x: x \geq 0 \text{ and } x^2 < 9\}$
(f) $\{x: x^2 < 9\}$
(g) $\{x: x > 0 \text{ and } x^3 < 8\}$
(h) $\{x: \sqrt{x} \leq 5\}$
(i) $\{x: x^2 < x\}$
(j) $\{x: x^2 \leq x\}$
(k) $\{x: 0 < x < x^2\} \cap \{x: |x| < 2\}$
(l) $\{x: 0 \leq x < x^2\} \cap \{x: |x| < 2\}$

9. In each case, give an example of an open interval that contains a but not x.
(a) $a = 1, \quad x = 2$
(b) $a = 2, \quad x = 1$
(c) $a = 1, \quad x = 1.1$
(d) $a = 1, \quad x = 0.99$

10. In Problem 9, give an example of a *closed* interval that contains a but not x.

11. In each case, give an example of a closed interval that contains a but excludes both x_1 and x_2.
(a) $a = 1.6, \quad x_1 = 1.5, \quad x_2 = 1.65$
(b) $a = -3.1, \quad x_1 = -3.12, \quad x_2 = -3.08$
(c) a not specified, $\quad x_1 = a - \frac{1}{10}, \quad x_2 = a + \frac{1}{100}$

12. In Problem 11, give an example of a closed interval having a as its *midpoint* and excluding both x_1 and x_2.

13. (a) Find two open intervals each of length 1 whose union contains the closed interval $[0, 1]$.
(b) Find eleven open intervals each of length 1 whose union contains the closed interval $[0, 10]$.

14. Let (a_1, b_1) and (a_2, b_2) be open intervals having a common point.
(a) Prove that $a_1 < b_2$ and $a_2 < b_1$.
(b) Show that $(a_1, b_1) \cup (a_2, b_2)$ is one of the open intervals

$$(a_1, b_1), \quad (a_1, b_2), \quad (a_2, b_1), \quad (a_2, b_2).$$

(c) Show that $(a_1, b_1) \cap (a_2, b_2)$ is also one of these open intervals.

15. Prove that each interior point of an interval has interior points of the interval on both sides of it.

Answers to problems (§1E)

8. (a) [3, 4]   (e) [0, 3)

§1F. Functions

1F1. Functions. Calculus studies real-valued *functions*. A real-valued function—for short, a *function*—is a rule for associating with each member of a given set some real number. Symbols are convenient: A function, f, is a rule for associating with each member x of a given set A some real number y. (In this context the letters x and y are called *variables*—x the "independent" variable, y the "dependent" variable.)

For each $x \in A$ the associated number y is called the *value* of f at x. We write

(1) $$y = f(x) \qquad (y = \text{"}f \text{ of } x\text{"}).$$

You will need to be thoroughly familiar with this notation.

A function may be thought of as a black box. In goes x, out comes $f(x)$ (Figure 1).

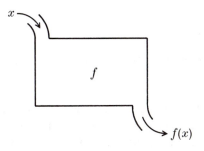

FIGURE 1

The set A is called the *domain* of f. We say that f is "defined" on A (and on each subset of A). The set of values of f is called the *image* of A under f, or, for short, the image of f.

Of course, other letters may be used for the function, the domain, and the variables.

Example 1. Let s be the function defined by

$$s(x) = \sqrt{x} \qquad (x \geq 0).$$

(The entry in the right-hand margin indicates the domain.) Then:

$$s(9) = 3,$$
$$s(u) = \sqrt{u} \qquad (u \geq 0),$$
$$s(x + 1) = \sqrt{x + 1} \qquad (x \geq -1),$$
$$s(t^2) = |t| \qquad (t \in \mathscr{R}).$$

The image of this function is the interval $x \geq 0$. For, in the first place, all values of s lie in this set. And, secondly, every number $x \geq 0$ is a value of s, since x^2 lies in the domain of s.

Example 2. Let h be the function defined by the formula

$$h(x) = \frac{x}{x + 1} \qquad (x \neq -1).$$

Then:

$$h(2) = \tfrac{2}{3}$$

$$h(-x) = \frac{x}{x - 1} \qquad (x \neq 1)$$

$$h(-2) = 2$$

$$h\left(\frac{1}{x}\right) = \frac{1}{x + 1} \qquad (x \neq 0, -1)$$

$$h(z) = \frac{z}{z + 1} \qquad (z \neq -1)$$

$$h(h(x)) = \frac{x}{2x + 1} \qquad (x \neq -1, -\tfrac{1}{2})$$

$$h(x + 1) = \frac{x + 1}{x + 2} \qquad (x \neq -2)$$

We often say for short, "the function $f(x) = x^2$," or even just "the function x^2."

1F2. Domains. For the most part we will consider functions whose domains are, as in the examples, sets of real numbers. We then stick to intervals or simple unions of intervals. The latter can also be described as intervals from which occasional points or other intervals have been removed. The following domains are as complicated as any we will encounter:

(i) $\{x : x \neq 0\}$ (the domain of the reciprocal function),
(ii) all x except odd multiples of $\pi/2$ (the domain of the tangent),
(iii) $\{x : |x| \geq 1\}$ (the domain of the inverse secant).

When a function is defined by an algebraic formula, its domain is understood to be its "natural" domain, that is, the set of *all* values for which the formula is meaningful—unless indicated otherwise. For instance, the domain of $s(x) = \sqrt{x}$ is understood to be the set of all $x \geq 0$ and the domain of $h(x) = x/(x + 1)$ all $x \neq -1$. The entries in the margin in Examples 1 and 2 are there for emphasis rather than necessity.

In the next example, the domain of the function under consideration is smaller than the natural domain.

The *graph of a function f* is the graph of the equation $y = f(x)$, that is, it is the set $\{(x, y): y = f(x)\}$.

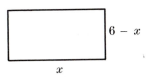

FIGURE 2

Example 3. Let $F(x)$ denote the area of a rectangle of perimeter 12 and base x (Figure 2). Clearly, x can be any number > 0 and < 6, but cannot be any other number. The domain of the function F is the open interval $(0, 6)$, and

$$F(x) = x(6 - x) \qquad\qquad (0 < x < 6).$$

The graph of this quadratic is the parabola $y = x(6 - x)$, that is,

$$y - 9 = -(x - 3)^2,$$

restricted to the interval $(0, 6)$ (Figure 3).

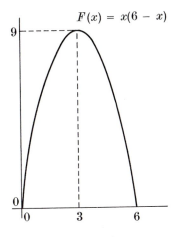

FIGURE 3

1F3. Sets of values. Let f be defined on a set I. The image of I under f is the set of values $f(x)$, for $x \in I$. (See §1F1.) Often we do not need to know this image exactly, but only that it is contained in some particular set J.

In this situation—when the values of f on I all lie in J—we say that f *takes I into J*.

Example 4. In Example 3, the values of F are all positive. Therefore F takes the interval $I = (0, 6)$ into the interval $J = \{x: x > 0\}$. (The reader can check that the image of I is the half-open interval $(0, 9]$.) If I' is any subset of I (for example, $I' = (0, 3)$), then f also takes I' into J.

1F4. Bounded functions. We say that f is *bounded* on I if there are two numbers a and b such that

$$a \le f(x) \le b$$

for every $x \in I$. In other words, f is bounded on I if there is a closed interval $J = [a, b]$ such that f takes I into J.

Example 5. The function F of Example 3 is bounded on its domain, $(0, 6)$.

The function $h(x) = 1/x \ (x > 0)$ is bounded on the interval $x \ge 1$, but it is *unbounded* on the interval $0 < x < 1$. Draw a picture.

1F5. Monotonic functions. A function f defined on a set I is said to be *increasing on I* provided that

(2)   if $x_1 < x_2$ in I then $f(x_1) < f(x_2)$.

We say it is *decreasing on I* provided that

(3)   if $x_1 < x_2$ in I then $f(x_1) > f(x_2)$.

A function that is either increasing on I or decreasing on I is said to be *monotonic on I*.

Example 6. The function F of Example 3 is increasing on the interval $(0, 3]$, and decreasing on the interval $[3, 6)$. It is not monotonic on the interval $(0, 6)$, as it is not increasing on $(0, 6)$ and it is not decreasing on $(0, 6)$.

Example 7. The linear function

(4) $l(x) = mx + b$ $(x \in \mathscr{R})$

is increasing in case $m > 0$ and decreasing in case $m < 0$. When $m \neq 0$, its image is all of $\mathscr{R}$, as is seen by solving the equation $l(x) = y$ for x:

$$x = \frac{1}{m} (y - b),$$

and then substituting in (4). When $m = 0$ this conclusion is false. The function is constant and, far from assuming all real values, assumes only one.

1F6. Operations with functions. The operations of algebra are performed on functions. The sum $f + g$ is that function whose value at x is $f(x) + g(x)$. This requires that x belong to the intersection of the domains of f and g. We assume that this intersection is an interval or simple union of intervals as described in §1F2. The functions $f - g$, fg, and f/g are defined correspondingly, with the extra restriction for f/g that its domain must exclude all x for which $g(x) = 0$. Expressions such as cf (c a constant), $|f|$, and f^n also have their obvious meanings.

Problems (§1F)

1. State the natural domain of each of the following functions.

   (a) $f(x) = x^2$

   (b) $g(x) = x^3$

   (c) $h(x) = \dfrac{x^2}{x}$

   (d) $k(x) = \dfrac{x - 1}{x + 1}$

   (e) $p(x) = \dfrac{x^2 - 1}{x + 1}$

   (f) $q(x) = \sqrt{1 - x^2}$

   (g) $r(x) = \dfrac{1}{\sqrt{1 - x^2}}$

   (h) $F(x) = 1 + \dfrac{1}{x}$

   (i) $G(x) = \dfrac{1}{1 - 1/x}$

   (j) $H(x) = \dfrac{1}{1 - 1/x^2}$

2. If $l(x) = 3x + 4$, find:

   (a) $l(3)$

   (b) $l(0)$

   (c) $l(-2)$

   (d) $l(-x)$

   (e) $l(2x)$

   (f) $l(1 + x)$

   (g) $l(3x + 4)$

   (h) $l(x^2)$

   (i) $l\left(\dfrac{x - 4}{3}\right)$

   (j) $l(l(x))$

3. If $f(x) = x^2$, find:

   (a) $f(4)$

   (b) $f(t)$

   (c) $f(\sqrt{2})$

   (d) $f(u^{1/4})$

   (e) $\dfrac{f(x)}{x}$

   (f) $f(a + b)$

   (g) $f(a) + f(b)$

4. If $q(x) = (x - 2)^2 + 5$, find:
   (a) $q(t)$
   (b) $q(2)$
   (c) $q(0)$
   (d) $q(x + 2)$
   (e) $q(x - 2)$
   (f) $q(x^2)$
   (g) $q(-x)$
   (h) $q(2 - x)$
   (i) $q(2 + \sqrt{x - 5})$
   (j) $q(2 - \sqrt{x - 5})$

5. If $g(x) = x + 1/x$ $(x \neq 0)$, find:
   (a) $g(1)$
   (b) $g(2x)$
   (c) $g\left(\dfrac{1}{x}\right)$
   (d) $3g(3)$
   (e) $g(t^2)$
   (f) $\dfrac{1}{g(u)}$
   (g) $xg(x) - 1$
   (h) $\dfrac{1}{g(\frac{1}{2})}$

6. If $f(x) = \dfrac{x - 1}{x + 1}$ $(x \neq -1)$, then:
   (a) $f(0) =$
   (b) $f(1) =$
   (c) $f(s) =$
   (d) $f\left(\dfrac{1}{x}\right) =$
   (e) $\dfrac{1}{f(x)} =$
   (f) $f(3x + 4) =$
   (g) If $f(x) = 2$, $x =$
   (h) If $f(x) = 0$, $x =$
   (i) If $f(x) = -1$, $x =$
   (j) If $f(x) = 3$, $x =$

7. Let $p(x) = 1/x - 1$ $(x \neq 0)$. Verify that:
   (a) $p(x) + p(-x) = -2$
   (b) $p(2x) = \frac{1}{2}[p(x) - 1]$
   (c) $p(1 - x) = \dfrac{1}{p(x)}$
   (d) $\dfrac{-1}{p(x + 1)} = p(x) + 2$
   (e) $\dfrac{1}{p(x) + 1} = p\left(\dfrac{1}{x}\right) + 1$

8. Let $l(x) = 2x - 3$ and let I be the interval $[1, 2]$. Which of the following intervals does l take I into?

$$[-3, 0] \quad [-2, 1] \quad [-1, 2] \quad [0, 3] \quad [1, 4]$$

9. Let $q(x) = x^2 + x$ and let I be the interval $[-1, \frac{1}{2}]$. Which of the following intervals does q take I into?

$$[-1, 0] \quad [-\tfrac{3}{4}, \tfrac{1}{2}] \quad [-\tfrac{1}{2}, \tfrac{3}{4}] \quad [-\tfrac{1}{4}, 1] \quad [0, \tfrac{3}{2}]$$

10. If g takes I into the interval $(1, 4)$, which of the following intervals does $-g$ necessarily take I into?

$$(1, 4) \qquad (0, 4) \qquad (-1, 4) \qquad (-4, 0) \qquad (-3, 3)$$

11. If h takes I into the interval $(1, 4)$, which of the following intervals does $1/h$ necessarily take I into?

$$(1, 4) \qquad (0, 4) \qquad (-4, 0) \qquad (\tfrac{1}{2}, 2) \qquad (\tfrac{1}{100}, 1)$$

12. Find an open interval I that contains the given point a and that f takes into the given interval J.

(a) $f(x) = \tfrac{1}{3}x + 1, \qquad a = -1, \quad J = (0, 1)$
(b) $f(x) = x^2 - 2x, \qquad a = 3, \quad J = (0, 5)$
(c) $f(x) = x^2 - 2x, \qquad a = -1, \quad J = (0, 5)$
(d) $f(x) = x^3, \qquad\quad a = -1, \quad J = (-2, 2)$

(e) $f(x) = \dfrac{1}{x} \; (x > 0), \qquad a = 1, \quad J = (\tfrac{1}{10}, 2)$

(f) $f(x) = \dfrac{x - 1}{x + 1} \; (x > -1), \qquad a = 2, \quad J = (0, 1)$

13. If f takes I into $(0, 5)$ and g takes I into $(-5, 10)$, which of the following intervals does $f + g$ necessarily take I into?

$$(0, 5) \qquad (-5, 10) \qquad (0, 10) \qquad (-5, 15) \qquad (0, 15)$$

14. If f takes I into $(0, 5)$ and g takes I into $(5, 10)$, which of the following intervals does fg necessarily take I into?

$$(5, 10) \qquad (5, 50) \qquad (0, 50) \qquad (-50, 0) \qquad (-25, 25)$$

15. If f takes I into $(0, 5)$ and g takes I into $(-5, 10)$, which of the following intervals does fg necessarily take I into?

$$(-5, 10) \qquad (-5, 50) \qquad (0, 50) \qquad (-50, 25) \qquad (-25, 50)$$

16. Decide whether the following functions are bounded on the intervals mentioned.

(a) $k(x) = \dfrac{1}{x}$, on $(-1, 0)$ \qquad (e) $h(x) = \dfrac{1 + x}{1 - x}$, on $(0, 1)$

(b) $s(x) = \sqrt{x}$, on $\{x: x \geq 0\}$ \qquad (f) $F(x) = \dfrac{1 - x}{1 + x}$, on $(-1, 1)$

(c) $t(x) = \dfrac{1}{1 + x^2}$, on $\mathcal{R}$ \qquad (g) $G(x) = \dfrac{1 - x^2}{1 + x}$, on $(-1, 1)$

(d) $g(x) = \dfrac{x}{1 + x^2}$, on $\mathcal{R}$

17. Determine on what intervals the following functions are increasing and on what intervals they are decreasing.

(a) x^2

(b) x^3

(c) $\sqrt{x}$

(d) $|x|$

(e) $x + |x|$

(f) $1 - \dfrac{1}{x}$ $(x > 0)$

18. Determine on what intervals the following functions are increasing and on what intervals they are decreasing.

(a) $x^2 + \dfrac{1}{x^2}$ $(x > 0)$ [*Hint*. Where is $x - 1/x$ increasing?]

(b) $x + \dfrac{1}{x}$ $(x > 0)$

19. (a) If f is increasing, is $2f$ necessarily increasing?

(b) If f is increasing, is $-f$ necessarily decreasing?

(c) If f is increasing, is f^2 necessarily increasing?

(d) If f is increasing and is never zero, is $1/f$ necessarily decreasing?

20. Let f and g be functions defined on the same interval.

(a) If f and g are both increasing, is $f + g$ necessarily increasing?

(b) Give an example in which f is increasing, g is decreasing, and $f + g$ is increasing.

(c) Give an example in which f and g are both monotonic but $f + g$ is not monotonic.

21. A function f has the following property:

(5)
$$f(x + 1) = f(x) + f(1) + 1 \qquad (x \in \mathscr{R}).$$

(a) What is $f(0)$?

Assume further that $f(1) = 1$. Find:

(b) $f(2)$

(c) $f(3)$

(d) $f(-1)$

22. A function g has the following property:

(6)
$$g(x + y) = g(x) + g(y) \qquad (x \in \mathscr{R}, y \in \mathscr{R}).$$

(a) What is $g(0)$?

Prove that:

(b) $g(-x) = -g(x)$ $(x \in \mathscr{R})$

(c) $g(2x) = 2g(x)$ $(x \in \mathscr{R})$

Assume further that $g(1) = 1$. Find:

(d) $g(2)$

(e) $g(3)$

(f) $g(\tfrac{1}{2})$

23. Prove that these all say the same thing:
    (i) f is bounded on I.
    (ii) There is a number M such that

$$|f(x)| \le M \qquad \text{for all} \qquad x \in I.$$

(iii) The graph of f over I lies between two horizontal lines.

24. Prove that:
    (a) All constant functions are bounded.
    (b) If f is bounded on I then so is $-f$.
    (c) If f and g are bounded on I then so is $f + g$.
    (d) If f and g are bounded on I then so is fg. [*Hint.* See Problem 23.]

25. What are the images of the functions in Problem 1?

Answers to problems (§1F)

1. (a) $\mathscr{R}$ (d) $\{x : x \ne -1\}$ (h) $\{x : x \ne 0\}$
2. (a) 13   (e) $6x + 4$
3. (a) 16   (c) 2
4. (a) $(t - 2)^2 + 5$ (c) 9   (g) $(x + 2)^2 + 5$
5. (a) 2   (c) $x + \dfrac{1}{x}$
6. (a) -1 (d) $\dfrac{1 - x}{1 + x}$ (g) -3
8. $[-2, 1], [-1, 2]$
11. $(0, 4), (\frac{1}{100}, 1)$
13. $(-5, 15)$
16. (c) yes
17. (a) decreasing on $x \le 0$, increasing on $x \ge 0$

§1G. Composite and inverse

1G1. Composite functions. Consider the function

$$h(x) = (x + 1)^2 \qquad\qquad (x \in \mathscr{R}).$$

To get from x to $h(x)$ we perform two operations: first add 1, then square. First we operate with

$$g(x) = x + 1,$$

then with

$$f(x) = x^2.$$

In brief,

(1) $$h(x) = f(g(x)).$$

The h-box consists of a g-box followed by an f-box (Figure 1).

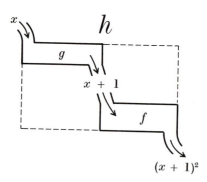

FIGURE 1

In general, if f and g are functions such that the image of g is contained in the domain of f, then the *composite* function

(2) $$f(g(x))$$

is defined for each x in the domain of g.

Example 1. Let

$$f(x) = \frac{1}{x} \quad (x \neq 0) \qquad \text{and} \qquad g(x) = 1 + x.$$

Then

$$f(f(x)) = x \quad (x \neq 0), \qquad\qquad g(g(x)) = 2 + x,$$

$$f(g(x)) = \frac{1}{1 + x} \quad (x \neq -1), \qquad g(f(x)) = 1 + \frac{1}{x} \quad (x \neq 0).$$

1G2. *Inverse functions.* Let f be a function with domain I and image J. It can happen that there are two points in I having the same functional value (Figure 2)—or else no such pair exists (Figure 3). In the latter case, for each number $y \in J$ there is *exactly one* number $x \in I$ such that $f(x) = y$. We call it $f^{-1}(y)$ ("f inverse" of y). We thus define a function f^{-1} with domain J and image I:

(3) $$f^{-1}(y) = x \qquad\qquad (y \in J)$$

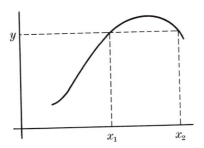

FIGURE 2

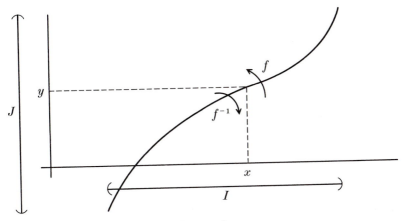

FIGURE 3

means

(4) $$f(x) = y \qquad (x \in I).$$

The relation between f and f^{-1} can be described as follows. Putting $y = f(x)$ in (3) we get

(5) $$f^{-1}(f(x)) = x \qquad (x \in I);$$

and putting $x = f^{-1}(y)$ in (4) we get

(6) $$f(f^{-1}(y)) = y \qquad (y \in J).$$

If we apply both f and f^{-1}, in either order, we return to where we started. The functions f and f^{-1} are said to be *inverses* of one another.

The letter y was a helpful symbol during the course of the preceding discussion, but there is nothing magic about that letter in formula (6). Instead, we ordinarily write

(7) $$f(f^{-1}(x)) = x \qquad (x \in J).$$

Let f be a function with an inverse. If a point (a, b) lies on the graph of f, then (b, a) is a point on the graph of f^{-1}. A little reflection shows that the graph of f^{-1} is the mirror image, through the line $y = x$, of the graph of f. See Figure 4.

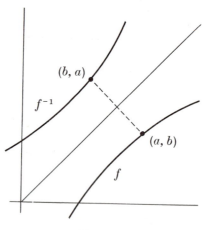

FIGURE 4

1G3. Monotonic functions. Clearly, if a function f is increasing, then it has an inverse. Moreover, f^{-1} is also an increasing function. For, given y_1 and y_2 in J, let $x_1 = f^{-1}(y_1)$, $x_2 = f^{-1}(y_2)$. If $x_2 \leq x_1$ then (since f is increasing) $y_2 \leq y_1$. Thus if $y_1 < y_2$ then $x_1 < x_2$. This shows that f^{-1} is increasing.

Similarly, if f is decreasing then f^{-1} exists and is decreasing.

1G4. Finding the inverse. Often f is given by a formula and we can find f^{-1} by solving the equation

$$f\big(f^{-1}(x)\big) = x.$$

Example 2. (See Figure 5.) Let $l(x) = mx + b$ $(m \neq 0)$. Then

$$l\big(l^{-1}(x)\big) = x,$$

i.e.,

$$ml^{-1}(x) + b = x.$$

Thus,

$$l^{-1}(x) = \frac{1}{m}(x - b).$$

If a car travels at constant speed m, the distance traveled is a linear function of time, and the slope of the straight-line graph is m. The graph of time as a

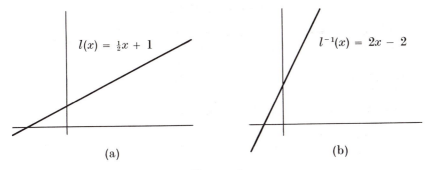

$l(x) = \tfrac{1}{2}x + 1$ $l^{-1}(x) = 2x - 2$

(a)   (b)

FIGURE 5

function of distance is also a straight line, this time with slope $1/m$. A car going 40 miles per hour is doing 1/40 hour per mile.

Example 3. (See Figure 6.) Let $F(x) = x^2 + 1$ $(x \geq 0)$. Then F is increasing and takes its domain $\{x: x \geq 0\}$ into the interval $\{x: x \geq 1\}$. In fact, F assumes every value ≥ 1. For, given $x \geq 1$, we can solve the equation

$$F(F^{-1}(x)) = x$$

for a unique value $F^{-1}(x)$ lying in the domain of F:

$$[F^{-1}(x)]^2 + 1 = x,$$
$$F^{-1}(x) = \sqrt{x - 1} \geq 0 \qquad\qquad (x \geq 1).$$

Since F is increasing, so is F^{-1}.

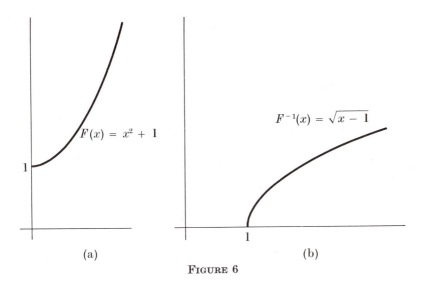

$F(x) = x^2 + 1$ $F^{-1}(x) = \sqrt{x - 1}$

(a)   (b)

FIGURE 6

Problems (§1G)

1. Let $f(x) = x^2$ and $g(x) = x^2 + 1$. Evaluate:
   (a) $f(0)$
   (b) $f(f(0))$
   (c) $f(f(f(0)))$
   (d) $g(0)$
   (e) $g(g(0))$
   (f) $g(g(g(0)))$
   (g) $f(g(2))$
   (h) $g(f(2))$

2. Let f and g be as in Problem 1. Verify that:

   (a) $g\left(\dfrac{1}{x}\right) = \dfrac{g(x)}{f(x)}$

   (b) $f(g(x)) = g(f(x)) + 2f(x)$

3. Let $s(x) = \sqrt{x}$ and $h(x) = x + 1$. Evaluate:
   (a) $s(h(0))$
   (b) $h(s(0))$
   (c) $s(h(8))$
   (d) $h(s(8))$

4. Let s and h be as in Problem 3. Verify that:
   (a) $(s(h(x)))^2 = h(x)$
   (b) $(h(s(x)))^2 = h(x) + 2s(x)$

5. Let s and h be as in Problem 3. Solve the equation
$$s(h(x)) = h(s(x)).$$

6. Compute $f(f(x)), f(g(x)), g(f(x))$, and $g(g(x))$ in each of the following cases.

   (a) $f(x) = x^2, \quad g(x) = \dfrac{1}{x}$

   (b) $f(x) = \sqrt{x}, \quad g(x) = x + 1$
   (c) $f(x) = x^2 + 1, \quad g(x) = \sqrt{x-1}$
   (d) $f(x) = 2x - 1, \quad g(x) = \frac{1}{2}(x + 1)$

7. Let
$$f(x) = \frac{x-1}{x+1}, \qquad g(x) = \frac{x+1}{x-1}, \qquad n(x) = -x, \qquad r(x) = \frac{1}{x}.$$
Verify that:

   (a) $f(n(x)) = r(f(x)) = g(x)$
   (b) $g(n(x)) = r(g(x)) = f(x)$
   (c) $f(r(x)) = n(f(x))$
   (d) $g(r(x)) = n(g(x))$
   (e) $f(g(x)) = r(x)$
   (f) $g(f(x)) = n(x)$
   (g) $f(f(x)) = r(n(x)) = n(r(x))$
   (h) $g(g(x)) = n(n(x)) = r(r(x))$

8. Express each of the functions below as composites of two or more of the following:
$$a(x) = x + 1 \qquad g(x) = x^3$$
$$b(x) = x - 2 \qquad h(x) = \frac{1}{x}$$
$$e(x) = 3x \qquad k(x) = \sqrt{x}$$
$$f(x) = x^2$$

(a) $3x + 1$

(b) $3x + 3$

(c) $3x^2$

(d) $9x^2$

(e) $(x^3 - 2)^2$

(f) $9x + 3$

(g) $\dfrac{1}{\sqrt{x^2 + 1}}$

(h) $x + 2$

(i) $x - 1$

(j) $x^2 - 1$

(k) $3x + 2$

(l) $\sqrt{x^3 + 1}$

(m) $\sqrt{x + 1}$

(n) $\sqrt{x} + 1$

(o) $\dfrac{1}{\sqrt{x}}$

(p) $\dfrac{1}{\sqrt{x - 2}}$

(q) $\dfrac{1}{\sqrt{x} - 2}$

(r) $x^{3/2}$

(s) $\frac{1}{3}x$

9. Specify a function g for which

   (a) $g(2x) = x$

   (b) $g(1 + x) = x$

   (c) $g(1 - x) = x$

   (d) $g(x^2) = x$

   (e) $g\left(\dfrac{1}{x}\right) = x$

10. Specify a function h such that

    (a) $h(1 + x) = 4x - 1$

    (b) $h(2x) = 4x - 1$

    (c) $h(x^2) = 4x - 1$

    (d) $h(1 - x) = 4x - 1$

    (e) $h\left(\dfrac{1}{x}\right) = 4x - 1$

11. Give an example of a function f for which

    (a) $f(f(x)) = x + 2$

    (b) $f(f(x)) = 4x + 3$

    (c) $f(f(x)) = x$

    (d) $f(f(x)) = 2x + 4$

12. Match the functions in the first column with their inverses in the second column.

$$f_1(x) = \frac{1}{x + 2} \qquad g_1(x) = \frac{x}{1 - x}$$

$$f_2(x) = \frac{x}{x - 1} \qquad g_2(x) = \frac{x}{x - 1}$$

$$f_3(x) = 3 + \frac{1}{x} \qquad g_3(x) = \frac{1}{x} - 2$$

$$f_4(x) = \frac{x}{2} - 2 \qquad g_4(x) = \frac{1}{x - 3}$$

$$f_5(x) = \frac{x}{x + 1} \qquad g_5(x) = 2x + 4$$

13. Find the inverses:

(a) $f(x) = x - 1$

(b) $f(x) = 3x - 1$

(c) $f(x) = \dfrac{1}{x + 1}$

(d) $f(x) = \dfrac{1}{x}$

(e) $f(x) = \dfrac{x + 13}{x + 14}$

(f) $f(x) = x^3$

(g) $f(x) = \sqrt{2x + 1}$

(h) $f(x) = x^2 + 4 \ (x \geq 0)$

14. Verify that each of the following functions is its own inverse.

(a) $f(x) = x$

(b) $f(x) = -x$

(c) $f(x) = \dfrac{1}{x}$

(d) $f(x) = \sqrt{1 - x^2} \ (x \geq 0)$

(e) $f(x) = \dfrac{x + 1}{x - 1}$

(f) $f(x) = 2 + \dfrac{1}{x - 2}$

(g) $f(x) = -\dfrac{x}{x + 1}$

(h) $f(x) = \dfrac{ax + b}{x - a}$

15. Verify that each of the following functions satisfies

$$f(f(f(x))) = x.$$

(a) $f(x) = 1 - \dfrac{1}{x}$

(b) $f(x) = 2 - \dfrac{1}{x - 1}$

(c) $f(x) = -\dfrac{1}{x + 1}$

(d) $f(x) = a - \dfrac{1}{x + b}$,   where $a + b = 1$.

16. Let f and g be defined "piecewise," as follows:

(8) $f(x) = 1 - \tfrac{1}{3}x \ (x < 0)$, $f(x) = 1 - x \ (x \geq 0)$

and

(9) $g(x) = 2x - x^2 \ (x < 1)$, $g(x) = x^2 - 2x + 2 \ (x \geq 1)$.

(a) Sketch the graphs.

Find:

(b) $f(g(-1))$

(c) $f(g(0))$

(d) $f(g(\frac{1}{2}))$

(e) $f(g(1))$

(f) $f(g(2))$

(g) $g(f(-1))$

(h) $g(f(0))$

(i) $g(f(\frac{1}{2}))$

(j) $g(f(1))$

(k) $g(f(2))$

(l) $f(f(-1))$

(m) $g(g(2))$

17. Let f and g be as in Problem 16. Solve for x:

(a) $f(x) = 2$

(b) $g(x) = 2$

(c) $f(x) = x$

(d) $g(x) = x$

(e) $f(x) = g(x)$

(f) $f(g(x)) = 2$

(g) $g(f(x)) = 2$

18. Let f and g be defined "piecewise," as follows:

(10) $$f(x) = |x| \quad (x < 1), \qquad f(x) = 2x - 1 \quad (x \geq 1)$$

and

(11) $$g(x) = 2 - x^2 \quad (x < 0), \qquad g(x) = x + 2 \quad (x \geq 0).$$

(a) Sketch the graphs.

Find:

(b) $f(g(0))$

(c) $f(g(1))$

(d) $f(g(-2))$

(e) $f(f(-1))$

(f) $f(f(-2))$

(g) $g(f(0))$

(h) $g(f(-1))$

(i) $g(f(-2))$

(j) $g(g(1))$

(k) $g(g(-1))$

19. Let f and g be as in Problem 18.

Solve for x:

(a) $f(x) = 0$

(b) $g(x) = 0$

(c) $f(x) = x$

(d) $g(x) = x$

(e) $f(x) = g(x)$

(f) $f(g(x)) = 1$

(g) $g(f(x)) = 1$

Next:

(h) Prove that $f(x) \geq 0$ for all x.

(i) Give an example of a value of x for which $g(x) < 0$.

(j) Prove that $g(f(x)) > 0$ for all x.

(k) Does f have an inverse?

(l) Does $g(f(x))$ have an inverse?

20. Let f and g be as in Problem 18.

(a) Write out a "piecewise" definition of $f(g(x))$.

(b) Sketch its graph.

Answers to problems (§1G)

1. (a) 0   (b) 0   (c) 0   (d) 1   (e) 2   (f) 5   (g) 25   (h) 17

3. (a) 1   (b) 1   (c) 3   (d) $2\sqrt{2}+1$

6. (a) x^4, $1/x^2$, $1/x^2$, x

8. (a) $a(e(x))$ (b) $e(a(x))$

13. (a) $f^{-1}(x) = x + 1$

16. (b) 2   (c) 1   (d) $\frac{1}{4}$ (e) 0   (f) -1 (g) $\frac{10}{9}$ (h) 1   (i) $\frac{3}{4}$
    (j) 0   (k) -3 (l) $-\frac{1}{3}$ (m) 2

17. (a) -3

18. (b) 3   (c) 5   (d) 2   (e) 1   (f) 3   (g) 2   (h) 3   (i) 4   (j) 5
    (k) 3

19. (a) 0

§1H. Completeness

There is a subtle property of the real line that we have not yet discussed, namely, the property of *completeness*. It expresses in mathematical terms the intuitive fact that the line is all one connected piece. Proofs involving this property tend to be sophisticated. Many teachers consider them too sophisticated for a first course. We have concentrated the sophistication into a single theorem, which we call the "Fundamental Lemma on Closed Intervals" (§1H4 below). (A "lemma" is a theorem of interest primarily for proving other theorems rather than for its own sake.) Its statement is somewhat abstract. However, the needed results all follow from it without effort.

1H1. Upper bounds. Let S be any set of real numbers (other than the "empty" set). A number u is called an *upper bound* of S if every member of S is $\le u$.

Some sets have upper bounds and some haven't. The set $\mathscr{R}$ has no upper bound. Neither has the set of all positive integers.

For the open interval $(0, 1)$, 1 and all numbers greater than 1 are upper bounds. For the closed interval $[0, 1]$, also, 1 and all greater numbers are upper bounds. There are no others; in each case, 1 is the *least* of all the upper bounds.

Not all sets have a least element. The open interval $(0, 1)$ has no least element. But the *set of upper bounds* of $(0, 1)$ does have a least element. The *Axiom of Completeness* asserts that sets of this type always have.

1H2. Axiom of Completeness. *If a set has an upper bound it has a least upper bound.*

1H3. Lower bounds. A number l is called a *lower bound* of a set S (not the "empty" set) if every member of S is $\geq l$. The following result is included for the sake of completeness.

If a set S has a lower bound it has a greatest lower bound.

Proof. The set L of all lower bounds of S has an upper bound (any member of S) and hence has a least upper bound, u.

u is a lower bound of S. For if not, some $x \in S$ satisfies $x < u$. Since x is an upper bound of L, u would not be the least.

u is the greatest lower bound of S. For, u is an upper bound of L, and, as we have just proved, $u \in L$.

1H4. Fundamental Lemma on Closed Intervals. The lemma refers to points and intervals within a given interval $[a, b]$. As a matter of fact, it speaks about a *set of intervals* in $[a, b]$.

We introduce an abbreviation. Let I be an interval in $[a, b]$ of more than one point, and let $x \in I$. We will say for short that I is an interval "about" x if either $x = a$, or $x = b$, or x belongs to the interior of I (Figure 1). We may then state the lemma as follows.

Let $\mathscr{I}$ be a set of closed intervals in $[a, b]$. Suppose that:

(i) *Whenever two members of $\mathscr{I}$ intersect, their union is also in $\mathscr{I}$; and*

(ii) *Each point of $[a, b]$ has an interval about it that belongs to $\mathscr{I}$.*

Then $[a, b]$ itself belongs to $\mathscr{I}$.

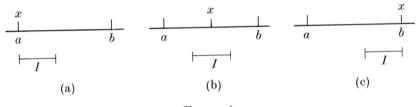

FIGURE 1

Proof. We look at all the intervals of the form $[a, x]$. Presumably, some belong to $\mathscr{I}$ and some don't. At least one does, since by (ii) the point a itself has an interval about it that belongs to $\mathscr{I}$; such an interval is of the form $[a, x_0]$ for some $x_0 > a$. (See Figure 2(a).)

Now consider the set S defined as follows: S is *the union of all the intervals $[a, x]$ that belong to $\mathscr{I}$.* In other words, a point belongs to S provided that some interval $[a, x]$ of $\mathscr{I}$ contains it. For instance, $x_0 \in S$.

The set S has an upper bound, namely, b. By the Axiom of Completeness, S has a *least* upper bound. Call the least upper bound u.

Since u is an upper bound of S, $x_0 \leq u$. Therefore $u \neq a$. Hence,

$$\text{either} \quad a < u < b \quad \text{or} \quad u = b.$$

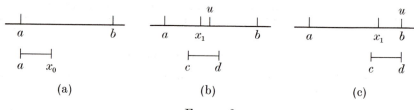

FIGURE 2

In any event, by (ii), there is an interval $[c, d]$ about u that belongs to $\mathscr{I}$. The two possibilities are shown in Figure 2, (b) and (c).

Let us show that in either case, $[a, d] \in \mathscr{I}$. Since $c < u$, c is not an upper bound of S (since u is the *least* upper bound). Consequently, some interval $[a, x_1]$ of $\mathscr{I}$ contains c (otherwise c *would* be an upper bound). Then $[a, x_1]$ and $[c, d]$ are intersecting members of $\mathscr{I}$. By (i), their union belongs to $\mathscr{I}$; that is, $[a, d] \in \mathscr{I}$.

Since $[a, d] \in \mathscr{I}$, $d \leq u$. This rules out the case $a < u < b$ (Figure 2(b)). Therefore $u = b$. Then $b = d$ (Figure 2(c)) and we have $[a, b] \in \mathscr{I}$. This is what was to be proved.

Problems (§1H)

1. Does the given set have a greatest element? If so, what is it?
    (a) the set of real numbers
    (b) the set of rational numbers
    (c) the set of positive integers
    (d) the set of negative integers
    (e) the set of positive rational numbers
    (f) the set of nonnegative rational numbers
    (g) the set of positive irrational numbers
    (h) the set of nonnegative irrational numbers
    (i) the set of irrational numbers between 1 and 2
    (j) the set $\{1, 3, 4, 6, 9\}$
    (k) the closed interval $[3, 4]$
    (l) the open interval $(3, 4)$
    (m) the half-open interval $[3, 4)$
    (n) the half-open interval $(3, 4]$

2. In Problem 1, does the given set have a least element? If so, what is it?

3. In Problem 1, does the given set have an upper bound? If so, what is its *least* upper bound?

4. In Problem 1, does the given set have a lower bound? If so, what is its *greatest* lower bound?

5. If 4 and 6 are both upper bounds of a given set,
    (a) is 10?   (b) is 7?   (c) is 5?

6. Can the given set be the set of all upper bounds of some set?
    (a) the set of positive integers
    (b) the set of nonnegative integers

    (c) the set of positive real numbers

    (d) the set of nonnegative real numbers

    (e) the open interval (0, 1)

    (f) the closed interval [0, 1]

    (g) the interval $\{x: x \geq 3\}$

7. Exhibit two sets with the following properties:

    (i) neither one has a greatest element,

    (ii) they have the same least upper bound, 1,

and

    (iii) they have no element in common.

Answers to problems (§1H)

1. (a) no   (i) no

3. (a) no   (i) yes, 2

6. (a) no   (d) yes

CHAPTER 2

Limits and Continuity

In this chapter, we define the notion of limit and the related concept of continuity, and establish their fundamental properties. The ideas here underlie all of calculus, but they are harder to understand than the topics in calculus proper. The problem of how to treat this material in a first course is a subject of never-ending controversy. We have included complete details so that each teacher can make his own choice.

Our own recommendation is to go over everything in the chapter that is not set in small type. This means learning definitions and statements of results, but not necessarily the proofs.

§2A. The problem of tangents

A central problem in which limits appear is the problem of tangents: to determine the line tangent to a given graph at a given point. This is not just an exercise in geometry. It is the common formulation into geometric terms of dozens of basic problems in the physical, mathematical, biological, and social sciences.

Before plunging into computation we have to decide what a tangent *is*. The old standbys—that the tangent meets the curve only once, that it does not cross the curve at the point of tangency—do not hold up. In Figure 1, the line *l* is surely tangent at *P*, yet it crosses the graph there and meets it elsewhere as well. We need a new idea.

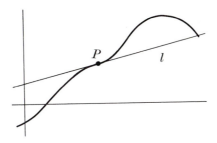

FIGURE 1

As a specific problem, let us ask for the tangent line to the parabola $y = x^2$ at the point (1, 1) (Figure 2). As we already know a point on the line, the problem is to decide on its slope. The new idea is to begin with *approximations* to the line we want. Let (x, y) be any point on the graph *other than*

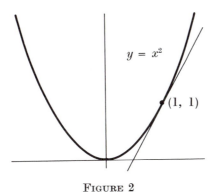

FIGURE 2

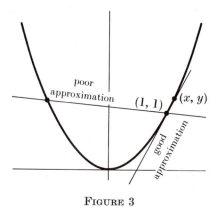

FIGURE 3

(1, 1). The *two* points determine a line. If (x, y) is close to $(1, 1)$ we expect the slope of the line to be close to the slope we are looking for (Figure 3). Our feelings are imprecise but compelling. The slope we seek will be a limiting value, in a sense yet to be defined, of the approximating slopes.

Since (x, y) is on the graph and different from $(1, 1)$, $x \neq 1$. The slope of the line through (x, y) and $(1, 1)$ is

$$(1) \qquad\qquad m = \frac{y - 1}{x - 1} \qquad\qquad (x \neq 1).$$

To learn something from this expression we put it in terms of x alone and simplify. Since $y = x^2$,

$$(2) \qquad\qquad m = \frac{x^2 - 1}{x - 1} = x + 1 \qquad\qquad (x \neq 1).$$

When x is close to 1 this is close to 2, so that 2 is apparently the limiting value we are after. It will turn out that 2 *is* the limiting value. The tangent line will then be, by decree, that line through $(1, 1)$ whose slope is 2.

The task for now is to put this concept on ice. This is the subject of the rest of the chapter. The application to tangents is resumed in Chapter 3.

Problems (§2A)

1. What is the equation of the tangent line in the example of the text?

2. Reasoning as in the text, decide on the slope of the tangent line to the given graph at the point indicated. Write the equation of the line and sketch the graphs.

(a) $y = x^2$,   at $(-1, 1)$ (g) $y = x^3 - x$,   at $(0, 0)$

(b) $y = x^2$,   at $(0, 0)$ (h) $y = \sqrt{x}$,   at $(1, 1)$

(c) $y = x^2$,   at $(\frac{1}{2}, \frac{1}{4})$ (i) $y = \sqrt{x}$,   at $(9, 3)$

(d) $y = x^2 + x$,   at $(1, 2)$ (j) $y = \dfrac{1}{x}$,   at $(1, 1)$

(e) $y = 2x^2 - 3x + 1$,   at $(0, 1)$ (k) $y = \dfrac{1}{x}$,   at $(2, \frac{1}{2})$

(f) $y = x^3$,   at $(0, 0)$ (l) $y = \dfrac{1}{x}$,   at $(\frac{1}{2}, 2)$

Answers to problems (§2A)

1. $y = 2x - 1$
2. (a) $y = -2x - 1$ (b) $y = 0$ (c) $y = x - \frac{1}{4}$ (d) $y = 3x - 1$
   (e) $y = -3x + 1$ (f) $y = 0$ (g) $y = -x$ (h) $y = \frac{1}{2}x + \frac{1}{2}$
   (i) $y = \frac{1}{6}x + \frac{3}{2}$ (j) $y = -x + 2$ (k) $y = -\frac{1}{4}x + 1$ (l) $y = -4x + 4$

§2B. Definition of limit

2B1. *The idea of a limit.* The intuitive idea of a limit is natural enough but the precise concept is hard to pin down. Let f be a function and let a be a point on the real line. We are going to define

$$(1) \qquad\qquad\qquad \lim_{x \to a} f(x),$$

the limit of $f(x)$ as x approaches a. The assertion that

$$(2) \qquad\qquad\qquad \lim_{x \to a} f(x) = b$$

will express in a precise way the rough idea that $f(x)$ is close to b when x is close to (but not equal to) a.

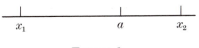

FIGURE 1

In order to talk about "x close to a" we assume that the domain of f includes an interval of the form (x_1, a) or (a, x_2) (Figure 1). Of course, it may include both, and much more besides.

We do not require, however, that the domain include the point a itself. The value of f at a, if there is one, has nothing to do with the question. In many important applications f is not defined at a. The following simple examples make this clear and should be thoroughly understood.

The slope function

$$f(x) = x + 1 \qquad\qquad (x \neq 1)$$

encountered in the preceding section was not defined at $x = 1$, but we were interested in $\lim_{x \to 1} f(x)$ and decided (correctly, as it will turn out) that $\lim_{x \to 1} f(x) = 2$. See Figure 2(a).

If g is the function defined (for some reason) by

$$g(x) = x + 1 \quad \text{if} \quad x \neq 1, \qquad g(1) = 3,$$

then g *is* defined at $x = 1$, but it will still turn out that $\lim_{x \to 1} g(x) = 2$, the value of $g(1)$ being irrelevant (Figure 2(b)).

Finally, if

$$h(x) = x + 1 \text{ for } all \ x,$$

then again $\lim_{x \to 1} h(x) = 2$; in this case, as it happens, h is defined at $x = 1$ and $h(1) = 2$ (Figure 2(c)).

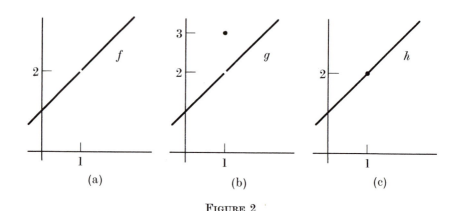

FIGURE 2

2B2. Interval cut at a point. Closeness can be expressed in terms of intervals. To be close to b means to be inside some small interval about b

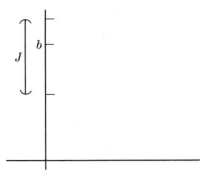

FIGURE 3

(Figure 3). Whether the endpoints are included or not is clearly of no consequence. For the rest of this chapter, the word "interval," unmodified, will mean *open* interval.

A set consisting of an interval minus one of its points will be called, for short, an interval *cut* at the point. An (open) interval I cut at a is a set of the form $(x_1, a) \cup (a, x_2)$ (Figure 4). To be close to but not equal to a means to be inside some small interval cut at a.

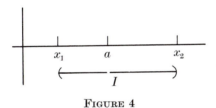

FIGURE 4

2B3. Definition of limit. We say that

$$\lim_{x \to a} f(x) = b$$

if, *given any (open) interval J about b, there is an (open) interval I cut at a that f takes into J* (Figure 5). In other words, given an interval J about b, there is an interval I about a such that

$$if \quad x \in I \quad then \quad f(x) \in J \qquad\qquad (x \neq a).$$

The definition is interpreted relative to the domain of f, which may include points on one side of a but not the other. When we say that f takes all points of I into J we understand this to mean all points of I at which f is defined.

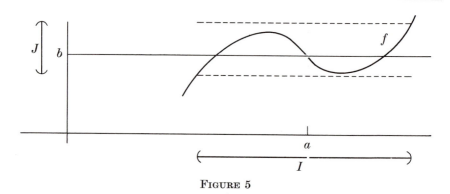

FIGURE 5

In the definition, J is a challenge and I a response. To verify that $\lim_{x \to a} f(x) = b$ one must be able to respond to every challenge. In addition, one must usually guess the value of the limit in the first place. The ability to make good guesses and test them depends upon experience and skill.

If there is no number b for which $\lim_{x \to a} f(x) = b$ we say that $\lim_{x \to a} f(x)$ does not exist.

2B4. Limit of a constant function. $\lim_{x \to a} c = c$ *(c a constant)*.

Proof. Given an interval J about c, let I be any interval at all cut at a (Figure 6). The constant function c takes I to the single value c and hence into J.

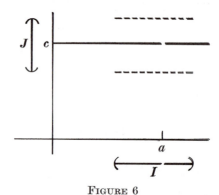

FIGURE 6

2B5. Limit of a linear function. Let $l(x) = mx + b$. Then $\lim_{x \to a} l(x) = ma + b$.

Proof. We assume $m > 0$. ($m < 0$ is similar and $m = 0$ is §2B4.) Given an interval $J = (y_1, y_2)$ about $ma + b$, choose $x_1 = (y_1 - b)/m$ and $x_2 = (y_2 - b)/m$, so that $l(x_1) = y_1$ and $l(x_2) = y_2$. Then $I = (x_1, a) \cup (a, x_2)$ is an interval cut at a that l takes into J (Figure 7).

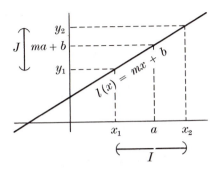

FIGURE 7

2B6. Notation. Another handy expression for $\lim_{x \to a} f(x) = b$ is

$$f(x) \to b \qquad \text{as} \qquad x \to a,$$

"$f(x)$ approaches b as x approaches a."

Example 1.

$$\lim_{x \to 4} 7 = 7, \qquad \lim_{x \to x_0} x = x_0, \qquad \lim_{x \to 0} 3x = 0, \qquad \lim_{x \to a} (-x) = -a,$$

$$\lim_{x \to 2} (4x + 3) = 11, \qquad 2x \to 6 \quad \text{as} \quad x \to 3.$$

Example 2. For each point (x, y) on the curve $y = x^2$ $(x > 0)$, let $T(x)$ and $P(x)$ be the areas of the triangle and parallelogram formed as shown in Figure 8. What happens to their ratio, $R(x) = P(x)/T(x)$, as $x \to 0$? What is your guess?

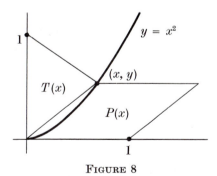

FIGURE 8

Comments. $R(x)$ is defined for $x > 0$ and nowhere else. Note that for $x = 0$ there are no areas to talk about.

It is obvious geometrically that $T(x) \to 0$ and $P(x) \to 0$ as $x \to 0$. So this problem, like the problem of tangents, asks for the limit of a certain fraction whose numerator and denominator are both approaching 0.

Solution. The triangle has base 1 and altitude x, and the parallelogram has base 1 and altitude $y = x^2$. Therefore

$$T(x) = \tfrac{1}{2}x, \qquad P(x) = x^2,$$

and

$$R(x) = \frac{P(x)}{T(x)} = 2x \rightarrow 0 \qquad (\text{as} \quad x \rightarrow 0),$$

as we know from §2B5.

Problems (§2B)

1. Evaluate the following limits.

   (a) $\lim\limits_{x \to 1} (6x - 7)$

   (b) $\lim\limits_{x \to 0} (57x + 2)$

   (c) $\lim\limits_{x \to 0} 4$

   (d) $\lim\limits_{t \to 6} (4t - 1)$

   (e) $\lim\limits_{y \to -1} (36y + 41)$

2. Evaluate the following limits.

   (a) $\lim\limits_{x \to 3} |x|$

   (b) $\lim\limits_{y \to -2} |y|$

   (c) $\lim\limits_{z \to 0} |z|$

   (d) $\lim\limits_{t \to 3} |t - 2|$

   (e) $\lim\limits_{t \to 0} \dfrac{|t - 2|}{t - 2}$

3. Evaluate the following limits.

   (a) $\lim\limits_{u \to 1} \dfrac{u^2 - 4}{u - 2}$

   (b) $\lim\limits_{u \to 2} \dfrac{u^2 - 4}{u - 2}$

   (c) $\lim\limits_{v \to 0} \dfrac{v^2 + v - 2}{v + 2}$

   (d) $\lim\limits_{v \to -2} \dfrac{v^2 + v - 2}{v + 2}$

4. Find

$$\lim_{x \to a} \frac{f(x) - f(a)}{x - a} \quad \text{and} \quad \lim_{t \to 0} \frac{f(a + t) - f(a)}{t}$$

(carrying through the calculations in both forms) for:

   (a) $f(x) = x^2, \quad a = 3$
   (b) $f(x) = x^2, \quad a = -\tfrac{1}{3}$
   (c) $f(x) = x^2 + 1, \quad a = 2$
   (d) $f(x) = 3x^2 - x, \quad a = 0$
   (e) $f(x) = \tfrac{1}{2}x^2 - 3x + 1, \quad a = 0$
   (f) $f(x) = (x - 3)^2 - 5, \quad a = 1$

5. (a) Can $\lim\limits_{x \to 2} f(x)$ exist if f is not defined at 2?

   (b) If $\lim\limits_{x \to 2} f(x) = 5$, what can we conclude about $f(2)$?

   (c) Can $\lim\limits_{x \to 2} f(x)$ be equal to $\lim\limits_{x \to 3} f(x)$?

   (d) Can it happen that a function g never assumes the value 6 and yet $\lim\limits_{x \to 3} g(x) = 6$?

   (e) Can one function be twice another and yet have the same limit at some point?

6. In each case, give an example of an interval cut at a that excludes both x_1 and x_2.

   (a) $a = 1$, $x_1 = 0$, $x_2 = 2$
   (b) $a = -1$, $x_1 = -2$, $x_2 = 0$
   (c) $a = 0$, $x_1 = -\frac{1}{3}$, $x_2 = \frac{1}{5}$
   (d) $a = 14.1$, $x_1 = 14.09$, $x_2 = 14.101$
   (e) a is unspecified, δ is an unspecified *positive* number, $x_1 = a - \delta$, $x_2 = a + \delta$

7. For each point (x, y) on the quarter-circle

$$x^2 + y^2 = 9 \qquad\qquad (x > 0, y > 0),$$

let $A(x)$ and $B(x)$ be the areas of the triangles formed as shown in Figure 9.

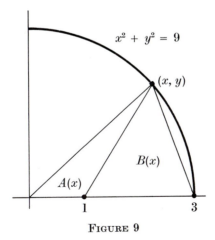

FIGURE 9

Find $\lim\limits_{x \to 3} \dfrac{B(x)}{A(x)}$.

8. Generalize Problem 7 to curves other than the quarter-circle.

9. Which of the following is equivalent to the assertion that

$$\lim_{x \to a} f(x) = b?$$

(*Interval* here means *open* interval.)

    (i)   If J is any interval about b and I is any interval cut at a, then f takes I into J.

    (ii)   There exist an interval J about b and an interval I cut at a such that f takes I into J.

    (iii)   If I is any interval cut at a, then there exists an interval J about b such that f takes I into J.

    (iv)   If J is any interval about b, then there exists an interval I cut at a such that f takes I into J.

    (v)   There exists an interval J about b such that, if I is any interval cut at a, then f takes I into J.

    (vi)   There exists an interval I cut at a such that, if J is any interval about b, then f takes I into J.

10. Let g be defined on $(0, 1)$ as follows:

$$g(x) = x \quad \text{except for} \quad x = \tfrac{1}{2}, \tfrac{1}{3}, \tfrac{1}{4}, \cdots;$$
$$g(x) = 0 \quad \text{for} \quad x = \tfrac{1}{2}, \tfrac{1}{3}, \tfrac{1}{4}, \cdots.$$

Find $\lim_{x \to 0} g(x)$.

Answers to problems (§2B)

1. (a) -1 (b) 2   (c) 4   (d) 23   (e) 5

2. (a) 3   (b) 2   (c) 0   (d) 1   (e) -1

3. (a) 3   (b) 4   (c) -1 (d) -3

4. (a) 6, 6   (b) $-\tfrac{2}{3}, -\tfrac{2}{3}$ (c) 4, 4   (d) $-1, -1$ (e) $-3, -3$
    (f) $-4, -4$

7. 2

10. 0

§2C. Implications of the definition

2C1. Boundedness near a point. *If* $\lim_{x \to a} f(x)$ *exists then there is an interval cut at a on which f is bounded.*

Proof. Call the limit b. Pick any interval $J = (u, v)$ about b. There is an interval I cut at a that f takes into J (Figure 1). Then f is bounded on I.

This is a result of great importance. Stated the other way around, it says that if f is unbounded on every interval cut at a, then $\lim_{x \to a} f(x)$ does not exist. More picturesquely:

(1)   *If f is unbounded near a then it has no limit at a.*

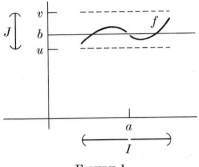

FIGURE 1

Example 1. The function $f(x) = 1/x$ $(x \neq 0)$ is not bounded near $x = 0$. For otherwise there would be a number $v > 0$ and an interval $I = (0, c)$ such that $f(x) \leq v$ for all $x \in I$. But this is not the case. For, we can find a number $x_0 > 0$ such that $x_0 < c$ and $x_0 < 1/v$. (See Figure 2.) Then $x_0 \in I$ and

$$f(x_0) = \frac{1}{x_0} > v.$$

Since f is not bounded near 0, it has no limit there. Thus,

$$\lim_{x \to 0} \frac{1}{x} \quad \text{does not exist.}$$

There are various ways that a function can fail to have a limit at a point, and this is one of them. In our work, it is the only one that will ever come up (except for an occasional example).

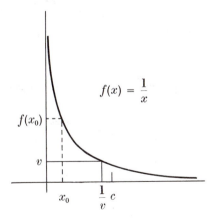

FIGURE 2

We will have many occasions to inquire whether a given function has a limit at a certain point. The thing to watch for in each case is whether it is bounded near the point. If not, then it cannot have a limit there.

2C2. Closeness of values near a point. *Let* $\lim\limits_{x \to a} f(x) = b$.

(a) *If* $b > r$ *there is an interval cut at* a *throughout which* $f(x) > r$.
(b) *If* $b < r$ *there is an interval cut at* a *throughout which* $f(x) < r$.

Proof. (a) Pick an interval J about b of the form (r, v). There is an interval I cut at a that f takes into J (Figure 3). Then $f(x) > r$ on I.

The proof of (b) goes the same way.

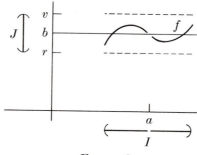

F<small>IGURE</small> 3

2C3. The squeeze theorem. *Suppose that* $f_1 \le g \le f_2$ *(on an interval cut at* a*) and that* $\lim\limits_{x \to a} f_1(x) = b$ *and* $\lim\limits_{x \to a} f_2(x) = b$. *Then* $\lim\limits_{x \to a} g(x) = b$.

Proof. Given an interval J about b, there is an interval I_1 cut at a that f_1 takes into J, and there is an interval I_2 cut at a that f_2 takes into J. The intersection,

$$I = I_1 \cap I_2,$$

is an interval cut at a that *both* f_1 and f_2 take into J (Figure 4). Since g lies between f_1 and f_2, it also takes I into J.

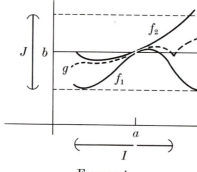

F<small>IGURE</small> 4

Problems (§2C)

1. The following functions are defined for $x \neq 1$ but not necessarily at $x = 1$. In each case, state whether the function has a limit as $x \to 1$. When the limit exists, evaluate it.

(a) $\dfrac{1}{x - 1}$

(b) $\dfrac{x + 1}{x - 1}$

(c) $\dfrac{x - 1}{x - 1}$

(d) $\dfrac{x^2 + 1}{x - 1}$

(e) $\dfrac{x^2 - 1}{x - 1}$

(f) $\dfrac{\sqrt{a^2}}{x - 1} - \dfrac{a}{x - 1}$

(g) $1 - \dfrac{1}{1 - x}$

(h) $\dfrac{(1 - x)^2 - 1}{x - 1}$

(i) $\dfrac{(1 - 2x)^2 - 1}{x - 1}$

2. Assume that $f \leq g$ on an interval cut at a and that $\lim_{x \to a} f(x)$ and $\lim_{x \to a} g(x)$ both exist. Prove that $\lim_{x \to a} f(x) \leq \lim_{x \to a} g(x)$.

3. Let sgn x denote the "sign" function, equal to 1 for $x > 0$, equal to 0 for $x = 0$, and equal to -1 for $x < 0$. Prove that $\lim_{x \to 0}$ sgn x does not exist.

4. Let h be a function that assumes only integers as values.
   (a) Can $\lim_{x \to 0.5} h(x)$ exist?
   (b) If so, what can be said about its value?

5. If two functions differ by a constant, can they have the same limit at some point?

6. Let g be a function about which we have the following information:
   (i) g is defined for $x \neq 2$ but not for $x = 2$,
   (ii) $g(x) > 3x + 4$ for all $x < 2$,

and

   (iii) $g(x) < 3x + 4$ for all $x > 2$.
What can be said about $\lim_{x \to 2} g(x)$?

7. Prove
   (a) If $\lim_{x \to a} f(x) = 0$ then $\lim_{x \to a} |f(x)| = 0$.
   (b) If $\lim_{x \to a} |f(x)| = 0$ then $\lim_{x \to a} f(x) = 0$

Answers to problems (§2C)

1. (a) no   (b) no   (c) 1   (d) no   (e) 2   (f) 0 $(a \geq 0)$,   no $(a < 0)$
   (g) no   (h) no   (i) 4

§2D. The algebra of limits

To do calculus you have to know the basic theorems for combining limits by the operations of algebra. The results are easy to remember, as they are just what one would guess.

The first two theorems are already familiar (§2B4 and §2B5).

2D1. Limit of a constant function. $\lim\limits_{x \to a} c = c.$

2D2. Limit of the function x. $\lim\limits_{x \to a} x = a.$

2D3. Limit of the negative of a function. If $\lim\limits_{x \to a} f(x) = b$, then $\lim\limits_{x \to a} [-f(x)] = -b.$

Proof. Let $J = (u, v)$ be an interval about $-b$. Then $J' = (-v, -u)$ is an interval about b and there is an interval I cut at a that f takes into J' (Figure 1). Then $-f$ takes I into J.

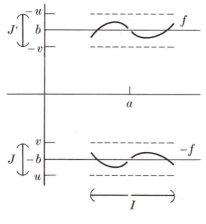

Figure 1

2D4. Limit of the reciprocal of a function. If $\lim\limits_{x \to a} f(x) = b$, where $b \neq 0$, then

$$\lim_{x \to a} \frac{1}{f(x)} = \frac{1}{b}.$$

Proof. We give the details for $b > 0$. (The proof for $b < 0$ is similar.) Let J be an interval about $1/b$. Since $1/b > 0$, there is an interval $J_0 \subset J$ that contains $1/b$ and has

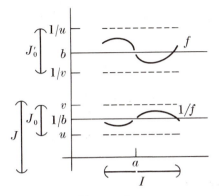

FIGURE 2

the form $J_0 = (u, v)$, $u > 0$. Then $J_0' = (1/v, 1/u)$ is an interval about b and there is an interval I cut at a that f takes into J_0' (Figure 2). Then $1/f$ takes I into J_0 and hence into J.

2D5. Limit of the sum of two functions. *If* $\lim\limits_{x \to a} f_1(x) = b_1$ *and* $\lim\limits_{x \to a} f_2(x) = b_2$, *then* $\lim\limits_{x \to a} [f_1(x) + f_2(x)] = b_1 + b_2$.

Proof. Let

(1) $$J = (b - h, b + k)$$

be an interval about b, $b = b_1 + b_2$. To find the desired interval I cut at a we "split the difference." Since

(2) $$J_1 = \left(b_1 - \frac{h}{2}, b_1 + \frac{k}{2}\right)$$

is an interval about b_1 there is an interval I_1 cut at a that f_1 takes into J_1. Since

(3) $$J_2 = \left(b_2 - \frac{h}{2}, b_2 + \frac{k}{2}\right)$$

is an interval about b_2 there is an interval I_2 cut at a that f_2 takes into J_2. Their intersection,

(4) $$I = I_1 \cap I_2,$$

is an interval cut at a that works for *both*: f_1 takes I into J_1 and f_2 takes I into J_2. Therefore $f_1 + f_2$ takes I into J.

Remarks. A particular consequence of the theorem is that *if* $\lim\limits_{x \to a} f(x) = b$ *then* $\lim\limits_{x \to a} [f(x) - b] = 0$, *and conversely.*

The theorem extends to any number of terms. For instance, if $f_1(x) \to b_1$, $f_2(x) \to b_2$, and $f_3(x) \to b_3$ (as $x \to a$), then

(5) $$f_1(x) + f_2(x) + f_3(x) \to b_1 + b_2 + b_3.$$

2D6. Limit of the difference of two functions. *If* $\lim\limits_{x\to a} f_1(x) = b_1$ *and* $\lim\limits_{x\to a} f_2(x) = b_2$, *then*

$$\lim_{x\to a} [f_1(x) - f_2(x)] = b_1 - b_2.$$

This follows from the sum theorem (§2D5) and the fact that $\lim\limits_{x\to a} [-f_2(x)] = -b_2$ (§2D3).

2D7. Limit of the product of two functions. *If* $\lim\limits_{x\to a} f_1(x) = b_1$ *and* $\lim\limits_{x\to a} f_2(x) = b_2$, *then* $\lim\limits_{x\to a} [f_1(x)f_2(x)] = b_1 b_2$.

Proof. We first treat two special cases.

(a)   If $f(x) \to 0$ *and* $g(x) \to 0$ *then* $f(x)g(x) \to 0$.

Given an interval J about 0, pick an interval $J_0 \subset J$ of the form

(6) $J_0 = (-l, l)$.

Then

(7) $J' = (-\sqrt{l}, \sqrt{l})$

is an interval about 0 and there are intervals I_1 and I_2 cut at a that f and g, respectively, take into J'. The intersection,

(8) $I = I_1 \cap I_2$,

is an interval cut at a that *both* f and g take into J'. Therefore fg takes I into J_0 and hence into J.

(b)   If $f(x) \to 0$ *then* $cf(x) \to 0$ (*c a constant*).

We give the proof for $c > 0$. ($c < 0$ is similar, and $c = 0$ is covered in §2D1.) Let

(9) $J = (-h, k)$

be any interval about 0. Then

(10) $J' = \left(-\dfrac{h}{c}, \dfrac{k}{c}\right)$

is an interval about 0 and there is an interval I cut at a that f takes into J'. Then cf takes I into J.

In the general case we have

(11) $f_1(x) - b_1 \to 0$ and $f_2(x) - b_2 \to 0$.

Hence, by (a) and (b),

(12) $f_1 f_2 - b_1 b_2 = (f_1 - b_1)(f_2 - b_2) + b_1 \cdot (f_2 - b_2) + (f_1 - b_1) \cdot b_2$

(13) $\to$ 0   +   0   +   0

                    $= 0$.

Therefore $f_1 f_2 \to b_1 b_2$.

Remarks. The theorem extends to any number of factors. For instance, if $f_1(x) \to b_1$, $f_2(x) \to b_2$, and $f_3(x) \to b_3$, then

(14) $$f_1(x)f_2(x)f_3(x) \to b_1 b_2 b_3.$$

For the case in which all the factors are the same, we get:

(15) $$\textit{If } \lim_{x \to a} f(x) = b \quad \textit{then} \quad \lim_{x \to a} [f(x)]^n = b^n.$$

Example 1.

$$\lim_{x \to 3} [(x - 1)^6] = [\lim_{x \to 3} (x - 1)]^6 = 2^6.$$

Example 2.

$$\lim_{x \to 2} [(3x^2 - 2x + 4)(x^2 + 5x - 3)]$$
$$= [\lim_{x \to 2} (3x^2 - 2x + 4)][\lim_{x \to 2} (x^2 + 5x - 3)]$$
$$= 12 \cdot 11$$
$$= 132.$$

2D8. Limit of the quotient of two functions. *If* $\lim_{x \to a} f_1(x) = b_1$ *and* $\lim_{x \to a} f_2(x) = b_2$, *where* $b_2 \neq 0$, *then*

$$\lim_{x \to a} \frac{f_1(x)}{f_2(x)} = \frac{b_1}{b_2}.$$

This follows from the product theorem (§2D7) and the fact that $\lim_{x \to a} [1/f_2(x)] = 1/b_2$ (§2D4).

2D9. Limit of a polynomial. *If* $P(x)$ *is a polynomial:*

$$P(x) = c_n x^n + c_{n-1} x^{n-1} + \cdots + c_1 x + c_0,$$

then $\lim_{x \to a} P(x) = P(a)$.

Proof. This follows from combining several of the results just developed. A typical term is $c_k x^k$. Since $c_k \to c_k$ and $x \to a$ we have $c_k x^k \to c_k a^k$, by the product theorem (§2D7). The final result then follows from the sum theorem (§2D5).

2D10. Limit of the square root of a function. *If* $f \geq 0$ *and* $\lim_{x \to a} f(x) = b$, *then* $\lim_{x \to a} \sqrt{f(x)} = \sqrt{b}$.

Proof. Since $f \geq 0$ we know that $b \geq 0$ (§2C2(b)).

Case 1. $b > 0$. Let J be any interval about $\sqrt{b}$. Since $\sqrt{b} > 0$, J contains an interval J_0 about $\sqrt{b}$ of the form $J_0 = (u, v)$, $u \geq 0$. Then $J_0' = (u^2, v^2)$ is an interval about b

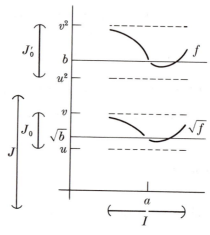

FIGURE 3

and there is an interval I cut at a that f takes into J_0' (Figure 3). Then $\sqrt{f}$ takes I into J_0 and hence into J.

Case 2. $b = 0$. Let $J = (-h, k)$ be an interval about 0. Then $J' = (-h, k^2)$ is an interval about 0 and there is an interval I cut at a that f takes into J'. (Since $f \geq 0$, f actually takes I into $[0, k^2)$.) Then $\sqrt{f}$ takes I into J.

2D11. *Limit of a rational power of a function.* *Let* r *be a rational number* > 0. *If* $f \geq 0$ *and* $\lim_{x \to a} f(x) = b$, *then* $\lim_{x \to a} [f(x)^r] = b^r$.

The proof is similar to the preceding.

Problems (§2D)

1. Evaluate the following limits.

(a) $\lim_{x \to 0} (3x^3 - 2x - 1)$

(b) $\lim_{x \to 1} (3x^3 - 2x - 1)$

(c) $\lim_{x \to -6} (x + 5)^8$

(d) $\lim_{x \to 4.7} [(x - 4.7)^3 (x + 2.8)^4]$

(e) $\lim_{x \to 4} \dfrac{x^2 - 3x + 2}{x^2 - 4x + 3}$

(f) $\lim_{x \to 1} \dfrac{x^2 - 3x + 2}{x^2 - 4x + 3}$

(g) $\lim_{x \to 1} \dfrac{(x - 1)^2}{x^2 - 1}$

(h) $\lim_{x \to 1} \dfrac{(x - 1)^2}{|x^2 - 1|}$

(i) $\lim\limits_{x \to 4} \dfrac{1 + \sqrt{x}}{1 - \sqrt{x}}$

(k) $\lim\limits_{x \to -2} \dfrac{1}{1 + \dfrac{1}{\sqrt{1 + x^2}}}$

(j) $\lim\limits_{x \to 1} (1 + x^{2/3})^{3/2}$

2. Find $\lim\limits_{x \to 2} f\big(g(x)\big)$ and $\lim\limits_{x \to 2} g\big(f(x)\big)$ for:

(a) $f(x) = \dfrac{1}{x + 1}$, $g(x) = x^2 + 1$

(b) $f(x) = \sqrt{x}$, $g(x) = x^2 + 1$

(c) $f(x) = \dfrac{x + 1}{x + 2}$, $g(x) = \dfrac{1}{x}$

3. Find $\lim\limits_{x \to 2} f^{-1}(x)$ for:

(a) $f(x) = \dfrac{x + 1}{x + 2}$

(b) $f(x) = \dfrac{1}{1 + \dfrac{1}{\sqrt{x}}}$

(c) $f(x) = x^{2/3} - 2 \ (x \geq 0)$

4. Evaluate the following limits.

(a) $\lim\limits_{t \to 0} \dfrac{(x + t)^3 - x^3}{t}$

(b) $\lim\limits_{u \to 0} \dfrac{1}{u} \left(\dfrac{1}{x + u} - \dfrac{1}{x} \right)$

(c) $\lim\limits_{h \to 0} \dfrac{1}{h} \left(\dfrac{1}{(x + h)^2} - \dfrac{1}{x^2} \right)$

5. Solve for h:

(a) $\lim\limits_{x \to 1} (3x^2 + h) = 6$

(c) $\lim\limits_{x \to 2} [3(x + h)^2 - 2(x + h)] = 0$

(b) $\lim\limits_{x \to 4} \dfrac{x + 2h}{x + h} = \dfrac{2}{3}$

(d) $\lim\limits_{x \to 3} \dfrac{x^2 + h^2}{x + h} = \dfrac{5}{2}$

6. (a) Let

$$F(k) = \lim\limits_{x \to 4} \dfrac{x^2 - 2x + 2k}{x^2 - 3x + k} \, .$$

Show that $F(k) = 2$ for all $k \neq -4$, but that $F(-4) = \frac{6}{5}$.

(b) Let
$$F(k) = \lim_{x \to -1} \frac{x^2 + 3x - 2k}{x^2 + 4x - 3k}$$

Show that $F(k) = \frac{2}{3}$ for all $k \neq -1$, but that $F(-1) = \frac{1}{2}$.

(c) Let
$$F(k) = \lim_{x \to 3} \frac{k^2(x^2 - 7) + 4k + x - 1}{kx + k(k - 1) + 1}.$$

Show that $F(k) = 2$ for all $k \neq -1$ but that $F(-1) = -7$.

7. Prove:
   (a) If f is bounded on an interval cut at a and if $\lim_{x \to a} g(x) = 0$ then $\lim_{x \to a} f(x)g(x) = 0$. [*Hint.* Use §2D7(b) and §2C3.]
   (b) If f has a limit as $x \to a$ and if $\lim_{x \to a} g(x) = 0$ then $\lim_{x \to a} f(x)g(x) = 0$.
   (c) If $\lim_{x \to a} h(x)/g(x)$ exists and $\lim_{x \to a} g(x) = 0$ then $\lim_{x \to a} h(x) = 0$.

8. Prove that these all say the same thing:
   (i) $\lim_{x \to a} f(x) = b$.
   (ii) Given any number $\epsilon > 0$, there is an interval I cut at a such that
   $$|f(x) - b| < \epsilon$$
   for all $x \in I$.
   (iii) Given any number $\epsilon > 0$, there is a number $\delta > 0$ such that
   $$|f(x) - b| < \epsilon \quad \text{whenever} \quad 0 < |x - a| < \delta.$$

Answers to problems (§2D)

1. (a) -1 (b) 0 (c) 1 (d) 0 (e) 2 (f) $\frac{1}{2}$ (g) 0 (h) 0
   (i) -3 (j) $2\sqrt{2}$ (k) $\dfrac{5 - \sqrt{5}}{4}$
2. (a) $\frac{1}{6}, \frac{10}{9}$ (b) $\sqrt{5}, 3$ (c) $\frac{3}{5}, \frac{4}{3}$
3. (a) -3 (b) not defined   (c) 8
4. (a) $3x^2$ (b) $-\dfrac{1}{x^2}$ (c) $-\dfrac{2}{x^3}$
5. (a) 3 (b) -1 (c) -2 or $-\frac{4}{3}$ (d) 1 or $\frac{3}{2}$

§2E. Continuity

The notion of continuity appeals strongly to our geometric intuition. One of the triumphs of mathematical thinking has been to put this notion into precise terms.

2E1. Definition of continuity. A function f is *continuous* at a point a provided that

(1)
$$\lim_{x \to a} f(x) = f(a).$$

That is, f is continuous at a provided that, given any interval J about $f(a)$, there is an interval I about a that f takes into J.

The definition states that f must be defined at a, it must have a limit as $x \to a$, and its value at a must be equal to its limit there.

Roughly speaking, if x is close to a, $f(x)$ is close to $f(a)$.

We say that f is continuous on a particular set if it is continuous at each point of the set. When the set need not be specified or when it can be identified from context, we say for short that f is continuous.

2E2. The role of continuity. This whole book deals with continuous functions. Intuitively, a function is continuous if you can draw its graph without lifting your pencil.

In many theorems, the requirement that a certain function be continuous appears as an explicit part of the hypothesis. You should greet this as a sign of reassurance rather than a cause for alarm. The spirit of the hypothesis is to show the generality of the theorem, not its restrictedness.

Suppose you have to apply the theorem. It would be a mistake to think you must run back each time to check whether a given function takes a certain interval into a certain other interval. On the contrary. We will build up a collection of results so that, at a glance, you will be able to recognize all sorts of functions as continuous on various intervals. (Of course, in *proving* these results we have to rely on precise definitions.) Generally, the only question will be whether a function is defined and continuous at an endpoint as well. And there the issue is usually whether the function is bounded near the point (§2C1).

2E3. The algebra of continuous functions. The theorems about limits have their counterparts about continuity. The following theorem summarizes, in terms of continuity, the results in §2D.

(a) *Constant functions are continuous, and the function $f(x) = x$ is continuous.*

(b) *Sums, differences, and products of continuous functions are continuous. In particular, polynomials are continuous.*

(c) *Quotients and rational powers of continuous functions (where defined) are continuous.*

2E4. Composition of continuous functions. The following basic result is often applied without explicit mention.

> *If g is continuous at a, and if f is continuous at g(a), then the composite function $f(g(x))$ is continuous at a.*

Proof. Given an interval J about $f(g(a))$ there is an interval I about $g(a)$ that f takes into J. In turn, there is an interval H about a that g takes into I. The composite function takes H into J.

2E5. The Intermediate-Value Theorem. *Let f be continuous on an interval $[a, b]$. For each number r between $f(a)$ and $f(b)$, there is a point $c \in [a, b]$ such that $f(c) = r$.*

Proof. Say $f(a) < r < f(b)$. Define

$$(2) \qquad\qquad g(x) = f(x) - r.$$

Then g is continuous, $g(a) < 0$, and $g(b) > 0$, and the theorem asserts that there is a point c for which $g(c) = 0$. Stated the other way around, it says:

(3)   *If a continuous function g is never 0 in $[a, b]$ then it does not change sign in $[a, b]$.*

To prove this we apply the Fundamental Lemma on Closed Intervals (§1H4). The set $\mathscr{I}$ of the lemma will be the set of all closed intervals in $[a, b]$ on which g does not change sign.

(i) *If g never changes sign on either of two intersecting intervals then it does not change sign on their union.* This is obvious.

(ii) *Each point has a closed interval about it on which g does not change sign.* This is true because g is continuous (§2C2).

By the Fundamental Lemma, g does not change sign on $[a, b]$.

The conclusion may certainly fail if f is not continuous; for instance, if $f(x) = 1$ for $x \geq 0$ and $f(x) = -1$ for $x < 0$ (Figure 1).

The hypothesis that f be continuous on an *interval* is also essential. The function $f(x) = 1/x$ $(x \neq 0)$ is continuous on its domain, positive at $x = 1$ and negative at $x = -1$, but not zero anywhere (Figure 2).

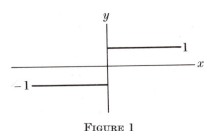

FIGURE 1

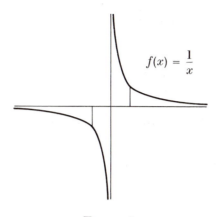

$$f(x) = \frac{1}{x}$$

FIGURE 2

2E6. Boundedness. *A function f continuous on a closed interval* $[a, b]$ *is bounded there.*

Proof. We apply the Fundamental Lemma on Closed Intervals (§1H4). The set $\mathscr{I}$ of the lemma will be the set of all closed intervals in $[a, b]$ on which f is bounded.

(i) *If f is bounded on each of two intersecting intervals, it is bounded on their union.* This is clear.

(ii) *Each point has a closed interval about it on which f is bounded.* This is true because f is continuous (§2C1).

By the Fundamental Lemma, f is bounded on $[a, b]$.

2E7. The Maximum-Value Theorem. *A function f continuous on a closed interval* $[a, b]$ *assumes a maximum value and a minimum value.*

Proof. By the preceding result (§2E6), the set of values of f has an upper bound. Let c denote its *least* upper bound (§1H2). We will prove that f assumes c as a value.

Suppose, on the contrary, that c is not a value of f. Then $c - f(x)$ is never zero but assumes arbitrarily small values. Then the function

$$g(x) = \frac{1}{c - f(x)}$$

is defined on $[a, b]$ and assumes arbitrarily large values—i.e., is unbounded. Since g is continuous (§2E3), this is a contradiction (§2E6).

It follows that c is a value of f. Clearly, it is the maximum value of f.

By a similar proof (or by applying the foregoing result to the function $-f$), we find that f assumes a minimum value.

The maximum and minimum values of f on a closed interval $[a, b]$ are denoted

$$\max_{[a,b]} f \quad \text{and} \quad \min_{[a,b]} f.$$

2E8. Image of a closed interval. *The image of a closed interval $[a, b]$ under a continuous function f is a closed interval.*

Proof. By the Maximum-Value Theorem (§2E7), f assumes a minimum at some point a_1 and a maximum at a point b_1. By the Intermediate-Value Theorem (§2E5), f also assumes all values between $f(a_1)$ and $f(b_1)$. Its image is therefore the closed interval $[f(a_1), f(b_1)]$.

2E9. Continuity of the inverse. *Let f be continuous on an interval. Then if f^{-1} exists, it is continuous.*

Proof. We show first that f is monotonic. Suppose not. Then there are points $x_1 < x_2 < x_3$ such that $f(x_2)$ does not lie between $f(x_1)$ and $f(x_3)$. Say $f(x_1) < f(x_3) < f(x_2)$ (Figure 3). By the Intermediate-Value Theorem (§2E5), the value $f(x_3)$ is duplicated somewhere in the interval $[x_1, x_2]$. But then f would not have an inverse.

Thus, f is monotonic.

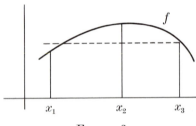

FIGURE 3

To show continuity of f^{-1} at a point b, consider any interval $J = (u, v)$ about $f^{-1}(b)$; we are to find an interval I about b that f^{-1} takes into J. We may assume that J is small enough so that u and v lie in the domain of f. Now, f is either increasing or decreasing; say increasing. Then

$$I = (f(u), f(v))$$

is an interval about b (Figure 4); moreover, the image of the closed interval $[u, v]$ is the closed interval $[f(u), f(v)]$ (§2E8). Then f^{-1} takes $[f(u), f(v)]$ back to $[u, v]$ and hence takes I into J

Remark. If a function f is monotonic then of course it has an inverse. In the course of the preceding proof we showed that, in the other direction:

(4)   *If f is continuous on an interval and has an inverse, then f is monotonic.*

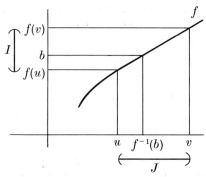

FIGURE 4

Problems (§2E)

1. Suppose f is continuous at $x = 3$.
   (a) Is f defined at $x = 3$?
   (b) If $f(3) = 6$, what can you say about $\lim_{x \to 3} f(x)$?

2. Give an example of a function on $[0, 1]$ that
   (a) is continuous on $(0, 1)$ but not at 0 nor at 1.
   (b) has exactly 11 points of discontinuity.

3. Show that the function $g(x) = |x|$ is continuous.

4. The following functions are *not defined* at $x = 2$ and are therefore not continuous there. Where possible, extend f by assigning a value at 2 in such a way that the resulting function *is* continuous at 2.

   (a) $f(x) = \dfrac{x}{x - 2}$ $\quad (x \neq 2)$

   (b) $f(x) = \dfrac{x^2 - 5x + 6}{x - 2}$ $\quad (x \neq 2)$

   (c) $f(x) = (x - 2)(x^2 - 1)\sqrt{x}$ $\quad (x \neq 2)$

   (d) $f(x) = \dfrac{x - 2}{x - 2}$ $\quad (x \neq 2)$

5. Which statement supports the assertion that $x \to 4$ as $\sqrt{x} \to 2$?
   (i) The square function is continuous
   (ii) The square-root function is continuous

6. Are there functions f and g, defined on $\mathscr{R}$ and *discontinuous* at a point a, for which
   (a) both $f + g$ and $f - g$ are continuous?
   (b) both $f + g$ and fg are continuous?

7. Let f be a continuous function. Quote appropriate theorems to show that each of the following functions is continuous.

(a) $g(x) = f((\sqrt{x} + 1)^{3/2})$

(b) $g(x) = \dfrac{1}{\sqrt{f(x^2 + 1)}}$

(c) $g(x) = f\left(\dfrac{1}{f(2x)}\right)$

8. Let f and g be defined on $\mathscr{R}$ and let $h(x) = f\big(g(x)\big)$. Consider a point a.

   (a) Can f be discontinuous at $g(a)$ while g and h are both continuous at a?

   (b) Can g be discontinuous at a while f is continuous at $g(a)$ and h is continuous at a?

   (c) Can g be discontinuous at a and f discontinuous at $g(a)$ and yet h continuous at a?

9. Show that the equation

$$x^7 + 3x - 1 = 0$$

has at least one root between 0 and 1.

10. Let f and g be continuous functions on $[0, 1]$ such that

$$f(0) = 1, \qquad g(0) = 0, \qquad \text{and} \qquad f(1) = g(1).$$

Let any number k between 0 and 1 be given. Prove that there is a point x such that the vertical chord between the graphs at x has length k (Figure 5).

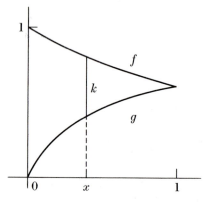

FIGURE 5

11. Let f be a continuous function that takes $[0, 1]$ into $[0, 1]$. Prove that f has a "fixed point," i.e., that there is a point x_0 such that $f(x_0) = x_0$. [*Hint.* Consider the function $f(x) - x$.]

12. Give an example of a function continuous on the non-closed interval $(0, 1]$ that is

    (a) not bounded.

    (b) bounded, but does not assume a maximum.

13. Give an example of a function discontinuous on the closed interval $[0, 1]$ that is

    (a) not bounded.

    (b) bounded, but does not assume a maximum.

14. (a) Specify a continuous function on $[3, 6]$ whose maximum value is 5, assumed at $x = 4$, and whose minimum value is 4, assumed at $x = 5$.

    (b) What is the image of this function?

    (c) Has this function an inverse?

15. Let f be continuous on a closed interval I and let I' be a closed interval contained in I. Which of the following statements is necessarily true?

    (i) $\max_{I} f \geq \max_{I'} f$ (iii) $\min_{I} f \geq \min_{I'} f$

    (ii) $\max_{I} f \leq \max_{I'} f$ (iv) $\min_{I} f \leq \min_{I'} f$

Answers to problems (§2E)

4. (a) not possible   (b) set $f(2) = -1$ (c) set $f(2) = 0$
    (d) set $f(2) = 1$

§2F. Partitions

The following specialized results are needed only in the construction of the integral in Chapter 5.

2F1. Partitions of an interval. Let I be a closed interval. By a *partition* of I is meant a set of abutting closed intervals whose union is I (Figure 1). We call these intervals the *segments* of the partition. A partition may have any number of segments: $1, 2, \cdots$.

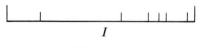

I

FIGURE 1

2F2. ϵ-partitions. Let f be a function defined on I, and let ϵ be a positive number.

Let P be a partition of I. We call P an "ϵ-partition" (with respect to f) if, within any one segment, f varies by less than ϵ. That is, P is an ϵ-partition if, whenever x_1 and x_2 belong to the same segment of P,

(1) $$|f(x_1) - f(x_2)| < \epsilon.$$

Suppose I_1 and I_2 are intersecting intervals in I that admit ϵ-partitions, say P_1 and P_2. Create a partition P_0 of $I_1 \cup I_2$ by taking as the points of subdivision all those

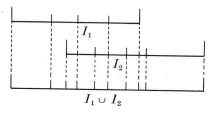

$$I_1 \cup I_2$$

FIGURE 2

that occur in either P_1 or P_2 (Figure 2). Each segment thus formed is a subset of one from P_1 or P_2; hence f varies within it by less than ϵ. Thus, P_0 is an ϵ-partition of $I_1 \cup I_2$.

2F3. Existence of ϵ-partitions. *Let f be continuous on a closed interval $[a, b]$, and let ϵ be any positive number.*

Then there is a partition of $[a, b]$ such that, within any one segment, f varies by less than ϵ. Briefly: $[a, b]$ admits an ϵ-partition.

Proof. We apply the Fundamental Lemma on Closed Intervals (§1H4). The set $\mathscr{I}$ of the lemma will consist of those closed intervals in $[a, b]$ that admit ϵ-partitions.

(i) *If each of two intersecting intervals admits an ϵ-partition, then so does their union.* This was noted above.

(ii) *Each point has a closed interval about it that admits an ϵ-partition.* For, let x_0 be any point of $[a, b]$. Because f is continuous, there is a closed interval I about x_0 such that

$$(2) \qquad\qquad |f(x) - f(x_0)| < \frac{\epsilon}{2}$$

for all $x \in I$. Then f varies within I by less than ϵ. Thus, the partition of I consisting of the single segment I itself is an ϵ-partition of I.

By the Fundamental Lemma, $[a, b]$ admits an ϵ-partition.

Problems (§2F)

1. In each case, specify an ϵ-partition for the function f on the interval I.

    (a) $f(x) = x$, $I = [0, 10]$, $\epsilon = 0.01$.
    (b) f is the constant function 3, $I = [0, 10]$, $\epsilon = 0.01$.
    (c) $f(x) = x^2$, $I = [0, 5]$, $\epsilon = 5$.

2. Let f be continuous on a closed interval I. Derive the fact that f is bounded on I from the existence of an ϵ-partition of I for $\epsilon = 1$.

Answers to problems (§2F)

1. (a) any partition in which the length of every segment is less than 0.01

CHAPTER 3

The Derivative

Now we start calculus.

This chapter is on the fundamentals of "differential" calculus. It begins with the definition of the derivative and develops all the standard formulas about derivatives of algebraic functions, plus the "chain rule." These formulas represent the basic technique of the subject and they come thick and fast. However, they are not difficult.

There are applications to geometry and motion.

§3A. The Δ-notation

3A1. The Δ-notation. Let f be a function and let x_0 be a point in its domain. Focusing on x_0, we express other values of x in the form

(1) $$x_0 + \Delta x,$$

using a new variable, Δx, to denote the difference or change in x. (The letter h is also commonly used. Δx is clumsier but more suggestive.)

The corresponding change in the value of the function is

(2) $$f(x_0 + \Delta x) - f(x_0).$$

It is often convenient to introduce a variable,

(3) $$y = f(x).$$

We then *define* Δy as the change in the value of the function:

(4) $$\Delta y = f(x_0 + \Delta x) - f(x_0).$$

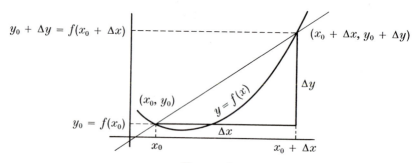

FIGURE 1

Note well. The symbol Δy is handy but contains less information than the longer expression it stands for. It is not just any change in y, but the one that results when x is changed from x_0 by the amount Δx.

Another natural abbreviation is

(5) $$y_0 = f(x_0).$$

Then by (4), $y_0 + \Delta y = f(x_0 + \Delta x)$. The relations are clarified in Figure 1.

Example 1. Let $y = f(x) = x^2$, $x_0 = 2$. Then

$$y_0 = f(x_0) = f(2) = 4.$$

If $\Delta x = 1$ then $x_0 + \Delta x = 3$,

$$f(x_0 + \Delta x) = f(3) = 9,$$

$$\Delta y = f(x_0 + \Delta x) - f(x_0) = 9 - 4 = 5.$$

If $\Delta x = -\frac{1}{2}$ then $x_0 + \Delta x = 1\frac{1}{2}$,

$$f(x_0 + \Delta x) = f(1\frac{1}{2}) = 2\frac{1}{4},$$

$$\Delta y = f(x_0 + \Delta x) - f(x_0) = 2\frac{1}{4} - 4 = -1\frac{3}{4}.$$

Example 2. Let $u = g(x) = 1/x$ $(x \neq 0)$, $x_0 = 4$. Then

$$u_0 = g(x_0) = g(4) = \frac{1}{4}.$$

If $\Delta x = 1$ then $x_0 + \Delta x = 5$,

$$g(x_0 + \Delta x) = g(5) = \frac{1}{5},$$

$$\Delta u = g(x_0 + \Delta x) - g(x_0) = \frac{1}{5} - \frac{1}{4} = -\frac{1}{20}.$$

If $\Delta x = -\frac{1}{2}$ then $x_0 + \Delta x = 3\frac{1}{2}$,

$$g(x_0 + \Delta x) = g(3\frac{1}{2}) = \frac{2}{7},$$

$$\Delta u = g(x_0 + \Delta x) - g(x_0) = \frac{2}{7} - \frac{1}{4} = \frac{1}{28}.$$

3A2. Continuity. By definition, f is continuous at x_0 provided that

(6)
$$\lim_{x \to x_0} f(x) = f(x_0).$$

In the Δ-notation, this takes the form

(7)
$$\lim_{\Delta x \to 0} f(x_0 + \Delta x) = f(x_0).$$

This can also be written as:

(8)
$$\Delta y \to 0 \qquad \text{as} \qquad \Delta x \to 0.$$

Example 3. Let $y = f(x) = x^2$. Then f is continuous at $x_0 = 2$:

$$\lim_{x \to x_0} f(x) = f(x_0), \qquad \lim_{x \to 2} x^2 = 2^2;$$

$$\lim_{\Delta x \to 0} f(x_0 + \Delta x) = f(x_0), \qquad \lim_{\Delta x \to 0} (2 + \Delta x)^2 = 2^2;$$

$$\Delta y = (2 + \Delta x)^2 - 2^2 \to 0 \qquad \text{as} \qquad \Delta x \to 0.$$

Example 4. Let $u = g(x) = 1/x$ $(x \neq 0)$. Then g is continuous at $x_0 = 4$:

$$\lim_{x \to x_0} g(x) = g(x_0), \qquad \lim_{x \to 4} \frac{1}{x} = \frac{1}{4} \; ;$$

$$\lim_{\Delta x \to 0} g(x_0 + \Delta x) = g(x_0), \qquad \lim_{\Delta x \to 0} \frac{1}{4 + \Delta x} = \frac{1}{4} \; ;$$

$$\Delta u = \frac{1}{4 + \Delta x} - \frac{1}{4} \to 0 \quad \text{as} \quad \Delta x \to 0.$$

3A3. Average rate of change. Let $y = f(x)$. The number

$$\Delta y = f(x_0 + \Delta x) - f(x_0)$$

is the net change in y when x changes from x_0 by the amount Δx. The "difference quotient,"

(9) $$\frac{\Delta y}{\Delta x} = \frac{f(x_0 + \Delta x) - f(x_0)}{\Delta x} \; ,$$

is then the *average rate of change of y with respect to x*, i.e., the average rate of change of y per unit change in x (between x_0 and $x_0 + \Delta x$). It is equal to the slope of the line joining (x_0, y_0) and $(x_0 + \Delta x, y_0 + \Delta y)$ (Figure 1).

If Δx is close to 0, $\Delta y/\Delta x$ is the average rate over a very small interval. What happens as $\Delta x \to 0$? If f is a continuous function, then $\Delta y \to 0$ too. What does $\Delta y/\Delta x$ do?

Example 5. Let $y = f(x) = x^2$, $x_0 = 2$. Extending the results of Example 1, we get:

Δx	1	-1	0.5	-0.5	0.1	-0.1	0.01	-0.01
Δy	5	-3	2.25	-1.75	0.41	-0.39	0.0401	-0.0399
$\dfrac{\Delta y}{\Delta x}$	5	3	4.5	3.5	4.1	3.9	4.01	3.99

The figures suggest (as they should) that as $\Delta x \to 0$, $\Delta y \to 0$. They also suggest that $\Delta y/\Delta x$ is settling down, approaching a limit—namely, 4. (As we will see, this is also true.)

Example 6. Let $u = g(x) = 1/x$ $(x \neq 0)$, $x_0 = 4$. Extending the results of Example 2, we get:

Δx	1	-1	$\dfrac{1}{2}$	$-\dfrac{1}{2}$	$\dfrac{1}{10}$	$-\dfrac{1}{10}$	$\dfrac{1}{100}$	$-\dfrac{1}{100}$
Δu	$-\dfrac{1}{20}$	$\dfrac{1}{12}$	$-\dfrac{1}{36}$	$\dfrac{1}{28}$	$-\dfrac{1}{164}$	$\dfrac{1}{156}$	$-\dfrac{1}{1604}$	$\dfrac{1}{1596}$
$\dfrac{\Delta u}{\Delta x}$	$-\dfrac{1}{20}$	$-\dfrac{1}{12}$	$-\dfrac{1}{18}$	$-\dfrac{1}{14}$	$-\dfrac{1}{16.4}$	$-\dfrac{1}{15.6}$	$-\dfrac{1}{16.04}$	$-\dfrac{1}{15.96}$

The figures suggest (as they should) that as $\Delta x \to 0$, $\Delta u \to 0$. They also suggest that $\Delta u/\Delta x$ is settling down, approaching a limit—namely, $-1/16$. (As we will see, this is also true.)

Whether $\Delta y/\Delta x$ approaches a limit as $\Delta x \to 0$ depends on the function and on the point x_0. For the functions we deal with, the limit will exist with only occasional exceptions.

Problems (§3A)

1. Let $y = f(x) = x^2$.
   (a) Find Δy and $\Delta y/\Delta x$ for $x_0 = 3$ and each of the values $\Delta x = 1, -1, 0.5, -0.5, 0.1, -0.1, 0.01, -0.01$.
   (b) Find Δy and $\Delta y/\Delta x$ for $x_0 = 4$ and each of the values $\Delta x = 1, -1, 0.5, -0.5, 0.1, -0.1, 0.01, -0.01$.
   (c) Express Δy and $\Delta y/\Delta x$ in terms of Δx for each of the values $x_0 = 3, 4, -1, -2, 0$.
   (d) Express Δy and $\Delta y/\Delta x$ in terms of x_0 and Δx.

2. Let $u = g(x) = 1/x$.
   (a) Find Δu and $\Delta u/\Delta x$ for $x_0 = 2$ and each of the values $\Delta x = 1, -1, 0.5, -0.5, 0.1, -0.1, 0.01, -0.01$.
   (b) Find Δu and $\Delta u/\Delta x$ for $x_0 = 1$ and each of the values $\Delta x = 1, 0.5, -0.5, 0.1, -0.1, 0.01, -0.01$. Can we consider the value $\Delta x = -1$ here?
   (c) Find Δu and $\Delta u/\Delta x$ for $x_0 = 100$ and $\Delta x = 1, -1, 0.1, -0.1, 0.01, -0.01$.
   (d) Express Δu and $\Delta u/\Delta x$ in terms of Δx for $x_0 = 2, 1, 100, -1, -2$.
   (e) Express Δu and $\Delta u/\Delta x$ in terms of x_0 and Δx.

3. Let $v = F(x) = \sqrt{x}$.
   (a) Find Δv and $\Delta v/\Delta x$ for $x_0 = 5$ and each of the values $\Delta x = 1, -1, 0.1, -0.1$.
   (b) Find Δv and $\Delta v/\Delta x$ for $x_0 = 6$ and each of the values $\Delta x = 1, -1, 0.1, -0.1$.
   (c) Find Δv and $\Delta v/\Delta x$ for $x_0 = 0$ and $\Delta x = 1, 0.25, 0.01$. Can we consider the value $\Delta x = -1$ here?

(d) Express Δv and $\Delta v/\Delta x$ in terms of Δx for $x_0 = 5, 6, 0$.

(e) Express Δv and $\Delta v/\Delta x$ in terms of x_0 and Δx.

4. Let $s = h(t) = 1 + t$.

   (a) Find Δs and $\Delta s/\Delta t$ for $t_0 = 5$ and each of the values $\Delta t = 1, -0.5$.

   (b) Find Δs and $\Delta s/\Delta t$ for $t_0 = 0$ and each of the values $\Delta t = -1, 0.5$.

   (c) Express Δs and $\Delta s/\Delta t$ in terms of Δt for each of the values $t_0 = 5, -3$.

   (d) Express Δs and $\Delta s/\Delta t$ in terms of t_0 and Δt. Interpret the result in terms of the graph of the equation $s = 1 + t$.

5. Express the continuity of the following functions at the given points, using Example 3 as a model.

   (a) $y = h(x) = x^3, \quad x_0 = 3$ (c) $s = g(t) = \dfrac{1}{t^2 + 1}, \quad t_0 = 0$

   (b) $v = k(x) = x^2 - x, \quad x_0 = \frac{1}{2}$ (d) $u = F(x) = \sqrt{x^3 + 1}, \quad x_0 = 1$

6. If

$$\frac{f(2 + \Delta x) - f(2)}{\Delta x} = 3$$

for all $\Delta x \neq 0$, what can you say about the function f?

7. Let f be an increasing function. What can you say about the sign of

$$\frac{f(x_0 + \Delta x) - f(x_0)}{\Delta x}$$

   (a) for $\Delta x > 0$?

   (b) for $\Delta x < 0$?

Answers to problems (§3A)

1. (a) $(x_0 = 3)$

Δx	1	-1	0.5	-0.5	0.1	-0.1	0.01	-0.01
Δy	7	-5	3.25	-2.75	0.61	-0.59	0.0601	-0.0599
$\dfrac{\Delta y}{\Delta x}$	7	5	6.5	5.5	6.1	5.9	6.01	5.99

(b) $(x_0 = 4)$

Δx	1	-1	0.5	-0.5	0.1	-0.1	0.01	-0.01
Δy	9	-7	4.25	-3.75	0.81	-0.79	0.0801	-0.0799
$\dfrac{\Delta y}{\Delta x}$	9	7	8.5	7.5	8.1	7.9	8.01	7.99

(c)

x_0	3	4	-1	-2	0
Δy	$6\Delta x + (\Delta x)^2$	$8\Delta x + (\Delta x)^2$	$-2\Delta x + (\Delta x)^2$	$-4\Delta x + (\Delta x)^2$	$(\Delta x)^2$

$\dfrac{\Delta y}{\Delta x}$	$6 + \Delta x$	$8 + \Delta x$	$-2 + \Delta x$	$-4 + \Delta x$	Δx

(d) $\Delta y = 2x_0\,\Delta x + (\Delta x)^2$, $\dfrac{\Delta y}{\Delta x} = 2x_0 + \Delta x$

2. (a) $(x_0 = 2)$

Δx	1	-1	0.5	-0.5	0.1	-0.1	0.01	-0.01
Δu	$-\dfrac{1}{6}$	$\dfrac{1}{2}$	$-\dfrac{1}{10}$	$\dfrac{1}{6}$	$-\dfrac{1}{42}$	$\dfrac{1}{38}$	$-\dfrac{1}{402}$	$\dfrac{1}{398}$
$\dfrac{\Delta u}{\Delta x}$	$-\dfrac{1}{6}$	$-\dfrac{1}{2}$	$-\dfrac{1}{5}$	$-\dfrac{1}{3}$	$-\dfrac{1}{4.2}$	$-\dfrac{1}{3.8}$	$-\dfrac{1}{4.02}$	$-\dfrac{1}{3.98}$

(b) $(x_0 = 1)$

Δx	1	0.5	-0.5	0.1	-0.1	0.01	-0.01
Δu	$-\dfrac{1}{2}$	$-\dfrac{1}{3}$	1	$-\dfrac{1}{11}$	$\dfrac{1}{9}$	$-\dfrac{1}{101}$	$\dfrac{1}{99}$
$\dfrac{\Delta u}{\Delta x}$	$-\dfrac{1}{2}$	$-\dfrac{2}{3}$	-2	$-\dfrac{10}{11}$	$-\dfrac{10}{9}$	$-\dfrac{100}{101}$	$-\dfrac{100}{99}$

(c) $(x_0 = 100)$

Δx	1	-1	0.1	-0.1	0.01	-0.01
Δu	$-\dfrac{1}{10,100}$	$\dfrac{1}{9,900}$	$-\dfrac{1}{100,100}$	$\dfrac{1}{99,900}$	$-\dfrac{1}{1,000,100}$	$\dfrac{1}{999,900}$
$\dfrac{\Delta u}{\Delta x}$	$-\dfrac{1}{10,100}$	$-\dfrac{1}{9,900}$	$-\dfrac{1}{10,010}$	$-\dfrac{1}{9,990}$	$-\dfrac{1}{10,001}$	$-\dfrac{1}{9,999}$

(d)

x_0	2	1	100	-1	-2
Δu	$-\dfrac{\Delta x}{4 + 2\Delta x}$	$-\dfrac{\Delta x}{1 + \Delta x}$	$-\dfrac{\Delta x}{100(100 + \Delta x)}$	$-\dfrac{\Delta x}{1 - \Delta x}$	$-\dfrac{\Delta x}{4 - 2\,\Delta x}$
$\dfrac{\Delta u}{\Delta x}$	$-\dfrac{1}{4 + 2\,\Delta x}$	$-\dfrac{1}{1 + \Delta x}$	$-\dfrac{1}{100(100 + \Delta x)}$	$-\dfrac{1}{1 - \Delta x}$	$-\dfrac{1}{4 - 2\,\Delta x}$

(e) $\Delta u = -\dfrac{\Delta x}{x_0(x_0 + \Delta x)}$, $\dfrac{\Delta u}{\Delta x} = -\dfrac{1}{x_0(x_0 + \Delta x)}$

3. (a) $(x_0 = 5)$

Δx	1	-1	0.1	-0.1
Δv	0.21	-0.24	0.02	-0.03
$\dfrac{\Delta v}{\Delta x}$	0.21	0.24	0.2	0.3

(b) $(x_0 = 6)$

Δx	1	-1	0.1	-0.1
Δv	0.20	-0.21	0.02	-0.02

$\dfrac{\Delta v}{\Delta x}$	0.20	0.21	0.2	0.2

(c) $(x_0 = 0)$

Δx	1	0.25	0.01
Δv	1	0.5	0.1

$\dfrac{\Delta v}{\Delta x}$	1	2	10

No

(d)

x_0	5	6	0
Δv	$\sqrt{5 + \Delta x} - \sqrt{5}$	$\sqrt{6 + \Delta x} - \sqrt{6}$	$\sqrt{\Delta x}$
$\dfrac{\Delta v}{\Delta x}$	$\dfrac{1}{\sqrt{5 + \Delta x} + \sqrt{5}}$	$\dfrac{1}{\sqrt{6 + \Delta x} + \sqrt{6}}$	$\dfrac{1}{\sqrt{\Delta x}}$

(e) $\Delta v = \sqrt{x_0 + \Delta x} - \sqrt{x_0}, \quad \dfrac{\Delta v}{\Delta x} = \dfrac{1}{\sqrt{x_0 + \Delta x} + \sqrt{x_0}}$

4. (a) $(t_0 = 5)$

Δt	1	-0.5
Δs	1	-0.5

$\dfrac{\Delta s}{\Delta t}$	1	1

(b) $(t_0 = 0)$

Δt	-1	0.5
Δs	-1	0.5

$\dfrac{\Delta s}{\Delta t}$	1	1

(c)

t_0	5	-3
Δs	Δt	Δt

$\dfrac{\Delta s}{\Delta t}$	1	1

(d) $\Delta s = \Delta t, \quad \dfrac{\Delta s}{\Delta t} = 1$

§3B. The derivative

3B1. Definition of derivative. The *derivative* of f at x_0, denoted $f'(x_0)$, is defined by:

(1)
$$f'(x_0) = \lim_{\Delta x \to 0} \frac{f(x_0 + \Delta x) - f(x_0)}{\Delta x}.$$

We may express this briefly as

(2)
$$f'(x_0) = \lim_{\Delta x \to 0} \frac{\Delta y}{\Delta x},$$

provided it is understood that

(3)
$$\Delta y = f(x_0 + \Delta x) - f(x_0).$$

When f has a derivative at x_0 (that is, when the above limit exists), we say that f is *differentiable* at x_0. If f is differentiable at each point of a particular set we say it is differentiable on that set. When the set need not be specified or when it can be identified from context, we say for short that f is differentiable.

The process of finding derivatives is called *differentiation*. The study of derivatives is called *differential calculus*.

The derivative, $f'(x_0)$, is also known as the *instantaneous rate of change* of $f(x)$ with respect to x at the point x_0.

Example 1. What does $f'(3)$ mean?

Answer. By definition $\big(\S 3B(1)\big)$,

$$f'(3) = \lim_{\Delta x \to 0} \frac{f(3 + \Delta x) - f(3)}{\Delta x}.$$

Briefly,

$$f'(3) = \lim_{\Delta x \to 0} \frac{\Delta y}{\Delta x},$$

where, it is understood,

$$\Delta y = f(3 + \Delta x) - f(3).$$

Example 2. Let $f(x) = x^2$. What is $f'(3)$?

Solution. $f(3) = 3^2$ and $f(3 + \Delta x) = (3 + \Delta x)^2$.

Therefore,

$$f(3 + \Delta x) - f(3) = (3 + \Delta x)^2 - 3^2 = 6\Delta x + (\Delta x)^2,$$

so that

$$\frac{f(3 + \Delta x) - f(3)}{\Delta x} = 6 + \Delta x.$$

Hence

$$f'(3) = \lim_{\Delta x \to 0} \frac{f(3 + \Delta x) - f(3)}{\Delta x} = \lim_{\Delta x \to 0} (6 + \Delta x) = 6.$$

Example 3. Let $f(x) = x^2$. What is $f'(x_0)$?

Solution.

$$f(x_0 + \Delta x) - f(x_0) = (x_0 + \Delta x)^2 - x_0^2 = 2x_0\,\Delta x + (\Delta x)^2,$$

and therefore

$$f'(x_0) = \lim_{\Delta x \to 0} \frac{f(x_0 + \Delta x) - f(x_0)}{\Delta x} = \lim_{\Delta x \to 0} (2x_0 + \Delta x) = 2x_0.$$

Example 4. Let $g(x) = \sqrt{x}$ $(x \geq 0)$. What is $g'(5)$?

Solution.

$$\Delta y = g(5 + \Delta x) - g(5) = \sqrt{5 + \Delta x} - \sqrt{5},$$

and

$$\frac{\Delta y}{\Delta x} = \frac{\sqrt{5 + \Delta x} - \sqrt{5}}{\Delta x}.$$

This expression for $\Delta y/\Delta x$ is not in a manageable form for evaluating the limit. We apply an algebraic trick—"rationalizing the numerator." Multiplying numerator and denominator by $\sqrt{5 + \Delta x} + \sqrt{5}$, we get

$$\frac{\Delta y}{\Delta x} = \frac{1}{\sqrt{5 + \Delta x} + \sqrt{5}}.$$

Now we see that

$$g'(5) = \lim_{\Delta x \to 0} \frac{\Delta y}{\Delta x} = \lim_{\Delta x \to 0} \frac{1}{\sqrt{5 + \Delta x} + \sqrt{5}}$$

$$= \frac{1}{2\sqrt{5}}.$$

3B2. Tangent line and slope. Let f be a function and let $P_0 = (x_0, y_0)$ be a point on its graph. What is meant by the slope of the graph at the point P_0? What is meant by the tangent line to the graph at P_0?

If $P = (x, y)$ is any other point on the graph then the two points P and P_0 determine a line. Let $\Delta x = x - x_0$ and $\Delta y = y - y_0$ (Figure 1). The slope

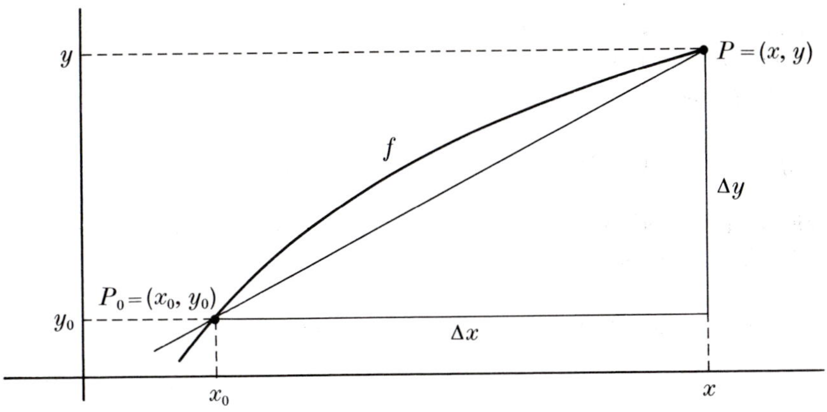

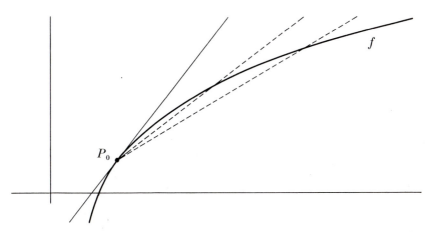

FIGURE 1

of the line PP_0 is $\Delta y/\Delta x$. We agreed in §2A that the slope of the tangent at P_0 should be the limiting value (if there is one) of $\Delta y/\Delta x$. At the time, we did not have a clear concept of limit.

3B3. Definition of tangent line and slope. The *tangent line* to the graph of f at a point $P_0 = (x_0, y_0)$ is the line through P_0 whose slope is $f'(x_0)$ (Figure 2). The number $f'(x_0)$ is also called the *slope of the graph* of f at P_0.

FIGURE 2

Example 5. Find the equation of the tangent line to the graph of $f(x) = x^2$ at the point $(3, 9)$.

Solution. As we saw in Example 2, $f'(3) = 6$. Hence the equation of the tangent line is

$$y - 9 = 6(x - 3).$$

3B4. *Sign of the derivative.* The derivative tells a great deal about the behavior of the function, as we will see at great length in Chapter 4. For now we point out a simple fact.

(a)   *If $f'(x_0) > 0$ then f is increasing at the point x_0.*

This is a technical phrase. It means that there is an interval I about x_0 such that, when $x_0 + \Delta x \in I$,

(4)
$$f(x_0 + \Delta x) > f(x_0) \quad if \quad \Delta x > 0 \qquad and$$
$$f(x_0 + \Delta x) < f(x_0) \quad if \quad \Delta x < 0.$$

See Figure 3(a).

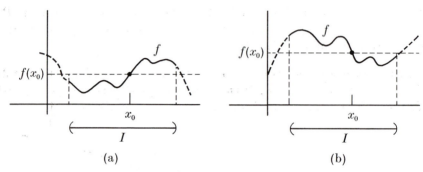

(a)   (b)

Figure 3

The proof is easy. By hypothesis,

(5)
$$f'(x_0) = \lim_{\Delta x \to 0} \frac{f(x_0 + \Delta x) - f(x_0)}{\Delta x} > 0.$$

By the theorem on closeness of values (§2C2), there is an interval I about x_0 on which

(6)
$$\frac{f(x_0 + \Delta x) - f(x_0)}{\Delta x} > 0 \qquad\qquad (\Delta x \neq 0).$$

Then Δx and $f(x_0 + \Delta x) - f(x_0)$ have the same sign. This is what (4) says. In similar fashion, we have:

(b)   *If $f'(x_0) < 0$ then f is decreasing at the point x_0.*

This means that there is an interval I about x_0 such that, when $x_0 + \Delta x \in I$,

(7)
$$f(x_0 + \Delta x) < f(x_0) \quad if \quad \Delta x > 0 \qquad and$$
$$f(x_0 + \Delta x) > f(x_0) \quad if \quad \Delta x < 0.$$

See Figure 3(b).

Caution. If $f'(x_0) = 0$, anything can happen. (See Problem 12.)

3B5. Velocity in linear motion. Suppose a point P is moving along a straight line. At each instant t, the coordinate s of P on the line depends only on t. That is, s is a function of the time t; say

$$s = f(t).$$

The motion of P is described by the properties of f. For example, to say that f is increasing is to say that P is moving in the positive direction.

The *average velocity* of P over a given period of time is equal to the net distance traveled by P divided by the total time. Consider the time interval between t_0 and $t_0 + \Delta t$. At time t_0, P is at $f(t_0)$; and at time $t_0 + \Delta t$, P is at $f(t_0 + \Delta t)$. The net distance traveled, Δs, is therefore given by

(8)
$$\Delta s = f(t_0 + \Delta t) - f(t_0).$$

Notice that Δs will be positive or negative according as P ends up at a point in the positive or in the negative direction from the starting point. Since the distance Δs has been covered in time Δt, the *average velocity* of P over the time interval is

(9)
$$\frac{\Delta s}{\Delta t} = \frac{f(t_0 + \Delta t) - f(t_0)}{\Delta t}.$$

This gives the average velocity as a difference quotient for f. It is natural to ask what happens as $\Delta t \to 0$. If f has a derivative at t_0, the *mathematical* answer is easy: as $\Delta t \to 0$, $\Delta s/\Delta t \to f'(t_0)$. Physically, we observe that for Δt near 0, (9) tells us how fast, and in what direction, P is moving over a small time interval starting at t_0 (or, if $\Delta t < 0$, ending at t_0). As the time interval becomes smaller we apparently acquire information about the motion of P at time t. For this reason, we shall *define* the *(instantaneous) velocity* of P at time t_0 to be the limit (if there is one) of $\Delta s/\Delta t$ as $\Delta t \to 0$. That is, *the velocity v of P at time t_0 is* defined as

(10)
$$v = \frac{ds}{dt} = f'(t_0).$$

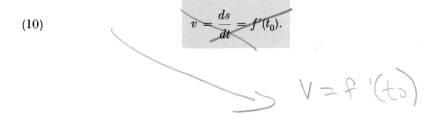

$$V = f'(t_0)$$

Problems (§3B)

1. Given $f(x) = 3x - 2$, find $f'(1), f'(0), f'(15), f'(x_0)$.
2. Given $h(t) = t^3$, find $h'(1), h'(-1), h'(0), h'(5), h'(t_0)$.
3. Given $F(t) = 1/t$, find $F'(1), F'(-1), F'(3), F'(-3), F'(t_0)$.
4. Let $g(u) = \sqrt{u}$.
   (a) Find $g'(1), g'(2), g'(3), g'(u_0)$.
   (b) Prove that g is not differentiable at $u = 0$.
5. Find the slope of the graph of $g(x) = x^2 - 1/x$ at $x = 1, x = 4, x = 11, x = x_0$.
6. Find the equation of the tangent line to the given curve at the point indicated.

   (a) $y = 3x^2 - 1$, $(2, 11)$ (c) $y = x - \sqrt{x}$, $(4, 2)$

   (b) $y = x^2 + 2x$, $(-1, -1)$ (d) $y = 4x + \dfrac{3}{x}$, $(3, 13)$

7. (a). How does the slope of the graph of $l(x) = mx + b$ at any point (as defined in §3B3) compare with the slope of the line $y = mx + b$ (as defined in §1B4)?
   (b) If f and g have the same derivative at a point x_0, what can one say about the tangent lines to their graphs at x_0?
   (c) Suppose f is differentiable and

   $$\frac{f(x_0 + 2) - f(x_0)}{2} = 3.$$

   What can one say about $f'(x_0)$?

   (d) Give an example of functions f and g for which

   $$f(1) < g'(1) < g(1) < f'(1).$$

8. The following functions give the position (in feet) after t seconds of a point P moving along a line. Find the average velocity of P over the time interval from the given value t_0 to $t_0 + \Delta t$.

   (a) $t + \dfrac{1}{t}$; $t_0 = 1$, $\Delta t = 1$

   (b) $2t^2 - 1$; $t_0 = 3$, $\Delta t = -1$

   (c) $\dfrac{t^2 - 1}{t^2 + 1}$; $t_0 = 2$, $\Delta t = 5$

   (d) $t^2 - |t|$; $t_0 = -1$, $\Delta t = 4$
   (e) $f(t) = 6$; $t_0 = 2$, $\Delta t = 2$
   (f) $f(t) = t^2$ $(t > 0)$, $f(t) = t^3$ $(t \le 0)$; $t_0 = -1$, $\Delta t = 2$

9. In Problem 8, in each case, find the instantaneous velocity at the given time t_0.

10. A body is thrown vertically from the ground. Its height after t seconds is $32t - 16t^2$ feet.
    (a) What is its initial velocity?
    (b) How long will it take to return to the ground?
    (c) What will be its velocity when it hits the ground?
    (d) How high will the body rise? [*Hint.* Note that its height at time t is $16 - 16(1 - t)^2$.]
    (e) What is its velocity at the highest point?

11. The position of a locomotive on a track is given by $s = t^4 - 8t^3 + 16t^2 + 3$.
    (a) What is its velocity at time 0?
    (b) At which values of t is the locomotive moving forward ($v > 0$) and at which in reverse ($v < 0$)?

12. Consider the functions $y = x^3, y = -x^3, y = x^2$ to show that if $f'(x_0) = 0$, f can be increasing at x_0, decreasing at x_0, or neither.

Answers to problems (§3B)

1. 3, 3, 3, 3
2. 3, 3, 0, 75, $3t_0^2$
3. $-1, -1, -\frac{1}{9}, -\frac{1}{9}, -1/t_0^2$
4. (a) $\frac{1}{2}, 1/2\sqrt{2}, 1/2\sqrt{3}, 1/2\sqrt{u_0}$
5. 3, $8\frac{1}{16}, 22\frac{1}{121}, 2x_0 + \dfrac{1}{x_0^2}$
6. (a) $y = 12x - 13$ (b) $y = -1$ (c) $y = \frac{3}{4}x - 1$ (d) $y = \frac{11}{3}x + 2$
8. (a) $\frac{1}{2}$ (feet/sec)   (b) 10   (c) $\frac{9}{125}$ (d) $\frac{3}{2}$ (e) 0   (f) 1
9. (a) 0   (b) 12   (c) $\frac{8}{25}$ (d) -1 (e) 0   (f) 3
10. (a) 32 feet/sec   (b) 2 sec   (c) -32 feet/sec   (d) 16 feet   (e) 0
11. (a) 0   (b) forward $0 < t < 2$ and $t > 4$,   reverse $t < 0$ and $2 < t < 4$

§3C. Differentiability

3C1. The derivative function. Let f be a function differentiable on an interval I. To each number $x \in I$, there is associated the number $f'(x)$, defined by

(1)
$$f'(x) = \lim_{\Delta x \to 0} \frac{f(x + \Delta x) - f(x)}{\Delta x}.$$

Thus, f' is a *function* on I. If, for example, $f(x) = x^2$, then $f'(x) = 2x$. The function f' is called the derivative of f.

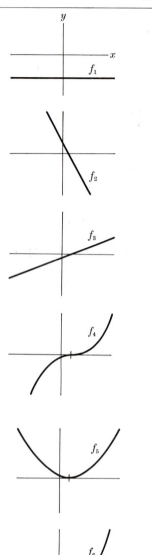

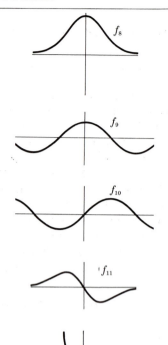

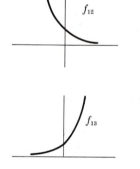

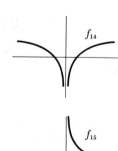

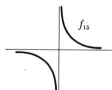

FIGURE 1(a)

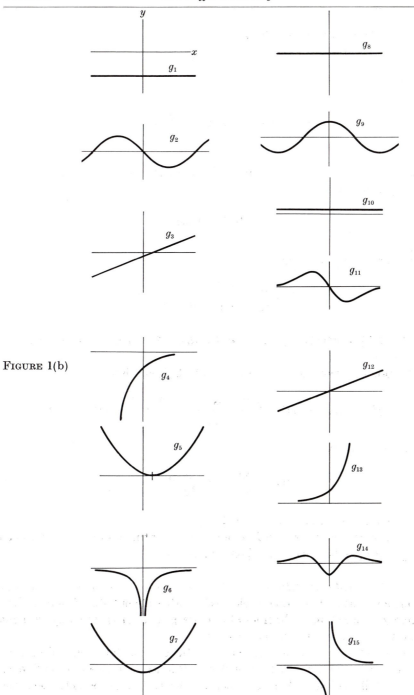

FIGURE 1(b)

There are other notations for $f'(x)$. If

$$y = f(x)$$

we write

.(2) $$y' = f'(x) \quad \text{and} \quad \frac{dy}{dx} = f'(x).$$

The symbol dy/dx is one piece, not a fraction. It is read "*d y d x.*" We also use the expression

$$\frac{d}{dx} f(x) = f'(x).$$

For instance, if $y = x^2$, then

$$y' = \frac{dy}{dx} = \frac{dx^2}{dx} = \frac{d}{dx} x^2 = 2x.$$

3C2. Differentiability and continuity. *A differentiable function is continuous.*

The proof is easy. At any point x at which $f'(x)$ exists,

(3) $$f(x + \Delta x) - f(x) = \frac{f(x + \Delta x) - f(x)}{\Delta x} \Delta x \rightarrow f'(x) \cdot 0 = 0$$

as $\Delta x \rightarrow 0$. Briefly,

(4) $$\Delta y = \frac{\Delta y}{\Delta x} \Delta x \rightarrow \frac{dy}{dx} \cdot 0 = 0 \qquad \text{(as } \Delta x \rightarrow 0\text{).}$$

This says that f is continuous at x $\big(\S 3A(8)\big)$.

The converse is not true. A function can be continuous at a point without being differentiable there. (See §3B, Problem 4.)

3C3. The role of differentiability. An immense amount of material throughout the rest of this book deals with differentiable functions. Intuitively, a function is differentiable if its graph turns smoothly without corners and has no vertical tangent lines.

We do not ordinarily run into curves with corners. But a vertical tangent line does crop up from time to time. This is the situation where the difference quotient fails to have a limit at a point because it is not bounded near the point. (See §3B, Problem 4.)

In many theorems, the requirement that a certain function be differentiable appears as an explicit part of the hypothesis. You should greet this as a sign of reassurance rather than a cause for alarm. The spirit of the hypothesis is to show the generality of the theorem, not its restrictedness.

Suppose you have to apply the theorem. It would be a mistake to think you must run back each time to evaluate the limit of some difference quotient. On the contrary. We will build up a collection of results so that, at a glance, you will be able to recognize all sorts of functions as differentiable on various intervals. (Of course, in *proving* these results we have to rely on precise definitions.) The exceptional points where there are vertical tangents can usually be identified by inspection.

Problems (§3C)

1. Match each function in Figure 1(a) ($f_1 - f_{15}$) with its derivative in Figure 1(b) ($g_1 - g_{15}$).

§3D. Derivatives of polynomials

To apply the derivative to problems in the mathematical, physical, biological, and social sciences, one must be able to *compute* derivatives with some facility. Fortunately, the necessary computations can be reduced to a rather mechanical process. We turn now to this matter, moving from the simple to the complex, always building on what we have learned. By the end of this section, we will be able to handle polynomials with ease, and, by the end of the chapter, algebraic expressions like

$$\frac{1}{\sqrt{x^2 + 1}}.$$

We will learn to differentiate other functions as they arise, so that by the end of Chapter 8 we will be able to handle just about anything we can write down, say $\log \sin \sqrt{x^2 + 1}$—without having to look at any limits at all.

Each formula is stated in both the f' and the d-notations. Recall the definition:

$$(1) \qquad f'(x) = \frac{d}{dx} f(x) = \lim_{\Delta x \to 0} \frac{f(x + \Delta x) - f(x)}{\Delta x}.$$

3D1. Derivative of a constant function. *If* $f(x) = c$ *(a constant), then* $f'(x) = 0$.

Briefly, $dc/dx = 0.$

Proof. For each point x, $f(x) = c$ and $f(x + \Delta x) = c$. Hence

(2) $$\frac{f(x + \Delta x) - f(x)}{\Delta x} = \frac{0}{\Delta x} = 0 \to 0 \qquad (\text{as } \Delta x \to 0).$$

3D2. Derivative of x. *If $f(x) = x$ then $f'(x) = 1$.*
Briefly, $dx/dx = 1.$

Proof. For each point x, $f(x) = x$ and $f(x + \Delta x) = x + \Delta x$. Hence

(3) $$\frac{f(x + \Delta x) - f(x)}{\Delta x} = \frac{\Delta x}{\Delta x} = 1 \to 1 \qquad (\text{as } \Delta x \to 0).$$

3D3. Derivative of x^n (Positive integral powers). *Let n be any positive integer.*
If $f(x) = x^n$ then $f'(x) = nx^{n-1}$ (where, it is understood, $n > 1$ in case $x = 0$).
Briefly, $(d/dx)x^n = nx^{n-1}.$

Before going into the proof, let us assemble some facts. We have just seen that the formula is true for $n = 1$:

$$\frac{d}{dx} x = 1 = 1 \cdot x^0 \qquad (x \neq 0).$$

We saw in §3B, Example 3, that it is true for $n = 2$:

$$\frac{d}{dx} x^2 = 2x = 2x^1.$$

Look back at that example and watch how the term $(\Delta x)^2$ disappears in the course of the derivation.

Now let us carry through the derivation for $n = 3$. For each x, $f(x) = x^3$ and

(4) $$f(x + \Delta x) = (x + \Delta x)^3 = x^3 + 3x^2 \Delta x + 3x(\Delta x)^2 + (\Delta x)^3.$$

This time two terms, the ones in $(\Delta x)^2$ and $(\Delta x)^3$, will disappear. We have

(5) $$f(x + \Delta x) - f(x) = (x + \Delta x)^3 - x^3$$
$$= 3x^2 \Delta x + 3x(\Delta x)^2 + (\Delta x)^3.$$

Hence

$$(6) \qquad \frac{f(x + \Delta x) - f(x)}{\Delta x} = 3x^2 + 3x \, \Delta x + (\Delta x)^2 \rightarrow 3x^2$$

$$(\text{as } \Delta x \rightarrow 0).$$

Proof of theorem. For each x,

$$(7) \qquad f(x) = x^n \quad \text{and} \quad f(x + \Delta x) = (x + \Delta x)^n.$$

Now we need the following fragment of the binomial theorem:

$$(8) \qquad (x + \Delta x)^n = x^n + nx^{n-1} \, \Delta x + (\Delta x)^2 P,$$

where P is a certain polynomial in x and Δx. With x fixed, P approaches some limit as $\Delta x \rightarrow 0$. Since

$$(9) \qquad f(x + \Delta x) - f(x) = nx^{n-1} \, \Delta x + (\Delta x)^2 P,$$

we have

$$(10) \qquad \frac{f(x + \Delta x) - f(x)}{\Delta x} = nx^{n-1} + (\Delta x)P \rightarrow nx^{n-1} \quad (\text{as } \Delta x \rightarrow 0).$$

Example 1. If $f(x) = x^6$ then $f'(x) = 6x^5$.
Briefly, $(d/dx)x^6 = 6x^5$.

3D4. Derivative of a constant times a function. *Let f be a differentiable function and let c be a constant.*
    If $g = cf$ then $g' = cf'$.
    Briefly, $(d/dx)(cy) = c(dy/dx)$.

    Proof. $g(x) = cf(x)$ and $g(x + \Delta x) = cf(x + \Delta x)$. Therefore

$$(11) \qquad \frac{g(x + \Delta x) - g(x)}{\Delta x} = c \frac{f(x + \Delta x) - f(x)}{\Delta x} \rightarrow cf'(x)$$

$$(\text{as } \Delta x \rightarrow 0).$$

Example 2.

    (a) $\dfrac{d}{dx} (3x^4) = 3 \dfrac{d}{dx} x^4 = 12x^3.$

    (b) $\dfrac{d}{dx} (4x^3) = 4 \dfrac{d}{dx} x^3 = 12x^2.$

3D5. Derivative of the negative of a function. *Let* f *be a differentiable function.*

If $g = -f$ then $g' = -f'$.

Briefly, $(d/dx)(-y) = -dy/dx.$

Proof. Take $c = -1$ in §3D4.

Example 3.

(a) $\dfrac{d}{dx}(-3x^4) = -12x^3.$

(b) $\dfrac{d}{dx}(-4x^3) = -12x^2.$

3D6. Derivatives of sums and differences. *Let* f *and* g *be differentiable.*

(a) *If* $h = f + g$ *then* $h' = f' + g'.$
(b) *If* $k = f - g$ *then* $k' = f' - g'.$

Briefly, $\dfrac{d}{dx}(u + v) = \dfrac{du}{dx} + \dfrac{dv}{dx}$, *and* $\dfrac{d}{dx}(u - v) = \dfrac{du}{dx} - \dfrac{dv}{dx}.$

Proof. (a). $h(x) = f(x) + g(x)$ and $h(x + \Delta x) = f(x + \Delta x) + g(x + \Delta x)$. Therefore,

(12)
$$\frac{h(x + \Delta x) - h(x)}{\Delta x} = \frac{f(x + \Delta x) - f(x)}{\Delta x} + \frac{g(x + \Delta x) - g(x)}{\Delta x}$$

$$\to f'(x) + g'(x) \qquad \text{(as } \Delta x \to 0\text{)}.$$

The proof of (b) goes the same way.

Remark. The theorem extends to any number of terms. For example,

(13)
$$\frac{d}{dx}(u + v - w) = \frac{du}{dx} + \frac{dv}{dx} - \frac{dw}{dx}.$$

3D7. Derivative of a polynomial. By combining the preceding results we can differentiate any polynomial.

Example 4. $(d/dx)(2x^4 - 5x^3 + 5x^2 - 6x + 7) = 8x^3 - 15x^2 + 10x - 6.$

Problems (§3D)

1. Find the following derivatives.

    (a) $\dfrac{d}{dx}(5x^8)$

    (c) $\dfrac{d}{ds}(13s^2)$

    (b) $\dfrac{d}{dx}(-3x^7)$

    (d) $\dfrac{d}{dt}(-15t^3)$

2. Find the following derivatives.

    (a) $\dfrac{d}{dx}(4x^5 - 5x^4 + 2x^3 - 3x^2)$

    (c) $\dfrac{d}{dx}[(x+1)(x-1)]$

    (b) $\dfrac{d}{dx}(1 + x + x^2 + x^3 + x^4)$

    (d) $\dfrac{d}{dx}[(3x+1)(x^2-2)]$

3. Find the following derivatives.

    (a) $\dfrac{d}{dt}(5t^{14} + 8t^{11} - 7t^{10} + 3)$

    (c) $\dfrac{d}{du}\dfrac{u^2 - 4}{u - 2}$

    (b) $\dfrac{d}{ds}(14s^5 + 11s^8 - 10s^7 + 3)$

    (d) $\dfrac{d}{dv}\dfrac{v^3 - 1}{v - 1}$

4. (a) Let $f(x) = 3x^3 + 2x^2 + 1$. Find $f'(-2)$.
   (b) Let $g(t) = 12t^5 - 7t^4 - 3$. Find $g'(1)$.
   (c) Let $h = f + g$, where $f(s) = 3s^5 - 2s^2 + 2$ and $g(u) = 4u^5 + 3u^3 - u$.
       Find $h'(1)$, $h'(-1)$, and $h'(0)$.

5. Find:

    (a) $\dfrac{d}{dx}\lim_{u \to 0}(x + 3u)^2$

    (b) $\dfrac{d}{dx}\lim_{v \to 1}(x + 3v)^2$

    (c) $\dfrac{d}{dx}\lim_{t \to 0}\left(x^3\sqrt{4 + t^2} - \dfrac{x^2}{2 + t^2}\right)$

    (d) $\dfrac{d}{dx}\lim_{\Delta x \to 0}\dfrac{(x + \Delta x)^3 - x^3}{\Delta x}$

    (e) $\dfrac{d}{dx}\lim_{t \to 0}\dfrac{(x + t)^4 - x^4}{t}$

6. Verify that $\dfrac{dy}{dx} = \dfrac{dy}{du}\dfrac{du}{dx}$, given:

    (a) $y = u^2$,   where $u = x^2$
    (b) $y = u^2$,   where $u = 3x + 1$
    (c) $y = 2u^3 - u$,   where $u = 1 - x$
    (d) $y = (2u - 1)^2$,   where $u = x^2 + 1$

7. Exhibit a function y for which:

    (a) $\dfrac{dy}{dx} = 3x^2$ (c) $\dfrac{dy}{dx} = 3x^2 - 2x + 1$

    (b) $\dfrac{dy}{dx} = 2x^2$ (d) $\dfrac{dy}{dx} = (x - 4)(x - 3)$

8. Compute $f(g(x))$, $f'(g(x))$, and $f'(g'(x))$ for:
    (a) $f(x) = 4x^3$, $g(x) = 3x^4$
    (b) $f(x) = 3x$, $g(x) = x^3 - x^2 + x - 1$
    (c) $f(x) = 3x^2 - 2x + 1$, $g(x) = 2x - 1$

9. Let $f(x) = 3x^2$. Find:

    (a) $f'(3x^2)$ (d) $\dfrac{d}{dx} f(6x)$

    (b) $\dfrac{d}{dx} f(3x^2)$ (e) $f'(x^3)$

    (c) $f'(6x)$ (f) $\dfrac{d}{dx} f(x^3)$

10. Let $g(u) = 4u^3$. Find:

    (a) $g'(4u^3)$ (d) $\dfrac{d}{du} g(12u^2)$

    (b) $\dfrac{d}{du} g(4u^3)$ (e) $g'(u^4)$

    (c) $g'(12u^2)$ (f) $\dfrac{d}{du} g(u^4)$

11. What are the equations of the horizontal tangents to the curve $y = x^3 - 2x^2 + x$?

12. Find the equation of the tangent line of slope 4 to the curve $y = x^4 + 2x^2 + 4x + 3$.

13. Find the point of intersection of the tangent lines to the curve $y = 4x^2 - 3x + 6$ of slopes 1 and -1.

14. Find c so that the line $y = 4x + 3$ is tangent to the curve $y = x^2 + c$.

15. The line $x = a$ cuts the curve $y = \frac{1}{3}x^3 + 4x + 1$ at a point P and the curve $y = 2x^2 + x - 1$ at a point Q. If the tangents to the curves at P and Q are parallel, what is a? What are the equations of the tangents?

16. Show that the curves $y = x^3 - 3x^2 + x - 1$ and $4y = x^2 - 16$ are tangent to one another—i.e., have a common tangent at a common point. What is the equation of the tangent line at that point?

17. A point moves in a straight line according to the law $s = t^3 - 9t^2 + 39t - 14$. At what instant is its velocity equal to 12?

18. A point P moves on the x-axis according to the law $x = t^3 + 2t^2 + 10t + 1$ and a point Q according to the law $x = t^2 + 11t$.

    (a) How far apart are they when their velocities are the same?

    (b) How far apart are they when the velocity of P is twice that of Q?

19. Is there a function equal to the square of its derivative?

20. What is $\lim\limits_{h \to 0} \dfrac{(1 + h)^{23} - 1}{h}$?

Answers to problems (§3D)

1. (a) $40x^7$
2. (c) $2x$ (d) $9x^2 + 2x - 6$
3. (c) 1 (d) $2v + 1$
4. (a) 28 (b) 32 (c) $39, 47, -1$
5. (a) $2x$ (b) $2x + 6$ (c) $6x^2 - x$ (d) $6x$ (e) $12x^2$
8. (a) $108x^{12}, 108x^8, 1728x^6$ (b) $3x^3 - 3x^2 + 3x - 3, 3, 3$
    (c) $12x^2 - 16x + 6, 12x - 8, 10$
9. (a) $18x^2$ (b) $108x^3$ (c) $36x$ (d) $216x$ (e) $6x^3$ (f) $18x^5$
10. (a) $192u^6$ (b) $2304u^8$ (c) $1728u^4$ (d) $41472u^5$ (e) $12u^8$ (f) $48u^{11}$
11. $y = 0$ and $y = \frac{4}{27}$
12. $y = 4x + 3$
13. $(\frac{3}{8}, \frac{43}{8})$
14. 7
15. $a = 1, y = 5x + \frac{1}{3}, y = 5x - 3;$ $a = 3, y = 13x - 17, y = 13x - 19$
16. $y = x - 5$
17. $t = 3$
18. (a) $t = \frac{1}{3} : \frac{22}{27};$ $t = -1 : 2$
    (b) $t = 2 : 11;$ $t = -2 : 1$

§3E. Derivatives of products and quotients

We continue to collect formulas about derivatives.

3E1. Derivative of a product. One might guess that the derivative of a product is the product of the derivatives. The simplest example shows that this is not true. For instance, if $f(x) = x$ and $g(x) = 1$, then $(fg)' = 1$, while $f'g' = 0$. The actual formula is a little more subtle and its derivation requires some care. The rule is:

Let f and g be differentiable.

If $h = fg$ then $h' = fg' + gf'$.

Briefly,
$$\frac{d}{dx}(uv) = u\frac{dv}{dx} + v\frac{du}{dx}.$$

To help make the mechanism clear, we give the proof in both notations.

Proof 1. We are interested in

(1) $$h'(x) = \lim_{\Delta x \to 0} \frac{h(x + \Delta x) - h(x)}{\Delta x}.$$

We write

(2) $$h(x + \Delta x) - h(x) = f(x + \Delta x)g(x + \Delta x) - f(x)g(x)$$

and get stuck. The trick is to subtract and add

$$f(x + \Delta x)g(x).$$

Then we group the terms as follows:

(3) $$h(x + \Delta x) - h(x) = f(x + \Delta x)[g(x + \Delta x) - g(x)]$$
$$+ [f(x + \Delta x) - f(x)]g(x).$$

(To check, multiply this out.) Dividing by Δx, we have

(4) $$\frac{h(x + \Delta x) - h(x)}{\Delta x} = f(x + \Delta x)\frac{g(x + \Delta x) - g(x)}{\Delta x}$$
$$+ \frac{f(x + \Delta x) - f(x)}{\Delta x}g(x).$$

Since f is differentiable, it is continuous (§3C2), so that $f(x + \Delta x) \to f(x)$ as $\Delta x \to 0$. Therefore,

(5) $$\frac{h(x + \Delta x) - h(x)}{\Delta x} \to f(x)g'(x) + f'(x)g(x) \qquad (\text{as } \Delta x \to 0).$$

Proof 2. Let $y = uv$, where u and v are differentiable functions of x. A change Δx results in changes Δu and Δv. These result in a change Δy. Then

(6) $$y + \Delta y = (u + \Delta u)(v + \Delta v)$$
$$= uv + u\,\Delta v + v\,\Delta u + \Delta u\,\Delta v.$$

Since $y = uv$ we may subtract it off to get

(7) $\Delta y = u\,\Delta v + v\,\Delta u + \Delta u\,\Delta v.$

Dividing by Δx, we have

(8) $\dfrac{\Delta y}{\Delta x} = u\,\dfrac{\Delta v}{\Delta x} + v\,\dfrac{\Delta u}{\Delta x} + \Delta u\,\dfrac{\Delta v}{\Delta x}.$

Since u is differentiable, it is continuous (§3C2), so $\Delta u \to 0$ as $\Delta x \to 0$. Therefore,

(9) $\dfrac{\Delta y}{\Delta x} \to u\,\dfrac{dv}{dx} + v\,\dfrac{du}{dx} + 0\,\dfrac{dv}{dx} = u\,\dfrac{dv}{dx} + v\,\dfrac{du}{dx}$ (as $\Delta x \to 0$).

Example 1. If $h(x) = (x^3 - 3x^2 + 7x + 4)(2x^2 - 5x + 3)$, then

$h'(x) = (x^3 - 3x^2 + 7x + 4)(4x - 5) + (2x^2 - 5x + 3)(3x^2 - 6x + 7).$

3E2. Derivative of the reciprocal of a function. *Let g be differentiable.*

If $h(x) = \dfrac{1}{g(x)}$ *then* $h' = -\dfrac{g'}{g^2}.$

Briefly, $\dfrac{d}{dx}\left(\dfrac{1}{v}\right) = -\dfrac{1}{v^2}\dfrac{dv}{dx}.$

Proof. Consider any x for which $g(x) \neq 0$. Since g is continuous (§3C2), there is an interval about x on which g is never 0 (§2C2), and we will keep Δx small enough so that $x + \Delta x$ stays in that interval. Then $g(x + \Delta x) \neq 0$. We have

(10) $h(x + \Delta x) - h(x) = \dfrac{1}{g(x + \Delta x)} - \dfrac{1}{g(x)} = \dfrac{g(x) - g(x + \Delta x)}{g(x + \Delta x)g(x)}.$

Dividing by Δx, we have

$\dfrac{h(x + \Delta x) - h(x)}{\Delta x} = -\dfrac{g(x + \Delta x) - g(x)}{\Delta x}\dfrac{1}{g(x + \Delta x)g(x)}.$

Since g is continuous, $g(x + \Delta x) \to g(x)$ as $\Delta x \to 0$. Therefore,

(11) $\dfrac{h(x + \Delta x) - h(x)}{\Delta x} \to -g'(x)\dfrac{1}{g(x)^2}$ (as $\Delta x \to 0$).

Example 2. An important special case is:

(12)
$$\frac{d}{dx}\left(\frac{1}{x}\right) = -\frac{1}{x^2}$$
$(x \neq 0)$.

Example 3.

$$\frac{d}{dx}\left(\frac{1}{3x^2 - x}\right) = -\frac{6x - 1}{(3x^2 - x)^2}$$
$(x \neq 0, \tfrac{1}{3})$.

3E3. Derivative of a quotient. *Let f and g be differentiable.*

If $h(x) = \dfrac{f(x)}{g(x)}$ *then* $h' = \dfrac{gf' - fg'}{g^2}$.

Briefly,
$$\frac{d}{dx}\left(\frac{u}{v}\right) = \frac{v\,\dfrac{du}{dx} - u\,\dfrac{dv}{dx}}{v^2}.$$

Proof. We write h in the form

$$h = f \cdot \frac{1}{g}$$

and apply the rules for the product (§3E1) and the reciprocal (§3E2). Thus,

$$h' = f \cdot \left(\frac{1}{g}\right)' + \frac{1}{g}f'$$

$$= f \cdot \left(-\frac{g'}{g^2}\right) + \frac{g}{g^2}f',$$

as desired.

Example 4. If $y = \dfrac{12x}{7x + 3}$ $(x \neq -\tfrac{3}{7})$ then $y = 12\,\dfrac{x}{7x + 3}$ and

$$\frac{dy}{dx} = 12\,\frac{(7x + 3)(1) - (x)(7)}{(7x + 3)^2} = \frac{36}{(7x + 3)^2}.$$

3E4. Derivative of x^m (Integral powers). *Let m be any integer—positive, negative, or zero.*
    If $f(x) = x^m$ *then* $f'(x) = mx^{m-1}$ *(where, it is understood, $m > 1$ in case $x = 0$).*

Briefly,
$$\frac{d}{dx}x^m = mx^{m-1}.$$

Proof. We have already handled the case $m > 0$ (§3D3). For $m = 0$ the result is obvious ($f = 1$ and $f' = 0$).

For $m < 0$ put $n = -m$. Then n is a positive integer and

$$x^m = \frac{1}{x^n}.$$

We can now combine what we know about reciprocals (§3E2) with the power rule for positive integers (§3D3). We have

(13)
$$\frac{d}{dx}x^m = \frac{d}{dx}\frac{1}{x^n} = -\left(\frac{1}{x^n}\right)^2\frac{d}{dx}x^n$$

$$= -\frac{1}{x^{2n}}nx^{n-1} = mx^{m-1},$$

as required.

Note that the special case $m = -1$ states that

(14)
$$\frac{d}{dx}\left(\frac{1}{x}\right) = -\frac{1}{x^2}.$$

This formula was previously noted in (12).

Problems (§3E)

1. Find the following derivatives.

(a) $\dfrac{d}{dx}[(4x + 5)(x^2 - 2x + 2)]$

(b) $\dfrac{d}{dx}[(6x - 1)(x + 2x + 3)]$

(c) $\dfrac{d}{dt}[(t^8 - 5t + 1)(2t^{-6} - 3)]$

(d) $\dfrac{d}{dy}[(5y + 4)(y^3 - 7y^2 + 3y + 1)]$

(e) $\dfrac{d}{dv}[(2v^2 - 3v - 3)(4v^2 + 5v - 2)]$

(f) $\dfrac{d}{ds}[(3s^2 + 8s + 9)(5s^2 - 9s + 7)]$

(g) $\dfrac{d}{dx}\left[(x^2 - 3x + 5)(2x^3 - x^2 + 4x - 1)\right]$

(h) $\dfrac{d}{dx}\left(4x^3 - 4x^2 + \dfrac{3}{x^2} - \dfrac{4}{x^3}\right)^2$

(i) $\dfrac{d}{du}\left[(u^2 + u - 1)(u^2 - 1)(u^2 - 2u - 1)\right]$ [*Hint. fgh = f(gh).*]

(j) $\dfrac{d}{ds}(2s^2 + 5s - 2)^3$

2. Find $F'(x)$ for each of the following functions.

(a) $F(x) = \dfrac{x^2 - 1}{x^2 + 1}$

(b) $F(x) = \dfrac{x^2 + 1}{x^2 - 1}$

(c) $F(x) = \dfrac{1}{x + \dfrac{1}{x}}$

(d) $F(x) = \dfrac{(x^2 + 1)(x^3 + 1)}{6x^{-4} + 1}$

(e) $F(x) = \dfrac{2x - 1}{(3x^3 + 4)(4x^3 + 3)}$

3. Find $G'(t)$ for each of the following functions.

(a) $G(t) = \dfrac{1}{3t^2 + 4t - 6}$

(b) $G(t) = \dfrac{1}{2t^5 - 5t^4 - 3t^{-2}}$

(c) $G(t) = \dfrac{1}{(4t^{-5} + 7)(2t^2 - t + 8)}$

4. Verify that $\dfrac{dy}{dx} = \dfrac{dy}{du}\dfrac{du}{dx}$, given:

(a) $y = \dfrac{u}{u + 1}$, where $u = x^2$

(b) $y = \dfrac{1}{u}$, where $u = x^2 + x + 1$

(c) $y = u^2 + u + 1$, where $u = \dfrac{1}{x}$

(d) $y = \dfrac{u + 1}{u - 1}$, where $u = \dfrac{x + 1}{x - 1}$

5. Exhibit a function y such that

(a) $\dfrac{dy}{dx} = (x^2 + 1) \cdot 3x^2 + (x^3 + 1) \cdot 2x$

(b) $\dfrac{dy}{dx} = (x^2 - x + 1)(2x - 1) + (x^2 - x + 2)(2x - 1)$

(c) $\dfrac{dy}{dx} = \dfrac{(x^3 - 1) \cdot 2x - (x^2 + 1) \cdot 3x^2}{(x^3 - 1)^2}$

6. Compute $f(g(x))$, $f'(g(x))$, and $f'(g'(x))$ for:

(a) $f(x) = x^3$, $g(x) = x^{-2}$

(d) $f(x) = \dfrac{1}{x + 1}$, $g(x) = x^2$

(b) $f(x) = \dfrac{1}{x}$, $g(x) = x^2 + 1$

(e) $f(x) = \dfrac{1}{x^2}$, $g(x) = \dfrac{1}{x^3}$

(c) $f(x) = x^2 + 2x$, $g(x) = \dfrac{1}{x^2}$

7. Let $F(x) = 1/x$. Find:

(a) $F'(\tfrac{1}{2})$

(e) $F'(-\tfrac{1}{4})$

(b) $F'\left(\dfrac{1}{x}\right)$

(f) $F'\left(-\dfrac{1}{x^2}\right)$

(c) $\dfrac{d}{dx} F\left(\dfrac{1}{x}\right)$

(g) $\dfrac{d}{dx} F\left(-\dfrac{1}{x^2}\right)$

(d) $\dfrac{d}{dt} F\left(\dfrac{1}{t}\right)$

(h) $\dfrac{d}{du} F\left(-\dfrac{1}{u^2}\right)$

8. Let $G = 1/H$, where H is differentiable. Prove that

$$\dfrac{H'(x)}{H(x)} = -\dfrac{G'(x)}{G(x)}.$$

9. Show that the tangents to the curves $y = (x^2 + 5)/x^2$ and $y = (x^2 - 4)/(x^2 + 1)$ at $x = 2$ are perpendicular to each other.

10. Let a be any positive number. Let l_1 be the line through the origin of slope a and let l_2 be the tangent line to the curve $y = 1/x$ ($x > 0$) of slope $-a$. Show that the area of the triangle cut off by l_1, l_2, and the x-axis is equal to 1.

11. Starting at time $t = 0$, point P moves up the y-axis according to the law $y = t/(t + 1)$ and point Q according to $y = t - \frac{1}{2}$. When does Q catch up to P, and how fast is Q overtaking P at that instant ?

12. Point P moves on a straight line according to the law $s = 3t^2/(1 + t^2) + (2 - t)f(t) + 3t$, while Q moves according to $s = 3t^2/(1 + t^2) + (2 - t)f(t) - 4t$, where f is a function that is differentiable but otherwise unknown. Prove that the difference between their velocities is constant.

13. (a) Is there a (nonconstant) function whose square is its derivative ?

(b) Is there a function for which the derivative of its reciprocal is the reciprocal of its derivative ?

Answers to problems (§3E)

1. (a) $(4x + 5)(2x - 2) + 4(x^2 - 2x + 2)$

(c) $(t^8 - 5t + 1)(-12t^{-7}) + (2t^{-6} - 3)(8t^7 - 5)$

(h) $2(4x^3 - 4x^2 + 3x^{-2} - 4x^{-3})(12x^2 - 8x - 6x^{-3} + 12x^{-4})$

(i) $[(u^2 + u - 1)(u^2 - 1)](2u - 2) + [(u^2 + u - 1)(u^2 - 2u - 1)](2u)$
$$+ [(u^2 - 1)(u^2 - 2u - 1)](2u + 1)$$

(j) $3(2s^2 + 5s - 2)^2(4s + 5)$

2. (a) $\dfrac{4x}{(x^2 + 1)^2}$ (b) $-\dfrac{4x}{(x^2 - 1)^2}$ (c) $\dfrac{1 - x^2}{(x^2 + 1)^2}$

(d) $\dfrac{(6x^{-4} + 1)[(x^2 + 1)(3x^2) + (x^3 + 1)(2x)] - (x^2 + 1)(x^3 + 1)(-24x^{-5})}{(6x^{-4} + 1)^2}$

(e) $\dfrac{(3x^3 + 4)(4x^3 + 3)(2) - (2x - 1)[(3x^3 + 4)(12x^2) + (4x^3 + 3)(9x^2)]}{[(3x^3 + 4)(4x^3 + 3)]^2}$

3. (a) $-\dfrac{6t + 4}{(3t^2 + 4t - 6)^2}$ (b) $-\dfrac{10t^4 - 20t^3 + 6t^{-3}}{(2t^5 - 5t^4 - 3t^{-2})^2}$

(c) $-\dfrac{(4t^{-5} + 7)(4t - 1) + (2t^2 - t + 8)(-20t^{-6})}{[(4t^{-5} + 7)(2t^2 - t + 8)]^2}$

6. (a) $x^{-6}, 3x^{-4}, 12x^{-6}$ (b) $\dfrac{1}{x^2 + 1}, -\dfrac{1}{(x^2 + 1)^2}, -\dfrac{1}{4x^2}$

(c) $\dfrac{2x^2 + 1}{x^4}, \dfrac{2x^2 + 2}{x^2}, \dfrac{2x^3 - 4}{x^3}$ (d) $\dfrac{1}{x^2 + 1}, -\dfrac{1}{(x^2 + 1)^2}, -\dfrac{1}{(2x + 1)^2}$

(e) $x^6, -2x^9, \frac{2}{27}x^{12}$

7. (a) -4 (b) $-x^2$ (c) 1 (d) 1 (e) -16 (f) $-x^4$ (g) $-2x$
(h) $-2u$

11. time $t = 1$, velocity $\frac{3}{4}$

§3F. The chain rule

3F1. Composite functions. The setting for the chain rule is a composite function. Let g be differentiable at a point x, and let f be differentiable at the point $g(x)$. Consider the composite function

$$(1) \qquad\qquad h(x) = f\bigl(g(x)\bigr).$$

The chain rule states that h is differentiable at x and gives a formula for h' in terms of f', g, and g'.

Let

$$(2) \qquad\qquad u = g(x), \qquad y = f(u) = h(x).$$

In this notation the chain rule is easily remembered as follows: $dy/dx = (dy/du)(du/dx)$. Its essential content is also easy: if a car is traveling 10 times as fast as a man and the man is walking twice as fast as a child, then the car is traveling 20 times as fast as the child.

3F2. The chain rule. *Let g be differentiable at a point x and let f be differentiable at $g(x)$.*

If $h(x) = f\bigl(g(x)\bigr)$ then $h'(x) = f'\bigl(g(x)\bigr)g'(x)$.

Briefly,
$$\frac{dy}{dx} = \frac{dy}{du}\frac{du}{dx}.$$

Proof. Let u and y be as in (2). A change Δx results in changes Δu and Δy:

$$(3) \qquad \begin{aligned} \Delta u &= g(x + \Delta x) - g(x), \\ \Delta y &= f(u + \Delta u) - f(u) = h(x + \Delta x) - h(x). \end{aligned}$$

By definition of derivative,

$$(4) \qquad \frac{du}{dx} = \lim_{\Delta x \to 0} \frac{\Delta u}{\Delta x}, \qquad \frac{dy}{du} = \lim_{\Delta u \to 0} \frac{\Delta y}{\Delta u}, \qquad \frac{dy}{dx} = \lim_{\Delta x \to 0} \frac{\Delta y}{\Delta x}.$$

Except for a subtlety discussed below we may now reason as follows. Since g is differentiable it is continuous, i.e., as $\Delta x \to 0$, $\Delta u \to 0$ (§3C2). Consequently, as $\Delta x \to 0$, $\Delta y/\Delta u \to dy/du$ and

$$(5) \qquad \frac{\Delta y}{\Delta x} = \frac{\Delta y}{\Delta u}\frac{\Delta u}{\Delta x} \to \frac{dy}{du}\frac{du}{dx},$$

which is what we want.

3F3. Elaboration of the proof. The subtlety referred to is that Δu may assume the value 0 as $\Delta x \to 0$. In this case (5) has no meaning. We handle this case (along with the other) by means of a trick.

Introduce a new variable, p:

(6) $$p = \frac{\Delta y}{\Delta u} \quad \text{when} \quad \Delta u \neq 0, \qquad p = \frac{dy}{du} \quad \text{when} \quad \Delta u = 0.$$

Then:

(7) $$p \to \frac{dy}{du} \qquad\qquad (as\ \Delta x \to 0)$$

(whether or not $\Delta u = 0$ in the process).

Next, when $\Delta u = 0$, $\Delta y = 0$, as we see from the second equation in (3). Therefore,

(8) $$\Delta y = p\,\Delta u \qquad\qquad (\text{for } all\ \Delta u).$$

From (8) and (7),

(9) $$\frac{\Delta y}{\Delta x} = p\,\frac{\Delta u}{\Delta x} \to \frac{dy}{du}\frac{du}{dx} \qquad (as\ \Delta x \to 0),$$

as required.

3F4. Applying the chain rule. The chain rule is the outstanding tool for reducing intricate differentiations to a routine. Your appreciation of its power should increase with each new application.

Example 1. Find

$$\frac{d}{dx}(x^2 + 1)^3.$$

Solution 1. We think of $(x^2 + 1)^3$ as $f(g(x))$, where

$$f(x) = x^3 \quad \text{and} \quad g(x) = x^2 + 1.$$

Then

$$f'(x) = 3x^2 \quad \text{and} \quad g'(x) = 2x.$$

Therefore,

$$\frac{d}{dx}(x^2 + 1)^3 = f'(g(x))g'(x)$$

$$= 3(x^2 + 1)^2 \cdot 2x = 6x(x^2 + 1)^2.$$

Solution 2 (d-notation). Let $y = (x^2 + 1)^3$. Thus,

$$y = u^3, \quad \text{where} \quad u = x^2 + 1.$$

Then

$$\frac{dy}{du} = 3u^2 \quad \text{and} \quad \frac{du}{dx} = 2x.$$

Hence

$$\frac{dy}{dx} = \frac{dy}{du}\frac{du}{dx} = 3u^2 \cdot 2x = 6x(x^2 + 1)^2.$$

Caution. Do not forget the factor du/dx. The temptation is to write

(10) $y = (x^2 + 1)^3, \qquad \dfrac{dy}{dx} = 3(x^2 + 1)^2.$ **X**

Almost every student makes this mistake at least once. Each of the authors of this book was a student and made the mistake at least once.

Example 2. Find $(d/dx)\big((x^2 + 1)/(x^2 - 1)\big)^4$.

Solution. We think of $\big((x^2 + 1)/(x^2 - 1)\big)^4$ as $f(g(x))$, where

$$f(x) = x^4 \qquad \text{and} \qquad g(x) = \frac{x^2 + 1}{x^2 - 1}.$$

Then

$$f'(x) = 4x^3 \qquad \text{and} \qquad g'(x) = \frac{(x^2 - 1)(2x) - (x^2 + 1)(2x)}{(x^2 - 1)^2}$$

$$= -\frac{4x}{(x^2 - 1)^2}.$$

Therefore,

$$\frac{d}{dx}\left(\frac{x^2 + 1}{x^2 - 1}\right)^4 = f'(g(x))g'(x)$$

$$= 4\left(\frac{x^2 + 1}{x^2 - 1}\right)^3 \frac{(-4x)}{(x^2 - 1)^2} = -16\,\frac{x(x^2 + 1)^3}{(x^2 - 1)^5}.$$

The symbols $g(x)$ or u help keep track of the computation. Most students use them at first, while the method is still new, but dispense with them as the process becomes automatic. The following example should be studied thoroughly as a model.

Example 3. If $h(x) = (x^3 + 1)^4$ then

$$h'(x) = 4(x^3 + 1)^3(3x^2).$$

Sometimes the chain rule has to be applied repeatedly.

Example 4. Let $y = [(x^3 + 1)^4 + 1]^3$. Then

$$\frac{dy}{dx} = 3[(x^3 + 1)^4 + 1]^2 \cdot 4(x^3 + 1)^3 \cdot 3x^2.$$

Problems (§3F)

1. Find $h'(x)$.
   (a) $h(x) = (x^2 + 1)^5$
   (b) $h(x) = (x^2 + 1)^{143}$
   (c) $h(x) = (x^3 - 1)^3$
   (d) $h(x) = (3x^5 - 4x^3 + 3x - 6)^2$
   (e) $h(x) = (3x^2 - 1)^{15}$
   (f) $h(x) = (x + 1/x)^{12}$

2. Find $f'(t)$.
   (a) $f(t) = (t^2 + 1)^4(t^2 + 2)^3$
   (b) $f(t) = (1 - 2t)^3(1 - 3t)^6$
   (c) $f(t) = (1 + t)^5(1 + 2t^2)^6$
   (d) $f(t) = (1 - t + t^2)^4(1 + t - t^2)^5$

3. Find $g'(u)$.

   (a) $g(u) = \dfrac{(1 + u)^5}{(1 + 2u^2)^6}$

   (b) $g(u) = \dfrac{(u^2 + 1)^4}{(u^2 + 2)^3}$

   (c) $g(u) = \dfrac{(1 - 2u)^3}{(1 - 3u)^6}$

   (d) $g(u) = \dfrac{(1 - u + u^2)^4}{(1 + u - u^2)^5}$

4. Find $F'(v)$.

   (a) $F(v) = \dfrac{(v^2 + v - 1)^2(v^3 + 3v^2 - 3)^3}{(4v^5 - 3v + 6)^4}$ $\left[\text{Hint. } F(v) = \dfrac{a^2b^3}{c^4}. \right]$

   (b) $F(v) = \dfrac{(v^3 + 3v^2 - 3)^3}{(v^2 + v - 1)^2(4v^5 - 3v + 6)^4}$

5. Let $H(x) = (x^2 + x + 1)^5$. Find:

   (a) $H'(x)$

   (b) $H'(1)$

   (c) $H'(2x)$

   (d) $H'(-x)$

   (e) $H'\left(\dfrac{1}{x}\right)$

6. Let y be a differentiable function of x. Express the following in terms of y and dy/dx.

   (a) $\dfrac{d}{dx}\, y^2$

   (b) $\dfrac{d}{dx}\, (y + 1)^2$

   (c) $\dfrac{d}{dx}\, (y^2 + 1)^2$

   (d) $\dfrac{d}{dx}\, (2y^3 - 4y^2 + 3y + 6)$

7. Express the following in terms of

$$u_1 = f'(x), \qquad u_4 = f'(f'(x)),$$
$$u_2 = f'(f(x)), \qquad u_5 = f'(f(f'(x))),$$
$$u_3 = f'(f(f(x))), \qquad u_6 = f''(x).$$

   (a) $\dfrac{d}{dx}\, f(x)$

   (b) $\dfrac{d}{dx}\, f'(x)$

   (c) $\dfrac{d}{dx}\, f(f(x))$

   (d) $\dfrac{d}{dx}\, f(f'(x))$

   (e) $\dfrac{d}{dx}\, f(f(f(x)))$

   (f) $\dfrac{d}{dx}\, f(f(f'(x)))$

8. Exhibit a function y for which:

(a) $\dfrac{dy}{dx} = 2(x^2 + 1) \cdot 2x$ (c) $\dfrac{dy}{dx} = 2(x^3 + 1) \cdot 3x^2$

(b) $\dfrac{dy}{dx} = 3(x^2 + 1)^2 \cdot 2x$ (d) $\dfrac{dy}{dx} = 3(x^3 + 1)^2 \cdot 3x^2$

9. Exhibit a function y for which:

(a) $\dfrac{dy}{dx} = 8x(x^2 + 1)^3$ (b) $\dfrac{dy}{dx} = 12x^2(x^3 - 1)^3$

Answers to problems (§3F)

1. (a) $10x(x^2 + 1)^4$ (b) $286x(x^2 + 1)^{142}$ (c) $9x^2(x^3 - 1)^2$
   (d) $2(3x^5 - 4x^3 + 3x - 6)(15x^4 - 12x^2 + 3)$ (e) $90x(3x^2 - 1)^{14}$

(f) $12\left(x + \dfrac{1}{x}\right)^{11}\left(1 - \dfrac{1}{x^2}\right)$

2. (a) $2t(7t^2 + 11)(t^2 + 1)^3(t^2 + 2)^2$ (b) $-6(4 - 9t)(1 - 2t)^2(1 - 3t)^5$

3. (a) $\dfrac{(1 + u^4)(5 - 24u - 14u^2)}{(1 + 2u^2)^7}$ (b) $\dfrac{2u(u^2 + 1)^3(u^2 + 5)}{(u^2 + 2)^4}$

4. With the abbreviations $a = v^2 + v - 1$, $b = v^3 + 3v^2 - 3$, $c = 4v^5 - 3v + 6$:

(a) $\dfrac{ab^2}{c^5}\,[9v(v + 2)ac + 2(2v + 1)bc - 4(20v^4 - 3)ab]$

(b) $\dfrac{b^2}{a^3c^5}\,[9v(v + 2)ac - 2(2v + 1)bc - 4(20v^4 - 3)ab]$

5. (a) $5(x^2 + x + 1)^4(2x + 1)$ (b) 1215 (c) $5(4x^2 + 2x + 1)^4(4x + 1)$

(d) $5(x^2 - x + 1)^4(-2x + 1)$ (e) $5\left(\dfrac{1}{x^2} + \dfrac{1}{x} + 1\right)^4\left(\dfrac{2}{x} + 1\right)$

6. (a) $2y\dfrac{dy}{dx}$

§3G. Rational powers; implicit differentiation

The power rule, which we have established for integral powers (§3D3 and §3E4), holds for fractional powers as well. The proof is a good example of the use of the chain rule.

3G1. Derivative of x^r (Rational powers). *Let r be any rational number—positive, negative, or zero.*

If $f(x) = x^r$ $(x \geq 0)$, then $f'(x) = rx^{r-1}$ (where, it is understood, $r > 1$ if $x = 0$).

Briefly,
$$\frac{d}{dx} x^r = rx^{r-1}.$$

Proof. We treat the case $x > 0$. (If $x = 0$ the result is obvious.)

Since r is a rational number, it is, by definition, a quotient of two integers. Say

$$(1) \qquad\qquad r = \frac{m}{n}.$$

Of course, we can choose m and n so that $n > 0$. We want the derivative of $x^{m/n}$.

We begin with the simpler function

$$(2) \qquad\qquad u = x^{1/n},$$

where $n > 0$. If we write

$$(3) \qquad\qquad \Delta u = (x + \Delta x)^{1/n} - x^{1/n}$$

and try to evaluate $\Delta u / \Delta x$ in some manageable form, we get stuck, because we don't know how to expand $(x + \Delta x)^{1/n}$.

So we turn the expression upside down. We have $x = u^n$ and

$$(4) \qquad\qquad x + \Delta x = (u + \Delta u)^n.$$

Hence

$$(5) \qquad\qquad \Delta x = (u + \Delta u)^n - u^n$$

and

$$(6) \qquad \frac{\Delta u}{\Delta x} = \frac{1}{(\Delta x / \Delta u)} = \frac{1}{[(u + \Delta u)^n - u^n]/\Delta u}.$$

Now, u is a continuous function of x $\left(\S 2\text{E}3(c)\right)$. So as $\Delta x \to 0$, $\Delta u \to 0$. Then the denominator on the right approaches $(d/du)u^n$. Hence

$$(7) \qquad \frac{du}{dx} = \frac{1}{(d/du)u^n} = \frac{1}{nu^{n-1}}.$$

Now we can handle $x^{m/n}$. Since

(8)
$$x^{m/n} = (x^{1/n})^m = u^m,$$

we can apply the chain rule. We have:

$$\frac{d}{dx} x^{m/n} = \frac{d}{dx} u^m = mu^{m-1} \frac{du}{dx}$$

(9)
$$= \frac{mu^{m-1}}{nu^{n-1}} = ru^{m-n}$$

$$= rx^{(1/n)(m-n)} = rx^{r-1}.$$

Example 1. An important special case is:

(10)
$$\frac{d}{dx} \sqrt{x} = \frac{1}{2\sqrt{x}} \qquad (x > 0).$$

Example 2. Find the equation of the tangent line to the circle $x^2 + y^2 = 1$ at a point (a, b) $(b > 0)$. See Figure 1.

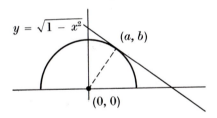

$y = \sqrt{1 - x^2}$ (a, b)

$(0, 0)$

FIGURE 1

Solution. The equation of the upper semicircle is $y = \sqrt{1 - x^2}$. By the chain rule,

$$\frac{dy}{dx} = \frac{1}{2\sqrt{1 - x^2}} (-2x) = -\frac{x}{y} \qquad (y > 0).$$

Hence the slope of the graph at (a, b) is $-a/b$. The equation of the tangent line is, therefore,

$$y - b = -\frac{a}{b} (x - a),$$

which simplifies to $ax + by = 1$. Note that the slope of the radius to (a, b) is b/a (if $a \neq 0$). Hence the tangent is perpendicular to the radius. This verifies that the tangent line to the circle as defined by calculus (§3B3) is the same as the one defined in geometry.

3G2. Implicit differentiation. In Example 2 we considered the equation

(11) $$x^2 + y^2 = 1 \qquad (y > 0)$$

and solved for y in terms of x in order to differentiate. The following variant is simpler. Just note mentally that we *can* solve for y as a differentiable function of x. Knowing this, differentiate in (11), using the chain rule. Observe that the right-hand side of this equation is the constant function 1. Thus:

(12) $$2x + 2y \frac{dy}{dx} = 0 \qquad \text{and so} \qquad \frac{dy}{dx} = -\frac{x}{y} \qquad (y > 0).$$

This procedure is called *implicit differentiation*. As the example suggests, the method is particularly useful in problems involving rational powers.

Implicit differentiation can be combined with the chain rule with respect to additional variables. Suppose that in (11), x and y are known to be differentiable functions of a third variable, say t. Then we may differentiate with respect to t directly in (11). Thus:

(13) $$2x \frac{dx}{dt} + 2y \frac{dy}{dt} = 0 \qquad \text{and so} \qquad \frac{dy}{dt} = -\frac{x}{y}\frac{dx}{dt} \qquad (y > 0).$$

Problems (§3G)

1. Find $g'(x)$.
   (a) $g(x) = x^{3/4}$
   (b) $g(x) = x^{4/3}$
   (c) $g(x) = x^{15/8}$
   (d) $g(x) = x^{-3/4}$
   (e) $g(x) = x^{-4/3}$
   (f) $g(x) = x^{-15/8}$

2. Find dv/dx.
   (a) $v = \sqrt{1 + x^2}$
   (b) $v = \sqrt{1 + x^{3/2}}$
   (c) $v = (1 + x^{3/2})^{2/3}$
   (d) $v = (1 + x^2)^{1/3}(1 + x^2)^{2/3}$

3. Find $h'(t)$.

   (a) $h(t) = \sqrt{\dfrac{1 + t}{1 - t}}$

   (b) $h(t) = \dfrac{t^{1/2} + t^{3/2}}{t^{1/2} - t^{3/2}}$

   (c) $h(t) = \dfrac{(t^2 - t + 3)^{5/2}(t^2 + t - 3)^{3/2}}{(3t^2 - 2t + 1)^{1/2}} \quad \left[Hint. \ h(t) = \dfrac{a^{5/2}b^{3/2}}{c^{1/2}}. \right]$

4. Exhibit a function y for which:

(a) $\dfrac{dy}{dx} = \dfrac{3}{2}\sqrt{x}$

(d) $\dfrac{dy}{dx} = 2x\sqrt{x^2 + 1}$

(b) $\dfrac{dy}{dx} = \dfrac{2}{3}\sqrt{x}$

(e) $\dfrac{dy}{dx} = x^2\sqrt{x^3 + 5}$

(c) $\dfrac{dy}{dx} = x\sqrt{x}$

5. Find dy/dx:

(a) $x^3 + y^3 = 1$

(d) $y^2 = \dfrac{x^2 + 1}{x^2 - 1}$

(b) $\sqrt{x} + \sqrt{y} = 1$

(e) $x^3y + x^2y^2 = 1 + x$

(c) $x^{2/3} + y^{2/3} = 1$

6. Is there a (nonconstant) function y for which
   (a) $2yy' = 1$? (b) $3y = 2xy'$?

7. Let a denote the x-intercept of the line tangent to the quarter-circle $x^2 + y^2 = 1$ $(x > 0, y > 0)$ at a point (x_1, y_1). Show that $ax_1 = 1$.

Answers to problems (§3G)

1. (a) $\frac{3}{4}x^{-1/4}$

2. (a) $\dfrac{x}{\sqrt{1 + x^2}}$ (b) $\dfrac{3}{4}\sqrt{\dfrac{x}{1 + x^{3/2}}}$ (c) $\dfrac{\sqrt{x}}{(1 + x^{3/2})^{1/3}}$ (d) $2x$

3. (a) $\dfrac{1}{(1 - t)\sqrt{1 - t^2}}$ (b) $\dfrac{2}{(1 - t)^2}$

(c) with the abbreviations $a = t^2 - t + 3$, $b = t^2 + t - 3$, $c = 3t^2 - 2t + 1$:

$$\frac{1}{2}\sqrt{\frac{a^3b}{c^3}}\,[3(2t + 1)ac + 5(2t - 1)bc - 2(3t - 1)ab]$$

5. (a) $-x^2/y^2$ (b) $-\sqrt{y/x}$ (c) $-\sqrt[3]{y/x}$ (d) $-\dfrac{2x}{x^4 - 1}$

$-4x\Big/2y(x^2 - 1)^2$

(e) $\dfrac{1 - 3x^2y - 2xy^2}{x^3 + 2x^2y}$

§3H. Related rates

3H1. Related rates. Interesting applications to problems of motion arise when two related variables are functions of time, t. In a typical problem, we are given the velocity of one of the variables at a given instant and asked to find the velocity of the other.

Example 1. Let $y = x^2$, where x is a differentiable function of t.
(a) Suppose that when $x = 4$, $dx/dt = 3$. What is dy/dt then?

Solution. $\dfrac{dy}{dt} = \dfrac{dy}{dx}\dfrac{dx}{dt} = 2x\dfrac{dx}{dt}$. Hence when $x = 4$, $\dfrac{dy}{dt} = 8\cdot 3 = 24$.

(b) Suppose that when $x = 3$, $dy/dt = 2$. What is dx/dt then?

Solution. $\dfrac{dy}{dt} = \dfrac{dy}{dx}\dfrac{dx}{dt} = 2x\dfrac{dx}{dt}$. Hence when $x = 3$, $2 = 6\dfrac{dx}{dt}$, $\dfrac{dx}{dt} = \dfrac{1}{3}$.

(c) Suppose that when $y = 4$ and $x < 0$, $dx/dt = 5$. What is dy/dt then?

Solution. $\dfrac{dy}{dt} = \dfrac{dy}{dx}\dfrac{dx}{dt} = 2x\dfrac{dx}{dt}$. When $y = 4$, $x = -2$. Hence when $y = 4$,

$dy/dt = -4\cdot 5 = -20$.

These very simple examples illustrate that the relations among the variables and their derivatives may be presented in a variety of forms. Notice that we never do learn what function x is of t. We don't even find out the value of t at which the derivatives are evaluated.

In applications we must seek the necessary relations from the data of the problem. Some of these facts are given *directly* in the statement of the problem. Others must be *derived*. The following general guide may be helpful.

Step 1. Identify the variables and assign symbols to them.

Step 2. State the problem in mathematical terms.

Step 3. Express relations among variables as equations. Bring in new variables and relations as needed, from the data of the problem.

Step 4. Obtain additional relations by differentiating. Expect to use the chain rule.

Step 5. Substitute numerical values and solve.

Example 2. A ladder 10 feet long is leaning against a wall. The foot of the ladder is moving away from the wall at the rate of 2 feet per second. How fast is the top of the ladder falling when the foot is 6 feet from the wall?

Solution. See Figure 1.

Step 1. Let x denote the distance from the corner to the foot of the ladder, and y the distance to the top.

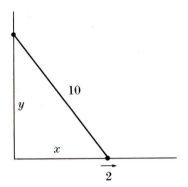

FIGURE 1

Step 2. We are given $dx/dt = 2$. We are to find dy/dt when $x = 6$. Note that dy/dt will be a negative number. The rate at which the ladder is *falling* is the corresponding positive number.

Step 3. The variables x and y are related by:

(1) $$x^2 + y^2 = 100.$$

Step 4. Differentiating implicitly (§3G2), we have

(2) $$2x\frac{dx}{dt} + 2y\frac{dy}{dt} = 0, \qquad \frac{dy}{dt} = -\frac{x}{y}\frac{dx}{dt} \qquad (0 < y \le 10).$$

Step 5. Substituting $x = 6$ (and hence $y = 8$) and $dx/dt = 2$, we get $dy/dt = -\frac{6}{8} \times 2 = -\frac{3}{2}$. When $x = 6$, the ladder is falling at $\frac{3}{2}$ feet per second.

Remark. According to (2), when the ladder is almost horizontal, the top is descending *very* fast. For instance, when $y = \frac{1}{100}$, $x^2 = 99.9999$. Then x is indistinguishable from 10, and (2) gives

$$\frac{dy}{dt} \approx -\frac{10}{\frac{1}{100}} \times 2 = -2000 \text{ ft./sec.}$$

(The symbol $\approx$ means "is approximately equal to.") Does this agree with your intuition?

Example 3. The volume of an expanding sphere is increasing at a rate of 10 cubic feet per second. How fast is the surface area increasing when the volume is 36π cubic feet?

Solution.

Step 1. Let V denote the volume and A the surface area.

Step 2. We are given $dV/dt = 10$. We are to find dA/dt when $V = 36\pi$.

Step 3. To connect dA/dt with dV/dt we bring in the new variable r, the radius. The formulas for A and V are

(3) $$A = 4\pi r^2, \qquad V = \tfrac{4}{3}\pi r^3.$$

(See §3H2 below.)

Step 4. We note that we *could* solve for r as a differentiable function of V. Hence r is a differentiable function of t, and we may differentiate implicitly in (3). We get

(4) $$\frac{dA}{dt} = 8\pi r \frac{dr}{dt}$$

and

(5) $$\frac{dV}{dt} = 4\pi r^2 \frac{dr}{dt}.$$

Step 5. Since $dV/dt = 10$, we get, from (5),

(6) $$10 = 4\pi r^2 \frac{dr}{dt}, \qquad \text{and therefore} \qquad \frac{dr}{dt} = \frac{5}{2\pi r^2}.$$

Hence in (4) we obtain

(7) $$\frac{dA}{dt} = 8\pi r \frac{5}{2\pi r^2} = \frac{20}{r}.$$

Finally, when $V = 36\pi$ we find $r = 3$ $\big(\text{from (3)}\big)$, and we have $dA/dt = \tfrac{20}{3}$. The area is increasing at $\tfrac{20}{3}$ square feet per second.

3H2. Formulas for area and volume. For use in problems now and later we borrow the following formulas.

(i) For a circle of radius r, area $= \pi r^2$ and circumference $= 2\pi r$.

(ii) For a sphere of radius r, volume $= \tfrac{4}{3}\pi r^3$ and surface area $= 4\pi r^2$.

(iii) For a (right circular) cylinder of radius r and height h, volume $= \pi r^2 h$ and surface area $= 2\pi rh$.

(iv) For a (right circular) cone of radius r and height h, volume $= \tfrac{1}{3}\pi r^2 h$ and surface area $= \pi r \sqrt{h^2 + r^2}$.

All of these geometric figures are bounded by *curved* lines or surfaces. Before their lengths or areas or volumes can be computed, the concepts must be made precise. These are matters to be decided by the methods of integral calculus. We will settle them in Chapter 6 and Chapter 8.

By methods of calculus it can be shown that π is an irrational number. Its decimal expansion begins

$$\pi = 3.14159\cdots.$$

A rational number that approximates π to 5 places is

$$3.14159.$$

A simple fraction that approximates π to 2 places is

$$\frac{22}{7} = 3.142857\cdots.$$

Problems (§3H)

1. Let $y = x^3$, where x is a differentiable function of t.
   (a) Suppose that when $x = 4$, $dx/dt = 3$. What is dy/dt then?
   (b) Suppose that when $x = 3$, $dy/dt = 9$. What is dx/dt then?
   (c) Suppose that when $y = 8$, $dx/dt = 2$. What is dy/dt then?

2. Let $y = (x^2 + 1)^2$ and $u = (x^2 - 1)^2$, where x is a differentiable function of t. If $dy/dt = \frac{1}{4}$ when $x = 2$, what is du/dt then?

3. Let $u = (x - 1)^3$ and $v = (x + 1)^3$, where x is a differentiable function of t. If $du/dt = 6$ when $dx/dt = \frac{1}{2}$, what is dv/dt then?

4. The length of a rectangle is increasing at 3 inches per second and the width is increasing at 2 inches per second. How fast is the area increasing when the length is 7 and the width 3?

5. The radius of a (right circular) cylinder is increasing at 3 inches per second and the height is increasing at 4 inches per second. When the radius is 6 and the height 10,
   (a) how fast is the volume increasing?
   (b) how fast is the surface area increasing?

6. The radius of a (right circular) cone is increasing at 4 inches per second while the height is increasing at 5 inches per second. When the radius is 7 and the height 15,
   (a) how fast is the volume increasing?
   (b) how fast is the surface area increasing?

7. (a) A gas is expanding according to the law $pV = $ constant, where p is the pressure and V is the volume. If the pressure is decreasing at 5 lbs. per square foot per hour, at what rate is the volume increasing when there are 2400 cubic feet of gas at a pressure of 500 lbs. per square foot?
   (b) A quantity of air is being compressed so that its volume is decreasing at 50 cubic feet per minute. The pressure p and volume V are related by the law

$$pV^{7/5} = \text{constant}.$$

   How fast is the pressure increasing when the volume is 1000 cubic feet and the pressure is 6000 lbs. per square foot?

8. A television camera is located on the ground 10 miles from the spot where a rocket is launched vertically. When the rocket is 10 miles high it is traveling 2 miles per second. At that instant, how fast is its distance from the camera increasing?

9. A stone dropped in a still pond produces a circular ripple.
   (a) If the radius increases at 2 feet per second, how fast is the enclosed area increasing when the radius is 12 feet?
   (b) If the area increases at 24 square feet per second, how fast is the radius increasing when the radius is 12 feet?

10. A rope connects a dock to the deck of a boat, 3 feet beneath the level of the dock.
    (a) If the rope is being reeled in at 6 feet per second, how fast is the boat approaching the dock when it is 4 feet away?
    (b) If the boat is drifting away at 2 feet per second, how fast is the rope leaving the dock when the boat is 12 feet away?

11. A man 6 feet tall is walking at 4 feet per second away from a pole 15 feet high. There is a light at the top of the pole.
    (a) How fast is the far end of the man's shadow moving?
    (b) How fast is the length of the shadow increasing?
    (c) Show that at each instant, the man's head is receding from the light faster than his feet are.
    (d) What are the rates in (c) when the man is 12 feet from the pole?

12. The length of a rectangle is increasing at 3 inches per second.
    (a) If the width decreases in such a way that the area remains constant at 16 square inches, how fast is the perimeter increasing when the length is 5 inches?
    (b) If the width decreases in such a way that the perimeter remains constant at 16 inches, how fast is the area increasing when the rectangle is a square?

13. One side of a right triangle is increasing at 2 units per second, while the other side is decreasing in such a way that the hypotenuse remains constant at 10. When the length of the first side is 6, is the area increasing or decreasing, and at what rate?

14. The radius of a cylinder is increasing at 3 inches per second, while the height decreases in such a way that the volume remains constant at 4π. When the radius is 2 and the height 1, at what rate is the surface area decreasing?

15. The height of a cylinder is increasing at 2 inches per second. When the radius is 6, the volume is increasing at 96π cubic inches per second. How fast is the surface area increasing at that instant?

16. The radius of a cone is increasing at 3 inches per second, while the height is decreasing in such a way that the volume remains constant at 12π. At what rate is the height decreasing when the radius is 3?

17. The height of a cone is increasing at 2 inches per second, while the radius is decreasing in such a way that the surface area remains constant at 15π square inches. At what rate is the volume changing when the height is 4 inches?

18. As a spherical mothball evaporates, its volume decreases at a rate proportional to its surface area. Show that the rate of decrease of the radius is constant.

19. A spherical balloon is expanding in such a way that the radius is increasing at a rate proportional to the surface area. Show that the surface area is increasing at a rate proportional to the volume.

20. A camera is closing in on a cube.
    (a) If the (apparent) volume increases at a constant rate of 10 cubic feet per second, how fast is the total surface area increasing when the volume is 8 cubic feet?
    (b) If the area of a face is increasing at 2 square feet per second, how fast is the length of an edge increasing when the area of a face is 12 square feet?
    (c) If the length of an edge is increasing at 2 feet per second, how fast is the volume increasing when the length of an edge is 6 feet?

21. A conical reservoir is 12 feet deep and the radius at the brim is 3 feet.
    (a) If water is being poured in at 2 cubic feet per minute, how fast is the surface level rising when the depth of the water is 4 feet? [*Hint.* How is the radius at any level related to the depth?]
    (b) If water is being poured in at 2 cubic feet per minute, how fast is the radius at the water level increasing when the volume of water in the reservoir is $9\pi/2$ cubic feet?
    (c) If water is being poured in in such a way that the surface level rises at a constant rate of 1 foot per minute, at what rate is water being added when the depth is 4 feet?

22. Ship A is 30 nautical miles west of point P and is headed toward P at 11 knots. (1 knot = 1 nautical mile per hour.) Ship B is 20 nautical miles north of P and headed toward P at 16 knots. At this instant, how fast are the ships approaching one another, i.e., how fast is the distance between them decreasing?

23. A point moves along the parabola $y = x^2$ with its x-coordinate increasing at 2 units per second.
    (a) What is dy/dt when $x = 3$?
    (b) At what point is $dy/dt = 8$?
    (c) Show that the slope of the curve at the point is increasing at a constant rate.
    (d) Another point moves along the line $y = x$ in such a way as to keep the y-coordinates of the two points the same. How fast are the points separating when they cross at $(1, 1)$?

24. A point is moving along the curve $y^2 = x^3 - x^2$. As it passes through the point $(2, 2)$, its x-coordinate is increasing at a rate of 3 units per second.
    (a) What is the rate of increase of its y-coordinate then?
    (b) How fast is the slope of the graph at the point changing at that instant?

25. A point Q starts at $Q_0 = (1, 0)$ and moves along the upper semicircle $x^2 + y^2 = 1$ with $dx/dt = -2$. When $x = 1/\sqrt{2}$:
    (a) At what rate is the distance from Q to Q_0 increasing?
    (b) At what rate is the slope of the curve at Q increasing?
    (c) Consider the tangent line to the curve at Q. At what rate is its x-intercept increasing?

Answers to problems (§3H)

1. (a) 144   (b) $\frac{1}{3}$ (c) 24
2. $\frac{3}{20}$
3. 24   or   0
4. 23 sq. in./sec.
5. (a) 504π (b) 108π
6. (a) $\dfrac{1085}{3}\pi$ (b) $\dfrac{1817}{\sqrt{274}}\pi$
7. (a) 24 cu. ft./hr.   (b) 420 lb./sq. ft./min.
8. $\sqrt{2}$ mi./sec.
9. (a) 48π (b) $1/\pi$
10. (a) $\frac{15}{2}$ (b) $8/\sqrt{17}$
11. (a) $\frac{20}{3}$ (b) $\frac{8}{3}$ (d) head $\frac{16}{5}$, feet $16/\sqrt{41}$
12. (a) $\frac{54}{25}$ in./sec.   (b) 0
13. increasing, $\frac{7}{2}$
14. 6π
15. 28π
16. 8
17. $\frac{6}{17}\pi$
20. (a) 20 sq. ft./sec.   (b) $1/\sqrt{12}$ ft./sec.   (c) 216 cu. ft./sec.
21. (a) $2/\pi$ (b) $2/9\pi$ (c) π
22. $5\sqrt{13}$ knots
23. (a) 12   (b) $x = 2$ (d) 2 units/sec.
24. (a) 6   (b) $\frac{3}{2}$
25. (a) $\dfrac{2}{\sqrt{2-\sqrt{2}}}$ (b) $4\sqrt{2}$ (c) 4

§3I. Higher derivatives

3I1. Higher derivatives. The derivative of f', denoted f'', is called the second derivative of f. Other notations are

$$y'' = \frac{d^2y}{dx^2} = f''(x) = \frac{d^2}{dx^2}f(x) \qquad \big(\text{where } y = f(x)\big).$$

The derivative of f'' is the third derivative, denoted f'''; etc. Notations for the fourth and higher derivatives are $f^{(4)}, f^{(5)}, \cdots$ (or $f^{\text{iv}}, f^{\text{v}}, \cdots$).

Example 1. If $f(x) = x^4$, then

$$f'(x) = 4x^3, \qquad f''(x) = 12x^2, \qquad f'''(x) = 24x,$$
$$f^{(4)}(x) = 24, \qquad f^{(5)}(x) = 0, \qquad f^{(6)}(x) = 0.$$

3I2. Acceleration. Consider the linear motion $s = f(t)$, with velocity

$$v = \frac{ds}{dt} = f'(t).$$

The *average acceleration* over the time interval between t and $t + \Delta t$ is equal to the net change in velocity, Δv, divided by the time Δt during which this change occurred. Therefore

$$\frac{\Delta v}{\Delta t} = \frac{f'(t + \Delta t) - f'(t)}{\Delta t}.$$

If f' has a derivative at t, the average acceleration approaches a limit as $\Delta t \to 0$; by definition, this is the *(instantaneous) acceleration at time t.* So the acceleration a at time t is given by

(1)
$$a = \frac{dv}{dt} = \frac{d^2s}{dt^2} = f''(t).$$

Example 2. A point P moves on a line in such a way that its position at the end of t seconds is

$$s = t^3 - 12t \qquad\qquad (t \geq 0),$$

where s is measured in feet. Analyze the motion.

Solution. The velocity of P at time t is

$$v = \frac{ds}{dt} = 3t^2 - 12 = 3(t - 2)(t + 2)$$

feet per second. When $t = 0$, $s = 0$ and so P is at the origin. The initial velocity (the velocity at $t = 0$) is -12, and $v = ds/dt$ remains negative for all $t < 2$. Hence at each instant $t \in [0, 2)$, s is decreasing and therefore P is moving in the negative direction. At each instant $t > 2$, ds/dt is positive, so that s is increasing. The point P reverses direction at $t = 2$; here $s = -16$. It passes through the origin again when $t^3 - 12t = 0$, i.e., at $t = \sqrt{12}$ seconds; at that instant its velocity is 24 feet per second.

The acceleration is

$$a = \frac{dv}{dt} = 6t.$$

At each instant $t > 0$, a is positive and so the velocity v is increasing. Since v is negative for $t < 2$ this means that $|v|$, the *speed* of P, is decreasing at each instant $t \in (0, 2)$.

Notice that the average velocity of P over the first $\sqrt{12}$ seconds is 0, although P actually travels 32 feet in that time (16 in the negative direction and 16 in the positive direction).

Sketch the graph of s vs. t.

Problems (§31)

1. Find the first four derivatives.
   (a) $f(x) = 15x - 4$
   (b) $g(x) = 4x^2 + 41x + 6$
   (c) $h(x) = 4x^3 - 2x^2 + 5x - 8$
   (d) $f(x) = x^4 + 11x^3 - 12x^2 + 2x - 3$

   (e) $g(x) = \dfrac{1}{x}$

   (f) $h(x) = \dfrac{x + 2}{x - 1}$

   (g) $F(x) = x^{7/2}$
   (h) $G(x) = x^{7/3}$
   (i) $H(x) = \sqrt{1 + x^2}$

2. (a) What is the 18th derivative of
$$x^{17} - 256x^{16} + \tfrac{54}{13}x^{11} - \tfrac{23}{19}x^8 + \sqrt{\pi}\, x^3 - 4\,?$$

   (b) What is the 17th derivative?

3. Exhibit a function y for which
   (a) $y'' = 3$ (b) $y'' = 3x$ (c) $y'' = 3x^2$

4. Find
$$\lim_{\Delta x \to 0} \frac{f'(x + \Delta x) - f'(x)}{\Delta x}$$

for:

   (a) $f(x) = 3x^2$ (c) $f(x) = \dfrac{1}{1 + x}$

   (b) $f(x) = 4x^3$ (d) $f(x) = \dfrac{1}{1 + x^2}$

5. Let $f(x) = x^2 + 1$. Find

   (a) $f'(x)$ (b) $f'(x^2 + 1)$

(c) $f''(x)$

(f) $\dfrac{d}{dx} f'(x^2 + 1)$

(d) $f''(x^2 + 1)$

(g) $\dfrac{d^2}{dx^2} f(x^2 + 1)$

(e) $\dfrac{d}{dx} f(x^2 + 1)$

6. Let $g(x) = x - \dfrac{1}{x}$. Find

(a) $g'(x)$

(e) $\dfrac{d}{dx} g\left(x - \dfrac{1}{x}\right)$

(b) $g'\left(x - \dfrac{1}{x}\right)$

(f) $\dfrac{d}{dx} g'\left(x - \dfrac{1}{x}\right)$

(c) $g''(x)$

(g) $\dfrac{d^2}{dx^2} g\left(x - \dfrac{1}{x}\right)$

(d) $g''\left(x - \dfrac{1}{x}\right)$

7. For $h = fg$ we have the formula
$$h' = fg' + f'g.$$
What is the corresponding formula
    (a) for h''?   (b) for h'''?

8. In Problem 8 of §3B, in each case, find the average acceleration over the indicated time interval.

9. In Problem 8 of §3B, in each case, find the instantaneous acceleration at the given time t_0.

10. A particle moves on a line so that its position at time t is $s = t^3 - 6t^2 + 9t$.
    (a) When is s increasing and when decreasing?
    (b) When is the velocity v increasing and when decreasing?
    (c) Find the net distance traveled by P in the first 6 seconds.
    (d) Find the *total* distance traveled by P in the first 6 seconds.

11. A point moves on a line so that $s = t^3 - 3t + 6$. Find its acceleration at those times when the velocity is 0.

12. Let u be a twice-differentiable function of x. Express y'' $(= d^2y/dx^2)$ in terms of u and its derivatives, for:
    (a) $y = (u^2 + 1)^2$ (b) $y = \sqrt{u^2 + 1}$

13. For $h(x) = f\big(g(x)\big)$, we have the formula
$$h'(x) = f'\big(g(x)\big)g'(x).$$
What is the corresponding formula
    (a) for h''?   (b) for h'''?

14. Let $f(x) = x^2$ $(x < 0)$, $f(x) = x^3$ $(x \geq 0)$.
   (a) Find $f'(0)$.   (b) Does $f''(0)$ exist?

Answers to problems (§3I)

1. (a) $15, 0, 0, 0$ (b) $8x + 41, 8, 0, 0$ (c) $12x^2 - 4x + 5, 24x - 4, 24, 0$
   (d) $4x^3 + 33x^2 - 24x + 2, 12x^2 + 66x - 24, 24x + 66, 24$
   (e) $-1/x^2, 2/x^3, -6/x^4, 24/x^5$
   (f) $-3/(x - 1)^2, 6/(x - 1)^3, -18/(x - 1)^4, 72/(x - 1)^5$
   (g) $\frac{7}{2}x^{5/2}, \frac{35}{4}x^{3/2}, \frac{105}{8}x^{1/2}, \frac{105}{16}x^{-1/2}$
   (h) $\frac{7}{3}x^{4/3}, \frac{28}{9}x^{1/3}, \frac{28}{27}x^{-2/3}, -\frac{56}{81}x^{-5/3}$

   (i) $\dfrac{x}{\sqrt{1 + x^2}}, \dfrac{1}{(1 + x^2)^{3/2}}, -\dfrac{3x}{(1 + x^2)^{5/2}}, \dfrac{3(4x^2 - 1)}{(1 + x^2)^{7/2}}$

4. (a) 6 (b) $24x$ (c) $\dfrac{2}{(x + 1)^3}$ (d) $\dfrac{6x^2 - 2}{(x^2 + 1)^3}$

5. (a) $2x$ (b) $2(x^2 + 1)$ (c) 2 (d) 2 (e) $4x(x^2 + 1)$ (f) $4x$
   (g) $12x^2 + 4$

6. (a) $1 + \dfrac{1}{x^2}$ (b) $1 + \dfrac{x^2}{(x^2 - 1)^2}$ (c) $-\dfrac{2}{x^3}$

   (d) $-\dfrac{2x^3}{(x^2 - 1)^3}$ (e) $1 + \dfrac{1}{x^2} + \dfrac{x^2 + 1}{(x^2 - 1)^2}$

   (f) $-2x\dfrac{x^2 + 1}{(x^2 - 1)^3}$ (g) $-\dfrac{2}{x^3} - 2x\dfrac{x^2 + 3}{(x^2 - 1)^3}$

7. (b) $fg''' + 3f'g'' + 3f''g' + f'''g$
8. (a) $\frac{3}{4}$ (b) 4 (c) $-7.72/125$ (d) $\frac{3}{2}$ (e) 0 (f) $-\frac{1}{2}$
9. (a) 2 (b) 4 (c) $-\frac{44}{125}$ (d) 2 (e) 0 (f) -6
10. (a) increasing for $t < 1$ and $t > 3$, decreasing for $1 < t < 3$
   (b) increasing for $t > 2$, decreasing for $t < 2$
   (c) 54 (d) 62
11. $a = 6$ when $t = 1$, $a = -6$ when $t = -1$
12. (a) $4[(u^3 + u)u'' + (3u^2 + 1)u'^2]$

   (b) $\dfrac{(u^3 + u)u'' + u'^2}{(u^2 + 1)^{3/2}}$

13. (b) $f'(g(x))g'''(x) + 3f''(g(x))g'(x)g''(x) + f'''(g(x))g'(x)^3$
14. (a) 0 (b) no

CHAPTER 4

Applications of the Derivative

In §§4A–4B we learn how to determine the maximum and minimum values of a function (if there are any) and locate the points where they are assumed. The derivative is used as a simple but penetrating tool to cut to the heart of a problem that might otherwise seem hopelessly complex. §4C is devoted to applications of the theory.

Similarly, by checking a few facts supplied by the derivative and the second derivative, we are able to analyze the qualitative behavior of a function. This problem is taken up in §4D.

The subject of §4E is the Mean-Value Theorem. This theorem is fundamental for the rest of the book and for results in more advanced mathematics as well. One of its applications is to antiderivatives, discussed in §4F.

Finally, §4G deals with the differentiability of the inverse function.

A word about hypotheses. Many of the theorems of the chapter start out like this: "Let f be continuous on an interval I and differentiable on its interior." These hypotheses deserve study. They may seem picky—but they aren't. Just the opposite. They are designed so that you *don't* have to worry when applying the theorem. Explicitly, you don't have to worry about a vertical tangent at an endpoint.

§4A. Maximum and minimum on a closed interval

In this and the next section (§4B) we undertake a systematic study of maxima and minima—whether a given function does attain maximum or minimum values, where they occur, and what the values are. Then, in §4C, we take up applications.

4A1. Maximum and minimum. The theory rests upon two basic facts. One has to do with continuity. The other concerns the derivative.

The first is that *a function continuous on a closed interval attains both a maximum and a minimum on the interval.*

This is the Maximum-Value Theorem (§2E7), which we have restated here for emphasis and convenience of reference.

4A2. Interior maximum or minimum; critical points. The second basic result is as follows. Let f be continuous on an interval I and differentiable on its interior.

If f attains a maximum or minimum at a point x_0 in the interior of I, then $f'(x_0) = 0$.

It is convenient to define a *critical* point of f to mean a point at which $f' = 0$. Then the theorem reads:

An interior point at which f attains either a maximum or a minimum is necessarily a critical point of f.

Proof of theorem. If $f'(x_0) > 0$ then f is increasing at the point x_0 (§3B4) and cannot have either a maximum or a minimum there (Figure 1). The relation $f'(x_0) < 0$ is ruled out the same way. Therefore, $f'(x_0) = 0$.

Remark. The hypothesis that x_0 be an *interior* point is essential. (For example, the conclusion is false for $f(x) = x$ on $[0, 1]$ and $x_0 = 1$.) Do you see how it is used in the proof?

Example 1. Consider the parabola

$$y = 3(x - 2)^2 + 5 \qquad\qquad (x \in \mathscr{R}).$$

We know that the lowest point on the graph is $(2, 5)$, the vertex. Since 2 is an interior point of $\mathscr{R}$ (what isn't?), the derivative must be zero there. And, indeed, we have $y' = 6(x - 2)$, which is 0 at $x = 2$.

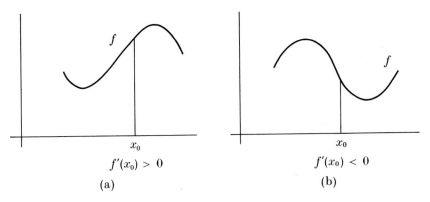

$$f'(x_0) > 0$$

(a)

$$f'(x_0) < 0$$

(b)

FIGURE 1

4A3. Maximum and minimum on a closed interval. Problems in maxima and minima vary in difficulty according to the character of the interval I on which the function is defined.

The simplest ones are those in which I is a *closed* interval. The guide to solving them is §4A1 and §4A2. Let f be continuous on a closed interval $[a, b]$ and differentiable on its interior. Then f attains both a maximum and a minimum (§4A1). Each of them occurs either at an endpoint or a critical point (§4A2). We compute the values of f at all these test points. The largest among them is the maximum of f on $[a, b]$; the smallest is the minimum.

Example 2. Find the maximum and minimum values of

$$f(x) = x^4 + 4x^3 - 20x^2 + 7$$

on the interval $[-1, 3]$.

Solution. The derivative is

$$f'(x) = 4x(x + 5)(x - 2).$$

Its zeros are $x = -5$, 0, and 2. We ignore -5, which falls outside the given interval. Hence our test points are -1, 0, 2, and 3. The values of f at these points are:

$$f(-1) = -16, \quad f(0) = 7, \quad f(2) = -25, \quad f(3) = 16.$$

Therefore the maximum value of f on $[-1, 3]$ is 16 (at $x = 3$) and the minimum is -25 (at $x = 2$). Sketch the graph.

Example 3. Find the maximum and minimum of $f(x) = x^3$ on the interval $[-1, 1]$.

Solution. The derivative is 0 at $x = 0$. But f is an increasing function. Its maximum is at $x = 1$ and its minimum at $x = -1$. Sketch the graph.

4A4. Monotonic functions. For later use we record the following corollary of the theorem on critical points (§4A2).

Let f be continuous on an interval I and differentiable on its interior.
If f has no critical points in the interior of I, then f is monotonic on I.
(Hence f has a continuous inverse on I.)

Proof. If f were not monotonic there would be three points $u < c < v$ in I with $f(c)$ either the largest or the smallest of the three values (Figure 2). Then either the maximum or the minimum of f on the closed interval $[u, v]$ would occur at an interior point of $[u, v]$ and that would be a critical point of f in the interior of I. The parenthetical remark follows from §2E9.

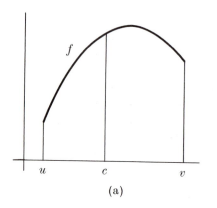
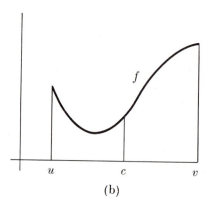

(a)   (b)

FIGURE 2

Problems (§4A)

1. Find the critical points and the maximum and minimum values of the following functions.

(a) $x^2 - 5x + 6$,   on $[0, 4]$ (d) $(1 - x)^3$,   on $[-2, 2]$

(b) $-2 + 3x - x^2$,   on $[0, 3]$ (e) $\sqrt{9 - 4x^2}$,   on $[-1, 1]$

(c) $(x^2 - 1)^2$,   on $[-1, 1]$ (f) $\dfrac{x^2}{x^4 + 1}$,   on $[-1, 2]$

Answers to problems (§4A)

1. (a) $x = \frac{5}{2}$;   max 6, min $-\frac{1}{4}$ (b) $x = \frac{3}{2}$;   max $\frac{1}{4}$, min -2

   (c) $x = -1, 0, 1$;   max 1, min 0   (d) $x = 1$;   max 27, min -1

   (e) $x = 0$;   max 3, min $\sqrt{5}$ (f) $x = 0, -1, 1$;   max $\frac{1}{2}$, min 0

§4B. Maximum or minimum on other intervals

Many important problems take the following form: *Let f be continuous on an interval I and differentiable on its interior; what is its maximum (or minimum) value?* When I is not a closed interval, the preceding guide is incomplete. There need not even be a maximum or minimum.

In practice these problems can usually be settled by directing the search to a suitable closed interval.

4B1. The special case: there is just one critical point. Let f be continuous on an interval I and differentiable on its interior, and suppose f has just one critical point, x_0. This is a common situation, and we will obtain a complete solution. Ordinarily, f turns out to have either a maximum or a minimum at the point.

Since x_0 is the only critical point, f is monotonic on the interval $x \le x_0$ and on the interval $x \ge x_0$ (§4A4). Hence we have the following three-point test for a maximum or minimum.

Compute the value of f at x_0 and at any two points on opposite sides of x_0.

(a) *If $f(x_0)$ is the largest of the three values then it is the maximum of f altogether* (Figure 1(a)).

(b) *If $f(x_0)$ is the smallest of the three values then it is the minimum of f altogether* (Figure 1(b)).

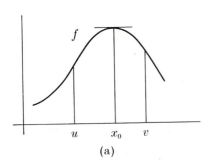 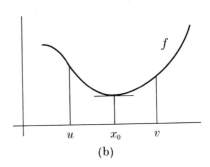

$$u \qquad x_0 \qquad v \qquad\qquad u \qquad x_0 \qquad v$$

(a)   (b)

FIGURE 1

If $f(x_0)$ lies between the other two values, then f is a monotonic function (Figure 2). In all cases, any other extreme values must occur at an endpoint.

Call the check points u and v:

$$u < x_0 < v.$$

Since we have freedom in choosing them, we try to choose them so that the computations are easy.

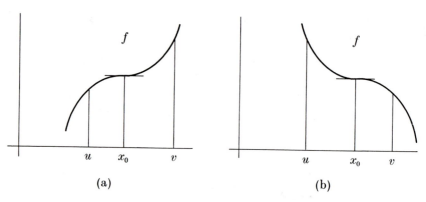

FIGURE 2

Example 1. What are the maximum and minimum values (if any) of

$$f(x) = 3x^2 - 2x + 4 \qquad\qquad (x \in \mathscr{R})\,?$$

Solution. We can analyze any quadratic function without calculus by completing the square; but we wish to illustrate the present methods. We have

$$f'(x) = 6x - 2.$$

This is 0 at $x = \frac{1}{3}$ and nowhere else. So $x_0 = \frac{1}{3}$ and

$$f(x_0) = f(\tfrac{1}{3}) = (3)(\tfrac{1}{9}) - (2)(\tfrac{1}{3}) + 4 = 3\tfrac{2}{3}.$$

Choose $u = 0$ and $v = 1$ (say). Then $u < x_0 < v$ and

$$f(u) = f(0) = 4 \qquad \text{and} \qquad f(v) = f(1) = 5.$$

Of the three values, $f(x_0)$ is the smallest. Therefore it is the minimum value of f. There is no maximum value, as $\mathscr{R}$ as no endpoints.

There is an alternative form of the tests in terms of the derivative.

(c) *If there are two points on opposite sides of x_0 where f' has opposite signs, then $f(x_0)$ is either a maximum or a minimum.*

It is clear how to tell which: if the signs are plus-minus, a maximum; if minus-plus, a minimum (Figure 3). Finally, if the signs are the same, f is a monotonic function.

The proofs are easy. Since we know that f is monotonic on the interval $x \le x_0$, all we have to do is decide which way. Consider a point $u < x_0$. If $f'(u) > 0$, then f is increasing at the point u (§3B4). Then f cannot be a decreasing function on the interval $x \le x_0$. So it must be increasing.

The other sections of the argument go the same way.

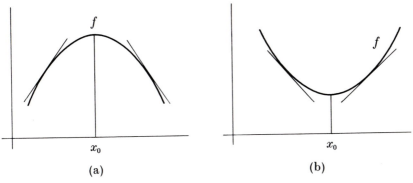

(a)   (b)

FIGURE 3

Example 2. In Example 1,

$$f'(x) = 6x - 2$$

and $x_0 = \frac{1}{3}$ is the only point where $f' = 0$. Pick $u = 0$ and $v = 1$. Then $u < x_0 < v$ and

$$f'(u) = f'(0) = -2 \qquad \text{and} \qquad f'(v) = f'(1) = 4.$$

Since the signs are minus-plus, $f(\frac{1}{3})$ is the minimum value of f.

4B2. The general case. Let f be continuous on an interval I and differentiable on its interior. We wish to determine whether and where f assumes a maximum or a minimum value. To do this we test certain points: the endpoints of I (if any) and the critical points of f (if any). Typically, there will be only a few test points. We call the leftmost test point a and the rightmost b. Then f is monotonic on the interval $x \leq a$ and on the interval $x \geq b$. Hence we have the following four-point test.

Compute f at a and b and at any two points $u < a$ and $v > b$.

(a) *If the values of f at a and b are larger than at u and v (resp.), then f assumes a maximum on I and it occurs in $[a, b]$.* (See Figure 4(a).)

(b) *If the values of f at a and b are smaller than at u and v (resp.), then f assumes a minimum on I and it occurs in $[a, b]$.* (See Figure 4(b).)

In each case, the problem is reduced to the closed interval $[a, b]$; and we know how to carry it on from there.

When a or b is an endpoint of I the rules are similar and simpler: just ignore the part about u or v (resp.) (and blot out the appropriate portion of Figure 4).

Again, there is an alternative test with the derivative

(c) *If there are two points $u < a$ and $v > b$ where f' has opposite signs, then f assumes either a maximum or a minimum on I, and it occurs in $[a, b]$.*

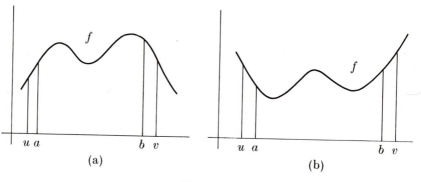

FIGURE 4

It is clear how to tell which: if the signs are plus-minus, f assumes a maximum; if minus-plus, a minimum (Figure 5).

This time, when f' has the same sign at the two check points, f need not be a monotonic function. We will not enter into a general discussion of this situation.

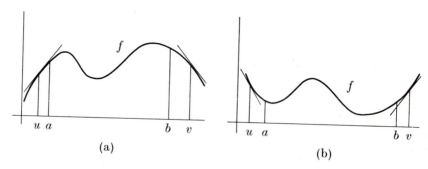

FIGURE 5

Example 3. Find the maximum and minimum values, if any, of

$$g(x) = \tfrac{1}{15}(3x^4 + 4x^3 - 12x^2 - 7) \qquad (x \in \mathcal{R}).$$

Solution. See Figure 6. We have

$$g'(x) = \tfrac{1}{15}(12x^3 + 12x^2 - 24x) = \tfrac{4}{5}x(x + 2)(x - 1).$$

Our test points are the zeros of g': -2, 0, and 1. Evidently, $g'(x) < 0$ for any $x < a = -2$ (since all three factors are negative) and $g'(x) > 0$ for all $x > b = 1$ (since all factors are positive). This is the pattern of Figure 5(b). Consequently, g attains a minimum, namely, its minimum on $[-2, 1]$.

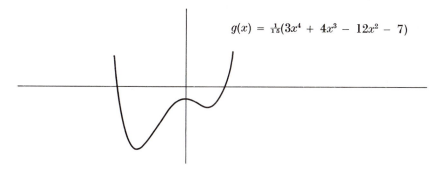

$$g(x) = \tfrac{1}{15}(3x^4 + 4x^3 - 12x^2 - 7)$$

FIGURE 6

To find the minimum on $[-2, 1]$ we compute g at the test points:

$$g(-2) = -\tfrac{13}{5}, \qquad g(0) = -\tfrac{7}{15}, \qquad g(1) = -\tfrac{4}{5}.$$

Therefore the minimum of g on $\mathscr{R}$ is $g(-2) = -\tfrac{13}{5}$.

By writing $g(x)$ in the form

$$g(x) = \tfrac{1}{15}[3x^4 + 4x^2(x - 3) - 7],$$

we see that g assumes arbitrarily large values and therefore has no maximum value.

Example 4. Find the maximum and minimum values, if any, of

$$f(x) = 2x^{3/2} - 9x + 12\sqrt{x} \qquad\qquad (x \geq 0).$$

Solution. See Figure 7. The domain of f is the interval $x \geq 0$, with left endpoint $a = 0$. We note that f is differentiable on the interior, and

$$f'(x) = 3\sqrt{x} - 9 + \frac{6}{\sqrt{x}} = \frac{3}{\sqrt{x}}(\sqrt{x} - 1)(\sqrt{x} - 2) \qquad (x > 0).$$

The zeros of f' are 1 and 4. Our test points are 0, 1, and 4.

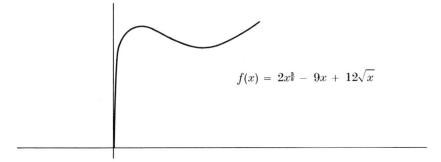

$$f(x) = 2x^{\frac{3}{2}} - 9x + 12\sqrt{x}$$

FIGURE 7

Since $f'(x) > 0$ for any $x > b = 4$, f attains a minimum, namely, its minimum value on $[0, 4]$ (Figure 5(b)). The values of f at the test points are:

$$f(0) = 0, \qquad f(1) = 5, \qquad f(4) = 4.$$

Hence the minimum value of f is $f(0) = 0$.

Writing $f(x)$ as

$$f(x) = x(2\sqrt{x} - 9) + 12\sqrt{x},$$

we see by inspection that f has no maximum value.

Problems (§4B)

1. Show that each of the following functions has a minimum, locate it, and find its value.

(a) $x + \dfrac{1}{x}$ $(x > 0)$

(b) $x^2 - 5x + 4$
(c) $2x^3 - 9x^2 + 12x + 1$ $(x > 1)$
(d) $(x^2 - 1)^4 + (x^2 + 1)^4$

(e) $\dfrac{1}{\sqrt{4 - x^2}} - \tfrac{1}{2}x^2$

2. Show that each of the following functions has a maximum, locate it, and find its value.

(a) $\dfrac{x}{1 + x^3}$ $(x > -1)$ (d) $3x^2 - x^3$ $(x > -1)$

(b) $5 - \sqrt{x} - \dfrac{2}{\sqrt{x}}$ (e) $x^2(4 - x^2)$

(c) $\dfrac{x}{1 + x^2}$ $(x \geq 0)$

3. According to the theory of probability, if p is the probability of success in a single trial of an experiment then P, the probability of exactly r successes in a series of n independent trials $(0 \leq r \leq n)$, is given by

$$P = C_{n,r}p^r(1 - p)^{n-r},$$

where $C_{n,r}$ depends on n and r but not on p (explicitly, $C_{n,r} = n!/r!\,(n - r)!$). P itself is a function of the three variables; but for fixed n and r it is a function of p alone. Prove that for fixed n and r, P is a maximum when $p = r/n$.

4. A calculus instructor permitted his class to choose a positive integer n with the understanding that any student with a grade of

$$g(n) = 100 \left(1 - \frac{12n}{10n^2 + 21} \right)$$

or more on the final exam would pass the course. What n should they choose? [*Hint.* Regard n as a continuous variable. *Caution.* Exert caution.]

Answers to problems (§4B)

1. (a) min 2 at $x = 1$ (b) min $-\frac{9}{4}$ at $x = \frac{5}{2}$ (c) min 5 at $x = 2$
   (d) min 2 at $x = 0$ (e) min $-\frac{1}{2}$ at $x = \sqrt{3}, -\sqrt{3}$
2. (a) max $2^{2/3}/3$ at $x = 1/2^{1/3}$ (b) max $5 - 2\sqrt{2}$ at $x = 2$
   (c) max $\frac{1}{2}$ at $x = 1$ (d) max 4 at $x = 2$
   (e) max 4 at $x = \sqrt{2}$ or $-\sqrt{2}$
4. $n = 2$

§4C. Applications of maxima and minima

A large variety of practical problems depend for their solution on finding the maximum or minimum of some quantity. Businessmen seek to minimize costs and maximize profits, airline pilots may try to minimize flight times, surgeons work to maximize the probability of recovery. Even Nature often appears to act in such a way that a particular quantity is maximized or minimized. With the help of calculus we can solve many such problems that would otherwise be intractable.

A problem ordinarily comes to us in descriptive language. Our first job is to recast it into the appropriate mathematical form. The form we want is:

(1)   *Find the maximum or minimum of the function f on the interval I.*

To put the problem into this form, proceed as follows.

Step 1. *Identify the variables* and assign them symbols.

Step 2. *Express the quantity* to be maximized or minimized *in terms of these variables.*

Step 3. *Identify relations among the variables,* from the data of the problem, so that you can *express all of them in terms of a particular one.*

Step 4. *Express the quantity* to be maximized or minimized *in terms of this single variable.*

Try to choose the variable in step 3 so that the functional expression in step 4 is as simple as possible.

Now check to see that the problem has been put into the form (1). Once the problem has been properly formulated, proceed as explained in §§4A and 4B.

Example 1. What is the most economical shape of a cylindrical can of given volume; that is, for which the surface area (including top and bottom) is a minimum?

Solution.

Step 1. Let r be the radius and h the height of the cylinder.

Step 2. The area of top and bottom are each πr^2 and the area of the wall is $2\pi rh$. The total surface area is

$$(2) \qquad S = 2\pi r^2 + 2\pi rh = 2\pi(r^2 + rh).$$

Step 3. The variables r and h are related by the equation

$$\pi r^2 h = \text{given volume};$$

or, simply,

$$(3) \qquad r^2 h = C \qquad \text{(a constant)}.$$

Step 4. We get a simpler expression by solving for h in terms of r:

$$(4) \qquad h = \frac{C}{r^2},$$

rather than the other way around. Then

$$(5) \qquad S = 2\pi \left(r^2 + \frac{C}{r}\right).$$

The mathematical problem is: Find the minimum value of S on the interval $r > 0$. We have

$$(6) \qquad S' = 2\pi \left(2r - \frac{C}{r^2}\right).$$

Solving $S' = 0$ we get

$$(7) \qquad r^3 = \frac{C}{2}.$$

As this equation has just one root, $r_0 = (C/2)^{1/3}$, the search is directed to this point (§4B1). Now, it is apparent from (6) that S' is increasing with r, so that $S' < 0$ for any $r < r_0$ and $S' > 0$ for $r > r_0$. Therefore S attains a minimum value at r_0 (§4B1(c)).

We can quickly get the ratio of h to r when $r = r_0$. From (3) and (7),

(8)
$$\frac{h}{r} = \frac{C}{(C/2)} = 2.$$

Thus, the optimal proportions call for the height to be twice the radius (Figure 1).

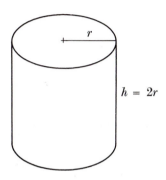

FIGURE 1

Example 2. Town B is 10 miles east of A and town C is 3 miles north of B. See Figure 2. An interstate highway is to be constructed from A to C. The cost along the existing roadbed from A to B is \$400,000/mile, while the cost elsewhere is \$500,000/mile. Where should the pivot point P be located so as to minimize the cost?

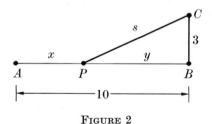

FIGURE 2

Solution.

Step 1. Let x, y, and s denote the distances as shown in the figure.

Step 2. The cost in hundreds of thousands of dollars is

$$C = 4x + 5s.$$

Step 3. The relations among x, y, and s are:

$$x + y = 10, \qquad y^2 + 9 = s^2.$$

We can express all variables in terms of x, in terms of y, or in terms of s.

Step 4. Our first inclination is to express C in terms of x, the distance from A to the pivot point. We have $y = 10 - x$ and

$$s = \sqrt{y^2 + 9} = \sqrt{(10 - x)^2 + 9}.$$

Then

$$C = 4x + 5\sqrt{(10 - x)^2 + 9}.$$

Alternatively, we can express C in terms of y:

(9) $$C = 4(10 - y) + 5\sqrt{y^2 + 9}.$$

This looks easier to work with.

Our mathematical problem is: Find the minimum value of C on the interval $0 \le y \le 10$. We have

(10) $$C' = -4 + 5\frac{y}{\sqrt{y^2 + 9}}.$$

Solving $C' = 0$ we get

(11) $$25y^2 = 16(y^2 + 9), \qquad y^2 = 16.$$

Of the two roots, $y = 4$ and $y = -4$, only the first lies within the domain of C. So we evaluate C at 0, 4, and 10:

$$\begin{aligned}
&\text{at} \quad y = 0, &&C = 55; \\
&\text{at} \quad y = 4, &&C = 49; \\
&\text{at} \quad y = 10, &&C = 5\sqrt{109} \approx 52.
\end{aligned}$$

The minimum cost is \$4,900,000, achieved at $y = 4$, i.e., by placing the turn 4 miles west of B (and 6 miles east of A) (Figure 3). This saves more than a quarter-million dollars over the straight-line route and over half a million over the L-shaped route through B.

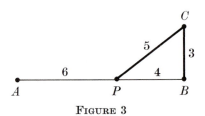

FIGURE 3

Problems (§4C)

1. The sum of two numbers is 12. Find the numbers if:
   (a) The sum of their cubes is a minimum.
   (b) The product of one of them by the cube of the other is a maximum.
   (c) Both are positive, and the product of one of them by the square of the other is a maximum.

2. If the difference of two numbers is 10, must their product be greater than -30?

3. A farmer wishes to select 10 acres of his land along a river and divide it into 8 small plots by means of a fence running parallel to the river and 9 fences perpendicular to it. Show that if the total amount of lumber is to be a minimum, the parallel fence should be as long as the 9 others combined.

4. A poster of 500 sq. in. is to have a margin of 6 in. at the top and 4 in. at each side and the bottom. What dimensions yield the largest printed area?

5. A sheet of metal 10 in. by 20 in. is to be made into an open box by cutting out a square at each corner and then folding up the four sides. What size square should be cut out in order that the resulting volume be a maximum, and what is the maximum volume?

6. A sheet of metal 12 inches wide is to be made into a gutter by turning up edges of equal length at right angles to the base. What should be the width of the gutter to maximize the volume the gutter can carry?

7. (a) A piece of wire 12 inches long is cut into two lengths, one of which is bent into a circle, the other into a square. Show that the sum of the areas has a *minimum* value of $36/(4 + \pi)$ square inches and that this is achieved when the side of the square is equal to the diameter of the circle.

   (b) A piece of wire 12 inches long is to be bent into the shape of a rectangle surmounted by a semicircle (Figure 4). Show that the *maximum* area is $72/(4 + \pi)$, achieved when the radius of the circle is equal to the smaller side of the rectangle.

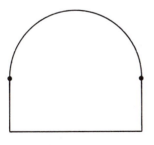

FIGURE 4

   (c) A piece of wire of length $9 + 4\sqrt{3}$ is cut into two lengths, one of which is formed into a square and the other into an equilateral triangle. What should the lengths be in order for the sum of the areas to be a minimum?

8. (a) Find the dimensions of the rectangle of area 25 whose perimeter is a minimum.

   (b) Which rectangle of perimeter 20 has the largest area?

9. A rectangular box is twice as long as wide.
   (a) What is its smallest possible surface area (top + bottom + 4 sides) if its volume is 72 cu. in.?
   (b) What is its largest possible volume if its surface area is 108 sq. in.?

10. A rectangle is to have one side along the x-axis and is to be inscribed in the semicircle $x^2 + y^2 = 1$ ($y > 0$). How should it be chosen so that
    (a) its perimeter is a maximum?
    (b) its area is a maximum?

11. A right circular cylindrical container open at the top is to hold a certain volume V_0. How should its dimensions be chosen so that the amount of material used is minimized?

12. If the strength of a rectangular beam is proportional to the width times the square of the depth, what relative dimensions should be chosen, when cutting a beam from a circular log, to achieve maximum strength?

13. Describe the isosceles triangle of maximum area if:
    (a) The perimeter is fixed.
    (b) Two sides have fixed length s.

14. At 4:00 p.m., automobile A is traveling at 60 mph toward a crossing 125 miles to the east, and car B is at the crossing headed north at 30 mph. At what time will the cars be closest to one another, and what will the distance between them then be?

15. A lakefront runs east-west. A man in a rowboat is 5 miles due north of a point A on the shore. He wishes to get to B, 10 miles due east of A, in the least time. He rows 3 mph. What is his minimum time if:
    (a) He walks 5 mph?   (b) He walks $3\frac{1}{3}$ mph?

16. Find the length of the longest horizontal chord cut off by the curves $y = \sqrt{x}$ and $y = x^2$ in the interval $[0, 1]$.

17. Locate the point $P = (x, y)$ on the quarter-circle

$$x^2 + y^2 = 1 \qquad\qquad (x \geq 0, y \geq 0)$$

so that the sum of the distances from P to the y-axis and to the point $(1, 0)$ shall be a maximum. What is the maximum?

18. Which line through $(1, 2)$ cuts off the triangle in the first quadrant having the least area?

19. What is the shortest distance from the point $(1, 2)$ to the parabola $y = x^2/4$?

20. Which line tangent to $y = 9 - x^2$ cuts off the smallest area in the first quadrant? [*Hint.* Denote the point of tangency by $(h, 9 - h^2)$ and express the x-intercept and y-intercept of the tangent line in terms of h.]

21. Find the point on the x-axis from which the sum of the distances to $(0, 4)$ and $(4, 2)$ is a minimum. [*Caution.* If $\sqrt{a} + \sqrt{b} = 0$ then $a = b$ but not conversely.]

22. Find the length of the shortest ladder from the ground to a vertical wall, passing over a shed 8 feet high that extends $3\frac{3}{8}$ feet from the wall. See Figure 5. [*Hint*. From similar triangles,

$$\frac{l}{b} = \frac{a + 3\frac{3}{8}}{a}.$$

Also, $b^2 = a^2 + 64$.]

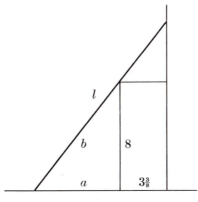

FIGURE 5

23. A manufacturer can produce soap flakes for $200 per ton over his weekly fixed costs of $10,000. If he sets his price at P dollars per ton, he can dispose of $500 - \frac{1}{2}P$ tons per week. At what selling price will his profit be maximized? [*Hint*. Regard the price as a continuous variable.]

24. An airline offers a charter flight at a fare of $100 per person if 50 to 100 passengers sign up. For each passenger beyond 100, the fare for *each* passenger is reduced by fifty cents. The plane has 200 seats. How many passengers will provide the largest total revenue? [*Hint*. Regard the number of passengers as a continuous variable.]

Answers to problems (§4C)

1. (a) 6 and 6   (b) 3 and 9   (c) 4 and 8
4. 20 × 25
5. side of square $5(1 - \frac{1}{3}\sqrt{3})$, maximum volume $\frac{1000}{9}\sqrt{3}$
6. 6 in.
7. (c) side of square $\sqrt{3}$, side of triangle 3
8. (a) 5 × 5   (b) 5 × 5
9. (a) 108   (b) 72
10. (a) corner at $(2/\sqrt{5}, 1/\sqrt{5})$ (b) corner at $(1/\sqrt{2}, 1/\sqrt{2})$
11. height = radius

12. depth $= \sqrt{2} \times$ width
13. (a) equilateral   (b) length of third side is $s\sqrt{2}$
14. 5:40 p.m., $25\sqrt{5}$ miles
15. (a) $3\frac{1}{3}$ hours, achieved by heading for the point $3\frac{3}{4}$ miles due east of A
    (b) $\frac{5}{3}\sqrt{5} \approx 3.7$ hours, achieved by rowing directly to B
16. $\dfrac{3}{4^{4/3}}$
17. max $\frac{3}{2}$ at $x = \frac{1}{2}$
18. $2x + y = 4$
19. $\sqrt{2}$
20. $y + 2\sqrt{3}x = 12$
21. $x = \frac{8}{3}$
22. $15\frac{5}{8}$ feet
23. $600
24. 150

§4D. Increase and decrease of functions

4D1. Introduction. In this section we study in more detail how the derivative gives information about the behavior of a function.

Suppose that $f'(x) > 0$ at each point x in a given interval. Is f an increasing function on that interval? That is, does $x_1 < x_2$ imply $f(x_1) < f(x_2)$?

Suppose $f'(x) = 0$ at each x in an interval. Is f constant on that interval? That is, if x_1 and x_2 are any two points, does $f(x_1)$ equal $f(x_2)$?

Each of these questions is somewhat deeper than it may appear. The hypothesis is about each individual point but the conclusion is about each pair of points. We will solve the first problem now and the second one in the next section (§4E).

4D2. Sign of the derivative. *Let f be continuous on an interval I and differentiable on its interior.*

(a) *If $f'(x) > 0$ throughout the interior of I, then f is an increasing function on I.*

(b) *If $f'(x) < 0$ throughout the interior of I, then f is a decreasing function on I.*

Proof. (a). Since f' is never 0 in the interior of I, we know from §4A4 that f is monotonic on I. Let x_0 be any point in the interior of I. Since $f'(x_0) > 0$, f is increasing at the point x_0 (§3B4). Hence f cannot be a decreasing function on I. Therefore it is an increasing function.

The proof of (b) goes the same way.

The converses are false. For example, x^3 is an increasing function on $\mathscr{R}$, but its derivative is not positive everywhere.

It is important that f' is assumed positive or negative only in the interior. In many interesting applications, f' is zero at the endpoints, or nonexistent (the curve having a vertical tangent). So the hypothesis is exactly the right one. The next examples illustrate this.

Example 1. Discuss the graph of

$$f(x) = \tfrac{1}{3}(x^3 - 3x^2 + 2) \qquad\qquad (x \in \mathscr{R}).$$

Discussion. The derivative is

$$f'(x) = x(x - 2).$$

For $x > 2$ both x and $x - 2$ are positive and therefore $f'(x) > 0$. By (a), f is increasing on the interval $x \geq 2$. Note that this is true even though $f'(2) = 0$. For $x < 0$ both x and $x - 2$ are negative, and again $f'(x) > 0$. By (a), f is increasing on the interval $x \leq 0$ (even though $f'(0) = 0$).

For $0 < x < 2$, x is positive and $x - 2$ is negative, and therefore $f'(x) < 0$. By (b), f is decreasing on the interval $0 \leq x \leq 2$ (even though f' is 0 at both endpoints). The graph is shown in Figure 1.

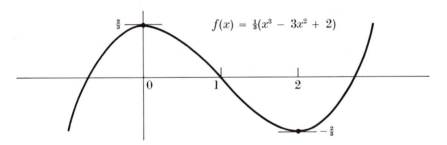

FIGURE 1

Example 2. The function $y = \sqrt{x}$ is continuous on the interval $x \geq 0$ and has a positive derivative $y' = 1/2\sqrt{x}$ on the interval $x > 0$. Therefore y is increasing on the interval $x \geq 0$. Note that y' does not exist at $x = 0$.

4D3. Local maximum and local minimum. We say that f achieves a *local maximum* at a point x_0 if on some interval about x_0, f attains a maximum at x_0. In Figure 2, f attains a local maximum at x_0. Although P_0 is not the highest point on the graph, it is king of its own hill. Another local maximum is at B; it has only a one-sided hill but is king of that side. Finally, P_1, which is the maximum point of the whole graph, is automatically a local maximum point.

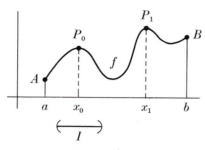

FIGURE 2

Local minimum is defined similarly.

Let f be continuous on an interval I and differentiable on its interior.

(1)   *If f has a local maximum or minimum at an interior point of I, then f' is zero there.*

Since a local maximum or minimum is a maximum or minimum on some interval, this follows from §4A2.

4D4. Concavity and inflection. Let f be differentiable on an interval I. We say that f or its graph are *concave up* on I if f' is increasing on I; and *concave down* if f' is decreasing. Where f is concave up, the slope of the tangent line is increasing with increasing x, so that as we move along the graph we swing upward (counterclockwise); see Figure 3(a). Where f is concave down we swing downward (clockwise); see Figure 3(b).

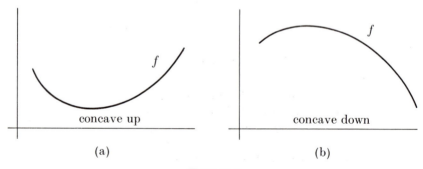

(a)	(b)

FIGURE 3

A point $\left(x_0, f(x_0)\right)$ is a *point of inflection* of f if there is an interval about x_0 on which f is concave one way on one side of x_0 and concave the other way on the other side. At a point of inflection we stop turning one way and start turning the other way (Figure 4).

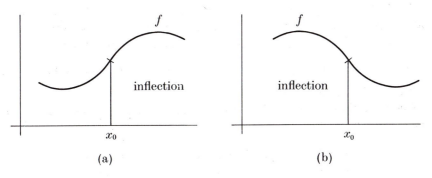

(a)   (b)

FIGURE 4

Example 3. Discuss the function

$$f(x) = \tfrac{1}{3}(x^3 - 3x^2 + 2).$$

Discussion. We have

(2) $$f'(x) = x(x - 2) = (x - 1)^2 - 1.$$

The zeros of f' are 0 and 2. We see that $f'(x) > 0$ for $x < 0$ and for $x > 2$, and $f'(x) < 0$ for $0 < x < 2$. Therefore, f is increasing on the interval $x \leq 0$, decreasing on the interval $0 \leq x \leq 2$, and increasing on $x \geq 2$. Hence there are a local maximum at $x = 0$ and a local minimum at $x = 2$.

From the second expression for f' in (2), it is clear that f' is decreasing on the interval $x \leq 1$ and increasing on the interval $x \geq 1$. Therefore f is concave down on the interval $x \leq 1$ and concave up on the interval $x \geq 1$. At $x = 1$ there is an inflection point.

Finally, from the expression

$$f(x) = \tfrac{1}{3}x^2(x - 3) + \tfrac{2}{3},$$

we verify that f assumes arbitrarily large values, both positive and negative, and so has no maximum and no minimum. The graph appears in Figure 1.

4D5. The second-derivative tests. Concavity and inflection are usually determined by a simple test with the second derivative. Local maxima can also be distinguished from local minima by means of the second derivative.

Let f be defined on an interval I and let f' be continuous on I and differentiable on its interior.

Let x_0 be a point in the interior of I.

LOCAL MAXIMUM OR MINIMUM

(a) *If $f'(x_0) = 0$ and $f''(x_0) > 0$ then f has a local minimum at x_0.*

Proof. Since $f''(x_0) > 0$, f' is increasing at the point x_0 (§3B4). Since its value there is 0, there is an interval about x_0 on which $f'(x) < 0$ for $x < x_0$ and $f'(x) > 0$ for $x > x_0$. Consequently, f attains a local minimum at x_0.

(b) *If $f'(x_0) = 0$ and $f''(x_0) < 0$ then f has a local maximum at x_0.*

The proof is similar to (a).

The rule is the opposite of the natural association of the words. A crude but helpful memory aid is:

positive:	yes, it holds water:	minimum
negative:	no, it spills water:	maximum

CONCAVITY

(c) *If $f''(x) > 0$ throughout the interior of I then f is concave up on I.*

Proof. By §4D2, f' is increasing on I. Hence, by definition, f is concave up.

(d) *If $f''(x) < 0$ throughout the interior of I then f is concave down on I.*

The proof is similar to (c).

INFLECTION

(e) *If f has an inflection point at x_0 then $f''(x_0) = 0$.*

Proof. By hypothesis, f' is increasing on one side of x_0 and decreasing on the other. Therefore f' attains either a local maximum at x_0 or a local minimum (depending on on which side of x_0 it is increasing). In either case, its derivative at the interior point x_0 is zero.

Example 4. Discuss the behavior of the function

$$g(x) = 3x^4 - 4x^3.$$

Discussion. We have

(3) $$g'(x) = 12x^2(x - 1), \qquad g''(x) = 36x(x - \tfrac{2}{3}).$$

We see that $g'(x) < 0$ on $x < 0$ and on $0 < x < 1$, and $g'(x) > 0$ on $x > 1$. Therefore g is decreasing on $x \le 1$ and increasing on $x \ge 1$. Consequently, g attains a minimum at $x = 1$, but has no local maximum.

Next, $g''(x) > 0$ for $x < 0$ and for $x > \tfrac{2}{3}$. Hence g is concave up on the intervals $x \le 0$ and $x \ge \tfrac{2}{3}$. Since $g''(x) < 0$ for $0 < x < \tfrac{2}{3}$, g is concave down on the interval $0 \le x \le \tfrac{2}{3}$. There are inflection points at $x = 0$ and $x = \tfrac{2}{3}$.

Finally, from the expression

$$g(x) = x^3(3x - 4),$$

we can check that g assumes arbitrarily large positive values. The graph is shown in Figure 5.

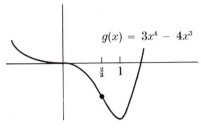

$$g(x) = 3x^4 - 4x^3$$

FIGURE 5

Example 5. Analyze the function

$$h(x) = \frac{1}{1 + x^2}.$$

Discussion. Note that $h(-x) = h(x)$. This means that the graph is symmetric with respect to the y-axis: the graph for $x < 0$ is the reflection, through the y-axis as mirror, of the graph for $x > 0$.

Evidently, $h(x) > 0$ for all x. Next, h is decreasing on the interval $x \geq 0$ and hence has a maximum at $x = 0$ (with value 1), but no minimum. Clearly, h assumes arbitrarily small positive values.

The derivatives are

$$h'(x) = -2\,\frac{x}{(1 + x^2)^2}, \qquad h''(x) = 6\,\frac{x^2 - \frac{1}{3}}{(1 + x^2)^3}.$$

We see that h is concave up on the intervals $x \leq -1/\sqrt{3}$ and $x \geq 1/\sqrt{3}$, and concave down on the interval

$$-\frac{1}{\sqrt{3}} \leq x \leq \frac{1}{\sqrt{3}}.$$

There are inflection points at $-1/\sqrt{3}$ and at $1/\sqrt{3}$. The value of the function at these points is $\frac{3}{4}$. The graph is shown in Figure 6.

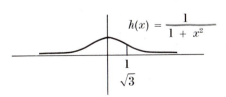

$$h(x) = \frac{1}{1 + x^2}$$

$$\frac{1}{\sqrt{3}}$$

FIGURE 6

Problems (§4D)

1. Analyze the following functions for increase, decrease, concavity, maximum and minimum, local maximum and minimum, and points of inflection, and sketch the graphs.

(a) $x^2 - 5x + 6$

(b) $x^3 - 8$

(c) $2x^3 - 9x^2 + 12x - 3$

(d) $(x - 1)(x + 1)^2$

(e) $x^4 - 4x + 1$

(f) $x^4 - 8x^2 + 1$

(g) $(x^2 - 1)^2$

(h) $x + \dfrac{1}{x + 1}$

(i) $x^2 + \dfrac{1}{x^2}$

(j) $\dfrac{1}{x^3 + 1}$

(k) $\dfrac{1}{x(x + 1)}$

(l) $\dfrac{x}{x^2 + 1}$

(m) $\dfrac{x^2}{x + 2}$

(n) $\sqrt{x^2 + 1}$

(o) $\sqrt{4 - x^2}$

(p) $x\sqrt{x + 1}$

(q) $x\sqrt{4 - x^2}$

(r) $x^2\sqrt{4 + x}$

(s) $2x^{3/2} - 9x + 12\sqrt{x}$

(t) $\dfrac{x^{3/2}}{1 + x^2}$

(u) $(x + 1)|x + 1|$

2. Give an example of a function f whose graph is concave up and for which
   (a) the graph of $1/f$ is concave up.
   (b) the graph of $1/f$ is concave down.

3. (a) Can a local maximum point be a point of inflection?
   (b) Can the first derivative be positive at a point of inflection?
   (c) Can the second derivative be zero at an interior local minimum?
   (d) Let f be defined on the interval $x > a$. If f is concave up on this interval, must it assume arbitrarily large positive values?

4. Find a and b, given that

$$y = x^3 + ax^2 - 3x + b$$

has a point of inflection at the point $(1, 1)$.

5. Show that if

$$f(x) = ax^3 + bx^2 + cx + d$$

has a point of inflection at the origin, then its graph is symmetric with respect to the origin—that is, for every x, $f(-x) = -f(x)$.

6. Show that whatever the coefficients a and b may be,

$$y = x^4 + ax + b$$

has no inflection points. Sketch the graph for $a = \frac{1}{2}, b = 1$.

7. Show that, if $ad \neq bc$,

$$f(x) = \frac{ax + b}{cx + d}$$

has no local maxima, no local minima, and no inflection points. Sketch the graph for $a = c = 1, b = -1, d = 0$.

8. Show that the function

$$f(x) = x^3 + ax^2 + bx + c$$

has no local maximum or minimum if $a^2 \leq 3b$, but has one local maximum and one local minimum if $a^2 > 3b$. [*Hint.* Do the cases $a^2 < 3b$ and $a^2 = 3b$ separately.]

9. The length of a rectangle is increasing at 3 inches per second, while the width decreases in such a way that the area remains constant at 20 square inches. On what interval is the perimeter increasing and on what interval decreasing?

10. A point moves in a straight line according to the law

$$s = t^4 - 12t^3 + 78t^2 - 28t + 6.$$

On what interval is the velocity increasing and on what interval decreasing?

11. Ship A is headed east at 11 knots (1 knot = 1 nautical mile per hour) and ship B is headed south at 16 knots. At time $t = 0$, A is 30 nautical miles west of a point P and B is 20 nautical miles north of P. On what interval is the distance between the ships increasing and on what interval is it decreasing?

12. Let g and h be continuous on an interval I and differentiable on its interior, and let a be a point of I. Prove that if $g'(x) > h'(x)$ throughout the interior, and if $g(a) \geq h(a)$, then $g(x) > h(x)$ for all $x > a$ in I. [*Hint.* Apply §4D2(a) to the function $f = g - h$.]

13. Show that if s is any rational number, $s > 1$, then

$$(1 + x)^s > 1 + sx \qquad \text{for all} \qquad x > 0.$$

[*Hint.* Apply Problem 12, with $g(x) = (1 + x)^s$, $h(x) = 1 + sx$, and $I = \{x : x \geq 0\}$. In verifying that $g'(x) > h'(x)$ for $x > 0$, make use of the following fact: if c is any number > 1 and t is any rational number > 0, then $c^t > 1$.]

14. Everybody "knows" that you get more from compound interest than simple interest and that the more often the interest is compounded the more you get. It is very easy to see that compounding 12 times a year gets you more than 2, or 3, or 4, or 6 times a year. But can you prove that it gets you more than 5, or 7, or 11 times a year?

Let r denote the rate of interest and n the number of times compounded. At the end of the year, each dollar has increased to $(1 + r/n)^n$ dollars. The problem is to show that

$$\left(1 + \frac{r}{n}\right)^n > \left(1 + \frac{r}{m}\right)^m \qquad \text{for } n > m.$$

Establish this by applying Problem 13, with $x = r/n$ and $s = n/m$.

Answers to problems (§4D)

1. (a) decreasing on $x \leq \frac{5}{2}$, increasing on $x \geq \frac{5}{2}$, concave up on $\mathscr{R}$, minimum $-\frac{1}{4}$ at $x = \frac{5}{2}$, no local maximum, no inflection points

   (b) increasing on $\mathscr{R}$, concave down on $x \leq 0$, concave up on $x \geq 0$, no local maximum or minimum, inflection point $(0, -8)$

   (c) increasing on $x \leq 1$ and on $x \geq 2$, decreasing on $1 \leq x \leq 2$, concave down on $x \leq \frac{3}{2}$, concave up on $x \geq \frac{3}{2}$, local maximum 2 at $x = 1$, local minimum 1 at $x = 2$, inflection point $(\frac{3}{2}, \frac{3}{2})$

   (d) increasing on $x \leq -1$ and on $x \geq \frac{1}{3}$, decreasing on $-1 \leq x \leq \frac{1}{3}$, concave down on $x \leq -\frac{1}{3}$, concave up on $x \geq -\frac{1}{3}$, local maximum 0 at $x = -1$, local minimum $-\frac{32}{27}$ at $x = \frac{1}{3}$, inflection point $(-\frac{1}{3}, -\frac{16}{27})$

   (e) decreasing on $x \leq 1$, increasing on $x \geq 1$, concave up on $\mathscr{R}$, minimum -2 at $x = 1$, no local maximum, no inflection points

   (f) decreasing on $x \leq -2$ and on $0 \leq x \leq 2$, increasing on $-2 \leq x \leq 0$ and on $x \geq 2$, concave up on $x \leq -2/\sqrt{3}$ and on $x \geq 2/\sqrt{3}$, concave down on $-2/\sqrt{3} \leq x \leq 2/\sqrt{3}$, minimum -15 at $x = 2$ and at $x = -2$, local maximum 1 at $x = 0$, inflection points $(-2/\sqrt{3}, -71/9)$ and $(2/\sqrt{3}, -71/9)$

   (g) decreasing on $x \leq -1$ and on $0 \leq x \leq 1$, increasing on $-1 \leq x \leq 0$ and on $x \geq 1$, concave up on $x \leq -1/\sqrt{3}$ and on $x \geq 1/\sqrt{3}$, concave down on $-1/\sqrt{3} \leq x \leq 1/\sqrt{3}$, minimum 0 at $x = 1$ and at $x = -1$, local maximum 1 at $x = 0$, inflection points $(1/\sqrt{3}, 4/9)$ and $(-1/\sqrt{3}, 4/9)$

   (h) increasing on $x \leq -2$ and on $x \geq 0$, decreasing on $-2 \leq x < -1$ and on $-1 < x \leq 0$, concave down on $x < -1$, concave up on $x > -1$, local maximum -3 at $x = -2$, local minimum 1 at $x = 0$, no inflection points

   (i) decreasing on $x \leq -1$ and on $0 < x \leq 1$, increasing on $-1 \leq x < 0$ and on $x \geq 1$, concave up on $x < 0$ and on $x > 0$, minimum 2 at $x = 1$ and at $x = -1$, no local maximum, no inflection points

   (j) decreasing on $x < -1$ and on $x > -1$, concave down on $x < -1$ and on $0 < x < 1/2^{1/3}$, concave up on $-1 < x < 0$ and on $x > 1/2^{1/3}$, inflection points $(0, 1)$ and $(1/2^{1/3}, \frac{2}{3})$

   (k) increasing on $x < -1$ and on $-1 < x \leq -\frac{1}{2}$, decreasing on $-\frac{1}{2} \leq x < 0$ and on $x > 0$, concave up on $x < -1$ and on $x > 0$, concave down on $-1 < x < 0$, local maximum -4 at $x = -\frac{1}{2}$, no local minimum, no inflection points

   (l) decreasing on $x \leq -1$ and on $x \geq 1$, increasing on $-1 \leq x \leq 1$, concave down on $x \leq -\sqrt{3}$ and on $0 \leq x \leq \sqrt{3}$, concave up on $-\sqrt{3} \leq x \leq 0$ and on $x \geq \sqrt{3}$, maximum $\frac{1}{2}$ at $x = 1$, minimum $-\frac{1}{2}$ at $x = -1$, inflection points $(0, 0)$, $(\sqrt{3}, \sqrt{3}/4)$, and $(-\sqrt{3}, -\sqrt{3}/4)$

(m) increasing on $x \le -4$ and on $x \ge 0$, decreasing on $-4 \le x < 2$ and on $-2 < x \le 0$, concave down on $x < -2$, concave up on $x > -2$, local maximum -8 at $x = -4$, local minimum 0 at $x = 0$, no inflection points

(n) decreasing on $x \le 0$, increasing on $x \ge 0$, concave up on $\mathscr{R}$, minimum 1 at $x = 0$, no local maximum, no inflection points

(o) increasing on $-2 \le x \le 0$, decreasing on $0 \le x \le 2$, concave down on $-2 \le x \le 2$, maximum 2 at $x = 0$, minimum 0 at $x = -2$ and at $x = 2$, no inflection points

(p) decreasing on $-1 \le x \le -\frac{2}{3}$, increasing on $x \ge -\frac{2}{3}$, concave up on $x \ge -1$, minimum $-\frac{2}{9}\sqrt{3}$ at $x = -\frac{2}{3}$, no local maximum, no inflection points

(q) decreasing on $-2 \le x \le -\sqrt{2}$ and on $\sqrt{2} \le x \le 2$, increasing on $-\sqrt{2} \le x \le \sqrt{2}$, concave up on $-2 \le x \le 0$, concave down on $0 \le x \le 2$, maximum 2 at $x = \sqrt{2}$, minimum -2 at $x = -\sqrt{2}$, inflection point $(0, 0)$

(r) increasing on $-4 \le x \le -\frac{16}{5}$ and on $x \ge 0$, decreasing on $-\frac{16}{5} \le x \le 0$, concave down on $-4 \le x \le -8(6 - \sqrt{6})/15$, concave up on $x \ge -8(6 - \sqrt{6})/15$, local maximum $512\sqrt{5}/125$ at $x = -\frac{16}{5}$, local minimum 0 at $x = -4$ and at $x = 0$, inflection point at $x = -8(6 - \sqrt{6})/15$

(s) increasing on $0 \le x \le 1$ and on $x \ge 4$, decreasing on $1 \le x \le 4$, concave down on $0 \le x \le 2$, concave up on $x \ge 2$, minimum 0 at $x = 0$, local maximum 5 at $x = 1$, local minimum 4 at $x = 4$, inflection point $(2, 16\sqrt{2} - 18)$

(t) increasing on $0 \le x \le \sqrt{3}$, decreasing on $x \ge \sqrt{3}$, concave up on $0 \le x \le \alpha$ and on $x \ge \beta$, where

$$\alpha = \left(\frac{13 - 4\sqrt{10}}{3}\right)^{1/2} \quad \text{and} \quad \beta = \left(\frac{13 + 4\sqrt{10}}{3}\right)^{1/2},$$

concave down on $\alpha \le x \le \beta$, maximum $3^{3/4}/4$ at $x = \sqrt{3}$, minimum 0 at $x = 0$, inflection points at $x = \alpha$ and $x = \beta$

(u) increasing on $\mathscr{R}$, concave down on $x \le -1$, concave up on $x \ge -1$, no local maximum or minimum, inflection point $(-1, 0)$

4. $a = -3, b = 6$

9. decreasing on $0 < l \le 2\sqrt{5}$, increasing on $l \ge 2\sqrt{5}$

10. increasing on $\mathscr{R}$

11. decreasing on $t \le \frac{50}{29}$, increasing on $t \ge \frac{50}{29}$

§4E. The Mean-Value Theorem

We turn now to the Mean-Value Theorem. It is a theorem that is easy to understand and easy to prove, yet that has many important theoretical consequences.

4E1. The Mean-Value Theorem. *Let f be continuous on a closed interval* $[a, b]$ *and differentiable on its interior.*

Then there is a point z in the interior such that

(1)
$$f(b) - f(a) = (b - a)f'(z).$$

Remark. Geometrically, this equation asserts that $f'(z)$ is the slope of the line joining $A = (a, f(a))$ and $B = (b, f(b))$:

(2)
$$f'(z) = \frac{f(b) - f(a)}{b - a} = \text{slope of } AB.$$

That is, the tangent line to the graph of f at z is to be parallel to AB (Figure 1).

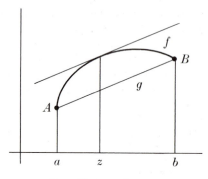

FIGURE 1

Proof of theorem. The line AB is the graph of a linear function, say g. The derivative of g at every point is constant, the slope of the line (2). We are looking for a point z in the interior of $[a, b]$ for which $f'(z) = g'(z)$. That is, the function $f - g$ is to satisfy

(3)
$$(f - g)'(z) = 0.$$

Now, evidently, $f - g$ is continuous on $[a, b]$ and differentiable on the interior. Hence if $f - g$ had no critical points in the interior, it would be monotonic on $[a, b]$ (§4A4). But $f - g$ cannot be monotonic, as it has the same value at a as at b (namely, 0). Therefore it has a critical point in the interior of $[a, b]$, i.e., a point z at which $(f - g)'$ is zero.

Remark. A novel feature of the theorem is that it does not tell us explicitly where z is, nor how to go about finding it.

The following special case of the Mean-Value Theorem is of interest.

4E2. Roots of equations (Rolle's Theorem). *Let f be continuous on an interval I and differentiable on its interior.*

Then between any two roots (in I) of the equation

$$f(x) = c \qquad\qquad (c \ any \ constant),$$

there lies a root of the equation

$$f'(x) = 0.$$

Proof. See Figure 2. If a and b are distinct roots, then by the Mean-Value Theorem there is a point z between them such that

$$f'(z) = \frac{f(b) - f(a)}{b - a} = 0.$$

Example 1. Show that the equation

$$x^7 - 14x + 23 = 0$$

has no more than three real roots.

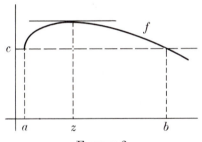

FIGURE 2

Solution. By Rolle's Theorem, any two roots enclose between them a root of

$$7x^6 - 14 = 0.$$

Since this equation has only two real roots (namely, $\pm 2^{1/6}$), the original cannot have more than three.

4E3. Constant functions. It is an elementary theorem that the derivative of a constant function is 0. With the help of the Mean-Value Theorem, we can now prove the converse.

Let f be continuous on an interval I and differentiable on its interior. If f'(x) = 0 throughout the interior of I, then f is constant on I.

(Hence $f'(x) = 0$ on all of I.)

Proof. We wish to show that, if a and b are any two points of I, then $f(a) = f(b)$. Now, by the Mean-Value Theorem, there is a point z between them for which

$$f(b) - f(a) = (b - a)f'(z).$$

Since $f'(z) = 0$, this shows that $f(b) = f(a)$.

Problems (§4E)

1. In each of the following circumstances, $f(b) - f(a) = 0$, yet there is no number $z \in (a, b)$ for which $f'(z) = 0$. In each case, reconcile this with the Mean-Value Theorem.

(a) $f(x) = \dfrac{1}{x^2}$, $[a, b] = [-2, 2]$

(b) $f(x) = 1 - |x|$, $[a, b] = [-1, 1]$

2. Show that the equation

$$x^{18} - 5x + 3 = 0$$

cannot have more than two real roots.

3. Show that the equation

$$x^3 + 6x^2 + 15x - 23 = 0$$

cannot have more than one real root.

4. Let

$$g(x) = x(x - 1)(x - 2)(x - 3).$$

How many real roots has the equation $g'(x) = 0$? For each, specify an interval containing that root and no others.

5. Let

$$g(x) = x^4 - 4x^3 - 2x^2 + 12x + 7.$$

Show that the equation $g'(x) = -8$ has a solution in the interval $(1, 3)$. [*Hint.* Consider the function $h(x) = g(x) + 8x$.]

6. Let f be continuous on $[a, b]$ and have a second derivative on (a, b). Show that if $f(x) = 0$ for at least three values of x in $[a, b]$, then $f''(x) = 0$ for at least one value of x in (a, b).

7. A man walks from point A to point B, 2 miles along a straight road, in 1 hour. Prove that there is at least one point at which his instantaneous velocity is exactly 2 miles per hour. (What conditions on the man's motion are you assuming?)

8. Show that there is a number x in the interval $(1, 3)$ at which the tangents to the graphs of

$$y = x^4 - x^3 - 5x + 9 \qquad \text{and} \qquad y = 6x^2 - 2x$$

are parallel.

9. Prove that if $xf'(x) = 0$ for all x in the interval $[0, 1]$, then f is constant on that interval.

10. (a) Let f be continuous on an interval I and differentiable on its interior. Prove that if $f'(x) \geq 0$ throughout the interior of I, then for any two points a and b of I, $a < b$, we have $f(a) \leq f(b)$. [*Hint.* Imitate the proof in §4E3.]

(b) Let g and h be continuous on an interval I and differentiable on its interior, and let a be a point of I. Prove that if $g'(x) \geq h'(x)$ throughout the interior of I, and if $g(a) \geq h(a)$, then $g(x) \geq h(x)$ for all $x > a$ in I. [*Hint.* Consider the function $f = g - h$.]

Answers to problems (§4E)

4. 3 real roots

§4F. Antiderivatives

4F1. The antiderivatives of a function. If f is the derivative of F we call F an *antiderivative* of f. For example, x^3 is an antiderivative of $3x^2$.

If F is an antiderivative of f then so is $F + C$, where C is any constant. For example, $x^3 + 5$ is an antiderivative of $3x^2$. So is $x^3 - 18$.

Thus, if a function has one antiderivative, it has many, obtained from the given one by adding constants. Moreover, as we shall now prove, we get *all* antiderivatives this way.

Let f be a function that has an antiderivative on an interval I.

(a) *Any two antiderivatives of f on I differ by a constant.*

Proof. If F_1 and F_2 are both antiderivatives of f then $F_1' = F_2'$ on I. Thus, the function $F_1 - F_2$ satisfies

$$(F_1 - F_2)'(x) = 0$$

for every $x \in I$. By §4E3, $F_1 - F_2$ is constant.

(b) *For each point $a \in I$ and for every number b, there is one and only one antiderivative F of f for which*

$$F(a) = b.$$

Proof. Clearly, all we have to do is pick any antiderivative and then adjust it (if necessary) by adding a constant so as to make the value at a equal to b. Thus, let F_0 be any antiderivative of f. Then F, the one we are after, satisfies

(1) $$F(x) = F_0(x) + C,$$

where C is a constant. (This is what we just proved in (a).) Now there is one and only one number C such that

(2) $$F(a) = F_0(a) + C = b$$

$\left(\text{namely, } C = b - F_0(a)\right)$. The function F is thus determined uniquely.

Example 1. Find the antiderivative of $f(x) = 2x$ whose value at $x = 2$ is 7.

Solution. We know that $d(x^2)/dx = 2x$. Therefore x^2 is an antiderivative of $2x$. If F is the antiderivative sought then

$$F(x) = x^2 + C.$$

To find C, evaluate the functions at $x = 2$, imposing the requirement $F(2) = 7$. This gives

$$7 = 4 + C, \qquad C = 3.$$

Therefore

$$F(x) = x^2 + 3.$$

4F2. Antiderivative formulas. To find antiderivatives we think of what we know about derivatives. Every differentiation formula gives us an antiderivative. For example, since $d(x^r)/dx = rx^{r-1}$, an antiderivative of rx^{r-1} is x^r. This is usually stated in the following form.

(a) *An antiderivative of* x^r *is* $\dfrac{1}{r+1} x^{r+1}$ *(where r is rational, $r \neq -1$, and $r > 0$ in case $x = 0$).*

The function $1/x$ $(x > 0)$ is not accounted for here. As it happens, it, too, has an antiderivative; but that is a deeper result. In fact, this question and its ramifications are the subject of an entire chapter (Chapter 7).

Other differentiation formulas are $(F + G)' = F' + G'$, $(F - G)' = F' - G'$, and $(cF)' = cF'$ (c a constant). These give:

If F and G are antiderivatives of f and g, then:

(b) $F + G$ *is an antiderivative of* $f + g$;
(c) $F - G$ *is an antiderivative of* $f - g$;
(d) cF *is an antiderivative of* cf (c *a constant*).

Combining these formulas, we can get the antiderivatives of all polynomials.

Example 2. What are the antiderivatives of

$$f(x) = x^3 - 3x^2 + 2x + 7?$$

Solution. One antiderivative is

$$F_0(x) = \tfrac{1}{4}x^4 - x^3 + x^2 + 7x.$$

The other antiderivatives are the functions of the form

$$F(x) = F_0(x) + C \qquad\qquad (C \text{ a constant}).$$

Example 3. A particle P moves along a line in such a way that its velocity at time t is given by

$$v = 4t^2 + 2t - 3.$$

At time $t = 1$, P is at the point 3. Where is P at time $t = 5$?

Solution. Since the velocity v is the derivative of the position s, s is an antiderivative of v. Therefore

$$s = \tfrac{4}{3}t^3 + t^2 - 3t + C,$$

where C is a suitable constant. To find it we refer to the condition that $s = 3$ when $t = 1$. This gives us

$$3 = \tfrac{4}{3} + 1 - 3 + C, \qquad C = 3\tfrac{2}{3}.$$

Therefore,

$$s = \tfrac{4}{3}t^3 + t^2 - 3t + 3\tfrac{2}{3}.$$

When $t = 5$, we get $s = 180\tfrac{1}{3}$.

Problems (§4F)

1. For each of the following functions, find the antiderivative whose value at 0 is 1.

(a) $x^2 - 3x + 4$

(b) $x^{3/2} + 2x^2$

(c) $5x^4 + 3x^2 + 2x + 1$

(d) $\dfrac{1}{(x + 1)^2}$

(e) $\dfrac{x}{\sqrt{x^2 + 1}}$

(f) $\dfrac{x^2}{\sqrt{x^2 + 1}} + \sqrt{x^2 + 1}$

2. Find the function F for which:
   (a) $F'(x) = x^2 + 2x - 4$, $F(2) = 1$
   (b) $F'(x) = 1/x^2$, $F(1) = 2$
   (c) $F'(x) = 2x - 1$, $F(5) = 6$

   (d) $F'(x) = \dfrac{1}{x^3} + x$, $F(3) = 1$

   (e) $F'(x) = 3(x - 1)^2(x + 1)^2 + 2(x - 1)(x + 1)^3$, $F(1) = 1$

   (f) $F'(x) = \dfrac{x}{(x^2 + 1)^2}$, $F(0) = 2$

3. Find the function G for which:
   (a) $G''(x) = x - 2$, $G'(1) = 2$, $G(2) = 3$
   (b) $G''(x) = 16$ for all x, $G'(0) = 0$, $G(0) = 0$
   (c) $G''(x) = 6x + 1$, $G(0) = 1$, $G(1) = 0$
   (d) $G''(x) = 1/x^3$, $G(1) = 1$, $G(2) = 0$
   (e) $G'''(x) = 2x$, $G''(0) = 1$, $G'(1) = \frac{4}{3}$, $G(0) = 2$
   (f) $G'''(x) = 2x$, $G'(0) = 1$, $G(0) = 1$, $G(1) = \frac{1}{12}$
   (g) $G'''(x) = 2x$, $G(0) = 0$, $G(1) = -\frac{1}{4}$, $G(2) = \frac{2}{3}$

4. Let f and g be functions such that $f'''(x) = g'''(x)$ for all x, $f''(0) = g''(0)$, $f'(0) = g'(0)$, and $f(0) = g(0)$. Show that $f(x) = g(x)$ for all x.

5. (a) Is there a function that has exactly one antiderivative?
   (b) How do the antiderivatives of the constant function $f(x) = \frac{1}{4}$ compare with those of the constant function $g(x) = -4$?
   (c) Two functions have identical derivatives on $\mathscr{R}$. Their values at $x = 0$ differ by 6. Compare their values at $x = 1$.
   (d) If two different functions have identical derivatives, in how many points can their graphs intersect?
   (e) If two different functions have identical second derivatives, in how many points can their graphs intersect?

6. What are the antiderivatives of $|x|$?

7. A particle P moves on a line so that its velocity at time t is $3t^2 - 2t + 1$. At time $t = 1$, it is at the point 2. Where is P at time $t = 3$?

8. The acceleration of a particle P moving on a line is given by $a = 3t^2 - 2t + 1$. At time $t = 1$, P is at the point 2 with a velocity of 4. Where is P at time $t = 3$ and what is its velocity then?

9. A particle at rest is subjected to a constant acceleration which moves it 60 feet in 4 seconds. What is the acceleration?

10. An automobile is traveling on a straight road at 88 feet per second (60 mph). What constant (negative) acceleration is required to stop the car in 40 feet?

11. A point P moves on a line so that its velocity at time t is $2t - 7$. The minimum value of the coordinate of P is $\frac{3}{4}$. What is the equation of motion of P?

12. A point P moves on a line so that its acceleration at time t is $3t^2 - 4t + 6$. Its coordinate is 0 at $t = 0$ and its coordinate is a minimum at $t = 1$. What is the minimum coordinate?

Answers to problems (§4F)

1. (a) $\frac{1}{3}x^3 - \frac{3}{2}x^2 + 4x + 1$
2. (a) $\frac{1}{3}x^3 + x^2 - 4x + \frac{7}{3}$
3. (a) $\frac{1}{6}x^3 - x^2 + \frac{7}{2}x - \frac{4}{3}$
7. 22
8. at $\frac{70}{3}$ with velocity 24
9. $7\frac{1}{2}$ ft./sec./sec.
10. -96.8 ft./sec./sec.
11. $s = t^2 - 7t + 13$
12. $-\frac{29}{12}$

§4G. The Inverse-Function Theorem

The result proved in this section will not be needed until the development of the exponential function (Chapter 7) and the trigonometric functions (Chapter 8).

Let $x = f(y)$ be differentiable on an interval I, and suppose that f' is never 0 in the interval.

Then f has a differentiable inverse, and

(1) $$(f^{-1})'(x_0) = \frac{1}{f'(y_0)}, \qquad where \qquad y_0 = f^{-1}(x_0).$$

Briefly,

(2) $$\frac{dy}{dx} = \frac{1}{dx/dy}.$$

Proof. Since f' is never 0 on I, f has a continuous inverse (§4A4). Therefore as $\Delta x \to 0$, $\Delta y \to 0$. Then

(3) $$\frac{\Delta x}{\Delta y} \to \frac{dx}{dy} \qquad\qquad (\text{as } \Delta x \to 0)$$

and we have

(4) $$\frac{\Delta y}{\Delta x} = \frac{1}{\Delta x/\Delta y} \to \frac{1}{dx/dy} \qquad\qquad (\text{as } \Delta x \to 0).$$

This is what we were to prove.

The thrust of the theorem is that $y = f^{-1}(x)$ is differentiable. To find dy/dx, one may write the equation $y = f^{-1}(x)$ in the form $f(y) = x$, differentiate implicitly with respect to x, and solve for dy/dx.

Example 1. Assuming the formula $d(x^2)/dx = 2x$, one can derive the formula $d\sqrt{x}/dx = 1/2\sqrt{x}$ as follows. The function

$$y = g(x) = \sqrt{x} \qquad\qquad (x \geq 0)$$

is the inverse of

$$x = f(y) = y^2 \qquad\qquad (y \geq 0);$$

and

$$\frac{dx}{dy} = 2y \neq 0 \qquad\qquad (y > 0).$$

By the theorem, dy/dx exists for $x > 0$ and

(5) $$\frac{dy}{dx} = \frac{1}{2y} = \frac{1}{2\sqrt{x}} \qquad\qquad (x > 0).$$

As mentioned above, the point of the theorem is that dy/dx exists. Knowing this, we may also write

$$y^2 = x$$

and differentiate implicitly to get

$$2y\,\frac{dy}{dx} = 1,$$

from which (5) follows.

CHAPTER 5

The Integral

Calculus is divided into *differential* calculus and *integral* calculus. We have been doing differential calculus. Now we start integral calculus. Its level of sophistication is a little higher.

The problem is introduced in §5A in terms of area. The integral itself is defined in §5B. The definition given is a departure from the usual one. (The *integral* is the same; just the form of the definition is different.) We "define" the integral by stating the properties we want it to have. This type of definition is probably new, so that it requires special attention.

§5C explains the relation between integrals and derivatives. This is the heart of the chapter. Be sure to learn and understand the results. Study the proofs, too. True, one could omit them and still learn how to do calculus problems. But we recommend that these particular proofs be studied very carefully—and for a very practical reason. *Studying these proofs helps learn what the results mean.* This is often the case in mathematics, and is particularly so here.

§§5D–5E are devoted to the construction of the integral and to a number of short but significant comments. The portions set in small type are technical and can be omitted with no loss of continuity.

§5F explains the method of integration by substitution, the principal procedure for transforming complicated integrals into simple ones.

§5G is a very brief introduction to numerical methods of evaluating integrals.

The final section, §5H, discusses other forms of the definition of the integral. It is set in small type and can be skipped without interrupting the main flow of ideas.

§5A. Area

5A1. The problem. We illustrate the idea of integration in an important case that is easy to picture. Let f be continuous and nonnegative on an interval $[a, b]$ (Figure 1). *What is the area of the region bounded by the graph of f, the x-axis, and the vertical lines at a and b?*

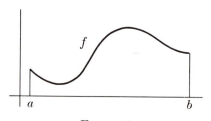

FIGURE 1

The question is far more difficult than it may appear. We cannot start in computing, because we don't know yet what to compute. We need a *definition*. In geometry, we handled areas of rectangles, then parallelograms, then triangles, then arbitrary polygons, and, finally (by a limiting process), circles—but never anything like this.

There are two things to be done: *define* the concept of area; *compute* the area. It is one of the triumphs of calculus that we are able to solve this problem in an extremely neat way, which includes an efficient method for computing areas of fairly complicated regions. Moreover, the underlying idea turns out to have an immense range of application.

5A2. Properties of area. The areas of familiar, simple regions have some obvious properties and we want the extended definition of area to be consistent with them. The fundamental property is that the area of a region cut into two parts is the sum of the areas of the parts. Another (which follows from it) is that the area of any region is greater than that of any within it. These properties will guide us in extending the definition.

5A3. Area as a function. We wish to define the area under the graph of f between a and b (Figure 1). So we stare at the figure. Progress occurs when we look at it in pieces and think about the area between *any* two points u and v ($u \leq v$). Let us call this area A_u^v (Figure 2). Area is now not just a number but a function of two variables, u and v.

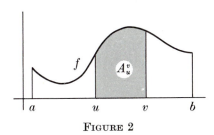

FIGURE 2

The area between a and u, plus the area between u and v, must be the area between a and v:

(A) $$A_a^u + A_u^v = A_a^v.$$

Property **(A)** is called *Additivity*. (See Figure 3.)

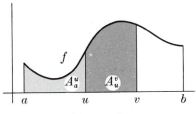

FIGURE 3

The area A_u^v is greater than the area of any rectangle erected on $[u, v]$ and lying under the graph of f. The largest of these is the one that just touches the graph at its lowest point (Figure 4). Its altitude is

$$\min_{[u,v]} f,$$

the minimum value of f on $[u, v]$ (§2E7). Its area is

$$(1) \qquad\qquad\qquad (v - u) \min_{[u,v]} f,$$

and this is $\leq A_u^v$. Similarly, the smallest rectangle enclosing the graph is the one with altitude

$$\max_{[u,v]} f,$$

the maximum value of f on $[u, v]$. Its area is

$$(2) \qquad\qquad\qquad (v - u) \max_{[u,v]} f,$$

and this is $\geq A_u^v$. Putting these facts together, we get:

$$\textbf{(B)} \qquad\qquad (v - u) \min_{[u,v]} f \leq A_u^v \leq (v - u) \max_{[u,v]} f.$$

Property **(B)** is called *Betweenness*.

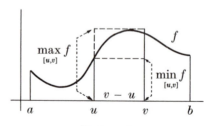

FIGURE 4

We are trying to define the area function A_u^v. Whatever it is, we want it to satisfy **(A)** and **(B)**. We will see later that there *is* a function A_u^v that does this; moreover, there is only one. Let us grant this for the moment. We might then decide to *define* A_u^v as *that function that satisfies* **(A)** *and* **(B)**.

This may seem an unorthodox way to "define" something. More to the point, it may seem futile, like saying to ourselves that the answer to our question is whatever it is that answers the question. As it happens, it is exactly how we are going to give the definition. And it will lead to an unexpectedly simple and powerful procedure for computing results.

Example 1. Let $f(x) = \frac{1}{3}x^2$ on $[0, 3]$. Verify that the function

$$A_u^v = \frac{1}{9}(v^3 - u^3)$$

satisfies properties **(A)** and **(B)**.

Solution. First we verify **(A)**. This takes the form:

(3) $A_0^u + A_u^v = A_0^v.$

Since $A_0^u = \frac{1}{9}u^3$, $A_u^v = \frac{1}{9}(v^3 - u^3)$, and $A_0^v = \frac{1}{9}v^3$, this is clear.

In preparation for **(B)** note that

$$\min_{[u,v]} f = \frac{1}{3}u^2 \qquad \text{and} \qquad \max_{[u,v]} f = \frac{1}{3}v^2,$$

since f is increasing. Thus we are to verify:

(4) $(v - u)(\frac{1}{3}u^2) \le \frac{1}{9}(v^3 - u^3) \le (v - u)(\frac{1}{3}v^2).$

For $u = v$ this is obvious. For $u < v$ we consider the equivalent inequalities obtained by clearing fractions and dividing through by $v - u$:

$$3u^2 \le v^2 + vu + u^2 \le 3v^2.$$

And these are clear from the fact that $0 \le u \le v$.

Comments. In the following sections we will see how the function $\frac{1}{9}(v^3 - u^3)$ was arrived at and why it is the only possible choice for A_u^v. According to this result, the area under the graph of $f(x) = \frac{1}{3}x^2$ between 0 and 3 is

$$A_0^3 = (\frac{1}{9})(3^3) = 3.$$

It is worth while to check this against a rough estimate. The areas of the triangle and trapezoid shown in Figure 5 are $(\frac{1}{2})(\frac{3}{2})(\frac{3}{4}) = \frac{9}{16}$ and $(\frac{1}{2})(\frac{3}{2})(\frac{3}{4} + 3) = \frac{45}{16}$, with total area $\frac{9}{16} + \frac{45}{16} = 3\frac{3}{8}$. This shows that the result $A_0^3 = 3$ is at least in the right ball park.

Problems (§5A)

1. In each of the following cases, verify that A_u^v satisfies **(A)** and **(B)**.

(a) $f(x) = 6$ on $[a, b] = [-1, 2]$, $A_u^v = 6v - 6u$

(b) $f(x) = 2x$ on $[a, b] = [2, 5]$, $A_u^v = v^2 - u^2$

(c) $f(x) = \dfrac{1}{x^2}$ on $[a, b] = [1, 2]$, $A_u^v = \dfrac{1}{u} - \dfrac{1}{v}$

(d) $f(x) = \sqrt{x}$ on $[a, b] = [1, 3]$, $A_u^v = \frac{2}{3}(v\sqrt{v} - u\sqrt{u})$ [*Hint.* The algebra in **(B)** is easier if u and v are replaced by s^2 and t^2.]

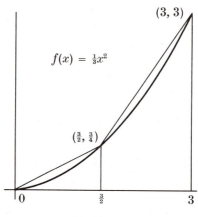

FIGURE 5

2. In each case in Problem 1, compute A_a^b. Then, as a check, estimate (or compute) the area under the graph of f between a and b by means of triangles, rectangles, or trapezoids, as in Example 1.

Answers to problems (§5A)

2. (a) 18   (b) 21   (c) $\frac{1}{2}$ (d) $2\sqrt{3} - \frac{2}{3}$

§5B. The integral

5B1. Definition of the integral of a continuous function. It is truly amazing how often the conditions **(A)** and **(B)** are relevant to problems in widely separate fields. The mathematical abstraction developed in response to these conditions is called the *integral*.

Let f be a continuous function defined on a closed interval $[a, b]$. (We no longer require that f be nonnegative.) To each pair of numbers u and v, where $a \le u \le v \le b$, we wish to associate a number,

(1)
$$\int_u^v f,$$

"the integral of f from u to v," in such a way that **(A)** and **(B)** are satisfied. That is, we wish to have:

(A) (Additivity)
$$\int_a^u f + \int_u^v f = \int_a^v f,$$

and

(B) (Betweenness) $\quad (v - u) \min_{[u,v]} f \leq \int_u^v f \leq (v - u) \max_{[u,v]} f.$

We will find that such an association can be made, and, moreover, in just one way. (This justifies our speaking of "the" integral.)

All this is pretty abstract. A good way to keep one's feet on the ground is to reread the discussion of area in §5A.

Let us recapitulate. We are introducing the integral as a function of the two variables u and v. The definition of the integral is this:

(2) $\qquad \int_u^v f \quad$ *is the unique function satisfying* **(A)** *and* **(B)**

(where f is continuous on an interval $[a, b]$ and $a \leq u \leq v \leq b$).

In the course of this chapter and the next, we will see that there *is* an integral and that there is only one, we will learn how to formulate problems in terms of integrals, and we will develop elegant and incisive methods for solving them.

5B2. *Terminology.* In the expression

$$\int_u^v f,$$

the symbol $\int$ is an "integral sign." It is a stylized letter "S," chosen because of the similarity (that will become apparent) between an integral and a *sum*. The function f is the *integrand*, the thing being *integrated*. The process of finding the integral is *integration*. The numbers u and v are the *limits of integration*; u is the lower limit, v the upper. (These have nothing to do with the *limit* of a function.)

Another notation for the integral, the "*d*-notation," is

(3) $\qquad\qquad\qquad\qquad \int_u^v f(x) \, dx.$

For example, if $f(x) = x^2/3$ then

(4) $\qquad\qquad\qquad \int_0^3 f \qquad$ and $\qquad \int_0^3 \tfrac{1}{3}x^2 \, dx$

mean the same thing. The x in (3) and (4) is called the "variable of integration." We could just as well use, say t:

$$\int_u^v f = \int_u^v f(x)\, dx = \int_u^v f(t)\, dt,$$

$$\int_0^3 \tfrac{1}{3}x^2\, dx = \int_0^3 \tfrac{1}{3}t^2\, dt.$$

The symbol dx (or dt) is a device for keeping track of the variable. Its significance will appear as we proceed.

5B3. Applications. It is truly amazing, we repeat, how often conditions **(A)** and **(B)** are relevant to problems in widely separate fields. Their relevance is decided *within the field of application*, whether mathematical or other. *Mathematics* then tells us that the function in question is an integral.

We give two illustrations here. Right now we cannot treat them adequately, as we have no techniques yet for *computing*. After we have acquired some, we will explore a wide range of applications.

5B4. Area under a graph. Let f be continuous and nonnegative on an interval $[a, b]$. We wish to define the area under the graph of f between a and b.

Our work has already been done. We saw in §5A3 that the area function A_u^v must satisfy the conditions **(A)** and **(B)**. Now, there is exactly one function—by definition, the integral, $\int_u^v f$—that satisfies **(A)** and **(B)**. This forces us to define the area from a to b as the integral:

(5)
$$A = \int_a^b f.$$

Memory aid. See Figure 1. The area between x and $x + \Delta x$ is approximately equal to the area of the rectangle of height $f(x)$ and width Δx, i.e., to

$$f(x)\, \Delta x.$$

The actual area from a to b is

$$\int_a^b f(x)\, dx.$$

Example 1. The area under the graph of $f(x) = \tfrac{1}{3}x^2$ between 0 and 3 is

$$\int_0^3 \tfrac{1}{3}x^2\, dx.$$

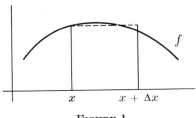

FIGURE 1

5B5. Work. Suppose a *constant* force acts along a line for a certain distance. By definition, the *work* done is the product of the force and the distance.

Now suppose that a force that varies *continuously* with position acts along a line between points a and b ($a \leq b$). The force at point x is, say, $f(x)$. How should the amount of work done be defined in this case?

Let W_u^v denote the work done between u and v (where $a \leq u \leq v \leq b$).

The work done between a and u, plus that between u and v, must equal the work done between a and v. This is an observation about *work*. In symbols,

(A) $$W_a^u + W_u^v = W_a^v.$$

The work done between u and v is at least as much as that done by a constant force with value $\min_{[u,v]} f$—namely, $(v - u) \min_{[u,v]} f$; and it is at most as much as that done by a constant force $\max_{[u,v]} f$—namely, $(v - u) \max_{[u,v]} f$. This insistence that a larger force operating over the same distance yields more work is, again, an observation about *work*. We have:

(B) $$(v - u) \min_{[u,v]} f \leq W_u^v \leq (v - u) \max_{[u,v]} f.$$

The requirements **(A)** and **(B)** force us to define the work done between a and b as the integral:

(6) $$W = \int_a^b f.$$

Memory aid. The work done between x and $x + \Delta x$ is approximately equal to that done by a constant force, with constant value $f(x)$, acting over the distance Δx; i.e., to

$$f(x)\, \Delta x.$$

The actual work done from a to b is

$$\int_a^b f(x)\, dx.$$

Example 2. Suppose a particle moves along the x-axis in such a way that at each point x it must overcome a force of $2x^2 + 1$ units. Then the total work done in moving from $x = 1$ to $x = 4$ is

$$\int_1^4 (2x^2 + 1)\, dx.$$

Problems (§5B)

1. Assume that f has an integral on $[1, 7]$ and that

$$\int_1^5 f = 3, \qquad \int_2^3 f = 1, \qquad \int_3^5 f = 1, \qquad \int_3^7 f = 6.$$

Evaluate:

(a) $\displaystyle\int_1^3 f$

(b) $\displaystyle\int_2^5 f$

(c) $\displaystyle\int_2^7 f$

(d) $\displaystyle\int_5^7 f$

(e) $\displaystyle\int_1^2 f$

(f) $\displaystyle\int_1^7 f$

2. Assume that g has an integral on $[1, 9]$ and that

$$\int_1^4 g = 1, \qquad \int_2^4 g(r)\, dr = 2, \qquad \int_2^6 g(s)\, ds = 0, \qquad \int_6^9 g(t)\, dt = 1.$$

Evaluate:

(a) $\displaystyle\int_1^2 g(u)\, du$

(b) $\displaystyle\int_2^9 g(v)\, dv$

(c) $\displaystyle\int_4^6 g(w)\, dw$

(d) $\displaystyle\int_1^9 g(x)\, dx$

(e) $\displaystyle\int_4^9 g(y)\, dy$

(f) $\displaystyle\int_1^6 g(z)\, dz$

3. Express each of the following as an integral (but do not attempt to evaluate it).
   (a) The area under the graph of $(x^2 + 1)^3$ between $x = 0$ and $x = 2$.
   (b) The area under the graph of $(t^2 + 1)^3$ between $t = 0$ and $t = 2$.
   (c) The area under the graph of $1/x$ between $x = 1$ and $x = 4$.
   (d) The work done in stretching a 12-inch spring to 15 inches, if the number of pounds of force to be overcome at any point is 50 times the number of inches of stretch at that point.

(e) The work done in moving a particle from 6 to 8 on the axis, if at every point the particle must overcome a force inversely proportional to the square of its distance from the point 2.

(f) The work done in moving a particle from 6 to 8 on the axis, if at every point the particle is repelled by a force inversely proportional to the square of its distance from the point 12.

4. Assume that f has an integral on an interval $[a, b]$. Prove that

$$\int_u^u f = 0$$

for each number u in $[a, b]$:

(a) By using (A).
(b) By using (B).

5. Assume that f and g each has an integral on $[a, b]$. Show that if

$$\max_{[a,b]} f \le \min_{[a,b]} g,$$

then $\int_a^b f \le \int_a^b g$.

6. Show that if f has the constant value k on $[a, b]$, then f has a unique integral on $[a, b]$ and

$$\int_a^b f = k(b - a).$$

Answers to problems (§5B)

1. (a) 2   (b) 2   (c) 7   (d) 5   (e) 1   (f) 8
2. (a) -1 (b) 1   (c) -2 (d) 0   (e) -1 (f) -1

3. (a) $\displaystyle\int_0^2 (x^2 + 1)^3 \, dx$ (b) $\displaystyle\int_0^2 (t^2 + 1)^3 \, dt \left[= \int_0^2 (x^2 + 1)^3 \, dx \right]$

(c) $\displaystyle\int_1^4 \frac{1}{x} \, dx$ (d) $\displaystyle\int_0^3 50 \, s \, ds$

(e) $\displaystyle\int_6^8 \frac{k}{(x - 2)^2} \, dx$ (f) $\displaystyle\int_6^8 \frac{k}{(12 - x)^2} \, dx$

§5C. Integrals and antiderivatives

5C1. Introduction. One of the milestones in the history of scientific thought was the realization that integration is related to differentiation and, moreover, in a very simple and important way: the processes are inverses of one another. In other words, integrals are essentially the same thing as antiderivatives (§4F). As a result, we can evaluate the integral of any continuous function of which we can find an antiderivative.

The precise relationship between antiderivatives and integrals is justly called the *Fundamental Theorem of Calculus*. We present it in two stages. In the present section (§5C) we show that *if* a continuous function has *either* an integral *or* an antiderivative, then it has both (and they are essentially the same); moreover, the integral is unique. In the following section (§5D) we show that *every* continuous function *does* have an integral (which is then also an antiderivative).

For convenience of reference we restate the conditions of *Additivity* and *Betweenness*:

(A)   (Additivity)
$$\int_a^u f + \int_u^v f = \int_a^v f$$

(B)   (Betweenness) $(v - u) \min_{[u,v]} f \leq \int_u^v f \leq (v - u) \max_{[u,v]} f$

(where f is continuous on $[a, b]$ and $a \leq u \leq v \leq b$).

5C2. The antiderivative as the integral. *Let f be continuous on $[a, b]$. If f has an antiderivative then it has an integral, namely,*

(1)
$$\int_u^v f = F(v) - F(u) \qquad (a \leq u \leq v \leq b),$$

where F is any antiderivative of f on $[a, b]$.

Proof. We verify that the function (1) satisfies **(A)** and **(B)**.
To verify **(A)** we are to check that

(2) $[F(u) - F(a)] + [F(v) - F(u)] = F(v) - F(a).$

This is obvious.
To verify **(B)** we must establish:

(3) $(v - u) \min_{[u,v]} f \leq F(v) - F(u) \leq (v - u) \max_{[u,v]} f.$

To do this we apply the Mean-Value Theorem (§4E1) to the function F on the interval $[u, v]$. It is applicable since F is differentiable there: $F' = f$, by hypothesis. We get

(4) $F(v) - F(u) = (v - u)F'(z),$

or, in other words,

$$(5) \qquad\qquad F(v) - F(u) = (v - u)f(z),$$

for some $z \in [u, v]$. Statement (3) is now obvious.

Remark. We have not yet shown that f has only one integral. But we can see quickly that all integrals obtained by this method, at least, are equal. For, if G is any other antiderivative of f then $G - F$ is constant (§4F1(a)), so that

$$(6) \qquad\qquad G(v) - F(v) = G(u) - F(u).$$

Therefore,

$$(7) \qquad\qquad G(v) - G(u) = F(v) - F(u).$$

Thus, (1) comes out the same no matter which antiderivative is used.

5C3. Applications. Before continuing with the proof of uniqueness of the integral in general, we pause for some examples.

The numerical value of the integral from a to b is given by

$$(8) \qquad\qquad \int_a^b f = F(b) - F(a)$$

(where F is any antiderivative of f). This formula enables us to evaluate the integral of any continuous function for which we know an antiderivative.

Example 1. What is the area under the graph of $f(x) = \frac{1}{3}x^2$ on the interval $[0, 3]$? (This example was previously discussed in §5A, Example 1 and §5B, Example 1.)

Solution. The function

$$F(x) = \tfrac{1}{9}x^3$$

is an antiderivative of f. Therefore the area is given by

$$\int_0^3 f = F(3) - F(0) = \tfrac{1}{9}(3^3 - 0^3) = 3.$$

Notation. The following convention is handy:

$$(9) \qquad\qquad F(x)\Big|_a^b = F(b) - F(a).$$

Example 2. In Example 1, the area is given by

$$\int_0^3 \tfrac{1}{3}x^2 \, dx \;=\; \tfrac{1}{9}x^3 \Big|_0^3 \;=\; \tfrac{1}{9}(3^3 - 0^3) \;=\; 3.$$

Example 3. How much work is done in moving a particle 4 units along the axis if the force applied at each point is proportional to the distance already traveled?

Solution. If we start at the origin, then at distance x, the force is kx, where k is the constant of proportionality. Therefore, the work done is

$$\int_0^4 kx \, dx \;=\; \tfrac{1}{2}kx^2 \Big|_0^4 \;=\; 8k.$$

5C4. *The integral as a function of its upper limit.* The integral $\int_u^v f$ is a function of the two variables u and v, its lower and upper limits. In order to establish the uniqueness of the integral, we have to study how the integral varies when we hold the lower limit fixed at a and vary the upper limit alone. For emphasis let us here denote the upper limit by x. Thus, we are to investigate

(10) $$\int_a^x f,$$

a function of the single variable x.

Let us try to guess how $\int_a^x f$ behaves. To keep things simple assume that f is a positive function so that the integral has a ready interpretation as an area. If x is increased by a small amount Δx, $\int_a^x f$ is increased by the area of the shaded column (Figure 1). Approximately, this column is a rectangle of height $f(x)$, so that its area is proportional to $f(x)$. This suggests that the rate

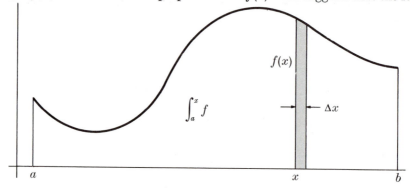

FIGURE 1

of change of $\int_a^x f$ is equal (or proportional) to $f(x)$. We now prove that this is always so (for any continuous f, not necessarily positive).

5C5. The integral as an antiderivative. *Let f be continuous on $[a, b]$. If f has an integral then it has an antiderivative, namely,*

$$(11) \qquad\qquad \int_a^x f \qquad\qquad (x \in [a, b]).$$

That is to say,

$$(12) \qquad\qquad \frac{d}{dx} \int_a^x f = f(x).$$

Proof. If we stare at **(A)** for a while we notice that two of the integrals there are of the form we are considering: $\int_a^x f$. This suggests solving for $\int_u^v f$ in **(A)** and substituting the result into **(B)**. When we do this we obtain

$$(13) \qquad (v - u) \min_{[u,v]} f \le \int_a^v f - \int_a^u f \le (v - u) \max_{[u,v]} f.$$

To bring this into sharper focus let us introduce

$$(14) \qquad\qquad F_0(x) = \int_a^x f.$$

Then (13) becomes

$$(15) \qquad (v - u) \min_{[u,v]} f \le F_0(v) - F_0(u) \le (v - u) \max_{[u,v]} f.$$

This holds for $u \le v$, but let us keep $u < v$. Then we can divide by $v - u$ so as to obtain a difference quotient:

$$(16) \qquad \min_{[u,v]} f \le \frac{F_0(v) - F_0(u)}{v - u} \le \max_{[u,v]} f.$$

To make it look more familiar, write $x = u$ and $\Delta x = v - u$. Then $v = x + \Delta x$ and we have

$$(17) \qquad \min_{[x,x+\Delta x]} f \le \frac{F_0(x + \Delta x) - F_0(x)}{\Delta x} \le \max_{[x,x+\Delta x]} f.$$

190 *The Integral* 5C6

Now as $\Delta x \to 0$ the two outside quantities both approach $f(x)$ (since f is continuous). The difference quotient between them is squeezed to the same limit (§2C3). Consequently, $F_0'(x) = f(x)$. This is what we wished to prove. (The notation here applies for $\Delta x > 0$. The case $\Delta x < 0$ is similar.)

5C6. Uniqueness of the integral. *Let f be continuous on $[a, b]$. If f has an integral it has only one.*

Proof. For any integral, $\int_u^v f$, the function

$$(18) \qquad F_0(x) = \int_a^x f$$

is an antiderivative of f, as we have just seen (§5C5). We shall now show that every integral leads to the *same* function $F_0(x)$. To do this we compute the number

$$(19) \qquad F_0(a) = \int_a^a f.$$

By **(A)**, as before,

$$(20) \qquad \int_u^v f = F_0(v) - F_0(u).$$

Therefore,

$$(21) \qquad F_0(a) = \int_a^a f = F_0(a) - F_0(a) = 0.$$

This shows that F_0 has the same value at a (namely, 0), no matter what integral we started with.

Thus, for every integral, F_0 is an antiderivative of f whose value at a is 0. As we know, these conditions determine the function F_0 uniquely (§4F1(b)). In turn, by (20), $\int_u^v f$ is determined uniquely.

Remark. We now know that if a continuous function has either an antiderivative or an integral, then it has both, and the integral is unique.

Problems (§5C)

1. Let $F(x) = 2x^2 - 3$ and $G(x) = 2x^2 + 1$. Evaluate:
   (a) $F(x)|_1^4$ and $G(x)|_1^4$
   (b) $F(x)|_0^5$ and $G(x)|_0^5$

2. Evaluate the following integrals.

(a) $\displaystyle\int_{-2}^{3} 2x \, dx$

(k) $\displaystyle\int_{1}^{3} (v^3 + 2) \, dv$

(b) $\displaystyle\int_{-3}^{2} x^2 \, dx$

(l) $\displaystyle\int_{1}^{2} (x^2 + x + 1 + x^{-2}) \, dx$

(c) $\displaystyle\int_{0}^{2} t^3 \, dt$

(m) $\displaystyle\int_{-4}^{6} 0 \, dx$

(d) $\displaystyle\int_{0}^{1} x^4 \, dx$

(n) $\displaystyle\int_{-1}^{1} 4x^3 \, dx$

(e) $\displaystyle\int_{4}^{9} r\sqrt{r} \, dr$

(o) $\displaystyle\int_{-1}^{1} 5x^4 \, dx$

(f) $\displaystyle\int_{2}^{5} \left(u + \frac{1}{u^2} \right) du$

(p) $\displaystyle\int_{2}^{2} (x^{31} - 3x^{1/12}) \, dx$

(g) $\displaystyle\int_{1}^{4} \frac{1}{\sqrt{s}} \, ds$

(q) $\displaystyle\int_{0}^{1} 5(x^2 + 1)^4 2x \, dx$

(h) $\displaystyle\int_{-3}^{2} 1 \, dx$

(r) $\displaystyle\int_{-7}^{7} 16(x^2 + 1)^{15} 2x \, dx$

(i) $\displaystyle\int_{3}^{10} 6 \, dy$

(s) $\displaystyle\int_{0}^{1} \left(\frac{x}{2\sqrt{x + 1}} + \sqrt{x + 1} \right) dx$

(j) $\displaystyle\int_{-3}^{2} (3x^2 + 2x + 1) \, dx$

3. If $F' = f$, and if

$$\int_{1}^{5} f = 3, \qquad \int_{2}^{6} f = 4, \qquad \int_{5}^{6} f = 5,$$

and $F(1) = 6$, what is $F(2)$?

4. If f' is continuous, and if the graph of f passes through the points $(1, 2)$ and $(3, 5)$, what is

$$\int_{1}^{3} f' \, ?$$

5. Evaluate

$$\int_0^1 \left[\frac{d}{dx} \sqrt{1 + x^3} \right] dx.$$

6. Find the areas under the graphs of the following functions.
   (a) $f(x) = 1/x^2$, on the interval $[1, 2]$
   (b) $f(x) = \sqrt{x}$, on $[1, 3]$
   (c) $f(x) = (x^2 + 1)^3$, on $[0, 2]$
   (d) $f(x) = x^4 + x^2$, on $[-3, 2]$
   (e) $f(x) = x^2(x^3 + 1)^4$, on $[-1, 1]$
   (f) $f(x) = (2x + 1)(x^2 + x + 2)^2$, on $[-2, 1]$ $[0, 1]$
   (g) $f(x) = (x + 1)(x^2 + 2x + 2)^2$, on $[-2, 1]$ $[-1, 1]$

7. Find the work done in stretching a 12-inch spring to 15 inches, if the number of pounds of force to be overcome at any point is 50 times the number of inches of stretch at that point.

8. Find the work done in moving a particle from 6 to 8 on the axis, if at every point the particle must overcome a force inversely proportional to the square of its distance:
   (a) from the point 2.
   (b) from the point 12.

9. To compress the spring on a certain railroad bumping post requires a force of 120 tons per foot of compression. How much work is done if the spring is depressed one inch?

10. Consider the force required to hold a thin rod, or spring, of natural length L at a stretched length $L + x$. According to *Hooke's law*, this force is equal to kx, where k is a constant (called the "modulus" of the rod or spring). (It is assumed that x does not exceed the "elastic limit" of the rod.)
   (a) A weight of 3 pounds stretches a spring 2 inches beyond its natural length. How much work is done in stretching the spring one foot beyond its natural length?
   (b) Suppose that holding a spring of natural length 10 feet at a length of 12 feet requires a force of 12 pounds. How much work is done in stretching the spring from a length of 12 feet to a length of 14 feet?

11. Two charged particles repel each other with a force equal to k/s^2, where s is the distance between them and k is a constant. How much work is done in moving the particles from a distance of 3 units to a distance of 1 unit?

12. An anchor weighing 100 pounds must be lifted 20 feet. How much work is done if the weight of the anchor cable is 1 pound per foot?

13. A cable weighing 2 pounds per foot is being unwound from a cylindrical drum under the force of gravity.
   (a) How much work is done by gravity during the first 50 feet? (Assume frictionless rotation of the drum.)
   (b) How much work is done during the *second* 50 feet?

14. An object of weight w, y units above the surface of the earth, is attracted toward the earth by a force of

$$w \left(\frac{r}{r + y} \right)^2 ,$$

where r is the radius of the earth. (We are regarding the earth as a perfect sphere.)

Find the work done in lifting a two-ton rocket from the earth's surface to a height of 100 miles. Take $r = 4000$ miles.

Answers to problems (§5C)

1. (a) 30 and 30   (b) 50 and 50
2. (a) 5   (b) $\frac{35}{3}$ (c) 4   (d) $\frac{1}{5}$ (e) $\frac{422}{5}$ (f) $10\frac{4}{5}$ (g) 2   (h) 5
      (i) 42   (j) 35   (k) 24   (l) $5\frac{1}{3}$ (m) 0   (n) 0   (o) 2   (p) 0
      (q) 31   (r) 0   (s) $\sqrt{2}$
3. 10
4. 3
5. $\sqrt{2} - 1$
6. (a) $\frac{1}{2}$ (b) $2\sqrt{3} - \frac{2}{3}$ (c) $47\frac{17}{35}$ (d) $66\frac{2}{3}$ (e) $\frac{32}{15}$ (f) $\frac{55}{3}$ (g) $\frac{19}{2}$ $\frac{62}{3}$
7. 225 in.-lbs.
8. (a) $k/12$ (b) $k/12$
9. $\frac{5}{12}$ ft.-tons
10. (a) 9 ft.-lbs.   (b) 36 ft.-lbs.
11. $\frac{2}{3}k$
12. 2200 ft.-lbs.
13. (a) 2500 ft.-lbs.   (b) 7500 ft.-lbs.
14. 195 ton-miles (approx.)

§5D. The Fundamental Theorem of Calculus

5D1. Introduction. So far, our knowledge of whether a given continuous function has an integral has depended on our knowledge of antiderivatives. When unable to come up with an antiderivative, we have been in the dark about the existence of the integral.

The *Fundamental Theorem of Calculus* asserts (in part) that *every* continuous function has an antiderivative and an integral. That is to say, there exists a function of which the given one is the derivative; or, equivalently, there is an associated function of two variables that satisfies conditions **(A)** and **(B)**.

Either way, the proof of the theorem will require penetrating insight and extraordinary patience.

5D2. The Fundamental Theorem of Calculus. *Let f be continuous on* [a, b]. *Then:*

(a)   *f has a unique integral.*

(b)
$$\frac{d}{dx}\int_a^x f = f(x)$$
(x ∈ [a, b]).

(c)
$$\int_a^b f = F(b) - F(a),$$

where F is any antiderivative of f on [a, b].

We know from §5C that if *f* has an integral, then (b) and (c) hold and the integral is unique. What we have to show now is that it does have one. To do this we roll up our sleeves and *construct* one.

5D3. Lower and upper sums. Consider a *partition* of an interval [u, v]. By this we mean a set of abutting closed intervals whose union is [u, v] (Figure 1). These intervals are called the *segments* of the partition. For each segment, [s, t], form the products

(1)
$$(t - s) \min_{[s,t]} f$$

and

(2)
$$(t - s) \max_{[s,t]} f$$

(Figure 1). Now add all the products (1), one for each segment of the partition. The result, L, is called a *lower sum* for *f* on [u, v]. The sum, U, of all the products (2) is an *upper* sum. Obviously, $L \leq U$.

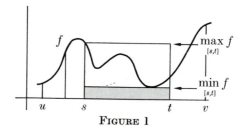

FIGURE 1

If a partition is "refined" by breaking given segments into smaller ones, the lower sum increases (or remains the same) and the upper sum decreases (Figure 2).

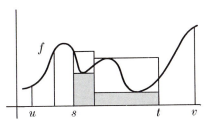

FIGURE 2

Let L and U be any lower and upper sums, not necessarily from the same partition. By pooling all the points of subdivision, from both the given partitions, we get a new partition which refines both. Let L' and U' denote the resulting lower and upper sums. Then, as just noted,

(3) $$L \leq L' \quad \text{and} \quad U' \leq U.$$

But $L' \leq U'$, since they are based on the same partition. It follows that $L \leq U$. We have proved:

(4)   *Every lower sum is $\leq$ every upper sum.*

On the other hand, as we shall now show, there are lower and upper sums as close together as we wish.

(5)   *Given any number $\epsilon > 0$, there exist lower and upper sums L and U such that $U - L < \epsilon$.*

Proof. We work with the number

(6) $$\epsilon' = \frac{\epsilon}{v - u}.$$

Pick a partition of $[u, v]$ such that on each segment, f varies by less than ϵ' (§2F3). Then on each segment, $[s, t]$,

(7) $$\max_{[s,t]} f - \min_{[s,t]} f < \epsilon'.$$

Therefore,

(8) $$(t - s) \max_{[s,t]} f - (t - s) \min_{[s,t]} f < \epsilon'(t - s).$$

Now, the sum of the terms $(t - s) \max_{[s,t]} f$, one for each segment $[s, t]$, is, by definition, U.

Similarly, summing $(t - s) \min_{[s,t]} f$ over all segments gives us L. On the right we get ϵ' times the sum of the lengths, i.e., $\epsilon'(v - u)$. So we have

$$(9) \qquad\qquad U - L < \epsilon'(v - u).$$

Thus, $U - L < \epsilon$. This is what we wanted to prove.

Remark. It follows easily from **(A)** and **(B)** that any integral must be $\geq$ every lower sum and $\leq$ every upper sum. By (5), there cannot be more than one such number. So we have another proof that the integral (if there is one) is unique.

5D4. Construction of the integral. There are many possible partitions of $[u, v]$. Each one yields a lower sum (and an upper sum). A lower sum is a number. Let $\mathscr{L}$ denote *the set of all lower sums* on $[u, v]$.

According to (4), any upper sum is an upper bound of $\mathscr{L}$. By the Axiom of Completeness (§1H2), $\mathscr{L}$ has a *least* upper bound. Denote it

$$(10) \qquad\qquad S_u^v.$$

We will show that S_u^v is the integral, $\int_u^v f$. That is, we will prove that S_u^v satisfies:

(A) $$S_a^u + S_u^v = S_a^v$$

and

(B) $$(v - u) \min_{[u,v]} f \leq S_u^v \leq (v - u) \max_{[u,v]} f.$$

Proof. If L is any lower sum on $[u, v]$, then $L \leq S_u^v$, since S_u^v is an upper bound of $\mathscr{L}$. If U is an upper sum then $S_u^v \leq U$, because S_u^v is the *least* upper bound of $\mathscr{L}$ (and hence is less than any other upper bound). So we have

$$(11) \qquad\qquad L \leq S_u^v \leq U$$

for every lower sum L and every upper sum U on $[u, v]$.

Now we can prove **(B)**. Consider the partition of $[u, v]$ consisting of the single segment $[u, v]$ itself. For this partition,

$$(12) \qquad L = (v - u) \min_{[u,v]} f \qquad \text{and} \qquad U = (v - u) \max_{[u,v]} f.$$

Putting these in (11), we get **(B)**.

To prove **(A)** we show that S_a^v and $S_a^u + S_u^v$ differ by less than any given positive number ϵ; hence they are equal.

Notice that if P_1 is any partition of $[a, u]$ and P_2 is a partition of $[u, v]$, then P_1 and P_2 together form a partition, say P_3, of $[a, v]$ (Figure 3). Moreover, the associated lower and upper sums satisfy

$$(13) \qquad L_1 + L_2 = L_3, \qquad U_1 + U_2 = U_3.$$

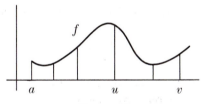

FIGURE 3

By (11),

(14) $$L_1 \le S_a^u \le U_1, \qquad L_2 \le S_u^v \le U_2,$$

and

(15) $$L_3 \le S_a^v \le U_3.$$

On the other hand, the two parts of (14), when added, give

(16) $$L_3 \le S_a^u + S_u^v \le U_3.$$

These relations hold no matter how P_1 and P_2 are chosen. Let us choose them so that $U_1 - L_1 < \epsilon/2$ and $U_2 - L_2 < \epsilon/2$ (see (5)). Then

(17) $$U_3 - L_3 < \epsilon.$$

Since S_a^v and $S_a^u + S_u^v$ both lie between L_3 and U_3, it follows that they differ from each other by less than ϵ. This completes the proof.

5D5. Consequences of the Fundamental Theorem. As a result of the Fundamental Theorem we know that

$$\sqrt{1 + x^3},$$

for example, has an antiderivative—although we have no formula for it. The fact can be useful just the same. We can speak meaningfully of the expression

$$\int_0^1 \sqrt{1 + x^3}\, dx,$$

and we can interpret it as an area, as an amount of work done, and in other ways.

More significantly, we now know that the function $1/x$ $(x > 0)$ has an antiderivative and an integral. This was the case omitted in the power rule. The new function so obtained is sufficiently important to be given a special name and studied in a separate chapter (Chapter 7).

5D6. A problem of notation. The Fundamental Theorem refers to

$$\int_a^x f,$$

a function of the upper limit, x. To express this in the d-notation we use a variable of integration different from x—for example, t:

$$\int_a^x f(t)\ dt.$$

By the Fundamental Theorem,

$$\frac{d}{dx}\int_a^x f(t)\ dt = f(x).$$

For example,

$$\frac{d}{dx}\int_0^x \sqrt{1 + t^3}\ dt = \sqrt{1 + x^3}; \qquad \frac{d}{dx}\int_1^x \frac{1}{t}\ dt = \frac{1}{x}.$$

Problems (§5D)

1. Differentiate with respect to x:

(a) $\displaystyle\int_1^x \sqrt{t^2 + 1}\ dt$

(f) $\displaystyle\int_{-3}^x (t^2 + 1)^{1/3}\ dt$

(b) $\displaystyle\int_0^x \frac{1}{t + 1}\ dt$

(g) $\displaystyle\int_3^x \frac{1}{s}\ ds$

(c) $\displaystyle\int_{-1}^x \frac{t}{(t^2 + 1)^{1/3}}\ dt$

(h) $\displaystyle\int_1^7 (t^3 + 3)^{3/5}\ dt$

(d) $\displaystyle\int_1^x \sqrt{t^4 + 1}\ dt$

(i) $\displaystyle\int_1^7 (x^3 + 3)^{3/5}\ dx$

(e) $\displaystyle\int_3^x \frac{1}{t}\ dt$

2. Evaluate the following derivatives. [*Hint.* Use the chain rule.]

(a) $\displaystyle\frac{d}{dx}\int_1^{x^2} \sqrt{t^8 + 1}\ dt$

(d) $\displaystyle\frac{d}{dt}\int_t^1 (x^2 + 1)\ dx$

(b) $\displaystyle\frac{d}{dx}\int_1^{x^2 - 3x + 1} \frac{1}{t^2 + 1}\ dt$

(e) $\displaystyle\frac{d}{dx}\int_x^{x^2} 3t^2\ dt$

(c) $\displaystyle\frac{d}{dx}\int_3^{\sqrt{x}} (t^3 + 1)^{3/2}\ dt$

(f) $\displaystyle\frac{d}{dy}\int_{y-1}^{y+1} (2x - 1)\ dx$

3. Compute the following integrals.

(a) $\displaystyle\int_0^1 \left[\frac{d}{dx}\sqrt{x^2+1}\right]dx$

(b) $\displaystyle\int_{-1}^1 \left[\frac{d}{dx}\sqrt{x^2+1}\right]dx$

(c) $\displaystyle\int_a^b f'(x)\,dx$ (assuming f' continuous)

4. Evaluate the following integrals.

(a) $\displaystyle\int_{-97}^{97} (x^5 - x^3 + x)\,dx$

(b) $\displaystyle\int_{-1}^2 |x|\,dx$

(c) $\displaystyle\int_0^1 g''(t)\,dt,$ where $\displaystyle g(t) = \int_1^t \sqrt{x^2+1}\,dx$

5. Find $f''(x)$ if

$$f(x) = \int_1^x \left[\int_1^t \frac{1}{y}\,dy\right]dt.$$

Answers to problems (§5D)

1. (a) $\sqrt{x^2+1}$ (b) $\dfrac{1}{x+1}$ (c) $\dfrac{x}{(x^2+1)^{1/3}}$ (d) $\sqrt{x^4+1}$ (e) $\dfrac{1}{x}$

(f) $(x^2+1)^{1/3}$ (g) $\dfrac{1}{x}$ (h) 0 (i) 0

2. (a) $2x\sqrt{x^{16}+1}$ (b) $\dfrac{2x-3}{(x^2-3x+1)^2+1}$ (c) $\dfrac{1}{2\sqrt{x}}(x^{3/2}+1)^{3/2}$

(d) $-(t^2+1)$ (e) $6x^5 - 3x^2$ (f) 4

3. (a) $\sqrt{2}-1$ (b) 0 (c) $f(b)-f(a)$

4. (a) 0 (b) $2\frac{1}{2}$ (c) $\sqrt{2}-1$

5. $\dfrac{1}{x}$

§5E. Properties of the integral; antiderivatives

5E1. Properties of the integral. We can derive properties of integrals via the Fundamental Theorem, from known properties of antiderivatives.

Let f and g be continuous on $[a, b]$.

(a) $\displaystyle\int_a^b (f + g) = \int_a^b f + \int_a^b g.$

Proof. If F and G are antiderivatives of f and g, then both sides are equal to $[F(b) - F(a)] + [G(b) - G(a)]$.

(b) $\displaystyle\int_a^b (f - g) = \int_a^b f - \int_a^b g.$

(c) $\displaystyle\int_a^b cf = c \int_a^b f$ (*c a constant*).

The proofs of these are similar to (a).

(d) *If* $f \geq 0$ *on* $[a, b]$ *then* $\displaystyle\int_a^b f \geq 0.$

Proof. The hypothesis implies that $\min\limits_{[a,b]} f \geq 0$. The result now follows from **(B)**.

(e) *If* $f \geq g$ *on* $[a, b]$ *then* $\displaystyle\int_a^b f \geq \int_a^b g.$

Proof. Since $f - g \geq 0$, we have $\int_a^b (f - g) \geq 0$, by (d). By (b), this is the same as

$$\int_a^b f - \int_a^b g \geq 0,$$

which is what we want.

Example 1. Show that $\int_0^1 \sqrt{1 + x^3}\, dx$ lies between 1 and $1\frac{1}{4}$.

Solution. We do not know an antiderivative of $\sqrt{1 + x^3}$ and so we cannot evaluate the integral by the Fundamental Theorem. But,

$$1 \leq \sqrt{1 + x^3} \leq 1 + x^3 \qquad (x \geq 0),$$

and therefore

$$\int_0^1 1\, dx \leq \int_0^1 \sqrt{1 + x^3}\, dx \leq \int_0^1 (1 + x^3)\, dx,$$

by (e). Since $\int_0^1 1\, dx = 1$ and $\int_0^1 (1 + x^3)\, dx = 1\frac{1}{4}$, this yields the required estimate.

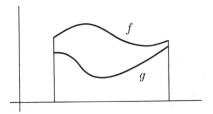

FIGURE 1

5E2. Area between two graphs. Suppose we want the area between two graphs. Let f and g be continuous on $[a, b]$, with $f \geq g$. If g is nonnegative, then the area between the graphs is simply the area under f minus the area under g (Figure 1). Hence $A = \int_a^b f - \int_a^b g$. In view of property (b) above, we may write this more simply as:

$$(1) \qquad A = \int_a^b (f - g).$$

In case g assumes negative values, we take advantage of the fact that, in any event, it has a minimum value on $[a, b]$. Say its minimum value is $-c$. Then we consider the functions $f + c$ and $g + c$ (Figure 2). Since $g + c \geq 0$, we are back to the first case. The area between the graphs is not affected by the upward shift, and we have

$$A = \int_a^b [(f + c) - (g + c)] = \int_a^b (f - g).$$

Thus we again get (1).

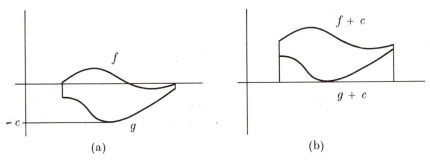

(a)   (b)

FIGURE 2

Example 2. Find the area bounded by the curves

$$y = f(x) = \sqrt{1 + x^3} + 2x - x^2$$

and

$$y = g(x) = \sqrt{1 + x^3} + x^2 - 3x + 2.$$

Solution. The curves intersect at $x = \frac{1}{2}$ and $x = 2$. Between these points, $f(x) - g(x) = -2x^2 + 5x - 2 > 0$. Hence the area in question is

$$A = \int_{1/2}^{2} (f - g) = \int_{1/2}^{2} (-2x^2 + 5x - 2)\, dx$$

$$= -\tfrac{2}{3}x^3 + \tfrac{5}{2}x^2 - 2x \Big|_{1/2}^{2} = \tfrac{9}{8}.$$

5E3. Interchange of limits of integration. Let f be continuous on $[a, b]$. It is often convenient to consider the expression

$$\int_{b}^{a} f,$$

even though the lower limit is greater than the upper. For consistency with the Fundamental Theorem we will want to have

(2) $$\int_{b}^{a} f = F(a) - F(b) \qquad\qquad (F' = f).$$

We accordingly *define*

(3) $$\int_{b}^{a} f = -\int_{a}^{b} f \qquad\qquad (a < b).$$

Then formula (2) holds. Another consequence is the following generalized property of additivity:

(4) $$\int_{a}^{b} f + \int_{b}^{c} f = \int_{a}^{c} f$$

for any numbers a, b, c, whatever their order, provided only that f is continuous on each of the three intervals they determine. (*Proof.* Both sides are equal to $F(c) - F(a)$.) For example, if f is continuous on the interval $[1, 3]$, then

$$\int_{1}^{3} f + \int_{3}^{2} f = \int_{1}^{2} f.$$

5E4. The integral in antiderivative problems. The Fundamental Theorem provides us with a handy notation for antiderivatives. Let f be continuous on $[a, b]$ and let c be any number. As we know, f has exactly one antiderivative, say F, whose value at a is c. Evidently,

$$(5) \qquad F(x) = c + \int_a^x f(t)\, dt.$$

To analyze the function F, or to evaluate $F(x)$ for a specific x, we may use all we know about integrals.

Example 3. Find the antiderivative of $f(x) = 2x$ whose value at $x = 2$ is 7.

Solution. The required function F is

$$F(x) = 7 + \int_2^x 2t\, dt = 7 + x^2 - 4 = x^2 + 3.$$

Compare the ease and directness of this solution with the one given in §4F1, Example 1.

Example 4. A particle P moves along a line in such a way that its velocity at time t is given by

$$v = 4t^2 + 2t - 3.$$

At time $t = 1$, P is at the point 3. Where is P at time $t = 5$?

Solution. The position function $s = F(t)$ is an antiderivative of v, and $F(1) = 3$. Hence

$$F(t) = 3 + \int_1^t (4x^2 + 2x - 3)\, dx$$

and

$$F(5) = 3 + \int_1^5 (4x^2 + 2x - 3)\, dx = 180\tfrac{1}{3}.$$

Again, compare this solution with the one in §4F2, Example 3.

Problems (§5E)

1. Evaluate the following integrals.

(a) $\displaystyle\int_2^1 x^2\, dx$

(b) $\displaystyle\int_1^0 (-x^3)\, dx$

(c) $\displaystyle\int_1^{-1} (3x^2 - 10x + 3)\, dx$

(d) $\displaystyle\int_1^2 \sqrt{x^2 - 1}\, dx + \int_2^1 \sqrt{x^2 - 1}\, dx$

2. State the relations among the numbers

$$a = \int_0^1 (1 - t^3)\, dt, \qquad b = \int_0^1 (t^3 - 1)\, dt,$$

$$c = \int_1^0 (1 - t^3)\, dt, \qquad d = \int_1^0 (t^3 - 1)\, dt.$$

3. Differentiate the following functions with respect to x.

   (a) $\displaystyle\int_x^1 \frac{1}{1 + t^2}\, dt$

   (c) $\displaystyle\int_x^x \frac{1}{\sqrt{1 + t^3}}\, dt$

   (b) $\displaystyle\int_x^0 \sqrt{1 - t^3}\, dt$

   (d) $\displaystyle\int_1^3 \sqrt{x^3 - 1}\, dx$

4. Let $F(x) = \displaystyle\int_x^1 f(t)\, dt$. What is $F'(x)$?

5. Show that

$$\left| \int_a^b f \right| \le \int_a^b |f|.$$

6. Using the fact that if $0 < t < 1$ then $0 < t < \sqrt{t} < 1$, show that:

   (a) $\dfrac{2}{3} \le \displaystyle\int_0^1 \sqrt{1 - x^2}\, dx \le 1$

   (b) $1 \le \displaystyle\int_0^1 \sqrt{1 + x^2}\, dx \le \dfrac{2}{3}(2\sqrt{2} - 1)$

7. Show that

$$\int_{147}^{149} (3x^4 - 2x^2 + 4)\, dx \ge 0.$$

8. Find the function F for which:
   (a) $F'(x) = x^2 + 2x - 4, \quad F(2) = 1$
   (b) $F'(x) = 1/x^2, \quad F(1) = 2$
   (c) $F'(x) = 2x - 1, \quad F(5) = 6$
   (d) $F'(x) = 1/x^3 + x, \quad F(3) = 1$
   (e) $F'(x) = 3(x - 1)^2(x + 1)^2 + 2(x - 1)(x + 1)^3, \quad F(1) = 1$
   (f) $F'(x) = x/(x^2 + 1)^2, \quad F(0) = 2$

9. Find the area of the region bounded by the given curves.
   (a) $y = 0, \quad x = 4, \quad$ and $y = -\sqrt{x}$
   (b) $y = 0 \quad$ and $\quad y = x^3, \quad$ on the interval $[-2, 1]$
   (c) $y = x \quad$ and $\quad y = \sqrt{x}$
   (d) $2x + y = 4 \quad$ and $\quad x^2 + y = 4$
   (e) $y = x^2 \quad$ and $\quad x^2 + y = 2$
   (f) $y = x^2 - 2 \quad$ and $\quad y = x$

(g) $y = 4 - x^2$ and $y = x + 2$
(h) $y = 4 - x^2$ and $y = 2x + 1$
(i) $y = x^2 - 1$ and $y = 2 - 2x^2$
(j) $y = x^4 - 4x^2$ and $y = 4x^2$
(k) $y = 3 - x^2$ and $y = 3 - 2x$
(l) $y = 1 - x^2$ and $y = x^3 - x$
(m) $y = 0$ and $y = x^3 - 5x^2 + 6x$
(n) $y = x$ and $y = x^3 - 5x^2 + 7x$

Answers to problems (§5E)

1. (a) $-\frac{7}{3}$ (b) $\frac{1}{4}$ (c) -8 (d) 0
2. $a = -b = -c = d$

3. (a) $-\dfrac{1}{1 + x^2}$ (b) $-\sqrt{1 - x^3}$ (c) 0   (d) 0

4. $-f(x)$
8. (a) $\frac{1}{3}x^3 + x^2 - 4x + \frac{7}{3}$
9. (a) $\frac{16}{3}$ (b) $\frac{17}{4}$ (c) $\frac{1}{6}$ (d) $\frac{4}{3}$ (e) $\frac{8}{3}$ (f) $\frac{9}{2}$ (g) $\frac{9}{2}$ (h) $\frac{32}{3}$
   (i) 4   (j) $\frac{512}{15}\sqrt{2}$ (k) $\frac{4}{3}$ (l) $\frac{4}{3}$ (m) $\frac{37}{12}$ (n) $\frac{37}{12}$

§5F. Integration by substitution; the indefinite integral

5F1. Discussion. To evaluate an integral by the Fundamental Theorem one looks for an antiderivative of the integrand. A complicated integrand may sometimes be rendered tractable by means of a change from the original variable, say x, to a new one, say u. There are two ways of considering the process: in the form $u = g(x)$ or in the form $x = h(u)$. In this section we concentrate on the first way.

The process rests upon the chain rule. The d-notation in the integral helps us keep track of the variables.

Suppose we are to evaluate

$$(1) \qquad \int_0^1 2x\sqrt{x^2 + 1}\; dx.$$

We want an antiderivative of $2x\sqrt{x^2 + 1}$. The key is to notice that the factor $2x$ is the derivative of $x^2 + 1$. The integral is therefore of the form

$$(2) \qquad \int_0^1 \sqrt{g(x)}\; g'(x)\; dx$$

(with $g(x) = x^2 + 1$). In terms of a new variable

$$(3) \qquad u = g(x),$$

this is

(4)
$$\int_0^1 \sqrt{u}\, \frac{du}{dx}\, dx.$$

Now as it turns out, we may replace the expression $\dfrac{du}{dx}\, dx$ in the integrand by du, just as though we had "canceled" the dx's—provided only that we adjust the limits to the new variable. In the present case, when $x = 0$, $u = x^2 + 1 = 1$; and when $x = 1$, $u = 2$. The original integral then reduces to

(5)
$$\int_1^2 \sqrt{u}\, du,$$

and this is something we can easily handle. We have:

(6)
$$\int_0^1 2x\sqrt{x^2 + 1}\, dx = \int_1^2 \sqrt{u}\, du \qquad (u = x^2 + 1)$$
$$= \tfrac{2}{3}(2^{3/2} - 1).$$

The following theorem justifies the procedure.

5F2. Integration by substitution. The setting is a function g defined on an interval $[a, b]$ and having a continuous derivative there. Keep in mind (1) and (2).

Next, $g(x)$ appears as part of a composite function, $f(g(x))$. In (2), f is the square root. In general, f is assumed to be defined on the image of g (which, by §2E8, is an interval), and to be continuous there.

The theorem is:

Let f, g, and $[a, b]$ be as above. Then

(7)
$$\int_a^b f(g(x))g'(x)\, dx = \int_{g(a)}^{g(b)} f(u)\, du.$$

Briefly: The substitution

(8)
$$u = g(x), \qquad du = g'(x)\, dx$$

is valid under the integral sign.

Proof. There is nothing hard about the proof except keeping track of the symbols. We have mentioned that f is defined on an interval. Let F be an antiderivative of f on this interval. By the Fundamental Theorem,

$$(9) \qquad \int_{g(a)}^{g(b)} f(u)\, du = F(u)\Big|_{g(a)}^{g(b)}.$$

Now, by the chain rule, $F\big(g(x)\big)$ is an antiderivative of $f\big(g(x)\big)g'(x)$. Therefore,

$$(10) \qquad \int_a^b f\big(g(x)\big)g'(x)\, dx = F\big(g(x)\big)\Big|_a^b,$$

also by the Fundamental Theorem.

The right-hand sides in (9) and (10) are the same thing. Therefore the left-hand sides are equal. This is what we wanted to prove.

Example 1. Evaluate

$$\int_0^2 \frac{x^2}{(x^3 + 2)^2}\, dx.$$

Solution. Trying $u = g(x) = x^3 + 2$, we get

$$du = g'(x)\, dx = 3x^2\, dx, \qquad \text{so that} \qquad x^2\, dx = \tfrac{1}{3}du.$$

Therefore,

$$(11) \qquad \int_0^2 \frac{x^2}{(x^3 + 2)^2}\, dx = \tfrac{1}{3}\int_{g(0)}^{g(2)} \frac{1}{u^2}\, du \qquad (u = g(x) = x^3 + 2)$$

$$= \tfrac{1}{3}\int_2^{10} \frac{1}{u^2}\, du = \tfrac{2}{15}.$$

With practice, we learn to write, more briefly: Set $u = x^3 + 2$; then $du = 3x^2\, dx$ and

$$(12) \qquad \int_0^2 \frac{x^2}{(x^3 + 2)^2}\, dx = \tfrac{1}{3}\int_2^{10} \frac{1}{u^2}\, du \qquad (u = x^3 + 2)$$

$$= \tfrac{2}{15}.$$

5F3. The indefinite integral. There are two steps in the above procedure: finding a suitable substitution, and adjusting the limits. The first requires thought, the second is mechanical. It is worth while, then, to set aside the question of limits and practice the first step alone.

The traditional symbol for an "integral without limits" is an integral without limits:

$$(13) \qquad \int f(x)\,dx$$

(called the "indefinite" integral). With cheerful ambiguity, it represents any antiderivative: by definition,

$$(14) \qquad \int f(x)\,dx = F(x) \qquad means \qquad \frac{d}{dx}F(x) = f(x).$$

For example,

$$(15) \qquad \int 2x\,dx = x^2;$$

or, when we want to indicate *all* possibilities,

$$(16) \qquad \int 2x\,dx = x^2 + C,$$

C denoting any constant (the "constant of integration"). Despite the ambiguity, the notation is convenient.

The rule for change of variable is

$$(17) \qquad \int f(u)\,\frac{du}{dx}\,dx = \int f(u)\,du$$

(where it is assumed that f and du/dx are continuous). *Proof.* If

$$(18) \qquad \frac{d}{du}F(u) = f(u),$$

then by the chain rule,

$$(19) \qquad \frac{d}{dx}F(u) = f(u)\,\frac{du}{dx}.$$

Example 2. Find

$$\int \frac{x^3}{\sqrt{1 + x^2}} \, dx.$$

Solution. Trying $u = 1 + x^2$, we get $du = 2x \, dx$ and

$$x^3 \, dx = x^2 \cdot x \, dx = (u - 1) \cdot \tfrac{1}{2} du.$$

Then

$$\int \frac{x^3}{\sqrt{1 + x^2}} \, dx = \tfrac{1}{2} \int \frac{u - 1}{\sqrt{u}} \, du = \tfrac{1}{2} \int \left(\sqrt{u} - \frac{1}{\sqrt{u}} \right) du$$

(20)

$$= \tfrac{1}{3} \sqrt{u}(u - 3) = \tfrac{1}{3} \sqrt{1 + x^2} \, (x^2 - 2).$$

As a check we can verify directly that

$$\frac{d}{dx} \left[\tfrac{1}{3} \sqrt{1 + x^2} \, (x^2 - 2) \right] = \frac{x^3}{\sqrt{1 - x^2}} \cdot$$

Do it.

ALWAYS CHECK BY DIFFERENTIATING!

Sometimes we proceed in stages.

Example 3. Find

$$F(x) = \int \frac{\sqrt{x} - 1}{\sqrt{x}(x - 2\sqrt{x} + 2)^2} \, dx.$$

Solution. The substitution $u = \sqrt{x}$ looks as though it may clean things up, so we try it. (This hopeful principle doesn't always work, however.) We get

$$du = \frac{1}{2\sqrt{x}} \, dx$$

and

(21) $$F(x) = \int \frac{2(u - 1)}{(u^2 - 2u + 2)^2} \, du.$$

If we next put $v = u^2 - 2u + 2$ we get $dv = (2u - 2) \, du$ and

(22) $$F(x) = \int \frac{1}{v^2} \, dv = -\frac{1}{v} = -\frac{1}{x - 2\sqrt{x} + 2} \cdot$$

Check by differentiating.

Problems (§5F)

1. Evaluate:

(a) $\int (-x)\sqrt{1 - x^2}\, dx$

(b) $\int 3x\sqrt{1 - x^2}\, dx$

(c) $\int (2x + 1)\sqrt{x^2 + x}\, dx$

(d) $\int \dfrac{x + 1}{\sqrt{x^2 + 2x + 3}}\, dx$

(e) $\int (x^3 + 1)^{17} x^2\, dx$

(f) $\int \dfrac{x^2}{(x^3 + 1)^2}\, dx$

(g) $\int \dfrac{(x^3 + 1)^2}{x^2}\, dx$

(h) $\int (x^3 + 1)^2\, dx$

(i) $\int (2x + 1)^{73}\, dx$

(j) $\int \dfrac{1}{\sqrt{3x - 2}}\, dx$

(k) $\int (2x - 3)^{1/3}\, dx$

(l) $\int \dfrac{1}{(x - 3)^3}\, dx$

(m) $\int \dfrac{1}{\sqrt{x}(\sqrt{x} + 1)^2}\, dx$

(n) $\int \dfrac{(1/x + 2)^{13}}{x^2}\, dx$

(o) $\int x\sqrt{x + 1}\, dx$

(p) $\int \dfrac{x^2 + \sqrt{x} + 1}{\sqrt{x}}\, dx$

(q) $\int \dfrac{2x(x^4 - 1)}{(x^4 + 3x^2 + 1)^2}\, dx$

[*Hint.* Set $t = x + 1/x$. Then $(x^2 - 1)\, dx = x^2\, dt$ and $x^4 + 3x^2 + 1 = (t^2 + 1)x^2$.]

2. Compute:

(a) $\int_2^5 \dfrac{1}{\sqrt{x - 1}}\, dx$

(b) $\int_0^1 (x - 2)^5\, dx$

(c) $\int_1^8 \sqrt{3x + 1}\, dx$

(d) $\int_{1/2}^1 (2x - 1)^{10}\, dx$

(e) $\int_0^2 \dfrac{x}{\sqrt{x^2 + 1}}\, dx$

(f) $\int_1^2 x^2(x^3 + 1)^5\, dx$

(g) $\int_0^1 x\sqrt{1 - x^2}\, dx$

(h) $\int_0^1 (2x - 1)\sqrt{x^2 - x}\, dx$

(i) $\int_0^1 x^3\sqrt{x^4 + 1}\, dx$

(j) $\int_0^2 (x^2 + x + 1)^3(2x + 1)\, dx$

(k) $\int_{\sqrt{6}}^5 \dfrac{x}{(x^2 + 2)^{1/3}}\, dx$

(l) $\int_1^2 \dfrac{x + 1}{\sqrt{x^2 + 2x - 1}}\, dx$

3. Express in terms of f and g alone:

(a) $\displaystyle\int [f(x)]^3 [g(x)]^3 [f(x)g'(x) + g(x)f'(x)]\, dx$

(b) $\displaystyle\int \frac{f(x)[f(x)g'(x) - g(x)f'(x)]}{[g(x)]^3}\, dx$

Answers to problems (§5F)

1. (a) $\frac{1}{3}(1 - x^2)^{3/2}$ (b) $-(1 - x^2)^{3/2}$ (c) $\frac{2}{3}(x^2 + x)^{3/2}$

(d) $\sqrt{x^2 + 2x + 3}$ (e) $\frac{1}{54}(x^3 + 1)^{18}$ (f) $-\dfrac{1}{3(x^3 + 1)}$

(g) $\frac{1}{5}x^5 + x^2 - \dfrac{1}{x}$ (h) $\frac{1}{7}x^7 + \frac{1}{2}x^4 + x$ (i) $\frac{1}{148}(2x + 1)^{74}$

(j) $\frac{2}{3}\sqrt{3x - 2}$ (k) $\frac{3}{8}(2x - 3)^{4/3}$ (l) $-\dfrac{1}{2(x - 3)^2}$

(m) $-\dfrac{2}{\sqrt{x + 1}}$ (n) $-\frac{1}{14}\left(\dfrac{1}{x} + 2\right)^{14}$ (o) $\frac{2}{15}(3x - 2)(x + 1)^{3/2}$

(p) $\frac{2}{5}x^{5/2} + x + 2x^{1/2}$ (q) $-\dfrac{1}{x^2 + 3 + 1/x^2}$

2. (a) 2   (b) $-\frac{21}{2}$ (c) 26   (d) $\frac{1}{22}$ (e) $\sqrt{5} - 1$
   (f) $\frac{1}{2}9^5 - \frac{1}{9}2^5$ (g) $\frac{1}{3}$ (h) 0   (i) $\frac{1}{3}\sqrt{2} - \frac{1}{6}$ (j) 600   (k) $\frac{15}{4}$
   (l) $\sqrt{7} - \sqrt{2}$

§5G. Approximating sums

In practical work, many integrals arise that cannot be handled by the Fundamental Theorem, so that to evaluate them one must resort to some method of numerical approximation. (Even when the Fundamental Theorem can be used, one may prefer to use an approximation.) This aspect is a subject in itself, with its own vast literature. Moreover, because of the development of electronic computers, methods that were in use only a few years ago are already out of date. In this book, we necessarily confine our discussion of the subject to a tiny scratch on the surface. The scratch follows. It can be omitted without loss of continuity.

5G1. Approximating sums. Let f be continuous on $[a, b]$. We know from §5D(5) that

(1) $$\int_a^b f$$

is the *unique* number that lies between every lower sum L on $[a, b]$ and every upper sum U.

It seems clear that the way to get U close to L is to choose partitions with sufficiently small segments. We now prove this. After that we take up its practical consequences.

By the *norm* of a partition is meant the length of its largest segment.

Darboux's Theorem. *Given* $\epsilon > 0$, *there exists* $\delta > 0$ *such that, for any partition of norm less than* δ, $U - L < \epsilon$.

Proof. We work with the number $\epsilon' = \epsilon/(b - a)$. Let P_0 be an $(\epsilon'/2)$-partition (§2F3). Choose δ to be one-half the length of its *smallest* segment.

Consider any partition P of norm $< \delta$. Each segment of P meets at most two (consecutive) segments of P_0. Therefore, f varies on each segment of P by less than ϵ'. As we saw in §5D3, this implies that $U - L < \epsilon$.

5G2. Control of the error. Let us now assume for simplicity that f is *increasing* on $[a, b]$. A similar discussion will apply in case f is decreasing. For a function that changes direction several times, the discussion can be applied separately to the intervals of increase and decrease.

Consider a partition of $[a, b]$ into n segments of *equal* length. This is also for simplicity. Call the points of subdivision $x_0, x_1, \cdots, x_{n-1}, x_n$, where $x_0 = a$ and $x_n = b$. The segments are $[x_0, x_1], \cdots, [x_{n-1}, x_n]$. Call their common length Δ_n:

$$(2) \qquad \Delta_n = \frac{1}{n}(b - a).$$

Let L_n and U_n denote the lower and upper sums.
Define

$$(3) \qquad \epsilon_n = U_n - L_n.$$

We notice that this is equal to the sum of the areas of the little rectangles enclosing the graph of f (Figure 1). As the figure shows,

$$(4) \qquad \epsilon_n = \Delta_n[f(b) - f(a)].$$

If f is a function we don't know how to "integrate," i.e., for which we do not know an antiderivative, we can estimate $\int_a^b f$ by computing L_n and U_n. Formula (4) shows how to choose n so as to achieve any desired degree of accuracy.

The formula for L_n itself is simple. Since f is an increasing function, its minimum on each segment occurs at the left endpoint. Therefore, the contribution to L_n from the first segment, $[x_0, x_1]$, is $\Delta_n f(x_0)$; the contribution

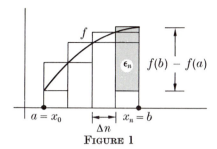

FIGURE 1

from the last segment, $[x_{n-1}, x_n]$, is $\Delta_n f(x_{n-1})$; and similarly in between. Thus,

$$(5) \qquad L_n = \Delta_n f(x_0) + \cdots + \Delta_n f(x_{n-1}).$$

Since Δ_n factors out, we usually first compute

$$(6) \qquad S_n = f(x_0) + \cdots + f(x_{n-1})$$

and then

$$(7) \qquad L_n = \Delta_n S_n.$$

Having found ϵ_n and L_n, we then compute U_n from (3).

Example 1. Compute $\int_0^1 \sqrt{1 + x^3}$ to within an error of 0.05.

Solution. See Figure 2. Here $[a, b] = [0, 1]$, $b - a = 1$, and $\Delta_n = 1/n$. Next, $f(x) = \sqrt{1 + x^3}$, $f(0) = 1.0$, and $f(1) = \sqrt{2} = 1.4$ (to one decimal place). Hence

$$(8) \qquad \epsilon_n = \frac{1}{n}[f(1) - f(0)] = \frac{0.4}{n}.$$

To achieve $\epsilon_n < 0.05$ we can take $n = 9$, but of course the computations are simpler if $n = 10$, since then the points of subdivision are $x_1 = 0.1$, $x_2 = 0.2$, etc. We get the following table to two places.

x	x^3	$1 + x^3$	$\sqrt{1 + x^3}$
0.0	0.00	1.00	1.00
0.1	0.00	1.00	1.00
0.2	0.01	1.01	1.00
0.3	0.03	1.03	1.01
0.4	0.06	1.06	1.03
0.5	0.12	1.12	1.06
0.6	0.22	1.22	1.11
0.7	0.34	1.34	1.16
0.8	0.51	1.51	1.22
0.9	0.73	1.73	1.31

$$S_{10} = 10.9$$

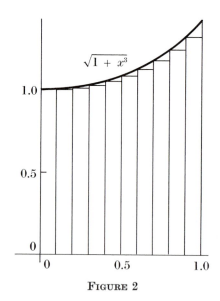

FIGURE 2

Since $\Delta_{10} = \frac{1}{10}$, $S_{10} = 10.9$, and $\epsilon_{10} = 0.04$,

$$L_{10} = \Delta_{10} S_{10} = 1.09$$

and

$$U_{10} = L_{10} + \epsilon_{10} = 1.13.$$

We have shown that

$$1.09 \leq \int_0^1 \sqrt{1 + x^3}\, dx \leq 1.13.$$

This improves on the result

$$1.00 \leq \int_0^1 \sqrt{1 + x^3}\, dx \leq 1.25$$

found (with less effort) in §5E1, Example 1.

Problems (§5G)

1. Compute the lower and upper sums for the function $f(x) = x^2$ on $[0, 3]$, using a partition into equal subintervals of length $\frac{1}{2}$. Compare the result with $\int_0^3 x^2\, dx$.

2. Compute the lower and upper sums for the function $f(x) = 1/x$ on $[1, 3]$, using a partition into equal subintervals of length $\frac{1}{4}$. Compare the result with 1.099, the known value of $\int_1^3 1/x\, dx$ (to three places).

3. Estimate $\int_0^1 \sqrt{1 - x^3}\, dx$ with an error of no more than 0.1 by computing upper and lower sums with a sufficiently large value of n. (Use a slide rule or a table of square roots.)

Answers to problems (§5G)

1. $L_6 = 6\frac{7}{8}$, $U_6 = 11\frac{3}{8}$, $\displaystyle\int_0^3 x^2\,dx = 9$

2. $L_8 = 1.02$, $U_8 = 1.19$

3. $0.78 = L_{10} < \displaystyle\int_0^1 \sqrt{1 - x^3}\,dx < U_{10} = 0.88$

§5H. The Riemann integral

Let f be continuous on a closed interval $[a, b]$. In §5B1 we defined the integral of f as a function of two variables satisfying the conditions **(A)** and **(B)**. This is a departure from the usual definitions of the integral as the *Darboux* integral or the *Riemann* integral. However, all three definitions are equivalent.

The *Darboux* integral is, by definition, the least upper bound of the set of all lower sums. We showed in §5D4 that this integral satisfies **(A)** and **(B)**. Therefore it is the same as the integral defined in §5B1.

To define the *Riemann* integral, we again start with a partition of $[a, b]$. Call its points of subdivision $x_0, x_1, \cdots, x_n$ (where $x_0 = a$ and $x_n = b$). We do not assume that the lengths of the segments are equal. The greatest of these lengths is called the *norm* of the partition.

In each segment, $[x_{i-1}, x_i]$, pick a number z_i:

$$(1) \qquad\qquad x_{i-1} \le z_i \le x_i \qquad\qquad (i = 1, \cdots, n).$$

Form the product

$$(2) \qquad\qquad f(z_i)(x_i - x_{i-1}) \qquad\qquad (i = 1, \cdots, n).$$

Add all these products:

$$(3) \qquad\qquad f(z_1)(x_1 - x_0) + \cdots + f(z_n)(x_n - x_{n-1}).$$

The expression (3) is called a *Riemann sum* associated with the partition. It is customarily abbreviated to

$$(4) \qquad\qquad \sum_{i=1}^{n} f(z_i)(x_i - x_{i-1}).$$

There are many Riemann sums associated with the same partition, because of the freedom in the choice of the numbers z_i.

Suppose, now, that there exists a number S with the following property: given any number $\epsilon > 0$, there is a number $\delta > 0$ such that, for every partition of norm $< \delta$, and for every Riemann sum R on any such partition, it is true that

$$(5) \qquad\qquad |S - R| < \epsilon.$$

Then the Riemann integral of f on $[a, b]$ is, by definition, the number S.

Let us show that the Riemann integral is the same as the others. We will show that the choice $S = \int_\omega^b f$ satisfies the definition just given.

Consider any partition of $[a, b]$. The lower and upper sums satisfy

(6)
$$L \leq \int_a^b f \leq U.$$

Now let R be any Riemann sum on the same partition. In forming the sum we pick $z_i \in [x_{i-1}, x_i]$. Then, obviously,

(7)
$$\min_{[x_{i-1}, x_i]} f \leq f(z_i) \leq \max_{[x_{i-1}, x_i]} f.$$

It follows that

(8)
$$L \leq R \leq U.$$

Thus, both R and the integral lie between L and U.

Now let $\epsilon > 0$ be given. By Darboux's Theorem (§5G1), there is a number $\delta > 0$ such that, for any partition of norm $< \delta$, $U - L < \epsilon$. Obviously, for every Riemann sum R on any such partition,

(9)
$$\left| \int_a^b f - R \right| < \epsilon.$$

This is what we wanted to prove.

CHAPTER 6

Applications of the Integral

In this chapter we see how to "set up" integrals, as a matter of routine, in a variety of standard applications. Each derivation is given in detail, to make it easier to skip around in the chapter.

Some of the work uses an extension of the definition of the integral. The chapter begins with this extension, called "The General Theorem for the Integral."

Each integral is set up according to an underlying pattern. We now wish to call attention to certain philosophical aspects of this pattern.

Let us return to the example of §5B5: work. Initially, work is defined (in physics) for a constant force only, so that the question of how much work is done by a varying force has no meaning. To give it meaning, we must first decide which of the properties of work done by a constant force are "essential." This decision may involve considerations from other parts of mathematics, or, as in this case, from outside mathematics.

The work done by a constant force, F, acting along a line for a distance s, is defined to be Fs. Behind this definition, though not a part of it, lies the idea of "getting something done," like rolling a ball up a hill, or compressing a spring. This idea leads to others. One can certainly "get something done" using a varying force, so we are naturally led to think about how work might be defined in that case.

The total amount "accomplished" by a force F acting over an interval $[a, u]$, added to that accomplished over $[u, v]$, seems evidently to be the amount accomplished over $[a, v]$. At least, so the physicists tell us. Next, with a greater force we expect greater accomplishment. The physicists tell us that, too. When these two observations about force are stated *mathematically*, they become conditions **(A)** and **(B)** for the function W_u^v being used to represent accomplishment. We conclude that the work done by the force F over the interval $[a, b]$ is $\int_a^b F$.

There are two subtle misconceptions to be avoided here. One is that we merely *define* the work to be $\int_a^b F$. The other, at the other extreme, is that we somehow *prove*, out of nothing, that the work is $\int_a^b F$. What we actually do is something in between. We represent the conditions presented by the physicists as *mathematical* conditions to be satisfied by a certain function of two variables, W_u^v. *From here on, the rest is mathematics.* Suppose we find from these assumptions (as we do) that the work function, W_u^v, satisfies **(A)** and **(B)** (with respect to F). We have then *proved*—from those assumptions—that $W_u^v = \int_u^v F$. We are then *forced* to define the work done over the interval $[a, b]$ as $\int_a^b F$.

But we are not forced to accept the initial assumptions. They come from physics. If a physicist friend informs us tomorrow that recent experiments suggest more force may result in less work, then our formula is in doubt. We must await further news about the characteristics of work.

Similar considerations govern all the applications. The integrals we derive are necessarily what they are only if the underlying assumptions are granted. It is always possible to question the assumptions. But, once granted, they force us to the definitions.

§6A. A general theorem for the integral

6A1. The problem. We have looked at two simple applications of the integral: area (§5B4) and work (§5B5). In each case we contemplate a function of the form S_u^v ($a \le u \le v \le b$) and decide quickly that it satisfies

(A) $$S_a^u + S_u^v = S_a^v.$$

In addition, a certain function f is present as part of the data of the problem, and we are able to verify the relations

(B) $$(v - u) \min_{[u,v]} f \leq S_u^v \leq (v - u) \max_{[u,v]} f.$$

We then have no choice but to define S_a^b to be $\int_a^b f$.

In more elaborate applications we may find ourselves with more than just one function f on our hands. Suppose the data are such that, in addition to **(A)**, we get:

(1) $$(v - u) \min_{[u,v]} (fg) \leq S_u^v \leq (v - u) \max_{[u,v]} (fg).$$

This is property **(B)** with respect to the function fg. We must define $S_a^b = \int_a^b (fg)$. This example is really not different from the others. Here we also deal with a single function, namely, fg. We might have called it h.

Suppose, though, we come up with the following variant:

(2) $$(v - u)(\min_{[u,v]} f)(\min_{[u,v]} g) \leq S_u^v \leq (v - u)(\max_{[u,v]} f)(\max_{[u,v]} g).$$

Can we conclude that $S_a^b = \int_a^b (fg)$? Not from anything we know so far. Expression (2) is *not* property **(B)** with respect to the function fg. That is stated in (1).

Example 1. Just to be sure about the distinction, look at the functions $f(x) = x$ and $g(x) = 1 - x$ on $[0, 1]$ (Figure 1). We have

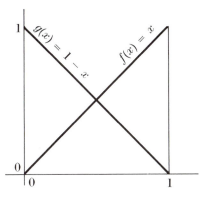

FIGURE 1

$$\max_{[0,1]} f = 1 \quad \text{and} \quad \max_{[0,1]} g = 1,$$

so that

$$(\max_{[0,1]} f)(\max_{[0,1]} g) = 1.$$

On the other hand, $f(x)g(x) = x(1 - x)$, and it is easy to see that the maximum of this function occurs at $x = \frac{1}{2}$, so that

$$\max_{[0,1]} (fg) = \tfrac{1}{4}.$$

Here is another illustration. Suppose the data of the problem lead, in addition to **(A)**, to

(3) $$\qquad (v - u) \min_{[u,v]} (f - g) \le S_u^v \le (v - u) \max_{[u,v]} (f - g).$$

This is property **(B)** with respect to the function $f - g$. We must define $S_a^b = \int_a^b (f - g)$. But suppose, instead, that the data lead to

(4) $$\quad (v - u)(\min_{[u,v]} f - \max_{[u,v]} g) \le S_u^v \le (v - u)(\max_{[u,v]} f - \min_{[u,v]} g).$$

This is *not* property **(B)**. Our course is in doubt.

Example 2. Let $f(x) = x$ and $g(x) = x^2$ on [0, 1] (Figure 2). Then

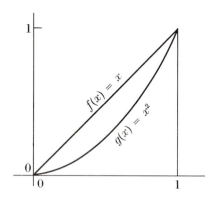

FIGURE 2

$$\max_{[0,1]} f = 1 \qquad \text{and} \qquad \min_{[0,1]} g = 0,$$

so that

$$\max_{[0,1]} f - \min_{[0,1]} g = 1.$$

On the other hand, $f(x) - g(x) = x - x^2$, so that

$$\max_{[0,1]} (f - g) = \tfrac{1}{4}.$$

It is clear what is at the root of the distinction. In (2) and (4) the bounds for S_u^v each involve not just one maximum or minimum but two, *applied separately* to two different functions.

Many applications lead to expressions of the same sort. As it turns out, we can handle them, and very easily. It happens that (2) actually is equivalent to (1), and (4) is equivalent to (3). The "rule" is simply to pretend we were not alert enough to notice the difference.

6A2. The General Theorem for the Integral. The ingredients are:

(i) One or more functions $f, g, \cdots$ continuous on an interval $[a, b]$. For simplicity, we state the theorem for the case of two functions, f and g.

(ii) Some algebraic combination of these functions, which we call $C(f, g)$. For instance, $C(f, g)$ might be fg, or $f - g$, or $\sqrt{f^2 + g^2}$, etc. We also have to consider $C\big(f(x_1), g(x_2)\big)$, in which f and g are evaluated at different points.

(iii) A function S_u^v defined for $u \leq v$ in $[a, b]$.

The theorem is:

Let $[a, b]$, f, g, C, and S_u^v be as above.
Suppose that

(A) $$S_a^u + S_u^v = S_a^v$$

and

(B) $$(v - u)C(\min_{[u,v]} f, \max_{[u,v]} g) \leq S_u^v \leq (v - u)C(\max_{[u,v]} f, \min_{[u,v]} g).$$

Then

$$S_a^b = \int_a^b C(f, g).$$

Remark. Hypothesis **(B)** may also be given with $\min_{[u,v]} f$ and $\min_{[u,v]} g$ on the left and $\max_{[u,v]} f$ and $\max_{[u,v]} g$ on the right; etc.

Proof of theorem. First we show that the function

(5) $$F_0(x) = S_a^x$$

is an antiderivative of $C(f(x), g(x))$. To do this we repeat the argument of §5C5, wherein it is shown that $\int_a^x f$ is an antiderivative of f. By **(A)**,

(6) $$S_u^v = F_0(v) - F_0(u).$$

We rewrite this in terms of x and Δx, putting $x = u$ and $\Delta x = v - u$. Then $v = x + \Delta x$ and we have

(7) $$S_u^v = F_0(x + \Delta x) - F_0(x).$$

We substitute this into **(B)** above and divide by Δx (assumed > 0). This gives:

(8) $$C(\min_{[x,x+\Delta x]} f, \max_{[x,x+\Delta x]} g) \leq \frac{F_0(x + \Delta x) - F_0(x)}{\Delta x} \leq C(\max_{[x,x+\Delta x]} f, \min_{[x,x+\Delta x]} g).$$

The key point is that as $\Delta x \to 0$, the two outside expressions approach $C(f(x), g(x))$. This is because f and g are continuous and C is an algebraic combination (made up of sums, products, etc.) The difference quotient is then squeezed to the same limit. Thus, $F_0'(x) = C(f(x), g(x))$.

Since F_0 is an antiderivative of $C(f, g)$, we know from §5C2 that $F_0(v) - F_0(u)$ is the integral, $\int_u^v C(f, g)$. But $F_0(v) - F_0(u)$ is S_u^v. So S_u^v is the integral.

Remark. Having proved that S_u^v is the integral, we know after the fact that it satisfies the original condition (B):

$$(9) \qquad \min_{[u,v]} C(f, g) \le S_u^v \le \max_{[u,v]} C(f, g)$$

—by definition of the integral.

Example 3. The area between two graphs. Let f and g be continuous on $[a, b]$, with $f \ge g$. We have observed that the area between their graphs is given by

$$(10) \qquad A = \int_a^b (f - g).$$

We got to this via $\int_a^b f - \int_a^b g$, essentially, the area under the graph of f minus the area under the graph of g (§5E2).

Now let us see how we might derive (10) directly from (A) and (B).

Let A_u^v denote the area between the graphs from $x = u$ to $x = v$. Certainly we are to have

$$\textbf{(A)} \qquad A_a^u + A_u^v = A_a^v.$$

Next, we argue for the following bounds:

$$\textbf{(B)} \qquad (v - u)(\min_{[u,v]} f - \max_{[u,v]} g) \le A_u^v \le (v - u)(\max_{[u,v]} f - \min_{[u,v]} g).$$

The quantity on the right is the area of the enclosing region shown in Figure 3. As for the left, either it is the area of an enclosed region (Figure 3) or else it is a negative number (Figure 4).

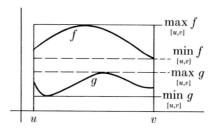

FIGURE 3

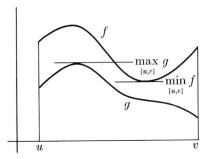

FIGURE 4

Because of **(A)** and **(B)**, we are forced by the General Theorem to define the area by (10).

Notice that the region between u and v is not enclosed by any rectangle of height $\max\limits_{[u,v]} (f - g)$. In Figure 4, the region does not enclose any rectangle of base $v - u$ and height $\min\limits_{[u,v]} (f - g)$. Hence we would have been hard put to recognize, *in advance*, the relations

(11) $$(v - u) \min_{[u,v]} (f - g) \le A_u^v \le (v - u) \max_{[u,v]} (f - g).$$

Problems (§6A)

1. In each case, S_u^v is defined for $u \le v$ in $[a, b]$ and satisfies **(A)** and the additional condition stated, where f and g are continuous on $[a, b]$. Express S_u^v as an integral.

(a) $(v - u) \min\limits_{[u,v]} f \min\limits_{[u,v]} g \le S_u^v \le (v - u) \max\limits_{[u,v]} f \max\limits_{[u,v]} g$

(b) $(v - u) \min\limits_{[u,v]} f \max\limits_{[u,v]} g \le S_u^v \le (v - u) \max\limits_{[u,v]} f \min\limits_{[u,v]} g$

(c) $(v - u)\pi (\min\limits_{[u,v]} f)^2 \le S_u^v \le (v - u)\pi (\max\limits_{[u,v]} f)^2$

(d) $(v - u)u \min\limits_{[u,v]} f \le S_u^v \le (v - u)v \max\limits_{[u,v]} f$

(e) $(v - u)(u^2 + v) \le S_u^v \le (v - u)(v^2 + u)$

(f) $u(u + 1)(v - u) \le S_u^v \le v(v + 1)(v - u)$

(g) $u(v + 1)(v - u) \le S_u^v \le v(u + 1)(v - u)$

(h) $(v^2 - u^2)(2u + v) \le S_u^v \le (v^2 - u^2)(2v + u)$

(i) $6u^2(v - u) \le S_u^v \le 6v^2(v - u)$

(j) $2(v - u)(2u^2 + v^2) \le S_u^v \le 2(v - u)(2v^2 + u^2)$

Answers to problems (§6A)

1. (a) $\int_u^v fg$ (b) $\int_u^v fg$ (c) $\pi \int_u^v f^2$ (d) $\int_u^v xf(x)\, dx$

(e) $\int_u^v (x^2 + x)\, dx$ (f) $\int_u^v (x^2 + x)\, dx$ (g) $\int_u^v (x^2 + x)\, dx$

(h) $6 \int_u^v x^2\, dx$ (i) $6 \int_u^v x^2\, dx$ (j) $6 \int_u^v x^2\, dx$

§6B. Volume

6B1. Properties of volume. To find the volume of any but the simplest solids raises problems like those for area. We have no *definition* of the volume of an arbitrary solid. As with area, we will state some principles to guide us to a definition. Again as with area, the fundamental property is that the volume of a solid cut into two parts is the sum of the volumes of the parts; and another (which follows from the first) is that the volume of any solid is greater than that of any within it.

In dealing with area, we used one additional principle: that the area of a rectangle is the product of its dimensions. A corresponding principle about volume is that the volume in a right cylinder is the area of the base times the height. (By definition, a right cylindrical solid with a given plane region as base is the set of all points lying in or above the region, up to a given height (Figure 1).) We shall adopt this principle for all bases (for which area is defined). That is, we will assume as an axiom:

(1)   *The volume enclosed by any right cylinder is the area of the base times the height.*

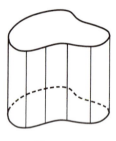

FIGURE 1

In particular, the volume enclosed by a right circular cylinder of radius r and height h is $\pi r^2 h$.

Example 1. Find the volume in a pyramid of height h whose base is a polygon of area A.

Solution. First we have to define the volume, then compute it.

It is convenient to measure from the vertex down, as shown in Figure 2. Consider the cross section at any level u. Call its area A_u. This cross section is a polygon, and it is easy to see that its sides are proportional to u. Now, the *area* of a polygon is proportional to the *squares* of its sides. Hence A_u is proportional to u^2. Since the area at level h is A itself, we have

(2) $$\frac{A_u}{A} = \frac{u^2}{h^2}.$$

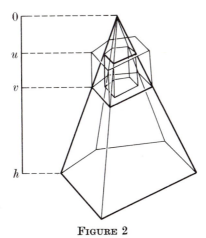

Let V_u^v denote the volume of the portion of the pyramid between levels u and v. Of course, we require

(A) $$V_0^u + V_u^v = V_0^v.$$

This portion of the solid lies between two prisms of height $v - u$: one with area of base A_u and one with area of base A_v. Therefore

(3) $$A_u(v - u) \le V_u^v \le A_v(v - u).$$

Substituting from (2), we get

(4) $$(v - u)\,\frac{A}{h^2}\,u^2 \le V_u^v \le (v - u)\,\frac{A}{h^2}\,v^2.$$

To put this in the form **(B)** we observe that $u^2 = \min_{[u,v]} y^2$ and $v^2 = \max_{[u,v]} y^2$. So we have

(B) $$(v - u)\,\min_{[u,v]}\frac{A}{h^2}\,y^2 \le V_u^v \le (v - u)\,\max_{[u,v]}\frac{A}{h^2}\,y^2.$$

Because of **(A)** and **(B)**, we are forced to define the total volume as

(5) $$V = \frac{A}{h^2}\int_0^h y^2\,dy = \tfrac{1}{3}Ah.$$

6B2. Solids of revolution. The solid obtained when a plane region is revolved in space about an axis lying to one side of it is known as a solid of revolution. Each point traces out a circle whose radius is its distance from the axis of revolution.

Sometimes it is handy to think of revolving just the boundary curve of the region, thus generating a surface, and to speak of the solid enclosed by the surface. For example, in Figure 3, the quarter-circular region revolves about the x-axis to generate a hemispherical ball; the quarter-circle itself generates the hemisphere enclosing the ball. (The figure shows one-fourth of the revolution.)

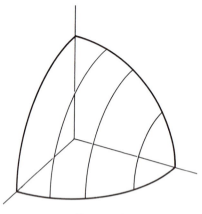

FIGURE 3

Usually, as here, the axis of revolution is one of the coordinate axes. If it is the x-axis, a point (x, y) traces out the circle of radius $|y|$, center $(x, 0)$ $\big($Figure 4(a)$\big)$. If it is the y-axis, (x, y) traces out the circle of radius $|x|$, center $(0, y)$ $\big($Figure 4(b)$\big)$.

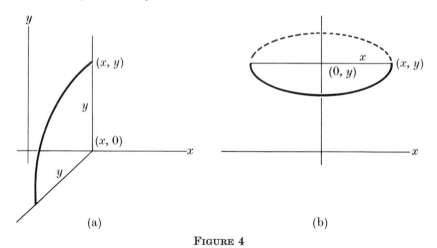

FIGURE 4

6B3. Volume of a solid of revolution: disc method. Let f be nonnegative and continuous on $[a, b]$. We wish to define and compute the volume enclosed when the graph of f is revolved about the x-axis.

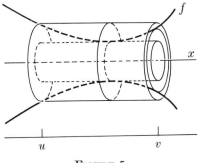

FIGURE 5

Let V_u^v denote the volume between u and v. Of course we require:

(A) $$V_a^u + V_u^v = V_a^v.$$

The portion of the solid between u and v is enclosed within a cylinder—look at it sideways—of radius $\max_{[u,v]} f$ and "height" $v - u$; and it encloses a cylinder of radius $\min_{[u,v]} f$ and height $v - u$. See Figure 5. Consequently,

(B) $$\pi(v - u)(\min_{[u,v]} f)^2 \leq V_u^v \leq \pi(v - u)(\max_{[u,v]} f)^2.$$

Because of **(A)** and **(B)**, we are forced to define the volume by

(6) $$V = \pi \int_a^b f^2.$$

Memory aid. The volume of the solid between x and $x + \Delta x$ is approximately equal to the volume of the cylinder of radius $f(x)$ and "height" Δx (Figure 6);

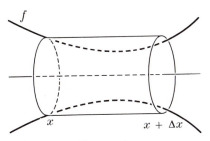

FIGURE 6

that is, to

$$\pi f(x)^2 \Delta x.$$

The actual volume of the solid between a and b is

$$\pi \int_a^b f(x)^2 \, dx.$$

Example 2. Find the volume of a sphere of radius r.

Solution. We obtain the right-hand hemisphere by revolving the quarter-circle $y = \sqrt{r^2 - x^2}$ $(0 \le x \le r)$ about the x-axis (Figure 3). By (6), the required volume is

$$V = 2\pi \int_0^r (r^2 - x^2)\, dx = 2\pi(r^3 - \tfrac{1}{3}r^3) = \tfrac{4}{3}\pi r^3.$$

Example 3. Find the volume generated when the region lying between the graphs of $y = x^2$ and $y = 2x$ is revolved about the x-axis.

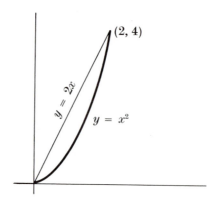

$(2, 4)$

$y = 2x$

$y = x^2$

FIGURE 7

Solution. The region is pictured in Figure 7. We subtract the volume generated by the graph of x^2 from that generated by the graph of $2x$. Thus,

$$V = \pi \int_0^2 (2x)^2\, dx - \pi \int_0^2 (x^2)^2\, dx = \tfrac{64}{15}\pi.$$

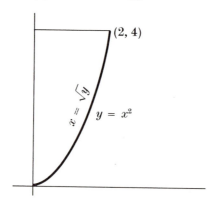

$(2, 4)$

$x = \sqrt{y}$

$y = x^2$

FIGURE 8

Example 4. Find the volume of the solid generated when the region bounded by the y-axis, the graph of $y = x^2$, and the line $y = 4$ is revolved about the y-axis.

Solution. The region is pictured in Figure 8. We consider discs perpendicular to the y-axis. The graph is the graph of $x = \sqrt{y}$. From (6), we have

$$V = \pi \int_0^4 (\sqrt{y})^2 \, dy = 8\pi.$$

Problems (§6B)

1. Find the volume generated when the given graph is revolved about the x-axis.
   (a) $y = x^3$ $(0 \le x \le 3)$
   (b) $y = x^{3/2}$ $(0 \le x \le 4)$
   (c) $x + y = 4$ $(0 \le x \le 4)$

2. Find the volume generated when each graph in Problem 1 is revolved about the y-axis.

3. Show that the volume of the *paraboloid of revolution* generated by revolving the curve $y^2 = x$ $(0 \le x \le a)$ about the x-axis is proportional to a^2.

4. (a) Show that revolving the graph of $9x^2 + 16y^2 = 144$ about the x-axis generates a volume of 48π.
   (b) Show, more generally, that the volume of the *ellipsoid of revolution* generated by revolving the *ellipse*

   $$\frac{x^2}{a^2} + \frac{y^2}{b^2} = 1 \qquad\qquad (a > 0)$$

   about the x-axis is $\frac{4}{3}\pi ab^2$.

5. Find the volume generated when the triangle with vertices $(0, 0)$, $(4, 0)$, and $(1, 3)$ is revolved about the x-axis.

6. Find the volume generated when the region bounded by the given curves is revolved about the x-axis.
   (a) $y = 5x - x^2 - 6$, the x-axis
   (b) $y = x^2$, $y = 4 - x^2$
   (c) $y = 1/x$, $y = 2$, $x = 2$
   (d) $y = x^2 + 2$, $y = 5$
   (e) $y = x^3$, $y = 1$, $x = 2$
   (f) $y = \sqrt{x}$, $x + y = 2$, $y = 0$
   (g) $y = \sqrt{16 - x^2}$, $x = -3$, $x = 2$, $y = 0$
   (h) $y = \sqrt{x}$, $x = 4 - y^2$, $y = 0$

7. Find the volume generated when the region bounded by the given curves is revolved about the y-axis.

(a) $y = x^5$, $x = 0$, $y = 32$
(b) $y = x^3$, $y = 0$, $x = 2$
(c) $y = x^2$, $y = x$
(d) $y = 2x$, $y = 0$, $x = 2$

8. The region bounded by the graph of $y = 4x - x^2$ and the x-axis is divided by the line $y = 2x$ into two parts. Find the volumes generated by revolving each part about the x-axis.

9. Find the volume in a right circular cone of height h and radius of base r.

10. Find the volume generated when the region bounded by the curves $xy = 4$ and $4x + 3y = 16$ is revolved about

(a) the x-axis.   (b) the y-axis.

11. Find the volume generated when the region bounded by the given curves is revolved about the line $x = 2$.

(a) $y = x^2$, $x = 2$, $y = 0$
(b) $y = x^2$, $x = 0$, $y = 4$

Answers to problems (§6B)

1. (a) $\frac{2187}{7}\pi$ (b) 64π (c) $\frac{64}{3}\pi$
2. (a) $\frac{729}{5}\pi$ (b) $\frac{384}{7}\pi$ (c) $\frac{64}{3}\pi$
5. 12π
6. (a) $\frac{1}{30}\pi$ (b) $\frac{64}{3}\pi\sqrt{2}$ (c) $\frac{9}{2}\pi$ (d) $\frac{152}{5}\pi\sqrt{3}$
    (e) $\frac{120}{7}\pi$ (f) $\frac{5}{6}\pi$ (g) $\frac{205}{3}\pi$ (h) 4π
7. (a) $\frac{640}{7}\pi$ (b) $\frac{64}{5}\pi$ (c) $\frac{1}{6}\pi$ (d) $\frac{32}{3}\pi$
8. $\frac{32}{5}\pi$, $\frac{416}{15}\pi$
9. $\frac{1}{3}\pi r^2 h$
10. (a) $\frac{128}{27}\pi$ (b) $\frac{32}{9}\pi$
11. (a) $\frac{8}{3}\pi$ (b) $\frac{40}{3}\pi$

§6C. Volume (continued)

6C1. Volume of a solid of revolution: shell method. Consider again the volume generated by revolving a region about the x-axis. If it is not the region under the graph of a function, the disc method for finding the volume may be awkward. In such cases the following approach is often useful.

We suppose that the region is bounded on the right and left by two curves, $x = f(y)$ and $x = g(y)$, where $f \geq g$ (Figure 1). It is assumed that f and g are continuous on an interval $[\alpha, \beta]$, where $\alpha \geq 0$. Let V_u^v denote the volume generated by the portion of the region between the levels $y = u$ and $y = v$. Certainly,

(A) $$V_\alpha^u + V_u^v = V_\alpha^v.$$

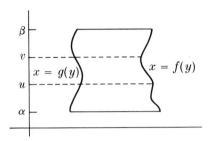

FIGURE 1

Next, the portion between u and v is enclosed in a rectangle with base $\max_{[u,v]} f - \min_{[u,v]} g$ and altitude $v - u$ (Figure 2). The solid swept out by this

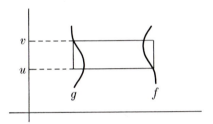

FIGURE 2

rectangle is a cylindrical shell of radii u and v and "height" $\max_{[u,v]} f - \min_{[u,v]} g$ (Figure 3). Its volume is the volume of the larger cylinder minus the volume

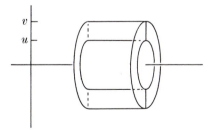

FIGURE 3

of the smaller. Therefore,

$$(1) \qquad V_u^v \le \pi v^2 (\max_{[u,v]} f - \min_{[u,v]} g) - \pi u^2 (\max_{[u,v]} f - \min_{[u,v]} g).$$

In the other direction, either the portion of the region between u and v

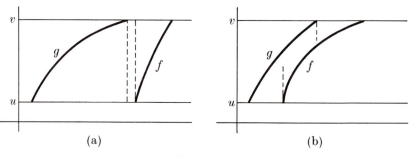

(a)   (b)

FIGURE 4

encloses a rectangle with base $\min\limits_{[u,v]} f - \max\limits_{[u,v]} g$ $\big($Figure 4(a)$\big)$, or else $\min\limits_{[u,v]} f -$
$\max\limits_{[u,v]} g$ is negative $\big($Figure 4(b)$\big)$. In either event,

$$(2) \qquad \pi v^2(\min_{[u,v]} f - \max_{[u,v]} g) - \pi u^2(\min_{[u,v]} f - \max_{[u,v]} g) \le V_u^v.$$

To put (1) and (2) in the form **(B)** we take advantage of the factor $v^2 - u^2$
that appears in each. This is equal to $(v - u)(v + u)$. For (2) we note that
$v + u \ge 2u = 2 \min\limits_{[u,v]} y$, and for (1) we note that $v + u \le 2v = 2 \max\limits_{[u,v]} y$.
So we have

$$2\pi(v - u)(\min_{[u,v]} y)(\min_{[u,v]} f - \max_{[u,v]} g)$$

(B) $$\qquad \le V_u^v$$

$$\le 2\pi(v - u)(\max_{[u,v]} y)(\max_{[u,v]} f - \min_{[u,v]} g).$$

Because of **(A)** and **(B)**, we are forced to define the volume as

$$(3) \qquad V = 2\pi \int_\alpha^\beta y[f(y) - g(y)]\, dy.$$

If we wish we may simplify this by putting

$$(4) \qquad w(y) = f(y) - g(y),$$

the width of the region at level y. Then

$$(5) \qquad V = 2\pi \int_\alpha^\beta y w(y)\, dy.$$

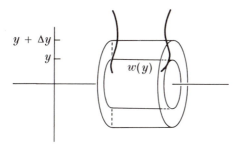

FIGURE 5

Memory aid. The volume of the shell between levels y and $y + \Delta y$ is approximately the same as the volume of a cylindrical shell of inner radius y, "height" $w(y)$, and thickness Δy (Figure 5). This is approximately equal to the area of the inner wall of the cylinder, $2\pi y w(y)$, times the thickness, Δy; that is, to

$$2\pi y w(y) \, \Delta y.$$

The actual volume between α and β is

$$2\pi \int_{\alpha}^{\beta} y w(y) \, dy.$$

Example 1. Find the volume obtained when the region bounded by the parabola $x = y(2 - y)$ and the lines $x = 3$, $y = 0$, and $y = 2$ is revolved about the x-axis (Figure 6).

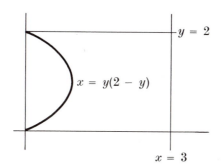

$y = 2$

$x = y(2 - y)$

$x = 3$

FIGURE 6

Solution. Here

$$w(y) = 3 - y(2 - y) = y^2 - 2y + 3.$$

The required volume is

$$V = 2\pi \int_{0}^{2} y w(y) \, dy = 2\pi \int_{0}^{2} (y^3 - 2y^2 + 3y) \, dy = \tfrac{28}{3}\pi.$$

Problems (§6C)

1. Find the volume generated when the region bounded by the curves indicated is revolved about the x-axis.

(a) $x = y(2 - y)$, $x = 2y(2 - y)$
(b) $x = \sqrt{y} - y^2$, $x = 0$
(c) $x = y^3 + y$, $x = 2$, $y = 0$

2. Consider the region in the first quadrant bounded by the coordinate axes and the curve $y = 4 - x^2$.

(a) Use both the disc method and the shell method to find the volume obtained when the region is revolved about the x-axis.
(b) Use both methods to find the volume obtained when the region is revolved about the y-axis.

3. Consider the region bounded by the graph of $y = x^3$, the line $x = 1$, and the line $y = 8$.

(a) Use both the disc method and the shell method to find the volume obtained when the region is revolved about the x-axis.
(b) Use both methods to find the volume obtained when the region is revolved about the y-axis.

4. The region bounded by the curves $g(y) = \frac{1}{2}(y - 1)^2(4 - y) + 1$ and $h(y) = \frac{1}{2}(y - 1)^2(y - 4) + 1$ and the lines $y = 0$ and $y = 4$, is revolved about the x-axis. Find the volume generated.

5. Derive the formula

(6)
$$V = 2\pi \int_\alpha^\beta (y - k)[f(y) - g(y)] \, dy$$

for the volume generated by revolving the region of Figure 1 about the line $y = k$, where $k \le \alpha$.

6. The region bounded by the graph of $x = y^2$, the lines $y = 2$ and $y = 3$, and the y-axis, is revolved about the line $y = -2$. Use formula (6) of Problem 5 to find the volume generated.

7. Find the volume generated when the region bounded by $y = \sqrt{1 + x^2}$, $x = 0$, $y = 0$, and $x = 1$, is revolved about the y-axis.

Answers to problems (§6C)

1. (a) $\frac{8}{3}\pi$ (b) $\frac{3}{10}\pi$ (c) $\frac{14}{15}\pi$
2. (a) $\frac{256}{15}\pi$ (b) 8π
3. (a) $\frac{321}{7}\pi$ (b) $\frac{58}{5}\pi$
4. $\frac{96}{5}\pi$
6. $\frac{347}{6}\pi$
7. $\frac{2}{3}\pi(2\sqrt{2} - 1)$

§6D. Length of arc

6D1. Properties of length. Let f be a continuous function on $[a, b]$. We wish to define the length of its graph.

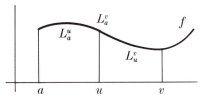

FIGURE 1

Let L_u^v denote the length between u and v (Figure 1). What properties should it have? Certainly we want it to satisfy

(A) $$L_a^u + L_u^v = L_a^v.$$

Next, we look for bounds for L_u^v. Let us begin with the observation that of two straight-line graphs with positive slopes, the one with the larger slope is the longer (Figure 2).

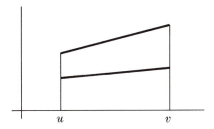

FIGURE 2

We extend this idea to graphs whose slopes are not necessarily constant. Imagine walking across a field, heading eastward but edging north as you go. Your friend does the same, keeping due north of you at all times and, at every instant, heading *more* northward than you (Figure 3). Obviously, he walks farther than you. (In fact, he has to walk faster, and you both complete the trip in the same time.)

In mathematical terms, this asserts that of two functions, f_1 and f_2, if $f_1'(x) \geq f_2'(x) \geq 0$ for each x, then the graph of f_1 is (as long as or) longer than the graph of f_2.

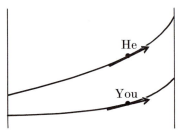

FIGURE 3

Since south is as good as north, we may say:

(1)   *If $|f_1'(x)| \geq |f_2'(x)|$ for each x, then the graph of f_1 is longer than the graph of f_2.*

We shall adopt this as a guiding principle. Actually, we need it only when either f_1 or f_2 is a straight line. We remark (see Figure 4) that the length of a segment of slope m, between u and v, is

(2) $$(v - u)\sqrt{1 + m^2}.$$

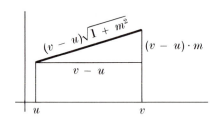

FIGURE 4

6D2. Definition of arc length. Our assumption (1) leads easily to a definition and formula for length of arc. We assume that f is differentiable on $[a, b]$ and that f' is continuous. We need this because our methods lead to an integral involving f'.

Consider the graph of f between u and v. The function $|f'|$ is continuous and so it assumes a minimum and maximum on $[u, v]$. By (1), the graph of f is longer than the segment l_1 with slope $\min_{[u,v]} |f'|$ and shorter than the segment l_2 with slope $\max_{[u,v]} |f'|$. See Figure 5. The lengths of these segments are given by (2), and we have

(B) $$(v - u)\sqrt{1 + (\min_{[u,v]} |f'|)^2} \leq L_u^v \leq (v - u)\sqrt{1 + (\max_{[u,v]} |f'|)^2}.$$

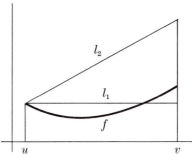

FIGURE 5

We were led to this from principle (1). We did not prove that principle but only assumed it. No one is obliged to accept it. But anyone who does assume (1) and also believes that arc length is additive is now forced to define the length of the graph of f as

(3)
$$L = \int_a^b \sqrt{1 + f'^2}.$$

Memory aid. See Figure 6. The length of arc between x and $x + \Delta x$ is approximately

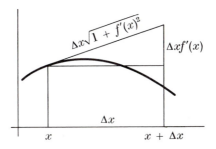

FIGURE 6

equal to the length of the tangent line at x,

$$\sqrt{1 + f'(x)^2}\, \Delta x.$$

The actual length of arc from a to b is

$$\int_a^b \sqrt{1 + f'(x)^2}\, dx.$$

Example 1. Find the length of the graph of

$$f(x) = x^{3/2} \qquad\qquad (0 \le x \le 4).$$

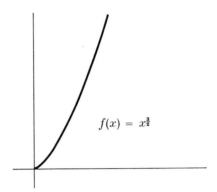

$$f(x) = x^{\frac{3}{2}}$$

FIGURE 7

Solution. (See Figure 7.) Here $f'(x) = \frac{3}{2}x^{1/2}$. The length of the graph is

$$L = \int_0^4 \sqrt{1 + \tfrac{9}{4}x}\, dx = \tfrac{8}{27}(1 + \tfrac{9}{4}x)^{3/2}\Big|_0^4$$

$$= \tfrac{8}{27}(10\sqrt{10} - 1).$$

Example 2. Find the length of the graph of

$$y = f(x) = x^{2/3} \qquad\qquad (0 \le x \le 8).$$

Solution. (See Figure 8.) Here

$$f'(x) = \tfrac{2}{3}x^{-1/3} \qquad\qquad (0 < x \le 8),$$

so that f' is continuous on $(0, 8]$; but f' does not exist at $x = 0$ and formula (3) does not apply. Looking at the graph sideways, however, we have

$$x = g(y) = y^{3/2} \qquad\qquad (0 \le y \le 4)$$

and therefore

$$L = \int_0^4 \sqrt{1 + \tfrac{9}{4}y}\, dy = \tfrac{8}{27}(10\sqrt{10} - 1)$$

(from Example 1).

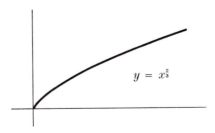

$$y = x^{\frac{2}{3}}$$

FIGURE 8

6D3. Arc length on the circle. Of course we want to test all this on a circle. Consider the circle

$$x^2 + y^2 = r^2,$$

with center $(0, 0)$ and radius r. Traditionally, arc length is measured from the point $Q_0 = (r, 0)$. But the function

(4) $$y = q(x) = \sqrt{r^2 - x^2}$$

is not differentiable there. (The tangent line is vertical.) So we work instead with x as a function of y (the same function!):

(5) $$x = q(y) = \sqrt{r^2 - y^2} \qquad (-r \le y \le r).$$

Its graph is the right-hand semicircle (Figure 9).

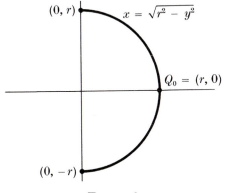

FIGURE 9

For $-r < y < r$ we have

(6) $$q'(y) = -\frac{y}{\sqrt{r^2 - y^2}} \qquad (-r < y < r).$$

Then we get

(7) $$\sqrt{1 + q'(y)^2} = \frac{r}{\sqrt{r^2 - y^2}} \qquad (-r < y < r).$$

The length of arc from Q_0 to a point (x, y) in the first quadrant, $y < r$, (Figure 10) is

(8) $$L_0^y = \int_0^y \sqrt{1 + q'^2} \qquad (0 \le y \le r).$$

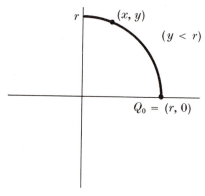

Because y appears as the upper limit of integration, let us use a different variable in the integrand, say t. Thus $\big($see $(7)\big)$,

$$(9) \qquad\qquad L_0^y = r \int_0^y \frac{1}{\sqrt{r^2 - t^2}}\, dt \qquad\qquad (0 \le y < r).$$

By the Fundamental Theorem of Calculus, this is a differentiable function of y for $0 \le y < r$. Therefore it is continuous.

We can't include $y = r$ in the formula, but we can get there by additivity. The length of arc from Q_0 to the point $(r/\sqrt{2},\, r/\sqrt{2})$ is

$$(10) \qquad\qquad r \int_0^{r/\sqrt{2}} \frac{1}{\sqrt{r^2 - t^2}}\, dt.$$

By symmetry, the length of arc between this point and $(0, r)$ is the same thing. (See Figure 11. Work with the function $y = q(x)$ and you get the same

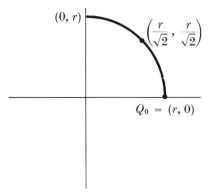

integral.) Hence the length of the quarter-circle is twice this integral. Similarly, we can measure arcs all the way around the circle.

6D4. The number π. This famous number is now *defined* as follows:

(11) π *is the length of the semicircle of radius* 1.

As (x, y) moves around this semicircle, the length of arc from $(1, 0)$ increases continuously from 0 to π. At the complete circle it reaches 2π. By the Intermediate-Value Theorem (§2E5), it also assumes every value in between.

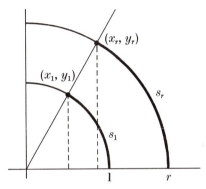

FIGURE 12

In Figure 12, the radial line into the first quadrant cuts the circles $x^2 + y^2 = 1$ and $x^2 + y^2 = r^2$ in the points (x_1, y_1) and (x_r, y_r). By similar triangles, $x_r = rx_1$ and $y_r = ry_1$. We show next that the corresponding arcs are related the same way:

(12) $s_r = rs_1.$

In fact, by (9),

(13)
$$s_r = r \int_0^{y_r} \frac{1}{\sqrt{r^2 - t^2}} \, dt = r \int_0^{y_1} \frac{1}{\sqrt{r^2 - r^2 u^2}} \, r \, du \qquad (t = ru)$$

$$= r \int_0^{y_1} \frac{1}{\sqrt{1 - u^2}} \, du = rs_1.$$

By additivity, this relation holds for the full circle as well. Hence the circumference of the circle of radius r is

$$C = 2\pi r.$$

Problems (§6D)

1. Find the length of the graph of:

(a) $y = \frac{1}{6}\left(x^3 + \frac{3}{x}\right)$ $(1 \leq x \leq 3)$

(b) $y = \frac{1}{24}\left(x^3 + \frac{48}{x}\right)$ $(2 \leq x \leq 4)$

(c) $y = \frac{1}{6}\sqrt{x}\,(4x - 3)$ $(1 \leq x \leq 9)$

(d) $y = \frac{1}{3}\sqrt{x}\,(x - 3)$ $(1 \leq x \leq 4)$

(e) $y = \frac{1}{8}\left(x^4 + \frac{2}{x^2}\right)$ $(1 \leq x \leq 3)$

(f) $y = \frac{1}{8}\left(2x^4 + \frac{1}{x^2}\right)$ $(1 \leq x \leq 2)$

(g) $y^3 = 8x^2$,   between the points $(1, 2)$ and $(8, 8)$
(h) $y^2 = x^3$,   between the points $(0, 0)$ and $(1, 1)$
(i) $x^{2/3} + y^{2/3} = 4$,   between the points $(2\sqrt{2}, 2\sqrt{2})$ and $(8, 0)$

2. Set up the integrals for the following lengths, but do not try to evaluate them.
   (a) The portion of $f(x) = -x^2 + 6x - 5$ lying above the x-axis
   (b) $x = y^2 + 1$ between $(1, 0)$ and $(2, 1)$
   (c) $f(x) = \sqrt{x}$ $(1 \leq x \leq 9)$
   (d) $f(x) = x^3$ $(0 \leq x \leq 2)$
   (e) $f(x) = \sqrt{9 - x^2}$ $(-2 \leq x \leq 2)$
   (f) *(Careful!)* $f(x) = \sqrt{9 - x^2}$ $(-3 \leq x \leq 3)$

3. In each case below, $P_1 = (x_1, y_1)$ and $P_2 = (x_2, y_2)$ are points on the circle $x^2 + y^2 = 1$, with neither coordinate zero. Express the length of the counterclockwise arc $\widehat{P_1 P_2}$ in terms of integrals of the form

$$H(u, v) = \int_u^v \frac{1}{\sqrt{1 - t^2}}\, dt \qquad (-1 < u < v < 1).$$

   (a) P_1 in the first quadrant, P_2 in the second
   (b) P_1 and P_2 both in the third quadrant, $y_1 < y_2$
   (c) P_1 in the second quadrant, P_2 in the fourth

Answers to problems (§6D)

1. (a) $\frac{14}{3}$ (b) $\frac{17}{6}$ (c) $\frac{55}{3}$ (d) $\frac{10}{3}$ (e) $\frac{92}{9}$ (f) $\frac{123}{32}$
   (g) $\frac{1}{27}(104\sqrt{13} - 125)$ (h) $\frac{1}{27}(13\sqrt{13} - 8)$ (i) 6

2. (a) $\displaystyle\int_{1}^{5} \sqrt{1 + 4(x - 3)^2}\, dx$

   (b) $\displaystyle\int_{0}^{1} \sqrt{1 + 4y^2}\, dy$ or $\displaystyle\int_{1}^{2} \sqrt{1 + \frac{1}{4(x - 1)}}\, dx$

   (c) $\displaystyle\int_{1}^{9} \sqrt{1 + \frac{1}{4x}}\, dx$ or $\displaystyle\int_{1}^{3} \sqrt{1 + 4y^2}\, dy$

   (d) $\displaystyle\int_{0}^{2} \sqrt{1 + 9x^4}\, dx$

   (e) $3\displaystyle\int_{-2}^{2} \frac{1}{\sqrt{9 - t^2}}\, dt$ or $6\displaystyle\int_{0}^{2} \frac{1}{\sqrt{9 - t^2}}\, dt$

   (f) $12\displaystyle\int_{0}^{3/\sqrt{2}} \frac{1}{\sqrt{9 - t^2}}\, dt$

3. (a) $H(x_2, x_1)$ (b) $H(x_2, x_1)$ or $H(y_1, y_2)$
   (c) $2H(0, y_1) + H(x_1, x_2)$ or $H(0, y_1) + \pi/2 + H(0, x_2)$

§6E. Surface area

6E1. Properties of surface area. Let f be a nonnegative, continuous function on $[a, b]$. When the graph of f is revolved about the x-axis, it generates a surface. We wish to define and compute the area of this surface.

The concept of surface area in general is very difficult. We will restrict ourselves for now to *surfaces of revolution*, as just described. Even these are not easy to analyze.

Of course, we suppose that surface area is additive (Figure 1). Thus, if

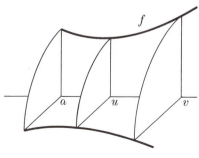

FIGURE 1

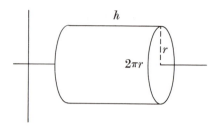

FIGURE 2

S_u^v denotes the area generated by the portion of the graph between u and v, then we assume:

(A)
$$S_a^u + S_u^v = S_a^v.$$

Next, we have to have information about some basic surface of revolution. Consider the simplest of all: that generated by a horizontal line segment. If the length of the segment is h and its distance above the x-axis is r, then it sweeps out a cylinder of radius r and "height" h (Figure 2). We can imagine cutting this surface along the segment and flattening it out. It forms a rectangle with sides h and $2\pi r$ and hence of area $2\pi rh$. Our assumption is that this is the area of the cylindrical surface:

(1)   *The area of a cylinder of radius r and height h is $2\pi rh$.*

Next we look for bounds on S_u^v. For this we must be able to *compare* areas. Obviously, in the case of the cylinder, a longer segment and larger radius yield a greater surface area. Our second assumption makes a similar comparison for any two graphs. Actually, we need it only when one or the other is a horizontal segment. Let us say for short that one graph lies *completely above* another if the minimum of the first is $\geq$ the maximum of the second, so that a horizontal segment can be interposed between them (Figure 3).

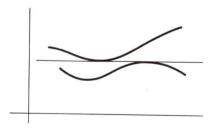

FIGURE 3

Our assumption is:

(2)   *A graph completely above and longer than another sweeps out a*
           greater area.

6E2. Definition of area of a surface of revolution. These assumptions
lead easily to a definition and formula for the area swept out by the graph of f.
Since they refer to the length of the graph, we assume that f has a continuous
derivative (§6D2).

According to §6D2, the length of the graph between u and v is a number
between

$$(3) \qquad\qquad (v - u) \min_{[u,v]} \sqrt{1 + f'^2}$$

and

$$(4) \qquad\qquad (v - u) \max_{[u,v]} \sqrt{1 + f'^2}$$

(see §6D(**B**) or §6D(3)). Place a horizontal segment l_1 of the shorter length
(3) at the distance $\min\limits_{[u,v]} f$; and another, l_2, of the greater length (4), at distance
$\max\limits_{[u,v]} f$. See Figure 4. According to principle (2), the area of the cylinder that
l_1 sweeps out is $\leq S_u^v$, and the area of the one l_2 sweeps out is $\geq S_u^v$. Hence,
by principle (1),

$$2\pi(v - u)(\min_{[u,v]} f) \min_{[u,v]} \sqrt{1 + f'^2}$$

(**B**) $$\leq S_u^v$$

$$\leq 2\pi(v - u)(\max_{[u,v]} f) \max_{[u,v]} \sqrt{1 + f'^2}.$$

We were led to this from the principles (1) and (2) about surface area and
the principle (1) of §6D about arc length. We did not prove any of these

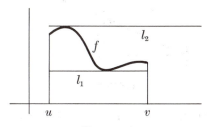

FIGURE 4

principles but only assumed them. No one is obliged to accept them. But anyone who does assume these principles and also believes that surface area is additive is now forced, by the General Theorem for the Integral, to define the surface area as follows:

(5)
$$S = 2\pi \int_a^b f\sqrt{1 + f'^2}.$$

Memory aid. See Figure 5. The area of the portion of the surface between x and $x + \Delta x$ is approximately the same as that swept out by the tangent line at x. The length of the

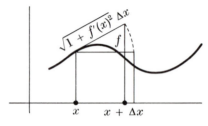

FIGURE 5

latter segment is $\sqrt{1 + f'(x)^2}\, \Delta x$. Its effect is approximately as if horizontal at height $f(x)$. It then sweeps out a cylinder of radius $f(x)$ and "height" $\sqrt{1 + f'(x)^2}\, \Delta x$, hence of area

$$2\pi f(x)\sqrt{1 + f'(x)^2}\, \Delta x.$$

The actual area of the surface between a and b is

$$2\pi \int_a^b f(x)\sqrt{1 + f'(x)^2}\, dx.$$

Example 1. Find the area of the surface of a right circular cone of radius r and altitude h (Figure 6).

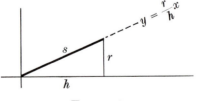

FIGURE 6

Solution. The cone is obtained by revolving the segment shown (Figure 6). The slope of the segment is r/h and its equation is $y = (r/h)x$. Therefore

$$S = 2\pi \int_0^h \frac{r}{h} x \sqrt{1 + (r/h)^2}\, dx = 2\pi \frac{r}{h} \sqrt{1 + (r/h)^2}\, (\tfrac{1}{2}h^2) = \pi r \sqrt{h^2 + r^2}.$$

6E3. x as a function of y. If f' is not continuous on $[a, b]$ the preceding results are not applicable. However, if f has an inverse, g, with a continuous derivative, we can proceed as follows. Here g and g' are defined on a closed interval $[\alpha, \beta]$ (namely, where $\alpha = \min_{[a,b]} f$ and $\beta = \max_{[a,b]} f$ (§2E8)). Consider that portion of the graph of g between $y = u$ and $y = v$. Let T_u^v denote the area of the surface generated by revolving this portion about the x-axis.

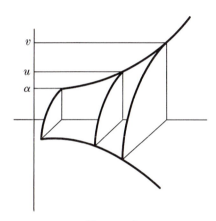

Figure 7

See Figure 7. Certainly, we demand

(A) $$T_\alpha^u + T_u^v = T_\alpha^v.$$

Reasoning as before, we observe that the length of the graph between $y = u$ and $y = v$ is a number between

(6) $l_1 = (v - u) \min_{[u,v]} \sqrt{1 + g'^2}$ and $l_2 = (v - u) \max_{[u,v]} \sqrt{1 + g'^2}.$

The minimum and maximum radii of revolution are $u = \min_{[u,v]} y$ and $v = \max_{[u,v]} y$. See Figure 8. Hence, by principles (1) and (2),

$$2\pi(v - u)(\min_{[u,v]} y) \min_{[u,v]} \sqrt{1 + g'^2}$$

(B) $$\leq T_u^v$$

$$\leq 2\pi(v - u)(\max_{[u,v]} y) \max_{[u,v]} \sqrt{1 + g'^2}.$$

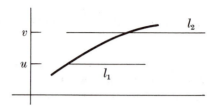

FIGURE 8

Because of **(A)** and **(B)** we are forced to define the surface area as

$$(7) \qquad T = 2\pi \int_\alpha^\beta y\sqrt{1 + g'(y)^2} \, dy.$$

Memory aid. See Figure 9. The area of the portion of the surface between y and $y + \Delta y$ is approximately as if swept out by the tangent line at y. The length of this segment is

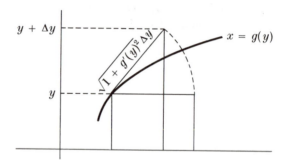

FIGURE 9

$\sqrt{1 + g'(y)^2} \, \Delta y$. Its effect is approximately as if horizontal at height y. It then sweeps out a cylinder of radius y and "height" $\sqrt{1 + g'(y)^2} \, \Delta y$, hence of area

$$2\pi y\sqrt{1 + g'(y)^2} \, \Delta y.$$

The actual area of the surface from α to β is

$$2\pi \int_\alpha^\beta y\sqrt{1 + g'(y)^2} \, dy.$$

Example 2. Find the area of a sphere of radius r.

Solution. The right-hand hemisphere is generated by revolving the quarter-circle $x^2 + y^2 = r^2$ ($x \geq 0, y \geq 0$) about the x-axis. Now, the function

$$y = q(x) = \sqrt{r^2 - x^2} \qquad (0 \leq x \leq r)$$

has no derivative at $x = r$; and its inverse,

$$x = q(y) = \sqrt{r^2 - y^2} \qquad (0 \le y \le r)$$

has no derivative at $y = r$. Consequently, neither of the formulas, (5) or (7), will be adequate alone.

To get S_0^r, the area of the hemisphere, we pick an intermediate point (x_1, y_1) $(0 < x_1 < r)$ and add:

$$S_0^r = S_0^{x_1} + T_0^{y_1}.$$

See Figure 10. Since both x and y will appear as variables, it is convenient to use

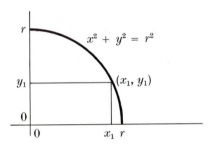

$$x^2 + y^2 = r^2$$

(x_1, y_1)

FIGURE 10

a neutral variable and write

$$q(t) = \sqrt{r^2 - t^2}.$$

We find that

(8) $$q'(t) = -\frac{t}{q(t)} \qquad (0 \le t < r)$$

and

(9) $$\sqrt{1 + q'(t)^2} = \frac{r}{q(t)} \qquad (0 \le t < r).$$

Consequently, from (5) and (9),

(10)
$$S_0^{x_1} = 2\pi \int_0^{x_1} q(t)\sqrt{1 + q'(t)^2}\, dt$$

$$= 2\pi \int_0^{x_1} r\, dt = 2\pi r x_1;$$

and from (7), (9), and (8),

$$T_0^{y_1} = 2\pi \int_0^{y_1} t\sqrt{1 + q'(t)^2}\, dt = 2\pi \int_0^{y_1} r\, \frac{t}{q(t)}\, dt$$

(11)

$$= -2\pi r \int_0^{y_1} q'(t)\, dt = 2\pi r[q(0) - q(y_1)] = 2\pi r(r - x_1).$$

Hence

(12) $$S_0 = S_0^{x_1} + T_0^{y_1} = 2\pi r x_1 + 2\pi r(r - x_1) = 2\pi r^2$$

and the area of the sphere is

(13) $$S = 2S_0^r = 4\pi r^2.$$

Problems (§6E)

1. Find the area of the surface generated when the given curve is revolved about the x-axis.

(a) $y = x^3 \quad (0 \le x \le \frac{2}{3})$

(b) $y = \frac{1}{6}\left(x^3 + \dfrac{3}{x}\right) \quad (1 \le x \le 3)$

(c) $y = \frac{1}{24}\left(x^3 + \dfrac{48}{x}\right) \quad (2 \le x \le 4)$

(d) $y = \frac{1}{6}\sqrt{x}(4x - 3) \quad (1 \le x \le 9)$

(e) $y = \frac{1}{3}\sqrt{x}\, |x - 3| \quad (1 \le x \le 4)$

(f) $y = \frac{1}{8}\left(x^4 + \dfrac{2}{x^2}\right) \quad (1 \le x \le 3)$

(g) $y = \frac{1}{8}\left(2x^4 + \dfrac{1}{x^2}\right) \quad (1 \le x \le 2)$

(h) $x^{2/3} + y^{2/3} = 1 \ (0 \le x \le \frac{3}{8}\sqrt{3})$ (Careful!)

2. A nonvertical line segment of length 1, with midpoint $(0, 1)$, is revolved about the x-axis to generate a surface. Show that the area of the surface is equal to 2π, independently of the slope of the segment.

Answers to problems (§6E)

1. (a) $\frac{98}{729}\pi$ (b) $\frac{208}{9}\pi$ (c) $\frac{47}{4}\pi$ (d) $\frac{2654}{9}\pi$ (e) $\frac{23}{9}\pi$ (f) $\frac{8429}{81}\pi$
   (g) $\frac{16911}{1024}\pi$ (h) $\frac{93}{80}\pi$

§6F. Average value; moments

6F1. The average value of a function. In a calculus class of five students, the final grades were 70, 80, 80, 90, 100. The average grade was then 84: if all grades were equal to the average, the total (420) would be the same as before.

Let f be a continuous function on $[a, b]$. We wish to define $\underset{[a,b]}{\text{avg}} f$, the average value of f on $[a, b]$. According to the general notion of average, it is the number such that a constant function on $[a, b]$ with that value will yield the same total result as before. But what do we mean here by the total result?

In case f is a velocity function over the time interval $[a, b]$, we have a ready answer. We agreed in grade school and §3B5 that the average velocity is the net distance divided by the time. Now, distance is the antiderivative of velocity (with respect to time); hence, by the Fundamental Theorem, the net distance is $\int_a^b f$. Since the total time is $b - a$,

$$(1) \qquad \underset{[a,b]}{\text{avg}} f = \frac{1}{b - a} \int_a^b f \qquad (a < b).$$

Using (1) we get

$$(2) \qquad \int_a^b \underset{[a,b]}{\text{avg}} f = (\underset{[a,b]}{\text{avg}} f)(b - a) = \int_a^b f.$$

Thus a constant function (velocity) on $[a, b]$ with value $\underset{[a,b]}{\text{avg}} f$ yields the same integral (distance) as before.

Guided by this example, *we take* (1) *as the definition* of $\underset{[a,b]}{\text{avg}} f$ $(a < b)$ in general. (We also define $\underset{[a,a]}{\text{avg}} f = f(a)$.)

If we compare the second equation in (2) with the condition **(B)** in the definition of integral, we see that

$$(3) \qquad \underset{[a,b]}{\min} f \le \underset{[a,b]}{\text{avg}} f \le \underset{[a,b]}{\max} f.$$

This is what we expect of an average value.

6F2. Functions and variables. The expression "average velocity" was used above as an abbreviation for average velocity *with respect to time*, or, more simply and unambiguously, $\underset{[a,b]}{\text{avg}} f$, the *average of the function f*. The average with respect to *distance* is the average of a different function.

Example 1. Suppose you travel 30 miles at 30 miles per hour and return at 60 miles per hour. Was your average speed 45 miles per hour? Not according to what we agreed in grade school and §3B5. The trip out took an hour and the return trip took a half-hour. You went 60 miles in an hour and a half. You averaged 40 miles per hour.

Figure 1(a) shows speed plotted against distance traveled, and Figure 1(b) shows speed as a function of time. In the first case, the speeds 30 and 60 are associated with equal numbers of units on the x-axis; while in the second, speed 30 holds for twice as many units as speed 60. So we expect the first average to be greater than the second. In fact, 45 is the average speed with respect to distance, and 40, as we have seen, the average with respect to time.

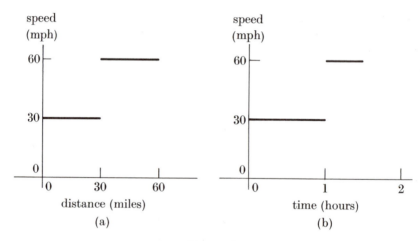

FIGURE 1

Example 2. Suppose $s = t^3$, where s is distance and t is time. Then v, the velocity, is

$$v = 3t^2 = 3s^{2/3}.$$

Velocity is a function of time and a different function of distance. Say

$$v = f(t) = 3t^2 \qquad \text{and} \qquad v = g(s) = 3s^{2/3}.$$

Consider the interval, say, $0 \le t \le 2$. Then $0 \le s \le 8$, and the average velocities are

$$\operatorname*{avg}_{[0,2]} f = \tfrac{1}{2} \int_0^2 3t^2 \, dt = 4$$

(the average velocity with respect to time), and

$$\operatorname*{avg}_{[0,8]} g = \tfrac{1}{8} \int_0^8 3s^{2/3} \, ds = \tfrac{36}{5}$$

(the average velocity with respect to distance). As a qualitative check we notice that both averages lie between 0 and 12, the minimum and maximum values of v for $0 \leq t \leq 2$.

6F3. Moments: finite case. *Moments* are closely related to averages. In mechanics a moment is, roughly, a turning effect. Consider n masses,

$$m_1, \cdots, m_n$$

situated at points

$$x_1, \cdots, x_n$$

along the x-axis. The *first moment about the origin* is by definition the number

(4) $$M_0 = m_1 x_1 + \cdots + m_n x_n.$$

The total mass is, of course,

(5) $$M = m_1 + \cdots + m_n.$$

The *center of mass* (or center of gravity) of the system, denoted $\bar{x}$, is defined by

(6) $$\bar{x} = \frac{M_0}{M} = \frac{m_1 x_1 + \cdots + m_n x_n}{m_1 + \cdots + m_n}.$$

If the total mass were concentrated at the center of mass, the first moment would come out the same. For, by (4), the first moment of the concentrated distribution is $M\bar{x}$. By (6), this is equal to the first first moment.

The same expressions came up in school in discussions of the seesaw. In (4), the fulcrum is at 0, and m_1 is the weight of the person sitting at location x_1, etc. The seesaw balances when the center of mass, $\bar{x}$, is at 0; equivalently, when the first moment is equal to 0.

In statistics, $\bar{x}$ is called the *weighted average* of the x's. If m_1 students got a test grade of x_1, etc., then the class average on the test was $\bar{x}$.

Example 3. What are the first moment and center of mass of three particles distributed on the line as follows:

$$\text{mass 3 at 1,} \qquad \text{mass 3 at 2,} \qquad \text{mass 2 at 3 ?}$$

Solution. The first moment is

$$M_0 = (3)(1) + (3)(2) + (2)(3) = 15.$$

The total mass is

$$M = 3 + 3 + 2 = 8.$$

The center of mass is

$$\bar{x} = \frac{M_0}{M} = \frac{15}{8}.$$

6F4. Moments: continuous case. Imagine, now, a *continuous* distribution of mass on an interval $[a, b]$. That is, assume that there is a continuous density function, ρ (rho), so that for each $x \in [a, b]$, the mass between a and x is $\int_a^x \rho$. We wish to define the first moment about the origin and the center of mass of this distribution.

Let M_u^v denote the contribution to the first moment that comes from the interval $[u, v]$. Physicists assure us that (in analogy with the finite case) we are to have:

(A)
$$M_a^u + M_u^v = M_a^v.$$

Now we want bounds on M_u^v. To get an upper bound we take the maximum density throughout and concentrate the resulting mass at the maximum x. (This assumption about moments also comes to us from the physicists.) The mass on $[u, v]$ with constant density $\max_{[u,v]} \rho$ is just $(v - u) \max_{[u,v]} \rho$; and the moment is the product of this with $\max_{[u,v]} x$ (as we know from the finite case). We argue similarly for a lower bound. Altogether, we get

(B)
$$(v - u)(\min_{[u,v]} x)(\min_{[u,v]} \rho) \leq M_u^v \leq (v - u)(\max_{[u,v]} x)(\max_{[u,v]} \rho).$$

Because of **(A)** and **(B)**, we are forced, by the General Theorem, to define the first moment about the origin as

(7)
$$M_0 = \int_a^b x\rho(x)\, dx.$$

The total mass is given by

(8)
$$M = \int_a^b \rho,$$

and the center of mass is defined by

(9)
$$\bar{x} = \frac{M_0}{M} = \frac{\int_a^b x\rho(x)\,dx}{\int_a^b \rho(x)\,dx}.$$

Memory aid. The mass between x and $x + \Delta x$ is approximately $\rho(x)\,\Delta x$. Therefore its contribution to the first moment is approximately

$$x\rho(x)\,\Delta x.$$

The total moment is actually

$$\int_a^b x\rho(x)\,dx.$$

Again, if the total mass were concentrated at the center of mass, the first moment would come out the same. For, by the finite case, it would be equal to $M\bar{x}$, and hence, by (9), to M_0.

Example 4. Let $\rho(x) = 1$ (constant) on $[0, 1]$ (Figure 2). Find the first moment and center of mass.

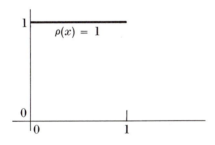

FIGURE 2

Solution. Obviously, $\bar{x} = \frac{1}{2}$. Since the mass is 1, $M_0 = \frac{1}{2}$. This checks with the formula:

$$M_0 = \int_0^1 x\,dx = \frac{1}{2}.$$

Example 5. Let $\rho(x) = \sqrt{x}$ on $[0, 1]$ (Figure 3). Find the first moment and center of mass.

Solution.

$$M = \int_0^1 \sqrt{x}\,dx = \frac{2}{3}, \qquad M_0 = \int_0^1 x^{3/2}\,dx = \frac{2}{5},$$

and

$$\bar{x} = \frac{\frac{2}{5}}{\frac{2}{3}} = \frac{3}{5}.$$

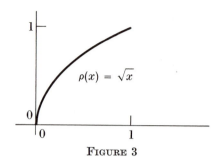

FIGURE 3

Example 6. Let $\rho(x) = x^2$ on $[0, 1]$ (Figure 4). Find the first moment and center of mass.

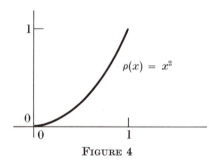

FIGURE 4

Solution.

$$M = \int_0^1 x^2 \, dx = \tfrac{1}{3}, \qquad M_0 = \int_0^1 x^3 \, dx = \tfrac{1}{4},$$

and

$$\bar{x} = \frac{\tfrac{1}{4}}{\tfrac{1}{3}} = \frac{3}{4}.$$

Example 7. Let $\rho(x) = x^2$ on $[-1, 1]$ (Figure 5). Find the first moment and center of mass.

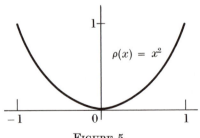

FIGURE 5

Solution. By symmetry, they are both 0. This checks with the formula:

$$M_0 = \int_{-1}^{1} x^3 \, dx = 0.$$

The mass, incidentally, is $\frac{2}{3}$ (twice that of Example 6).

Problems (§6F)

1. (a) Show that if f is nonnegative and continuous on $[a, b]$, then $\underset{[a,b]}{\text{avg}} f$ is the height of the rectangle with base $b - a$ whose area is the same as the area under the graph of f.
   (b) Deduce from this that the shaded areas in Figure 6 are equal.
   (c) Draw the corresponding diagrams for Example 1.
   (d) Draw the corresponding diagrams for Example 2.

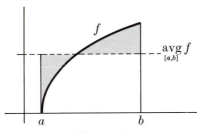

FIGURE 6

2. Find the average value of the following functions, and draw illustrative diagrams as in Figure 6.
   (a) $f(x) = 9 - x^2$ $(-3 \le x \le 3)$
   (b) $f(x) = \dfrac{1}{x^2}$ $(1 \le x \le 4)$
   (c) $f(x) = x^3$ $(-2 \le x \le 2)$
   (d) $f(x) = x\sqrt{4 - x^2}$ $(-2 \le x \le 2)$

3. A freely falling object, starting from rest, falls according to the law

$$s = \tfrac{1}{2}gt^2,$$

where t is the time in seconds, s is the distance in feet, and g is the gravitational constant.
   (a) Express the velocity as $v = F(t)$ and as $v = G(s)$.
   (b) Show that the average velocity with respect to time during the first 4 seconds is $2g$ feet per second. Illustrate with a diagram as in Figure 6.
   (c) Show that the average velocity with respect to distance during the same interval is $8g/3$ feet per second. Illustrate with a diagram as in Figure 6.

4. Let $f(x) = x^2$. Find:

(a) $\operatorname*{avg}_{[1,4]} f - 1$

(b) $\operatorname*{avg}_{[1,4]} (f - 1)$

(c) $2 \operatorname*{avg}_{[1,4]} f$

(d) $\operatorname*{avg}_{[1,4]} (2f)$

(e) $\operatorname*{avg}_{[1,4]} f^2$

(f) $\left(\operatorname*{avg}_{[1,4]} f\right)^2$

5. Let $f(x) = x^3$. Find:

(a) $\operatorname*{avg}_{[1,3]} f - 1$

(b) $\operatorname*{avg}_{[1,3]} (f - 1)$

(c) $2 \operatorname*{avg}_{[1,3]} f$

(d) $\operatorname*{avg}_{[1,3]} (2f)$

(e) $\operatorname*{avg}_{[1,3]} f^2$

(f) $\left(\operatorname*{avg}_{[1,3]} f\right)^2$

6. Find the total mass, first moment, and center of mass of the distributions given by the following density functions.

(a) $\rho(x) = 3x \quad (0 \le x \le 32)$

(b) $\rho(x) = 2x^2 \quad (0 \le x \le 4)$

(c) $\rho(x) = 3\sqrt{x} \quad (0 \le x \le 4)$

(d) $\rho(x) = \sqrt{x + 1} \quad (0 \le x \le 15)$

(e) $\rho(x) = 1/\sqrt{1 - x^2} \quad (0 \le x \le 1/\sqrt{2})$. [*Hint.* See §6D(10).]

Answers to problems (§6F)

2. (a) 6   (b) $\frac{1}{4}$ (c) 0   (d) 0

4. (a) 6   (b) 6   (c) 14   (d) 14   (e) $68\frac{1}{5}$ (f) 49

5. (a) 9   (b) 9   (c) 20   (d) 20   (e) $156\frac{1}{7}$ (f) 100

6. (a) $\frac{3}{2} \times 32^2$, 32^3, $21\frac{1}{3}$

(b) $\frac{128}{3}$, 128, 3

(c) 16, $\frac{192}{5}$, $2\frac{2}{5}$

(d) 42, $367\frac{1}{5}$, $8\frac{26}{35}$

(e) $\dfrac{\pi}{4}$, $1 - \dfrac{1}{\sqrt{2}}$, $\dfrac{4}{\pi}\left(1 - \dfrac{1}{\sqrt{2}}\right)$

§6G. Fluid force

6G1. Fluid force on a horizontal surface. Consider a tank containing a fluid weighing k pounds per cubic foot. We learned in school that the pressure exerted by the fluid at a depth of h feet is kh pounds per square foot, and that this pressure is the same in all directions, depending only on k and h.

The *force* on a surface of area A square feet, on which is exerted a constant pressure of p pounds per square foot, is pA pounds. Thus the force of the fluid on a horizontal surface of area A at depth h is khA pounds (equal to the weight of a column of fluid of height h over the surface).

If the surface is not horizontal, then the pressure is not the same at all points, and defining and computing the force become a problem in integration.

6G2. Fluid force on a vertical surface. Consider a vertical surface submerged in a fluid of constant density (weight per unit volume) k. We wish to define and compute the force of the fluid against the surface.

For convenience we use a coordinate system in which the positive y-axis is directed downward, so that the depth increases with y. Let α denote the minimum depth of the surface, and β the maximum depth. We assume that the width of the surface is a continuous function, w:

(1) $\qquad\qquad w(y) = $ width of surface at depth $y \qquad (\alpha \le y \le \beta)$.

In fact, we will assume somewhat more: that the surface is bounded by two curves, $x = f(y)$ and $x = g(y)$, where f and g are continuous on $[\alpha, \beta]$ and

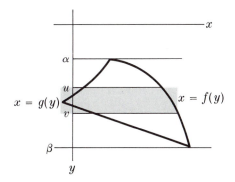

FIGURE 1

$f \ge g$ (Figure 1). Thus,

(2) $\qquad\qquad\qquad w(y) = f(y) - g(y).$

Consider the portion of the surface between depths u and v $(u \le v)$. Denote the force against this portion by F_u^v. We are assured by physicists that

(A) $\qquad\qquad\qquad F_\alpha^u + F_u^v = F_\alpha^v.$

To obtain an upper bound for F_u^v we look at a *greater* pressure against a *larger* area. The portion of the surface between u and v is enclosed in a rectangle of width

$$\max_{[u,v]} f - \min_{[u,v]} g$$

and height $v - u$, hence of area

$$(v - u)(\max_{[u,v]} f - \min_{[u,v]} g)$$

(Figure 1). The actual force F_u^v is less than if the surface had this larger area and were all at the maximum depth, $v = \max_{[u,v]} y$. (This, too, is an observation about *force*.)

Similarly, either the portion of the surface between u and v contains a rectangle of height $v - u$ and width

$$\min_{[u,v]} f - \max_{[u,v]} g$$

(Figure 2(a)), or else $\min_{[u,v]} f - \max_{[u,v]} g < 0$ (Figure 2(b)). In either case, S_u^v is

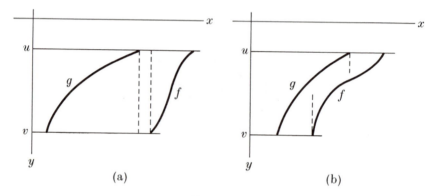

FIGURE 2

greater than $\min_{[u,v]} f - \max_{[u,v]} g$ multiplied by the force at the minimum depth $u = \min_{[u,v]} y$. Together with the above comparison, we have:

$$k(v - u)(\min_{[u,v]} y)(\min_{[u,v]} f - \max_{[u,v]} g)$$

(B)

$$\leq F_u^v$$

$$\leq k(v - u)(\max_{[u,v]} y)(\max_{[u,v]} f - \min_{[u,v]} g).$$

Because of **(A)** and **(B)**, we are forced, by the General Theorem, to define the force as

$$(3) \qquad F = k \int_\alpha^\beta y[f(y) - g(y)] \, dy = k \int_\alpha^\beta yw(y) \, dy.$$

Memory aid. The area of the strip between y and $y + \Delta y$ is approximately $w(y) \, \Delta y$ square feet (Figure 3(a)). The force against this strip is approximately as if the strip

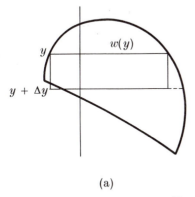

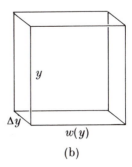

(a)   (b)

FIGURE 3

were horizontal at the constant depth y; then it would be supporting $yw(y) \, \Delta y$ cubic feet of fluid (Figure 3(b)). At k pounds per cubic foot, this is

$$kyw(y) \, \Delta y$$

pounds of force. The actual force against the surface from depth α to depth β is

$$k \int_\alpha^\beta yw(y) \, dy.$$

Example 1. What is the force against a vertical plate 2 feet square submerged in water with the top of the plate along the water's surface?

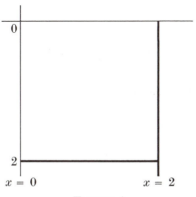

FIGURE 4

Solution. See Figure 4. The width function is constant: $w(y) = 2$. Hence the total force is

$$F = k \int_0^2 2y \, dy = 4k.$$

With $k = 62.5$ lbs., $F = 250$ lbs.

Example 2. What is the force against the portion above a diagonal of the square of Example 1?

Question. Do you expect it to be more than half that of Example 1, or less?

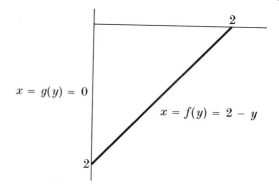

FIGURE 5

Solution. See Figure 5. Here

$$w(y) = 2 - y$$

and we have

$$F = k \int_0^2 y(2 - y) \, dy = \tfrac{4}{3}k,$$

one-third the other result.

Problems (§6G)

1. A trough with an isosceles trapezoidal cross section as shown (Figure 6) is filled with water weighing k pounds per cubic foot. What is the force against one side?

2. Plates in the form of the region bounded by the parabola $y = x^2$ and the line $y = 1$ (Figure 7) are submerged in a fluid in the positions shown in Figure 8, (a) and (b). Show that the force against the plate in the second position is two-and-a-half times that in the first.

3. A semicircular cylindrical tank of radius 10 feet is filled with water weighing k pounds per cubic foot (Figure 9). What is the force on one end of the tank?

4. Find the force on the submerged vertical plates pictured, if the liquid in which they are submerged has density k.
  (a) Figure 10
  (b) Figure 11
  (c) Figure 12
  (d) Figure 13
  (e) Figure 14

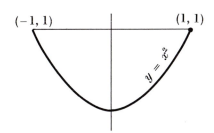

FIGURE 6

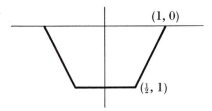

FIGURE 7

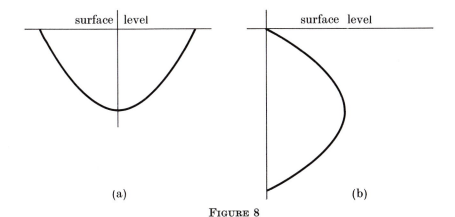

FIGURE 8

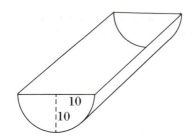

FIGURE 9

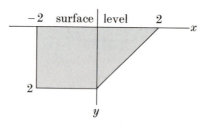

FIGURE 10

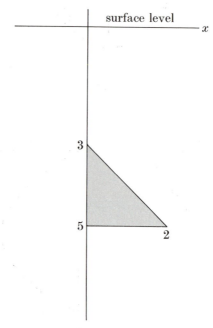

FIGURE 11

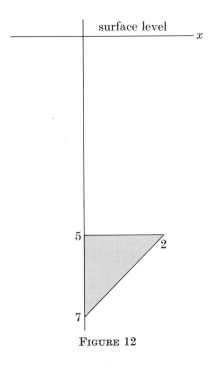

FIGURE 12

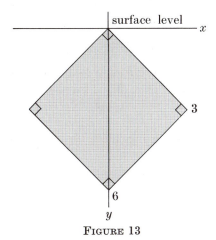

FIGURE 13

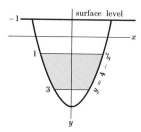

FIGURE 14

Answers to problems (§6G)

1. $\frac{2}{3}k$

3. $\frac{2000}{3}k$

4. (a) $\frac{16}{3}k$ (b) $\frac{26}{3}k$ (c) $\frac{34}{3}k$ (d) $54k$ (e) $\frac{8}{15}k(24\sqrt{3} - 11)$

§6H. Work

6H1. Work done in compressing a gas. A gas in a rigid container is compressed, say by the action of a piston. If p, the pressure (force per unit area), were constant from volume $V = b$ down to $V = a$, the amount of work done would be $p(b - a)$. In practice, the pressure increases as the volume diminishes. We wish to define the work done when the pressure is variable, say $p = f(V)$, where f is continuous.

Let W_u^v denote the work done when the volume is decreased from $V = v$ to $V = u$. Surely we must have

(A) $$W_u^v + W_a^u = W_a^v.$$

Next, we assume that greater pressure results in more work. (This, too, is an assumption about *work*.) Then W_u^v is greater than if the pressure were constant at the value $\min_{[u,v]} f$, and less than if constant at $\max_{[u,v]} f$. So we have

(B) $$(v - u) \min_{[u,v]} f \le W_u^v \le (v - u) \max_{[u,v]} f.$$

Because of **(A)** and **(B)**, we are forced to define the total amount of work done as

(1) $$W = \int_a^b f.$$

Memory aid. The work done in compressing volume $V + \Delta V$ down to volume V is approximately what would result from constant pressure $f(V)$, namely,

$$f(V)\,\Delta V.$$

The actual work done from volume b down to volume a is

$$\int_a^b f(V)\, dV.$$

The same formula gives the work done by an *expanding* gas in *increasing* a volume from a to b.

Note on units. To help keep track of units we may write (1) in the form

(2) $$W = \int_a^b p\, dV.$$

In this expression, p has units of pressure, and dV, representing V as it ranges between a and b, has units of volume. If p is in pounds per square foot and V is in cubic feet, then W is expressed in foot-pounds. If, as is common, p is in pounds per square inch and V is in cubic feet, then W is expressed in units of 144 foot-pounds.

Example 1. Air at pressure of 20 lb. per sq. in. is compressed from 200 cu. ft. to 50 cu. ft. Find the final pressure and the amount of work done if

(3) $$pV^{1.4} = C \qquad \text{(constant)}.$$

Solution. Here $a = 50$, $b = 200$, and

(4) $$p = f(V) = CV^{-1.4}.$$

Also, when $V = 200$, $p = 20$. This gives us C (from (3)):

(5) $$C = 20 \times 200^{1.4}.$$

Hence the final pressure is (from (4) and (5)):

(6) $$f(50) = C \times 50^{-1.4} = 20 \times 4^{1.4}.$$

It happens that $4^{1.4}$ is very close to 7. Granting this, we have

$$f(50) \approx 20 \times 7 = 140 \text{ lbs./sq. in.}$$

The work done is

(7)
$$W = C \int_{50}^{200} V^{-1.4}\, dV = \tfrac{5}{2} C(50^{-0.4} - 200^{-0.4})$$
$$= \tfrac{5}{2} \times 20 \times 200(4^{0.4} - 1)$$

(from (5)). Since $4^{0.4} \approx \tfrac{7}{4}$, we get

$$W \approx 10{,}000 \times \tfrac{3}{4}.$$

This is in units of 144 ft.-lbs. Hence, finally,

$$W \approx 10{,}000 \times \tfrac{3}{4} \times 144 = 1{,}080{,}000 \text{ ft.-lbs.}$$

6H2. Work done in pumping out a tank. A tank is filled with fluid, and the problem is to find how much work is done in pumping all the fluid out of the tank. We give the details for a tank in the form of a solid of revolution about the y-axis, from $y = 0$ to $y = \beta$. Other designs of tanks may be analyzed in analogous fashion.

It is convenient to direct the y-axis downward, as in Figure 1.

First the concept must be defined. Let k denote the weight of a cubic unit of fluid. Let $x = f(y)$ be the equation of the curve that is revolved to form the solid, where f is continuous.

Let W_u^v denote the work done in pumping out the fluid from the strip between levels u and v. Physicists assure us that

(A) $$W_0^u + W_u^v = W_0^v.$$

Next, we want bounds for W_u^v. The portion of the solid between levels u and v is enclosed in a cylinder of radius $\max\limits_{[u,v]} f$ and height $v - u$ (Figure 2). Hence its volume is less than

$$\pi(\max_{[u,v]} f)^2(v - u),$$

and therefore its weight is less than

$$k\pi(\max_{[u,v]} f)^2(v - u).$$

The work done is less than this weight times the maximum distance to be lifted, $v = \max\limits_{[u,v]} y$. (This, too, is an assumption about *work*.) This gives us an upper bound for W_u^v. There is a corresponding lower bound. Together, they give us

(B)
$$k\pi(v - u)(\min_{[u,v]} y)(\min_{[u,v]} f)^2$$
$$\leq W_u^v$$
$$\leq k\pi(v - u)(\max_{[u,v]} y)(\max_{[u,v]} f)^2.$$

Because of **(A)** and **(B)**, the General Theorem forces us to define the work done in emptying the tank as

(8) $$W = k\pi \int_a^\beta yf(y)^2 \, dy.$$

Memory aid. See Figure 3. The volume of fluid between levels y and $y + \Delta y$ is approximately

$$\pi f(y)^2 \, \Delta y;$$

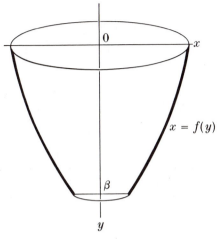

FIGURE 1

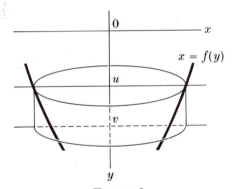

FIGURE 2

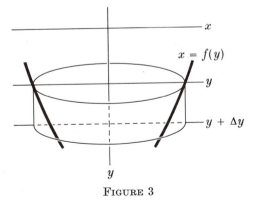

FIGURE 3

its weight,

$$k\pi f(y)^2 \, \Delta y.$$

It is lifted approximately the distance y. Hence the work done in lifting it is approximately

$$k\pi y f(y)^2 \, \Delta y.$$

The actual work done in emptying the tank is

$$k\pi \int_\alpha^\beta y f(y)^2 \, dy.$$

Example 2. A right circular conical tank of height 10 feet and radius 6 feet (at the top) is full of water (Figure 4). How much work is done in pumping all the water to the top of the tank?

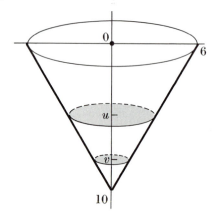

FIGURE 4

Solution. The equation of the line generating the cone is

$$x = \tfrac{3}{5}(10 - y).$$

The work done is

$$W = k\pi \int_0^{10} y(\tfrac{3}{5})^2 (10 - y)^2 \, dy = 300k\pi.$$

Since water weighs approximately $(200/\pi)$ lbs./cu. ft., this is about 60,000 foot-pounds.

Problems (§6H)

1. Assume that the pressure and volume of a gas are related by

$$pV^{1.4} = C,$$

where C is a constant.

(a) Find the work done in compressing 1024 cu. ft. of gas at 18 lbs./sq. in.
to 243 cu. ft. What is the final pressure?

(b) Show that the work done in compressing a gas of initial volume V_1 and
pressure p_1 to a volume V_2 is equal to

$$\tfrac{5}{2}p_1 V_1((V_1/V_2)^{2/5} - 1).$$

2. A tank in the shape of a *paraboloid of revolution*, obtained when the parabola
$y = x^2$ $(0 \le x \le \sqrt{6})$ is revolved about the y-axis, is full of water. How much
work is required to pump all the water over the top of the tank?

3. A right circular conical tank of height h and radius r is full of water. Show
that the work done in pumping all the water to the top is equal to

$$\tfrac{1}{12}k\pi r^2 h^2.$$

4. (a) A right cylinder with area of base A and height h is filled with fluid of
density k. Derive the formula

$$W = \tfrac{1}{2}kh^2 A$$

for the work done in pumping all the fluid to the top.

(b) Verify that this is the same as the work done in raising a weight equal
to the total weight of the fluid a distance half the height of the tank.

5. A tank in the shape of a rectangular box 10 feet high, with a base of area
10 square feet, is filled with water. How much work is done in pumping all the
water to the top of the tank?

6. A 10-foot-long water-filled trough has a semi-circular cross section of
diameter 3 feet. How much work is done in pumping the water out over the edge
of the trough?

7. A swimming pool is 75 feet long and 25 feet wide. It is 10 feet deep for its
first 25 feet; after that, the depth decreases linearly to 3 feet at the shallow end.
How much work is required to empty the pool by pumping all the water over the
edge?

Answers to problems (§6H)

1. (a) 5,160,960 ft.-lbs., $75\tfrac{23}{27}$ lbs./sq. in.

2. $36k\pi$

5. $500k$

6. $\tfrac{45}{2}k$

7. $60,208\tfrac{1}{3}k$

CHAPTER 7

Logarithm and Exponential

In calculus we study the logarithm as a function. The logarithmic and exponential functions are indispensable in many parts of mathematics and its applications.

In earlier courses where the logarithm is introduced, it is usually as the inverse of the exponential. That is, if $a^x = y$, then, by definition, $x = \log_a y$, the logarithm of y to the base a. This is correct but usually incomplete, as the definition of a^x for irrational x is ordinarily left out. It can be supplied directly, in terms of limits or least upper bounds. But then the work of verifying the laws of exponents is tedious and unexciting, and finding derivatives is very difficult.

What we shall do instead is take advantage of our knowledge of calculus. We have already developed a powerful theory of the integral (with the help of limits and least upper bounds). It is this that will be our point of departure.

A perhaps surprising feature is that we begin with the logarithm rather than the exponential.

§7A. The logarithm

7A1. How shall we define the logarithm? By a logarithm, in general, we mean a function L defined for all $x > 0$ in such a way that the fundamental relation

$$(1) \qquad\qquad L(\alpha\beta) = L(\alpha) + L(\beta) \qquad\qquad (\alpha > 0, \beta > 0)$$

is satisfied. Let us also require that the function be differentiable. (And exclude the constant function 0.) These properties will lead us to a definition.

To gain information we hold one variable fixed in (1) and vary the other. For emphasis, let us first rewrite (1) as

$$(2) \qquad\qquad L(\alpha x) = L(\alpha) + L(x) \qquad\qquad (\alpha > 0, \, x > 0).$$

Differentiating with respect to x, we get

$$(3) \qquad\qquad \alpha L'(\alpha x) = L'(x) \qquad\qquad (\alpha > 0, \, x > 0).$$

This holds for all values of $x > 0$. Setting $x = 1$, we have

$$(4) \qquad\qquad \alpha L'(\alpha) = L'(1) \qquad\qquad (\alpha > 0).$$

Equation (4) holds for all values of $\alpha > 0$. Shifting our emphasis, we rewrite it as

$$(5) \qquad\qquad L'(t) = \frac{1}{t} L'(1) \qquad\qquad (t > 0).$$

Thus, $L(t)$ will have a continuous derivative, proportional to $1/t$ (the number $L'(1)$ being the constant of proportionality). By the Fundamental Theorem of Calculus, it is the integral of its derivative. An initial value is

$$L(1) = 0,$$

obtained by setting $\beta = 1$ in (1). Consequently,

$$(6) \qquad\qquad L(x) = L'(1) \int_1^x \frac{1}{t} \, dt \qquad\qquad (x > 0).$$

We are free to choose the constant of proportionality, $L'(1)$, to suit ourselves. Sometimes the convenient choice is

$$(7) \qquad\qquad L'(1) = 0.43429448 \qquad\qquad \text{(approx.)}$$

(see §7E1 below). We will choose

$$L'(1) = 1.$$

7A2. The function log. We now define

$$(8) \qquad\qquad \log x = \int_1^x \frac{1}{t} \, dt \qquad\qquad (x > 0).$$

This function is called the *logarithm* (or the *natural logarithm*). By the Fundamental Theorem of Calculus,

$$(9) \qquad \frac{d}{dx} \log x = \frac{1}{x} \qquad (x > 0).$$

Thus, $\log x$ is defined for all $x > 0$ and has a continuous derivative there. Let us verify that it satisfies the characteristic property:

$$(10) \qquad \log (\alpha\beta) = \log \alpha + \log \beta \qquad (\alpha > 0, \beta > 0).$$

We have:

$$(11) \qquad \log (\alpha\beta) - \log \alpha = \int_{\alpha}^{\alpha\beta} \frac{1}{t} \, dt = \int_{1}^{\beta} \frac{1}{\alpha u} \, \alpha \, du \qquad (t = \alpha u)$$

$$= \int_{1}^{\beta} \frac{1}{u} \, du = \log \beta.$$

From (8) $\big($or, as before, from (10)$\big)$, we find that

$$(12) \qquad \log 1 = 0.$$

Setting $\beta = 1/\alpha$ in (10) yields the formula

$$(13) \qquad \log \frac{1}{\alpha} = -\log \alpha \qquad (\alpha > 0).$$

Finally,

$$(14) \qquad \log a^r = r \log a \qquad (a > 0, r \text{ rational}).$$

This can be derived from (10) by elementary means, as in earlier courses; or, more swiftly,

$$(15) \qquad \log a^r = \int_{1}^{a^r} \frac{1}{t} \, dt = \int_{1}^{a} \frac{1}{u^r} \, r u^{r-1} \, du \qquad (t = u^r)$$

$$= r \int_{1}^{a} \frac{1}{u} \, du = r \log a.$$

Example 1. Find $(d/dx) \log(x^2 + 3)$.

Solution. By the chain rule,

$$\frac{d}{dx}\log(x^2 + 3) = \frac{1}{x^2 + 3} \cdot 2x = \frac{2x}{x^2 + 3}.$$

Example 2. Find $(d/dx) \log(6x^2)$ $(x \neq 0)$.

Solution. By the chain rule,

$$\frac{d}{dx}\log(6x^2) = \frac{1}{6x^2} \cdot 12x = \frac{2}{x} \qquad (x \neq 0).$$

Alternatively,

$$\frac{d}{dx}\log(6x^2) = \frac{d}{dx}(\log 6 + \log x^2) = \frac{1}{x^2} \cdot 2x = \frac{2}{x} \qquad (x \neq 0).$$

If we know that $x > 0$ we can write, more simply,

$$\frac{d}{dx}\log(6x^2) = \frac{d}{dx}(\log 6 + 2\log x) = \frac{2}{x} \qquad (x > 0).$$

Example 3. Find $(d/dx) \log \sqrt{x^2 + 1}$.

Solution. By the chain rule,

$$\frac{d}{dx}\log \sqrt{x^2 + 1} = \frac{1}{\sqrt{x^2 + 1}} \cdot \frac{1}{2\sqrt{x^2 + 1}} \cdot 2x = \frac{x}{x^2 + 1}.$$

Alternatively,

$$\frac{d}{dx}\log \sqrt{x^2 + 1} = \frac{d}{dx}[\tfrac{1}{2}\log(x^2 + 1)] = \frac{1}{2} \cdot \frac{1}{x^2 + 1} \cdot 2x = \frac{x}{x^2 + 1}.$$

Example 4. Find $(d/dx) \log 1/x$ $(x > 0)$.

Solution. By the chain rule,

$$\frac{d}{dx}\log \frac{1}{x} = \frac{1}{1/x}\left(-\frac{1}{x^2}\right) = -\frac{1}{x} \qquad (x > 0).$$

Alternatively,

$$\frac{d}{dx}\log \frac{1}{x} = \frac{d}{dx}(-\log x) = -\frac{1}{x} \qquad (x > 0).$$

7A3. The antiderivative of 1/x. The gap in the power rule mentioned in §4F2 has now been filled. Since $d(\log x)/dx = 1/x$,

$$\int \frac{1}{x}\,dx = \log x \qquad\qquad (16) \qquad (x > 0).$$

Moreover, for $x < 0$ as well as for $x > 0$, we have

$$\int \frac{1}{x}\,dx = \frac{1}{2}\int \frac{1}{x^2}\cdot 2x\,dx$$

(17)

$$= \tfrac{1}{2}\log x^2 = \log\sqrt{x^2} = \log|x| \quad (x < 0 \text{ or } x > 0).$$

Thus,

$$\int \frac{1}{x}\,dx = \log|x|, \qquad \frac{d}{dx}\log|x| = \frac{1}{x} \quad (x < 0 \text{ or } x > 0).$$

(18)

Example 5. Find

$$\int \frac{2x}{x^2 + 3}\,dx.$$

Solution. This should be clear by inspection, since the numerator is the derivative of the denominator. But while the subject is new we proceed slowly and substitute $u = x^2 + 3$. Then $u > 0$ and $du = 2x\,dx$. Hence

$$\int \frac{2x}{x^2 + 3}\,dx = \int \frac{1}{u}\,du = \log u = \log(x^2 + 3).$$

Example 6. Find

$$\int \frac{1}{2x + 3}\,dx \qquad (x > -\tfrac{3}{2}).$$

Solution. Set $u = 2x + 3$. Then $u > 0$, $du = 2\,dx$, and

$$\int \frac{1}{2x + 3}\,dx = \frac{1}{2}\int \frac{1}{u}\,du = \tfrac{1}{2}\log u = \tfrac{1}{2}\log(2x + 3) \quad (x > -\tfrac{3}{2}).$$

Example 7. Find

$$\int \frac{x + 1}{x - 1}\,dx \qquad (x < 1).$$

Solution.

$$\int \frac{x+1}{x-1}\, dx = \int \left(1 + \frac{2}{x-1}\right) dx = x + 2 \log |x-1|$$

$$= x + \log\left[(x-1)^2\right] \quad (x < 1).$$

Problems (§7A)

1. Let $a = \log 2$, $b = \log 3$, $c = \log 5$. Express $\log x$ in terms of a, b, and c for each of the following.

(a) $x = 4, 6, 8, 9, 10, 12, 15, 16, 18, 20, 24, 25, 27, 30$

(b) $x = \frac{2}{3}, \frac{3}{4}, \frac{4}{5}, \frac{5}{6}, \frac{6}{5}, \frac{5}{4}, \frac{4}{3}, \frac{3}{2}$

(c) $x = 0.1, 0.2, 0.3, 0.4, 0.5, 0.6$

(d) $x = \sqrt{6}, \sqrt{8}, \sqrt{10}, 4^{1/3}, 12^{2/5}, 15^{4/3}$

2. State the natural domain of each of the following functions.

(a) $\log(x^2 + 1)$

(b) $\log \log x$

(c) $\dfrac{1}{\log 2x}$

(d) $\log(x + 1) - \log(x - 1)$

(e) $\log \dfrac{x+1}{x-1}$

(f) $\log \dfrac{1}{x}$

(g) $\log \left(1 + \dfrac{1}{x}\right)$

(h) $\log \left(1 + \dfrac{1}{1 + 1/x}\right)$

3. Differentiate the following.

(a) $x^2 \log x$

(b) $\log(x^2 + 2)$

(c) $\log \log x$

(d) $\sqrt{\log(3x - 1)}$

(e) $[\log(x^2 + 1)]^3$

(f) $\dfrac{1}{3 \log 2x}$

(g) $x(\log x - 1)$

(h) $\log \dfrac{x^2 - 7}{x^3 + 4x^2 - 1}$

(i) $\log \dfrac{x-1}{x+1}$

(j) $\log[x(x^2 + 1)^2]$

(k) $\log\sqrt{x^2 + 1}$

(l) $\log x^{2/3}$

4. Evaluate the following.

(a) $\displaystyle\int \frac{x}{3x^2 - 7}\, dx$

(k) $\displaystyle\int \frac{x}{1 - x}\, dx$

(b) $\displaystyle\int \frac{3x - 2}{x}\, dx$

(l) $\displaystyle\int \frac{x}{(1 - x)^2}\, dx$

(c) $\displaystyle\int \frac{x}{2x + 1}\, dx$

(m) $\displaystyle\int \frac{2x + 5}{(x + 3)(x + 2)}\, dx$

(d) $\displaystyle\int \frac{x^2}{x^3 + 1}\, dx$

(n) $\displaystyle\int \frac{(\sqrt{x} + 1)^{2/3}}{\sqrt{x}}\, dx$

(e) $\displaystyle\int \frac{\log x}{x}\, dx$

(o) $\displaystyle\int \frac{2x + 3}{3x + 5}\, dx$

(f) $\displaystyle\int \frac{(\log x)^2}{x}\, dx$

(p) $\displaystyle\int \frac{3x^2 - 2x + 7}{x + 1}\, dx$

(g) $\displaystyle\int \frac{1}{x \log x}\, dx$

(q) $\displaystyle\int \frac{x^2}{(x^3 + 1)^2}\, dx$

(h) $\displaystyle\int \frac{\log \sqrt{x}}{x}\, dx$

(r) $\displaystyle\int \frac{1}{x} \log \left(\frac{1}{x}\right) dx$

(i) $\displaystyle\int \frac{\log \log x}{x \log x}\, dx$

(s) $\displaystyle\int \frac{1}{x^2 - 1}\, dx$

(j) $\displaystyle\int \frac{1}{x + 1/x}\, dx$

[*Hint.* $1 = \frac{1}{2}[(x + 1) - (x - 1)]$.]

5. In isothermal expansion of a gas, the pressure and volume are related by the equation $pV = C$, where C is a constant. Show that under these conditions, the work done in expanding the gas from volume a to volume b (against a piston head, for example) is proportional to $\log(b/a)$.

6. Prove that the line through $(0, -1)$ and $(1, 0)$ is tangent to the graph of $y = \log x$.

7. (a) Prove that there is a number x_1 between 1 and 9 such that the slope of the tangent line to the graph of $y = \log x$ at $x = x_1$ is equal to $\frac{8}{9}$.
    (b) Prove that there is a number x_2 between 1 and 9 such that the slope of the tangent line to the graph of $y = \log x$ at $x = x_2$ is equal to $\frac{1}{4} \log 3$.

8. What is $\lim\limits_{x \to 0} (1/x) \log(1 + x)$? [*Hint.* Consider $\log'(1)$.]

Answers to problems (§7A)

1. (a) $2a$, $a + b$, $3a$, $2b$, $a + c$, $2a + b$, $b + c$, $4a$, $a + 2b$, $2a + c$, $3a + b$, $2c$, $3b$, $a + b + c$

 (b) $a - b$, $b - 2a$, $2a - c$, $c - (a + b)$, $a + b - c$, $c - 2a$, $2a - b$, $b - a$

 (c) $-(a + c)$, $-c$, $b - (a + c)$, $a - c$, $-a$, $b - c$

 (d) $\frac{1}{2}(a + b)$, $\frac{3}{2}a$, $\frac{1}{2}(a + c)$, $\frac{2}{3}a$, $\frac{2}{5}(2a + b)$, $\frac{4}{3}(b + c)$

2. (a) $\mathscr{R}$ (b) $x > 1$ (c) $0 < x < \frac{1}{2}$ or $x > \frac{1}{2}$ (d) $x > 1$

 (e) $x < -1$ or $x > 1$ (f) $x > 0$ (g) $x < -1$ or $x > 0$

 (h) $x < -1$ or $x > -\frac{1}{2}$

3. (a) $x + 2x \log x$ (b) $\dfrac{2x}{x^2 + 2}$ (c) $\dfrac{1}{x \log x}$

 (d) $\dfrac{3}{2(3x - 1)\sqrt{\log(3x - 1)}}$ (e) $\dfrac{6x}{x^2 + 1}\,[\log(x^2 + 1)]^2$

 (f) $-\dfrac{1}{3x[\log 2x]^2}$ (g) $\log x$ (h) $\dfrac{2x}{x^2 - 7} - \dfrac{3x^2 + 8x}{x^3 + 4x^2 - 1}$

 (i) $\dfrac{2}{x^2 - 1}$ (j) $\dfrac{1}{x} + \dfrac{4x}{x^2 + 1}$ (k) $\dfrac{x}{x^2 + 1}$ (l) $\dfrac{2}{3x}$

4. (a) $\frac{1}{6} \log |3x^2 - 7|$ (b) $3x - 2 \log |x|$ (c) $\frac{1}{2}x - \frac{1}{4} \log |2x + 1|$

 (d) $\frac{1}{3} \log |x^3 + 1|$ (e) $\frac{1}{2}(\log x)^2$ (f) $\frac{1}{3}(\log x)^3$ (g) $\log |\log x|$

 (h) $\frac{1}{4}(\log x)^2$ (i) $\frac{1}{2}(\log \log x)^2$ (j) $\frac{1}{2} \log(x^2 + 1)$ (k) $-x - \log |1 - x|$

 (l) $\log |1 - x| + \dfrac{1}{1 - x}$ (m) $\log |x^2 + 5x + 6|$ (n) $\frac{6}{5}(\sqrt{x} + 1)^{5/3}$

 (o) $\frac{2}{3}x - \frac{1}{9} \log |3x + 5|$ (p) $\frac{3}{2}x^2 - 5x + 12 \log |x + 1|$

 (q) $-\dfrac{1}{3(x^3 + 1)}$ (r) $-\frac{1}{2}(\log x)^2$ (s) $\frac{1}{2} \log \left|\dfrac{x - 1}{x + 1}\right|$

8. 1

§7B. The logarithm (continued)

7B1. Behavior of log x. Let us gather more information about the logarithm function. It has a positive derivative:

$$(1) \qquad\qquad\qquad \log'(x) = \frac{1}{x} > 0 \qquad\qquad (x > 0),$$

and therefore is an increasing continuous function. It is negative for $x < 1$, zero at $x = 1$, and positive for $x > 1$. Since the derivative of log x is *decreasing*, the graph of $y = \log x$ is concave down (Figure 1).

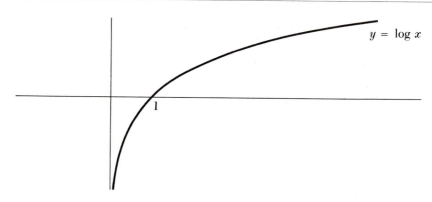

$$y = \log x$$

FIGURE 1

Let us show that $\log x$ assumes every real value. First, let any *positive* number s be given. We know that, say, $\log 2 > 0$. Pick an integer n such that

$$n > \frac{s}{\log 2}.$$

Then $n \log 2 > s > 0$; that is,

$$\log 2^n > s > \log 1.$$

By the Intermediate-Value Theorem (§2E5), $\log x$ assumes every value between $\log 2^n$ and $\log 1$ and hence assumes the value s. Thus, $\log x$ assumes all positive values. It follows from (13) of §7A2 that it assumes all negative values as well.

A table of values of $\log x$ (computed by advanced methods) appears at the back of the book.

7B2. The number e. What is the "base" of the logarithm? That is, what is the number whose logarithm is 1? We know there *is* such a number, of course. (Why?)

The base of the logarithm is called e. That is, the number e is *defined* by the equation

(2) $$\log e = 1.$$

What is the numerical value of e? As Figure 2 shows, $\log 2 < 1 < \log 4$. Consequently, $2 < e < 4$.

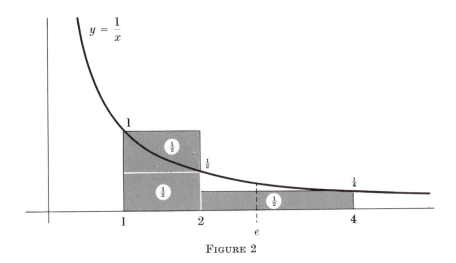

FIGURE 2

It can be shown that e is an irrational number. There are advanced methods for obtaining its value to any degree of accuracy. Mathematicians studying patterns in the decimal expansion have calculated e on computers to more than 100,000 places. In numerical work one rarely needs more than 3 places. The first 12 are:

$$(3) \qquad\qquad e = 2.7\ 1828\ 1828\ 459 \cdots$$

(This is easy to remember: 1828 is the year after Beethoven died, and $4 + 5$ is 9.)

Problems (§7B)

1. Is the function $\log x$ a polynomial?

2. Evaluate the following.

   (a) $\log e^2$

   (b) $\log \dfrac{1}{e}$

   (c) $\log \sqrt{e}$

   (d) $\displaystyle\int_1^{e^3} \frac{1}{x}\, dx$

   (e) $\displaystyle\int_1^{e^3} \frac{1}{x} \log x\, dx$

3. Find the minimum value of
   (a) $y = x \log x$ (b) $y = x^2 \log x$

4. (a) Where is the maximum value of $(\log x)/x$ attained, and what is the value?
   (b) Of all the lines through the origin that intersect the graph of $y = \log x$, which has the greatest slope?

5. The graph of $y = 1/\sqrt{x}$ ($1 \le x \le e$) is revolved about the x-axis. Find the volume enclosed.

6. Find the length of the graph of

$$y = \frac{x^2}{2} - \frac{\log x}{4}$$

from $x = 1$ to $x = e$.

Answers to problems (§7B)

1. no
2. (a) 2   (b) -1 (c) $\frac{1}{2}$ (d) 3   (e) $\frac{9}{2}$
3. (a) $-1/e$ (b) $-1/2e$
4. (a) max $1/e$ at $x = e$ (b) $y = \frac{1}{e}x$
5. π
6. $\frac{1}{4}(2e^2 - 1)$

§7C. Integration by parts

What is $\int \log x \, dx$?

Continuous function have antiderivatives (Fundamental Theorem). Many of these are expressible in terms of known functions and it behooves us to acquire a respectable knowledge of them. That is, we should learn how to "integrate" a good collection of functions.

We shall now show how to find $\int \log x \, dx$ and many other antiderivatives by the method of "integration by parts." This is a method based on the product rule for derivatives:

(1) $$(fg)' = fg' + gf',$$

or, in antiderivative form,

(2) $$fg = \int fg' + \int gf'.$$

The formula for integration by parts is simply equation (2), solved for $\int fg'$:

(3) $$\int f(x)g'(x) \, dx = f(x)g(x) - \int g(x)f'(x) \, dx.$$

The left side is the problem and the right side is the solution. That is, we look for functions f and g such that:

(i) *the given integrand has the form $f(x)g'(x)$*

AND

(ii) *we know how to integrate $g(x)f'(x)$.*

This program may seem almost impossible; but a little practice goes a long way.

Let us find $\int \log x \, dx$. We try

$$f(x) = \log x, \qquad g'(x) = 1.$$

Then

$$f'(x) = \frac{1}{x}, \qquad g(x) = x.$$

Thus,

(i) $\int \log x \, dx$ takes the form $\int f(x)g'(x) \, dx$

AND

(ii) $\int g(x)f'(x) \, dx = \int 1 \, dx$, which we know how to integrate.

Therefore the guess works. We have

$$\int \log x \, dx = f(x)g(x) - \int g(x)f'(x) \, dx$$

$$= x \log x - \int 1 \, dx = x \log x - x.$$

Thus,

(4) $$\int \log x \, dx = x \log x - x.$$

Check by differentiating!

In terms of variables $u = f(x)$ and $v = g(x)$, formula (3) becomes

(5) $$\int u \, dv = uv - \int v \, du.$$

Example 1. In this notation, the above substitutions are

$$u = \log x, \qquad dv = dx,$$

from which

$$du = \frac{1}{x}\, dx, \qquad v = x.$$

Then

$$\int \log x \, dx = \int u \, dv = uv - \int v \, du$$

$$= x \log x - \int 1 \, dx = x \log x - x.$$

Example 2. Find $\int x \log x \, dx$.

Solution. We try

$$u = \log x, \qquad dv = x \, dx.$$

Then

$$du = \frac{1}{x}\, dx, \qquad v = \tfrac{1}{2}x^2,$$

and

$$\int \overset{u}{(\log x)} \overset{dv}{(x\, dx)} = \overset{u}{(\log x)} \overset{v}{(\tfrac{1}{2}x^2)} - \int \overset{v}{(\tfrac{1}{2}x^2)} \overset{du}{\left(\frac{1}{x}\, dx\right)}$$

$$= \tfrac{1}{2}x^2 \log x - \tfrac{1}{4}x^2.$$

When the work is laid out as in these examples, it is easy to keep track of what we are doing.

Problems (§7C)

1. Evaluate the following.

(a) $\int x^2 \log x \, dx$

(b) $\int \sqrt{x} \log x \, dx$

(c) $\int \frac{\log \log x}{x} \, dx$

(d) $\int \log \sqrt{x-3} \, dx$

(e) $\int x \log \sqrt{x+1} \, dx$

(f) $\int (\log x)^2 \, dx$

(g) $\int x \log x^2 \, dx$

(h) $\int \frac{1}{x \log x} \, dx$

(i) $\int \log \frac{1}{x} \, dx$

(j) $\int \frac{\log(1/x)}{x^2} \, dx$

(k) $\int \log(x^2 - 1) \, dx$

(l) $\int (x+1) \log x \, dx$

2. Evaluate the following.

(a) $\displaystyle\int \log(x + c)\, dx$ (c) $\displaystyle\int x^2 \log(x + c)\, dx$

(b) $\displaystyle\int x \log(x + c)\, dx$

Answers to problems (§7C)

1. (a) $\frac{1}{3}x^3 \log x - \frac{1}{9}x^3$ (b) $\frac{2}{3}x^{3/2} \log x - \frac{4}{9}x^{3/2}$

(c) $\log x \log \log x - \log x$ (d) $\frac{1}{2}(x - 3) \log(x - 3) - \frac{1}{2}x$

(e) $\frac{1}{4}\left[(x^2 - 1) \log(x + 1) - \dfrac{x^2}{2} + x \right]$

(f) $x(\log x)^2 - 2x \log x + 2x$ (g) $x^2 \log |x| - \frac{1}{2}x^2$ (h) $\log |\log x|$

(i) $x - x \log x$ (j) $\dfrac{1 + \log x}{x}$ (k) $x \log(x^2 - 1) - 2x - \log \dfrac{x - 1}{x + 1}$

(l) $\frac{1}{2}x^2 \log x - \frac{1}{4}x^2 + x \log x - x$

2. (a) $(x + c) \log(x + c) - x$ (b) $\frac{1}{2}(x^2 - c^2) \log(x + c) - \frac{1}{4}(x - c)^2$

(c) $\frac{1}{3}(x^3 + c^3) \log(x + c) - \frac{1}{9}(x - c)^3 - \frac{1}{6}cx^2$

§7D. The exponential

7D1. The function exp. We have seen that the logarithm, defined for $x > 0$, is an increasing function whose image is all of $\mathscr{R}$. It therefore has an inverse defined on $\mathscr{R}$, whose image is the set of all positive numbers. This inverse is the *exponential*, denoted exp (Figure 1). By definition,

(1) $y = \exp x$ *means* $x = \log y$.

Since $\log e = 1$,

(2) $\exp 1 = e$.

7D2. Definition of e^x. We know from formula (14) of §7A2 that for each *rational* number r,

(3) $\log e^r = r$.

By definition of exp,

(4) $e^r = \exp r$ (*r* rational).

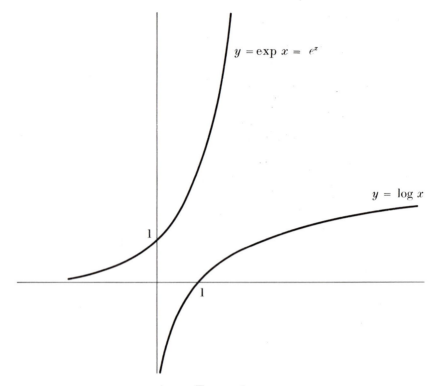

$y = \exp x = e^x$

$y = \log x$

FIGURE 1

We take this as our cue and *define* e^x for irrational x by

(5) $$e^x = \exp x.$$

Thus, (5) is true for every real number x.
    Recapitulating,

(6) $$y = e^x \qquad means \qquad x = \log y.$$

For emphasis:

(7) $$\log e^x = x \qquad and \qquad e^{\log y} = y.$$

The definition (5) is fully justified by the fact that e^x then satisfies the characteristic property:

(8) $$e^\alpha e^\beta = e^{\alpha+\beta} \qquad\qquad (\alpha \in \mathcal{R},\ \beta \in \mathcal{R}).$$

Proof. $\log(e^\alpha e^\beta) = \log e^\alpha + \log e^\beta = \alpha + \beta$, and taking exponentials yields the result.

A table of exponentials (computed by advanced methods) appears at the back of the book.

7D3. Derivative of e^x. Since the derivative of the *logarithm* is never zero, we know from the Inverse-Function Theorem (§4G) that the exponential is differentiable. We look at the equation $\log e^x = x$ and apply the chain rule, obtaining

(9)
$$\frac{1}{e^x} \frac{d}{dx} e^x = 1.$$

Therefore,

(10)
$$\frac{d}{dx} e^x = e^x.$$

The function e^x is equal to its own derivative. The antiderivative form of (10) is

(11)
$$\int e^x \, dx = e^x.$$

Example 1. $(d/dx)\, 2e^x = 2e^x$.

Example 2. By the chain rule:

(a) $\dfrac{d}{dx} e^{3x} = 3e^{3x}$.

(b) $\dfrac{d}{dx} e^{-x} = -e^{-x}$.

(c) $\dfrac{d}{dx} e^{x^2} = 2xe^{x^2}$.

Example 3. By the chain rule applied twice:

$$\frac{d}{dx} \sqrt{e^{4x} + 1} = \frac{1}{2\sqrt{e^{4x} + 1}} \frac{d}{dx} (e^{4x} + 1) = \frac{2e^{4x}}{\sqrt{e^{4x} + 1}}.$$

Example 4. Find $\int x^2 e^{x^3+1} \, dx$.

Solution. Set $u = x^3 + 1$. Then $du = 3x^2 \, dx$ and

$$\int x^2 e^{x^3+1} \, dx = \tfrac{1}{3} \int e^u \, du = \tfrac{1}{3} e^u = \tfrac{1}{3} e^{x^3+1}.$$

Example 5. Find $\int x e^x \, dx$.

Solution. We integrate by parts. Put

$$u = x, \qquad dv = e^x \, dx.$$

Then

$$du = dx, \qquad v = e^x,$$

and

$$\int x e^x \, dx = \int u \, dv = uv - \int v \, du$$

$$= x e^x - \int e^x \, dx = x e^x - e^x.$$

Sometimes we perform repeated integrations by parts.

Example 6. Find $\int x^2 e^x \, dx$.

Solution. Set

$$u = x^2, \qquad dv = e^x \, dx.$$

Then

$$du = 2x \, dx, \qquad v = e^x,$$

and $v \, du = 2x e^x \, dx$, which we know how to integrate by parts. We have

$$\int x^2 e^x \, dx = x^2 e^x - \int 2x e^x \, dx.$$

By Example 5, $\int 2x e^x \, dx = 2x e^x - 2e^x$. Hence, finally,

$$\int x^2 e^x \, dx = (x^2 - 2x + 2)e^x.$$

7D4. Exponential growth or decay; a differential equation. Various quantities change at rates proportional to their size. For instance, a population or a bank account may increase at a steady rate of 5 percent per year. Consider such a quantity, $y = f(t)$, which we assume to be always positive. Let y_0 denote its value at time $t = 0$. The assumption that y' is proportional to y is expressed by the *differential equation*

$$(12) \qquad\qquad y' = ky, \qquad \text{or} \qquad \frac{y'}{y} = k \qquad\qquad (y > 0),$$

where k is a constant. To solve the equation we note that for any time t,

$$(13) \qquad\qquad \int_0^t \frac{f'(x)}{f(x)} \, dx = \int_0^t k \, dx.$$

This gives us

(14) $$\log f(x)\Big|_0^t = kt, \qquad \text{that is,} \qquad \log \frac{f(t)}{f(0)} = kt.$$

Hence, finally,

(15) $$\log \frac{y}{y_0} = kt, \qquad \text{or} \qquad y = y_0 e^{kt}.$$

Example 7. A colony of bacteria increases in such a way that at each instant, the rate of increase per hour is equal to twice the size of the colony at that instant. How big will the colony be at the end of 1 hour?

Query. What is your guess?

Solution. We are given $k = 2$, i.e., $y' = 2y$. Hence

$$y = y_0 e^{2t}.$$

When $t = 1$, $y = y_0 e^2 \approx 10 y_0$. Thus, the colony will be 10 times its original size.

Problems (§7D)

1. Simplify the following expressions.

   (a) $\log e^{-x}$

   (b) $e^{-\log x}$

   (c) $e^{x + \log x}$

   (d) $\log(xe^{2x})$

   (e) $e^{-\log x^2}$

   (f) $e^{\log e^x}$

   (g) $\log(e^{\log x})$

   (h) $e^{2 \log 3}$

   (i) $\log(x^2 e^x)$

   (j) $\log(e^{x^2 - 2x})$

   (k) $e^{\log(e^x)}$ $\log e\, (e^x)$

2. Differentiate the following functions.

   (a) e^{2x-1}

   (b) e^{x^2+x}

   (c) e^{1-x}

   (d) $e^{x \log x}$

   (e) $e^{1/x}$

   (f) $\exp \sqrt{x^2 + 1}$

   (g) $\int_2^x e^{s^2}\, ds$

   (h) $\log(x^2 e^x)$

   (i) $\exp\left(\dfrac{x^2}{x^2 + 1}\right) \exp\left(\dfrac{1}{x^2 + 1}\right)$

   (j) $\exp \dfrac{x}{x - 1}$

   (k) $e^x \log x$

   (l) $\log(e^x + 1)$

   (m) $\dfrac{e^x - e^{-x}}{e^x + e^{-x}}$

   (n) xe^{-x^2}

   (o) $\dfrac{1}{e^x}$

   (p) $\log \dfrac{1}{e^x}$

3. Evaluate the following.

(a) $\int x^3 e^{x^4-1}\, dx$

(b) $\int e^x \sqrt{e^x + 1}\, dx$

(c) $\int \dfrac{e^{\sqrt{x}}}{\sqrt{x}}\, dx$

(d) $\int \dfrac{e^{1/x}}{x^2}\, dx$

(e) $\int e^{(e^x)} e^x\, dx$

(f) $\int \dfrac{e^x}{e^x - 1}\, dx$

(g) $\int \sqrt{e^x}\, dx$

(h) $\int e^{x^2} x^3\, dx$

(i) $\int x e^{x+2}\, dx$

(j) $\int e^{(e^x)} e^{2x}\, dx$

(k) $\int x e^{(x^2+1)}\, dx$

(l) $\int \sqrt{e^x + 1}\, dx$

[*Hint.* Set $u = \sqrt{e^x + 1}$.] 1.1

$\sqrt{e^x+1}$

(m) $\int e^{-\log x}\, dx$

(n) $\int x e^{-x}\, dx$

(o) $\int x e^{x^2}\, dx$

(p) $\int x e^{-x^2}\, dx$

(q) $\int x(x + 1) e^x\, dx$

(r) $\int e^x \left(\log x + \dfrac{1}{x} \right)\, dx$

[*Hint.* First try $\int e^x \log x\, dx$.]

(s) $\int e^x (f(x) + f'(x))\, dx$

(where f' is continuous)

4. (a) Show, by the method indicated in Example 6, that

$$\int x^n e^x\, dx = x^n e^x - n \int x^{n-1} e^x\, dx$$

(where n is a positive integer).

(b) Using (a) (as often as necessary), find

$$\int x^4 e^x\, dx.$$

5. Evaluate the following.

(a) $\int_0^1 e^{-x}\, dx$

(b) $\int_0^{\sqrt{\log 2}} x e^{x^2}\, dx$

(c) $\int_0^1 x e^{-x}\, dx$

6. Describe and sketch the graph between $x = -2$ and $x = 1$ of
   (a) $y = xe^x$ (b) $y = e^{3x} - e^x$

7. (a) What function is the negative of its derivative?
   (b) What function is twice its derivative?
   (c) What function satisfies the formula $f(2x) = f(x)^2$?

8. Find the volume generated when the region bounded by the given curves is revolved about the given line.
   (a) $y = e^{-x}$, $y = 1$, $x = 1$;   line $x = 1$
   (b) $y = \sqrt{x}\, e^{-x^2}$, $y = 0$, $x = 1$; x-axis

9. Find the length of the graph of

$$y = e^x \qquad (\tfrac{1}{2}\log 3 \le x \le \tfrac{1}{2}\log 24).$$

   [*Hint.* Let $u = \sqrt{1 + e^{2x}}$.]

10. (a) Show that the area under the graph of

$$y = \frac{e^x + e^{-x}}{2} \qquad\qquad (a \le x \le b)$$

   is numerically equal to the length of the graph over $[a, b]$.
   (b) Find the area of the surface generated when the portion of this graph over $[0, \log 2]$ is revolved about the x-axis.

11. A colony of bacteria increases at a constant relative rate so that it doubles in number every hour. How long does it take to multiply itself
   (a) by a factor of 10?   (b) by a factor of 16?

12. A particle moves along the x-axis so that its velocity at each point is proportional to the coordinate of the point. What is its equation of motion:
   (a) if $x = 1$ when $t = 0$ and $x = 3$ when $t = 1$?
   (b) if $x = 2$ when $t = 0$ and $x = 1$ when $t = 1$?

13. A particle moves along the x-axis so that its velocity at each point is equal to the coordinate of the point.
   (a) If $x = 2$ when $t = 0$, what is x when $t = 2$?
   (b) If $x = 1$ when $t = 1$, what is x when $t = 2$?

14. The *half-life* of a particular sample of radium is 1620 years. That is, the substance decays at a constant relative rate such that any given amount is reduced after 1620 years to half that amount. How long is required for a reduction of 25 percent?

15. Which statement supports the assertion that $x \to e^2$ as $\log x \to 2$?
   (i) The logarithm is continuous
   (ii) The exponential is continuous

16. Find $\lim\limits_{x \to 0} \dfrac{e^x - 1}{x}$ [*Hint.* Consider $\exp'(0)$.]

17. (a) By examining the graph of $1/t$ for $t \geq 1$, show that for $x > 0$,

$$\frac{x}{1 + x} < \log(1 + x) < x.$$

(b) Deduce that for each positive integer n,

$$\left(1 + \frac{1}{n}\right)^n < e < \left(1 + \frac{1}{n}\right)^{n+1}.$$

[*Hint.* Take $x = 1/n$, and apply formula (14) of §7A2.]

(c) By taking $n = 10$ in (b), check that
$$2.59 < e < 2.86.$$

Answers to problems (§7D)

1. (a) $-x$ (b) $1/x$ (c) xe^x (d) $2x + \log x$ (e) $1/x^2$ (f) e^x
   (g) $\log x$ (h) 9 (i) $x + 2\log|x|$ (j) $x^2 - 2x$ (k) e^x

2. (a) $2e^{2x-1}$ (b) $(2x + 1)e^{x^2+x}$ (c) $-e^{1-x}$ (d) $e^x \log x(1 + \log x)$

(e) $-\dfrac{1}{x^2}e^{1/x}$ (f) $\dfrac{x}{\sqrt{x^2 + 1}}\exp\sqrt{x^2 + 1}$ (g) e^{x^2} (h) $\dfrac{2}{x} + 1$

(i) 0 (j) $-\dfrac{1}{(x - 1)^2}\exp\dfrac{x}{x - 1}$ (k) $\dfrac{1}{x}e^x + e^x \log x$ (l) $\dfrac{e^x}{e^x + 1}$

(m) $\dfrac{4}{(e^x + e^{-x})^2}$ (n) $e^{-x^2}(1 - 2x^2)$ (o) $-e^{-x}$ (p) -1

3. (a) $\frac{1}{4}e^{x^4-1}$ (b) $\frac{2}{3}(e^x + 1)^{3/2}$ (c) $2e^{\sqrt{x}}$ (d) $-e^{1/x}$ (e) e^{e^x}
   (f) $\log|e^x - 1|$ (g) $2\sqrt{e^x}$ (h) $\frac{1}{2}e^{x^2}(x^2 - 1)$ (i) $xe^{x+2} - e^{x+2}$

(j) $e^{e^x}(e^x - 1)$ (k) $\frac{1}{2}e^{x^2+1}$ (l) $2\sqrt{e^x + 1} + \log\dfrac{\sqrt{e^x + 1} - 1}{\sqrt{e^x + 1} + 1}$

(m) $\log x$ (n) $-xe^{-x} - e^{-x}$ (o) $\frac{1}{2}e^{x^2}$ (p) $-\frac{1}{2}e^{-x^2}$
(q) $x^2 e^x - xe^x + e^x$ (r) $e^x \log x$ (s) $e^x f(x)$

4. (b) $(x^4 - 4x^3 + 12x^2 - 24x + 24)e^x$

5. (a) $1 - 1/e$ (b) $\frac{1}{2}$ (c) $1 - 2/e$

6. (a) decreasing $-2 \leq x \leq -1$, increasing $-1 \leq x \leq 1$, min $-1/e$ at $x = -1$, concave up on $[-2, 1]$

   (b) decreasing $-2 \leq x \leq -\frac{1}{2}\log 3$, increasing $-\frac{1}{2}\log 3 \leq x \leq 1$, min $-\frac{2}{9}\sqrt{3}$ at $x = -\frac{1}{2}\log 3$, concave down $x \leq -\log 3$, concave up $x \geq -\log 3$, inflection at $x = -\log 3$

8. (a) $\pi\left(1 - \dfrac{2}{e}\right)$ (b) $\dfrac{\pi}{4}\left(1 - \dfrac{1}{e^2}\right)$

9. $3 + \frac{1}{2}\log 2$

10. (b) $\pi(\frac{15}{16} + \log 2)$

11. (a) $\dfrac{\log 10}{\log 2} \approx 3.3$ hours   (b) 4 hours

12. (a) $x = e^{t \log 3}$ (b) $x = 2e^{-t \log 2}$

13. (a) $2e^2$ (b) e

14. $1620\,\dfrac{\log 4 - \log 3}{\log 2} \approx 680$ years

16. 1

§7E. Other bases

7E1. Logarithm to the base a. According to §7A1, a logarithm is a differentiable function L satisfying the fundamental relation

$$(1) \qquad L(\alpha\beta) = L(\alpha) + L(\beta) \qquad (\alpha > 0, \beta > 0)$$

(and other than the constant function 0). Now, if k is any constant, then kL satisfies the same equation; hence kL is also a logarithm function (if $k \neq 0$). There are no others; for, as we know from equation (6) of §7A1, any two logarithms are proportional.

The *base* of a logarithm is that number whose logarithm is 1. For each positive number a other than 1, there is a logarithm with base a—namely, the function $\log_a$ defined by

$$(2) \qquad \log_a x = \frac{1}{\log a} \log x \qquad (a \neq 1; x > 0).$$

The base of the *natural* logarithm is e (by definition of e): $\log_e x = \log x$. The choice (7) in §7A1 is the number $1/(\log 10)$ and hence yields the "common" logarithm, to base 10.

For the derivative of $\log_a x$ we have, obviously,

$$(3) \qquad \frac{d}{dx} \log_a x = \frac{1}{\log a}\frac{1}{x}.$$

7E2. Exponential to base a. At this juncture we are still in the curious position of knowing what e^x means for every real number x but having no definition yet for, say 2^x, or 10^x. For positive a different from e we have defined a^r for rational r only.

It would be good to have a^x satisfy

(4)
$$\log a^x = x \log a,$$

as we already know this for rational x ($\S 7A(14)$). It would also be good to have

(5)
$$\log_a a^x = x,$$

so that the logarithm to base a and exponential to base a are inverses. As it happens, (4) and (5) say the same thing (in view of the definition (2)). With them in mind, we define

(6)
$$a^x = e^{x \log a}.$$

(Taking logarithms then yields (4).) The characteristic property

(7)
$$a^\alpha a^\beta = a^{\alpha + \beta}$$

now follows. Moreover, we can now establish the formula

(8)
$$(a^\alpha)^\beta = a^{\alpha \beta} \qquad\qquad (\alpha > 0, \beta > 0)$$

(which until now was not defined, not even for $a = e$).

Proof. By (6),

(9)
$$(a^\alpha)^\beta = e^{\beta \log a^\alpha} = e^{\beta \alpha \log a} = a^{\beta \alpha}.$$

Finally, the derivative of a^x is

(10)
$$\frac{d}{dx} a^x = e^{x \log a} \cdot \log a = a^x \log a.$$

7E3. The number e. Here is a famous formula for e:

(11)
$$e = \lim_{x \to 0} (1 + x)^{1/x}.$$

To establish it we look at the definition of $\log'(1)$:

(12)
$$\log'(1) = \lim_{x \to 0} \frac{\log(1 + x) - \log 1}{x} = \lim_{x \to 0} \frac{1}{x} \log(1 + x).$$

But we know what $\log'(1)$ is: $\log'(1) = 1/1 = 1$. Thus,

(13) $$\frac{1}{x} \log(1 + x) \to 1 \qquad\qquad \text{(as } x \to 0).$$

In other words (see (4)),

(14) $$\log[(1 + x)^{1/x}] \to 1 \qquad\qquad \text{(as } x \to 0).$$

Taking exponentials, we get

(15) $$(1 + x)^{1/x} \to e \qquad\qquad \text{(as } x \to 0)$$

(since the exponential is continuous). This is what we were to prove.

7E4. Powers. We have defined

$$a^s$$

when a is a positive number and s is any real number. If we hold a fixed and vary s, which we then call x, we are dealing with the *exponential* function

$$a^x;$$

as we have just seen,

(16) $$\frac{d}{dx} a^x = a^x \log a. \quad \frac{d}{du} \uparrow a^u = a^u\, du \log a$$

If on the other hand we hold s fixed and vary a—which we then call x—we are dealing with the *power* function

$$x^s.$$

Let us prove that, for $x > 0$,

(17) $$\frac{d}{dx} x^s = sx^{s-1}$$

for all real numbers s. (We saw in §3G that this holds for *rational* s.)

   Proof. $x^s = e^{s \log x}$, and therefore

(18) $$\frac{d}{dx} x^s = e^{s \log x} \cdot s \cdot \frac{1}{x} = x^s \cdot s \cdot \frac{1}{x} = sx^{s-1}.$$

Be sure you understand the difference between (16) and (17).

7E5. Logarithmic differentiation. As we know, if y is a differentiable function of x then, by the chain rule, so is $\log y$ (assuming $y > 0$):

$$(19) \qquad \frac{d}{dx}\log y = \frac{1}{y}y' \qquad (y > 0).$$

Conversely, let us observe that if $\log y$ is a differentiable function of x, then so is y; this follows from the equation

$$(20) \qquad y = e^{\log y}$$

and the chain rule. Explicitly,

$$(21) \qquad y' = e^{\log y}\frac{d}{dx}(\log y) = y\frac{d}{dx}(\log y).$$

These ideas can be useful in simplifying or organizing involved computations. We do not memorize formula (21), but write out the computation step by step, the first step being to take logarithms.

Example 1. If $y = \sqrt{(x^2 - 1)/(x^2 + 3)}$ (where $x > 1$), find y'.

Solution. Here it is clear in the first place that y is differentiable. At any rate,

$$\log y = \tfrac{1}{2}[\log(x^2 - 1) - \log(x^2 + 3)].$$

Therefore,

$$\frac{y'}{y} = \frac{1}{2}\left[\frac{2x}{x^2-1} - \frac{2x}{x^2+3}\right], \qquad y' = x\sqrt{\frac{x^2-1}{x^2+3}}\left[\frac{1}{x^2-1} - \frac{1}{x^2+3}\right].$$

Example 2. If $y = x^x$, find y'.

Solution. $\log y = x\log x$, $y'/y = 1 + \log x$, $y' = y(1 + \log x) = x^x(1 + \log x)$.

Problems (§7E)

1. (a) Show that $\log_a b$ and $\log_b a$ are reciprocals.
   (b) If $\log_b a = 4$ and $\log_c b = \tfrac{1}{3}$, what is $\log_a c$?
2. Simplify the following.
   (a) $[\log_2(\tfrac{1}{2})]^{151}$
   (b) $\log_2[4^{\log_3 9}]$
   (c) $(\log_2 5)(\log_2 4)(\log_5 2)$
   (d) $2^{\log_2[2^{\log_3 27}]}$
   (e) $[\sqrt{2}^{\sqrt{2}}]^{\sqrt{2}}$
   (f) $e^{(\log x)/x}$

3. Find dy/dx in each case.

(a) $y = 3^{x^2}$

(b) $y = x^{\log x}$

(c) $y = (x + 1)^{x^2}$

(d) $y = (x^2 + 1)^x$

(e) $y = (x^2)^{x^2}$

(f) $y = (\log x)^x$

(g) $y = (\log x)^{\log x}$

(h) $y = e^e$

(i) $y = \log_x 2$

(j) $y = x^{\sqrt{2}}$

4. Evaluate the following.

(a) $\displaystyle\int x \cdot 3^{x^2}\, dx$

(b) $\displaystyle\int 2^x\, dx$

(c) $\displaystyle\int 3^{2x-1}\, dx$

(d) $\displaystyle\int \frac{3^{1/x}}{x^2}\, dx$

(e) $\displaystyle\int \frac{2^{\sqrt{x}}}{\sqrt{x}}\, dx$

(f) $\displaystyle\int x^{\sqrt{2}}\, dx$

5. Find the minimum value

(a) of x^x

(b) of $x^{-1/x}$

6. (a) Prove that

$$\lim_{x \to 0} (1 + 2x)^{1/x} = e^2 \quad \text{and} \quad \lim_{x \to 0} (1 - x)^{1/x} = \frac{1}{e}.$$

(b) What about e^α?

Answers to problems (§7E)

1. (b) $\frac{3}{4}$

2. (a) -1 (b) 4 (c) 2 (d) 8 (e) 2 (f) $x^{1/x}$

3. (a) $x \cdot 3^{x^2} \log 9$ (b) $2x^{\log x - 1} \log x$

(c) $(x + 1)^{x^2} \left[\dfrac{x^2}{x + 1} + 2x \log(x + 1) \right]$

(d) $(x^2 + 1)^x \left[\log(x^2 + 1) + \dfrac{2x^2}{x^2 + 1} \right]$ (e) $2x^{2x^2+1}(1 + \log x^2)$

(f) $(\log x)^x \left(\dfrac{1}{\log x} + \log \log x \right)$

(g) $\dfrac{1}{x} (\log x)^{\log x}(1 + \log \log x)$ (h) 0

(i) $-\dfrac{1}{x(\log x)^2} \log 2$ (j) $\sqrt{2}\, x^{\sqrt{2}-1}$

4. (a) $\dfrac{1}{\log 9} 3^{x^2}$ (b) $\dfrac{1}{\log 2} 2^x$ (c) $\dfrac{1}{6 \log 3} 9^x$ (d) $-\dfrac{1}{\log 3} 3^{1/x}$

(e) $\dfrac{1}{\log 2} 2^{1+\sqrt{x}}$ (f) $\dfrac{1}{1 + \sqrt{2}} x^{1+\sqrt{2}}$

5. (a) $e^{-1/e}$ (b) $e^{-1/e}$

CHAPTER 8

The Trigonometric Functions

The trigonometric functions as studied in calculus are functions of real numbers.

In §8A, we define the sine and cosine in terms of arc length (§6D), and then find their derivatives via the integral for arc length. §§8B-8C continue the discussion and introduce the other trigonometric functions.

§8D treats applications to geometry, including the trigonometric ratios in a right triangle.

§8E is on integrals and §8F introduces the inverse trigonometric functions. We discover that the integrals of some innocent-looking algebraic expressions involve logarithms or inverse trigonometric functions. The latter form the basis for applying the method of trigonometric substitution, explained in §8G.

§8A. The sine and cosine

8A1. Definitions of the functions sin and cos. The most important trigonometric functions are the sine and cosine. Their definitions are in terms of *directed* arc length on the unit circle, $x^2 + y^2 = 1$. The arcs will start at the point $Q_0 = (1, 0)$. The directed length is the length itself for counter-clockwise arcs, the negative of the length for clockwise.

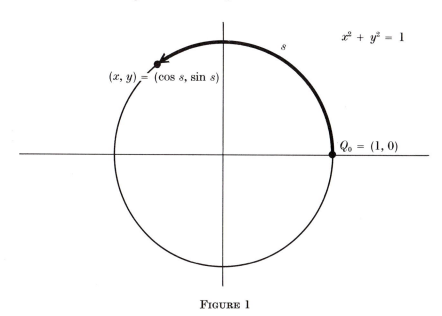

FIGURE 1

The sine function is denoted *sin* and the cosine is denoted *cos*. First we define sin s and cos s for $0 \leq s \leq 2\pi$. Let s be any number in this interval. There exists an arc from Q_0 of length s (see §6D4); let (x, y) denote its terminal point (Figure 1). Then, by definition,

(1) $$\sin s = y, \qquad \cos s = x.$$

In other words, the coordinates of the terminal point are, by definition, $(\cos s, \sin s)$.

Example 1. See Figure 2. Some special values of the sine and cosine are:

$$\sin 0 = 0, \qquad \cos 0 = 1,$$

$$\sin \frac{\pi}{2} = 1, \qquad \cos \frac{\pi}{2} = 0,$$

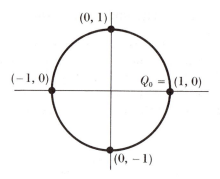

FIGURE 2

$$\sin \pi = 0, \qquad \cos \pi = -1,$$

$$\sin \frac{3\pi}{2} = -1, \qquad \cos \frac{3\pi}{2} = 0,$$

$$\sin 2\pi = 0, \qquad \cos 2\pi = 1.$$

Before extending the definitions to all real numbers we first describe the process pictorially. Imagine that an arc can be of any length s at all, perhaps even longer than the circle itself. For instance, picture winding a string of length s around and around, counterclockwise for s positive, clockwise for s negative. If we start at Q_0, then, as before, we call the terminal point (x, y) and define $\sin s$ and $\cos s$ by (1). Of course, two arcs from Q_0 that differ by a whole number of revolutions have the same terminal point; consequently, two numbers that differ by an integral multiple of 2π will have the same sine and cosine.

The formal definitions take the following concise form:

(2) $$\sin(s + 2n\pi) = \sin s, \qquad \cos(s + 2n\pi) = \cos s,$$

where n is any integer (positive, negative, or zero).

Example 2. See Figure 2.

$$\sin 5\pi = \sin(-\pi) = \sin \pi = 0,$$

$$\cos 5\pi = \cos(-\pi) = \cos \pi = -1.$$

The following famous formula is evident from the definitions:

(3) $$\sin^2 s + \cos^2 s = 1.$$

There are many other formulas relating sines and cosines. For now we mention the following:

$$(4) \qquad \sin s = \cos\left(s - \frac{\pi}{2}\right), \qquad \cos s = -\sin\left(s - \frac{\pi}{2}\right).$$

In other words, if (x_2, y_2) lies a quarter-turn (counterclockwise) beyond (x_1, y_1), then $y_2 = x_1$ and $x_2 = -y_1$. This is easy to check from Figure 3.

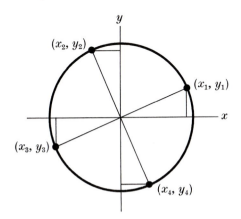

FIGURE 3

8A2. Derivatives of sin and cos. The formulas are:

$$(5) \qquad \frac{d}{ds}\sin s = \cos s, \qquad \frac{d}{ds}\cos s = -\sin s.$$

Proof. We first treat the case in which $-\pi/2 < s < \pi/2$ (Figure 4). According to formula (9) of §6D3,

$$(6) \qquad s = \int_0^y \frac{1}{\sqrt{1 - t^2}}\, dt \qquad\qquad (-1 < y < 1).$$

(The formula is stated there only for $y \geq 0$; by symmetry, it holds as well for $y \leq 0$.) Note that this equation expresses s in terms of $y = \sin s$. By the Fundamental Theorem of Calculus,

$$(7) \qquad \frac{ds}{dy} = \frac{1}{\sqrt{1 - y^2}} = \frac{1}{x} > 0 \qquad\qquad (-1 < y < 1).$$

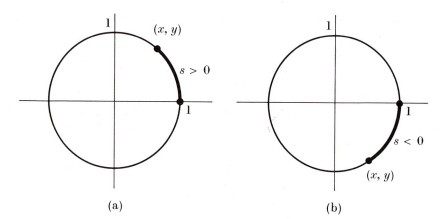

(a)   (b)

FIGURE 4

By the Inverse-Function Theorem (§4G), $dy/ds = x$; that is,

$$(8) \qquad \frac{d}{ds} \sin s = \frac{dy}{ds} = x = \cos s \qquad (-\pi/2 < s < \pi/2).$$

Next, from $\sin^2 s + \cos^2 s = 1$, we get, differentiating implicitly,

$$(9) \qquad 2 \sin s \cos s + 2 \cos s \frac{d}{ds} \cos s = 0.$$

Since $\cos s \neq 0$ we may divide. This gives us

$$\frac{d}{ds} \cos s = -\sin s.$$

We thus get the theorem for $-\pi/2 < s < \pi/2$. To extend the results to all values of s we may proceed in steps of size $\pi/2$, with the help of (4). Suppose we know the results for $s - \pi/2$. Then we can obtain them for s:

$$(10) \qquad \frac{d}{ds} \sin s = \frac{d}{ds} \cos \left(s - \frac{\pi}{2} \right) = -\sin \left(s - \frac{\pi}{2} \right) = \cos s;$$

and similarly for $d(\cos s)/ds$.

Example 3.

(a) $\dfrac{d}{ds} \sin^2 s = 2 \sin s \dfrac{d}{ds} \sin s = 2 \sin s \cos s.$

(b) $\dfrac{d}{ds} \cos(3s - 1) = -\sin(3s - 1) \dfrac{d}{ds} (3s - 1) = -3 \sin(3s - 1).$

(c) $\dfrac{d}{ds} \log \sin s = \dfrac{1}{\sin s} \dfrac{d}{ds} \sin s = \dfrac{\cos s}{\sin s}$ $(\sin s > 0).$

Problems (§8A)

1. Write out the *definition* of sin 1 and of cos 1.

2. Which of the following numbers are positive, which are negative, which are zero: sin 1, cos 1, sin 2, cos 2, sin(−1), cos(−1), sin(−2), cos(−2)?

3. Solve for s $(0 \le s \le 2\pi)$:
   (a) $2 \cos^2 s - 3 \cos s + 1 = 0$
   (b) $\sin^2 s + 3 \sin s + 2 = 0$
   (c) $\sin s \cos s + \cos s - \sin s - 1 = 0$
   (d) $\sin s = \cos s$
   (e) $\sin^2 s = \cos^2 s$

4. Prove that the sine function is not a polynomial.

5. Differentiate with respect to s:
   (a) $\cos^2 s$ (i) $\log e^{\sin s}$
   (b) $\sin^2 s + \cos^2 s$ (j) $\log \cos s$
   (c) $\sin 2s$ (k) $\cos(as + b)$
   (d) $\cos 2s$ (l) $e^{\log \cos s}$
   (e) $\cos^2 s - \sin^2 s$ (m) $\sin \log s$
   (f) $2 \sin s \cos s$ (n) $\log \sin s$
   (g) $e^{\sin s}$ (o) $\sin s^2$
   (h) $\sin e^s$ (p) $\cos \sin s$

6. State formulas for the nth derivatives of the sine and cosine.

Answers to problems (§8A)

2. positive: sin 1, cos 1, sin 2, cos(−1)
   negative: cos 2, sin(−1), sin(−2), cos(−2)

3. (a) $0, \pi/3, 5\pi/3, 2\pi$ (b) $3\pi/2$ (c) $0, 3\pi/2, 2\pi$ (d) $\pi/4, 5\pi/4$
   (e) $\pi/4, 3\pi/4, 5\pi/4, 7\pi/4$

5. (a) $-2 \cos s \sin s$ (b) 0   (c) $2 \cos 2s$ (d) $-2 \sin 2s$
   (e) $-4 \sin s \cos s$ (f) $2(\cos^2 s - \sin^2 s)$ (g) $\cos s\, e^{\sin s}$
   (h) $e^s \cos e^s$ (i) $\cos s$ (j) $-(\sin s)/(\cos s)$ (k) $-a \sin(as + b)$

   (l) $-\sin s$ (m) $\dfrac{1}{s} \cos \log s$ (n) $(\cos s)/(\sin s)$ (o) $2s \cos s^2$

   (p) $-\cos s \sin \sin s$

§8B. Further properties of the sine and cosine

8B1. The addition formulas. The following formulas are basic:

(1) $$\cos(s - t) = \cos s \cos t + \sin s \sin t,$$

(2) $$\sin(s - t) = \sin s \cos t - \cos s \sin t,$$

(3) $$\cos(-t) = \cos t, \qquad \sin(-t) = -\sin t.$$

We shall derive them quickly. Consider the points $A = (\cos s, \sin s)$ and $B = (\cos t, \sin t)$ of Figure 1(a). If h is the distance between them, then

(4) $$h^2 = (\cos s - \cos t)^2 + (\sin s - \sin t)^2$$
$$= 2 - 2 \cos s \cos t - 2 \sin s \sin t$$

(since $\sin^2 + \cos^2 = 1$). With the axes chosen so that B is the point $(1, 0)$, the coordinates of A become $(\cos(s - t), \sin(s - t))$ (Figure 1(b)). Then

(5) $$h^2 = [\cos(s - t) - 1]^2 + [\sin(s - t)]^2$$
$$= 2 - 2 \cos(s - t).$$

Equating the expressions for h^2, we get (1).

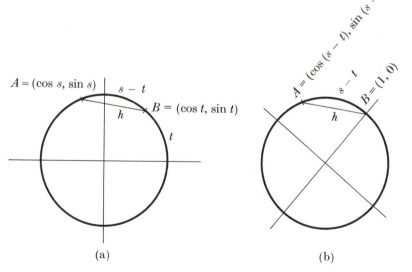

(a)   (b)

FIGURE 1

Since (1) is true for all s and t, we may hold t fixed and watch s vary. Differentiating with respect to s, and then changing signs, we get (2).

Finally, putting $s = 0$ in (1) and (2), we obtain (3).

Formulas (1), (2), and (3) lead to many others. If we replace t by $-t$ in (1) and (2), and take account of (3), we find that

(6)
$$\cos(s + t) = \cos s \cos t - \sin s \sin t,$$

and

(7)
$$\sin(s + t) = \sin s \cos t + \cos s \sin t.$$

Putting $s = \pi/2$ in (1) and (2), we get

(8)
$$\cos\left(\frac{\pi}{2} - t\right) = \sin t, \qquad \sin\left(\frac{\pi}{2} - t\right) = \cos t,$$

Similarly, we can get expressions for the sine and cosine of $\pi - t$, $\pi/2 + t$, etc.

The case $s = t$ in (6) and (7) is of special importance:

(9)
$$\cos 2s = \cos^2 s - \sin^2 s$$
$$= 2 \cos^2 s - 1 = 1 - 2 \sin^2 s,$$

(10)
$$\sin 2s = 2 \sin s \cos s.$$

Example 1. Find $\sin(\pi/4)$ and $\cos(\pi/4)$.

Solution. By (8), $\cos(\pi/4) = \sin(\pi/4)$. Since $\sin^2(\pi/4) + \cos^2(\pi/4) = 1$, we have $2 \sin^2(\pi/4) = 1$, and therefore

$$\sin\frac{\pi}{4} = \cos\frac{\pi}{4} = \frac{1}{\sqrt{2}}.$$

8B2. Graphs of the sine and cosine. The graph of a function f is customarily referred to an xy-coordinate system, where it appears as the graph of the equation $y = f(x)$. Accordingly, we now examine the graphs of the equations

(11)
$$y = \sin x \qquad \text{and} \qquad y = \cos x.$$

Thus, x and y no longer retain their specialized meanings in the definitions of the sine and cosine. The switch to the notation (11) takes a little getting used to, but everyone seems to manage it eventually.

The graph of $y = \sin x$ is formed by taking the portion over $0 \le x \le 2\pi$ and replicating it periodically, with period 2π.

We know from the definition that $\sin x$ increases from the value 0 at $x = 0$ to its maximum value of 1 as x increases to $\pi/2$; then decreases to its minimum value of -1 as x increases to $3\pi/2$, passing through 0 at $x = \pi$; then increases to 0 as x increases to 2π.

Next,

$$y'' = -\sin x,$$

so that the graph is concave down on the interval $0 \le x \le \pi$ and concave up on $\pi \le x \le 2\pi$. There are inflection points at $x = 0$ and $x = 2\pi$, where the slope has its maximum value of 1, and at $x = \pi$, with the slope at its minimum value of -1.

Since

$$\cos\left(x - \frac{\pi}{2}\right) = \sin x,$$

the graph of the cosine is the same as the graph of the sine but shifted to the left a distance $\pi/2$. The curves are shown in Figure 2.

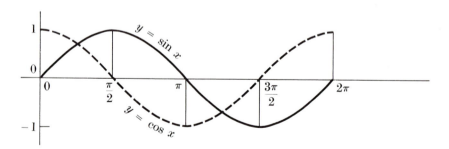

FIGURE 2

8B3. Limit of (sin x)/x. Here is a famous limit:

(12)
$$\lim_{x \to 0} \frac{\sin x}{x} = 1.$$

Its proof is immediate from considering the definition of $d(\sin x)/dx$ at $x = 0$:

$$(13) \qquad \lim_{x \to 0} \frac{\sin x}{x} = \lim_{x \to 0} \frac{\sin(0 + x) - \sin 0}{x} = \sin'(0) = \cos 0 = 1.$$

Example 2.

$$\lim_{x \to 0} \frac{x - \sin x}{x + \sin x} = \lim_{x \to 0} \frac{1 - (\sin x)/x}{1 + (\sin x)/x} = \frac{0}{2} = 0.$$

Example 3.

$$\lim_{x \to 0} \frac{1 - \cos x}{x} = \lim_{x \to 0} \left(\frac{1}{1 + \cos x} \right) \left(\frac{1 - \cos^2 x}{x} \right)$$

$$= \frac{1}{2} \lim_{x \to 0} \frac{\sin x}{x} \cdot \sin x = \tfrac{1}{2} \times 1 \times 0 = 0.$$

Example 4.

$$\lim_{x \to 0} \frac{1 - \cos x}{x^2} = \lim_{x \to 0} \left(\frac{1}{1 + \cos x} \right) \left(\frac{1 - \cos^2 x}{x^2} \right)$$

$$= \frac{1}{2} \lim_{x \to 0} \left(\frac{\sin x}{x} \right)^2 = \tfrac{1}{2} \times 1^2 = \tfrac{1}{2}.$$

Problems (§8B)

1. Express in terms of $\sin x$ and $\cos x$:

   (a) $\sin 3x$

   (b) $\cos 3x$

   (c) $\sin \left(\dfrac{\pi}{4} + x \right)$

   (d) $\cos \left(\dfrac{\pi}{4} + x \right)$

2. Solve for x $(0 \le x \le 2\pi)$:
   (a) $\sin x = \sin 2x$
   (b) $\cos x = \cos 2x$
   (c) $\cos x = \sin 2x$

3. If $f(x) = \sin 4x \cos 3x - \cos 4x \sin 3x$, what is $f'(\pi/3)$?

4. Use formulas (8) and (10) to find the sine and cosine of $\pi/3$ and $\pi/6$.

5. Use Example 1 and Problem 4 to find the sine and cosine of $\pi/12$.

6. (a) Find the equation of the line tangent to the graph of $y = \sin x$ at $x = \pi/4$.
   (b) What is the area of the triangle formed by this line with the lines $y = 0$ and $x = \pi/2$?

7. Give an example of a function f for which

$$f(2x) = 2f(x)f'(x).$$

8. Evaluate the following limits.

(a) $\lim\limits_{x \to 0} \dfrac{1 - \sin x}{1 - x}$

(g) $\lim\limits_{x \to \pi/2} \dfrac{\cos x}{1 + \sin x}$

(b) $\lim\limits_{x \to 0} \dfrac{\sin x}{1 + \cos x}$

(h) $\lim\limits_{x \to \pi/2} \dfrac{1 - \sin x}{\cos x}$

(c) $\lim\limits_{x \to 0} \dfrac{1 - \cos x}{\sin x}$

(i) $\lim\limits_{x \to 0} \dfrac{1 - \cos x}{x \sin x}$

(d) $\lim\limits_{x \to 0} \dfrac{\sin^2 x}{x^2}$

(j) $\lim\limits_{x \to \pi} \dfrac{\sin x}{1 - \cos x}$

(e) $\lim\limits_{x \to 0} \dfrac{\cos^2 x}{\sin^2 x + \cos^2 x}$

(k) $\lim\limits_{x \to \pi} \dfrac{1 + \cos x}{\sin x}$

(f) $\lim\limits_{x \to 2\pi} \dfrac{\sin x}{x}$

(l) $\lim\limits_{x \to \pi} \dfrac{2 \sin x \cos x}{\sin^2 x + \cos^2 x}$

9. Establish the following formulas.

(14)   (a) $\sin ax \sin bx = \frac{1}{2} \cos(ax - bx) - \frac{1}{2} \cos(ax + bx)$

(15)   (b) $\cos ax \cos bx = \frac{1}{2} \cos(ax - bx) + \frac{1}{2} \cos(ax + bx)$

(16)   (c) $\sin ax \cos bx = \frac{1}{2} \sin(ax - bx) + \frac{1}{2} \sin(ax + bx)$

Answers to problems (§8B)

1. (a) $3 \sin x - 4 \sin^3 x$ (b) $4 \cos^3 x - 3 \cos x$

(c) $\dfrac{1}{\sqrt{2}} (\sin x + \cos x)$ (d) $\dfrac{1}{\sqrt{2}} (\cos x - \sin x)$

2. (a) $0, \pi/3, 5\pi/3, 2\pi$ (b) $0, 2\pi/3, 4\pi/3, 2\pi$ (c) $\pi/6, \pi/2, 5\pi/6, 3\pi/2$

3. $\frac{1}{2}$

4. $\sin(\pi/6) = \cos(\pi/3) = 1/2$, $\cos(\pi/6) = \sin(\pi/3) = \sqrt{3}/2$

5. $\sin(\pi/12) = \frac{1}{4}\sqrt{2}(\sqrt{3} - 1)$, $\cos(\pi/12) = \frac{1}{4}\sqrt{2}(\sqrt{3} + 1)$

6. (a) $\sqrt{2}y = x + 1 - \dfrac{\pi}{4}$ (b) $\frac{1}{4}\sqrt{2} \left(1 + \dfrac{\pi}{4} \right)^2$

8. (a) 1   (b) 0   (c) 0   (d) 1   (e) 1   (f) 0   (g) 0   (h) 0
(i) $\frac{1}{2}$ (j) 0   (k) 0   (l) 0

§8C. Tangent, secant, cotangent, and cosecant

8C1. The functions tan and sec. The tangent, denoted *tan*, and the secant, denoted *sec*, are defined in terms of the sine and cosine:

$$(1) \qquad \tan x = \frac{\sin x}{\cos x}, \qquad \sec x = \frac{1}{\cos x} \qquad (\cos x \neq 0).$$

These functions are defined for all x except odd multiples of $\pi/2$. Note that

$$(2) \qquad \sec^2 x = 1 + \tan^2 x.$$

The tangent is periodic with period π; that is,

$$(3) \qquad \tan(x + \pi) = \tan x.$$

This is true since

$$\tan(x + \pi) = \frac{\sin(x + \pi)}{\cos(x + \pi)} = \frac{-\sin x}{-\cos x} = \tan x.$$

On the interval $(-\pi/2, \pi/2)$, the tangent assumes all real values. For, if m is any real number, then the line $y = mx$ cuts the unit circle in an arc s for which $\tan s = m$ (Figure 1). The derivative is

$$(4) \qquad \frac{d}{dx} \tan x = \sec^2 x.$$

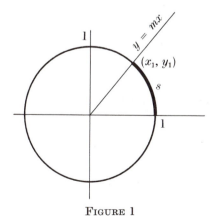

FIGURE 1

This follows directly from (1): by the quotient rule,

(5) $\quad \dfrac{d}{dx}\dfrac{\sin x}{\cos x} = \dfrac{(\cos x)(\cos x) - (\sin x)(-\sin x)}{\cos^2 x} = \dfrac{1}{\cos^2 x} = \sec^2 x.$

The graph of the tangent is indicated in Figure 2.

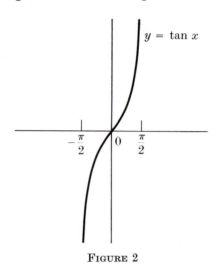

FIGURE 2

The secant is periodic with period 2π; that is, $\sec(x + 2\pi) = \sec x$. This is so because the cosine has the same property. On $[0, \pi/2)$, the secant assumes every value ≥ 1, since the cosine assumes every positive value ≤ 1. The derivative is

(6) $\qquad\qquad\qquad \dfrac{d}{dx} \sec x = \sec x \tan x.$

This follows directly from (1):

(7) $\quad \dfrac{d}{dx}\dfrac{1}{\cos x} = -\dfrac{1}{\cos^2 x}(-\sin x) = \dfrac{1}{\cos x}\dfrac{\sin x}{\cos x} = \sec x \tan x.$

The graph of the secant is indicated in Figure 3.

Example 1.

$$\dfrac{d}{dx}\tan^2 x = 2 \tan x \sec^2 x,$$

$$\dfrac{d}{dx}\sec^2 x = 2 \sec^2 x \tan x.$$

Query. Why are these equal?

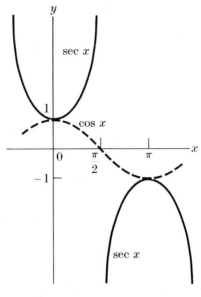

FIGURE 3

Example 2.

$$\frac{d}{dx} \log \sec x = \frac{\sec x \tan x}{\sec x} = \tan x \qquad (\sec x > 0),$$

$$\frac{d}{dx} \log \cos x = \frac{-\sin x}{\cos x} = -\tan x \qquad (\cos x > 0).$$

Query. Why is one the negative of the other?

8C2. *The functions* cot *and* csc. The cotangent and cosecant are referred to least often. They are denoted *cot* and *csc* and are defined by

(8) $$\cot x = \frac{\cos x}{\sin x}, \qquad \csc x = \frac{1}{\sin x} \qquad (\sin x \neq 0).$$

Evidently, $\cot(\pi/2 - x) = \tan x$, $\csc(\pi/2 - x) = \sec x$, and $\csc^2 x = 1 + \cot^2 x$. The derivatives are easily found to be

(9) $$\frac{d}{dx} \cot x = -\csc^2 x, \qquad \frac{d}{dx} \csc x = -\csc x \cot x.$$

Problems (§8C)

1. Evaluate the following.

   (a) $\tan 0$

   (b) $\sec 0$

   (c) $\tan \pi$

   (d) $\sec \pi$

   (e) $\tan \dfrac{\pi}{4}$

   (f) $\sec \dfrac{\pi}{4}$

   (g) $\tan\left(-\dfrac{\pi}{4}\right)$

   (h) $\sec\left(-\dfrac{\pi}{4}\right)$

2. Solve for x $(0 \le x \le 2\pi)$:

   (a) $\tan x = \sin x$

   (b) $\tan x = \sec x$

   (c) $\sin x = \sec x$

3. Find the first 4 derivatives of

   (a) $\tan x$

   (b) $\sec x$

4. Analyze the following graphs for concavity and inflection:

   (a) $y = \tan x$ $(-\pi/2 < x < \pi/2)$

   (b) $y = \sec x$ $(0 \le x < \pi/2)$

5. Derive the following formulas.

   (a) $\tan(-x) = -\tan x$

   (b) $\sec(-x) = \sec x$

   (c) $\tan(s + t) = \dfrac{\tan s + \tan t}{1 - \tan s \tan t}$

6. Differentiate the following with respect to x:

   (a) $\sec^2 x \tan x$

   (b) $\sec x \tan^2 x$

   (c) $\sec x \cos x$

   (d) $\sec^3 x$

   (e) $\tan^3 x$

   (f) $\tan x \cos x$

   (g) $\sec x \sin x$

   (h) $\sec^2 x + \tan^2 x$

   (i) $\sec^2 x - \tan^2 x$

   (j) $\cot x \csc x$

   (k) $\cot x \tan x$

   (l) $\cot x \sin x$

   (m) $\csc x \sin x$

   (n) $\csc x \tan x$

   (o) $\csc^2 x$

   (p) $\cot^2 x$

   (q) $\cot^2 x - \csc^2 x$

   (r) $\csc x \sin^2 x$

   (s) $\tan 2x \cot 3x$

   (t) $\sec^3 2x$

   (u) $\cot^2 \dfrac{\pi}{4}$

   (v) $\sin \tan x$

   (w) $\tan \sin x$

   (x) $\log \tan x$

   (y) $e^{\cot x}$

   (z) $\sin \dfrac{1}{x}$

7. Evaluate the following limits.

(a) $\lim\limits_{x\to 0} \dfrac{\tan x}{x}$

(e) $\lim\limits_{x\to 0} x \sec x$

(b) $\lim\limits_{x\to 0} \dfrac{\sec x - 1}{x}$

(f) $\lim\limits_{x\to 0} x \cot x$

(c) $\lim\limits_{x\to 0} \dfrac{\sin x - \tan x}{x}$

(g) $\lim\limits_{x\to 0} \dfrac{\tan^2 3x}{x \sin x}$

(d) $\lim\limits_{x\to 0} x \csc x$

8. Prove the following inequalities for $0 < x < \pi/2$.
   (a) $\sec x > \tan x > \sin x$
   (b) $\tan x > x > \sin x$ [*Hint.* See §4D, Problem 12.]

Answers to problems (§8C)

1. (a) 0   (b) 1   (c) 0   (d) -1 (e) 1   (f) $\sqrt{2}$ (g) -1 (h) $\sqrt{2}$
2. (a) $0, \pi, 2\pi$ (b) $\pi/2$ (c) no solution
3. (a) $\sec^2 x$, $2\sec^2 x \tan x$, $2(\sec^4 x + 2\sec^2 x \tan^2 x)$,
      $8(2\sec^4 x \tan x + \sec^2 x \tan^3 x)$
   (b) $\sec x \tan x$, $\sec^3 x + \sec x \tan^2 x$, $5\sec^3 x \tan x + \sec x \tan^3 x$,
      $5\sec^5 x + 18\sec^3 x \tan^2 x + \sec x \tan^4 x$
4. (a) concave down on $-\pi/2 < x \le 0$,   concave up on $0 \le x < \pi/2$,
      inflection point at $x = 0$
   (b) concave up on $0 \le x < \pi/2$
6. (a) $\sec^4 x + 2\sec^2 x \tan^2 x$ (b) $2\sec^3 x \tan x + \sec x \tan^3 x$ (c) 0
   (d) $3\sec^3 x \tan x$ (e) $3\tan^2 x \sec^2 x$ (f) $\cos x$ (g) $\sec^2 x$
   (h) $4\tan x \sec^2 x$ (i) 0   (j) $-\cot^2 x \csc x - \csc^3 x$ (k) 0
   (l) $-\sin x$ (m) 0   (n) $\sec x \tan x$ (o) $-2\csc^2 x \cot x$
   (p) $-2\csc^2 x \cot x$ (q) 0   (r) $\cos x$
   (s) $-3\tan 2x \csc^2 3x + 2\cot 3x \sec^2 2x$ (t) $6\sec^3 2x \tan 2x$ (u) 0
   (v) $\sec^2 x \cos \tan x$ (w) $\cos x \sec^2 \sin x$ (x) $\sec x \csc x$
   (y) $-\csc^2 x \, e^{\cot x}$ (z) $-\dfrac{1}{x^2} \cos \dfrac{1}{x}$
7. (a) 1   (b) 0   (c) 0   (d) 1   (e) 0   (f) 1   (g) 9

§8D. Applications to geometry

8D1. Angle; radians and degrees. An angle is the figure formed by two rays (the sides) emanating from a common point (the vertex). To assign it a direction, we think of it as generated by holding one ray fixed (the initial side) and rotating the other (the terminal side), the positive direction being counterclockwise. Its measure may then be any number at all, positive, negative, or zero.

The *radian* measure of an angle is the directed length of arc it sweeps out at radius 1 (Figure 1). One complete revolution, 360°, therefore has radian measure 2π. So:

$$2\pi \text{ radians } = 360°,$$

$$\pi \text{ radians } = 180°,$$

$$\pi/2 \text{ radians } = 90°,$$

etc.

Since an angle of θ radians sweeps out an arc of length θ on the unit circle, its arc on a circle of radius r is $r\theta$ (§6D(12)).

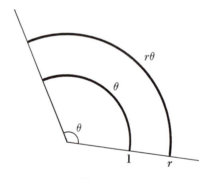

FIGURE 1

For simplicity, we often use the same symbol to denote a geometric angle, its vertex, and its measure. We speak of trigonometric functions of a number, of an angle, and of the measure of an angle. Thus, we say

$$\sin\frac{\pi}{6} = \frac{1}{2}, \qquad \sin\left(\frac{\pi}{6} \text{ radians}\right) = \frac{1}{2}, \qquad \sin 30° = \tfrac{1}{2},$$

and, if A is an angle of 30°,

$$\sin A = \tfrac{1}{2}.$$

Tables of trigonometric functions appear at the front of the book. They are calculated by advanced analytic methods.

8D2. Trigonometric ratios; area of a triangle. Suppose A is an *acute* angle in a *right* triangle. The trigonometric functions of A are easily expressed as ratios of sides of the triangle.

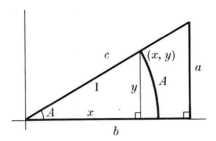

FIGURE 2

An angle of radian measure A sweeps out an arc of length A at radius 1. If axes are placed as shown in Figure 2 and the endpoint of the arc is (x, y), then by definition, $\sin A = y$, $\cos A = x$, and $\tan A = y/x$. From the similar triangles, we see that

(1) $$\sin A = \frac{a}{c}, \qquad \cos A = \frac{b}{c}, \qquad \tan A = \frac{a}{b}.$$

Here, c is the hypotenuse of the triangle and a is the side opposite angle A. Calling b the side "adjacent" to A, we have:

(2) $$\sin A = \frac{\text{opp}}{\text{hyp}}, \qquad \cos A = \frac{\text{adj}}{\text{hyp}}, \qquad \tan A = \frac{\text{opp}}{\text{adj}}.$$

See Figure 3. Remember that this means nothing unless A is an *acute* angle in a *right* triangle.

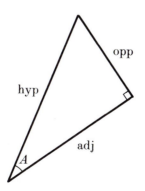

FIGURE 3

The following examples should be studied until fully understood.

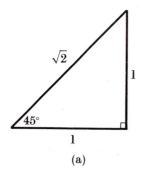

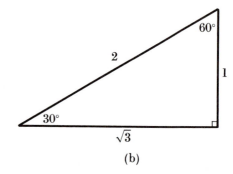

(a)   (b)

FIGURE 4

Example 1. See Figure 4.

$$\sin 45° = \frac{1}{\sqrt{2}} \qquad \sin 30° = \tfrac{1}{2} \qquad \sin 60° = \tfrac{1}{2}\sqrt{3}$$

$$\cos 45° = \frac{1}{\sqrt{2}} \qquad \cos 30° = \tfrac{1}{2}\sqrt{3} \qquad \cos 60° = \tfrac{1}{2}$$

$$\tan 45° = 1 \qquad \tan 30° = \frac{1}{\sqrt{3}} \qquad \tan 60° = \sqrt{3}$$

Example 2. See Figure 5.

$$\sin A = \frac{a}{c} = \frac{h}{b} = \frac{p}{a} \qquad \sin B = \frac{b}{c} = \frac{h}{a} = \frac{q}{b}$$

$$\cos A = \frac{b}{c} = \frac{q}{b} = \frac{h}{a} \qquad \cos B = \frac{a}{c} = \frac{p}{a} = \frac{h}{b}$$

$$\tan A = \frac{a}{b} = \frac{h}{q} = \frac{p}{h} \qquad \tan B = \frac{b}{a} = \frac{h}{p} = \frac{q}{h}$$

If A is a right or obtuse angle in a triangle we do not try to express its trigonometric functions as ratios of sides of the triangle. Instead, we reason as follows.

If A is a right angle then its radian measure is $\pi/2$ and we have $\sin A = 1$ and $\cos A = 0$; but $\tan A$ is not defined.

If A is an obtuse angle then $A' = \pi - A$ is acute and

(3) $\sin A = \sin A', \qquad \cos A = -\cos A', \qquad \tan A = -\tan A'.$

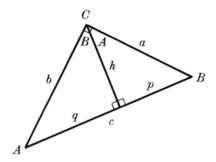

FIGURE 5

Example 3. In Figure 6,

$$\sin A = \sin A' = \frac{h}{c},$$

$$\cos A = -\cos A' = -\frac{k}{c},$$

$$\tan A = -\tan A' = -\frac{h}{k}.$$

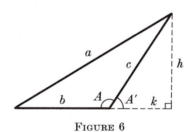

FIGURE 6

Example 4.

$$\sin 135° = \sin 45° = \frac{1}{\sqrt{2}}$$

$$\cos 150° = -\cos 30° = -\tfrac{1}{2}\sqrt{3}$$

$$\tan 120° = -\tan 60° = -\sqrt{3}$$

The area, S, of any triangle is given by the formula

(4) $$S = \tfrac{1}{2}bc \sin A.$$

Here, b and c are two of the sides and A is the included angle. The formula is immediate from Figure 7. Of course, we can also write (4) in two other forms, obtained by permuting the letters.

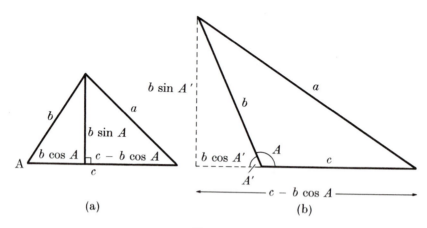

(a)   (b)

FIGURE 7

8D3. The Law of Cosines. The Law of Cosines is one of several formulas developed in numerical trigonometry for "solving" triangles, that is, finding unknown sides and angles when others are given. It is not our aim here to solve triangles. We will need the formula later on for theoretical purposes. For now we use it to help fix ideas about triangles. It states:

In a triangle with sides a, b, and c,

$$(5) \qquad a^2 = b^2 + c^2 - 2bc \cos A,$$

where A is the angle opposite a.

Proof. Notice that for $A = 90°$ this is just the Theorem of Pythagoras. If A is either acute or obtuse we get the diagrams shown in Figure 7. In either case,

$$(6) \qquad a^2 = (b \sin A)^2 + (c - b \cos A)^2,$$

and this simplifies to (5).

Of course, we can also write (5) in two other forms, obtained by permuting the letters.

Example 5. If two sides of a triangle are $b = 5$ and $c = 2$ and the included angle A is 60°, what is the third side? (See Figure 8.) What is the area?

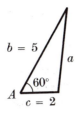

FIGURE 8

Solution.

$$a^2 = 5^2 + 2^2 - (2)(5)(2)(\tfrac{1}{2}) = 19, \qquad a = \sqrt{19} \approx 4.4.$$

$$S = \tfrac{1}{2} \cdot 5 \cdot 2 \cdot \tfrac{1}{2}\sqrt{3} = \tfrac{5}{2}\sqrt{3} \approx 4.3.$$

Example 6. The sides of a triangle are $a = 6$, $b = 7$, and $c = 10$ (Figure 9). What is angle C (opposite c)? What is the area?

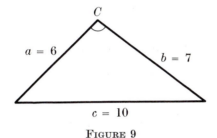

FIGURE 9

Solution.

$$10^2 = 6^2 + 7^2 - (2)(6)(7) \cos C,$$

$$\cos C = -\tfrac{5}{28} \approx -0.18, \qquad C \approx 100°.$$

For the area,

$$\sin C = \sqrt{1 - (\tfrac{5}{28})^2} \approx 0.97,$$

$$S = (\tfrac{1}{2})(6)(7)(0.97) = 20.4.$$

8D4. Applications of calculus to geometry. We can now use our knowledge of the calculus of trigonometric functions to solve problems involving angles.

Caution. In the derivation of the formulas of calculus, such as

$$\frac{d}{ds} \sin s = \cos s,$$

s was the length of arc along a circle of radius 1. One may therefore interpret s as the measure of an angle in radians. *But not in degrees!* That would be like reading your speedometer as 40 miles per hour and announcing your speed as 40 miles per minute.

When doing calculus with angles, use radian measure.

Example 7. An aircraft spotter observes a plane flying toward him at an altitude of 3750 feet. He notes that when the angle of elevation is 30°, it is increasing at the rate of 3° per second. What is the ground speed of the aircraft?

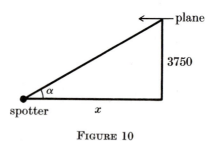

FIGURE 10

Solution. The situation is pictured in Figure 10. From the figure,

$$x = 3750 \cot \alpha.$$

Therefore,

$$\frac{dx}{dt} = 3750 \, (-\csc^2\alpha) \, \frac{d\alpha}{dt}.$$

We are given that $d\alpha/dt = \pi/60$ when $\alpha = 30°$. Since $\csc 30° = 2$,

$$\frac{dx}{dt} = -(3750)(4) \left(\frac{\pi}{60}\right) = -250\pi.$$

The plane is approaching at 250π feet per second. Since 88 feet per second is 60 miles per hour, this is

$$(\tfrac{60}{88})(250\pi) \approx 535 \text{ miles per hour.}$$

Example 8. Show that among all triangles with a given side and opposite angle, the one with the other two sides equal has the greatest area.

Solution. Let a and A be given. We may take $a = 1$. Angle B will be the variable. Figure 11 shows the picture in case A is acute. Denoting the area by S, we have, from the figure,

$$S = \tfrac{1}{2}(\sin B \cot A + \cos B) \sin B$$

$$= \frac{1}{2 \sin A} \sin(A + B) \sin B.$$

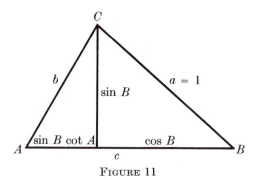

FIGURE 11

If A is a right angle or obtuse we get the same formula. Remembering that A is constant, B variable, we obtain

$$\frac{dS}{dB} = \frac{1}{2 \sin A} \sin(A + 2B).$$

This is positive for $A + 2B < \pi$, 0 at π, and negative thereafter. Hence S achieves a maximum at $A + 2B = \pi$. Since $A + B + C = \pi$ in any case, we get $B = C$. Therefore $b = c$.

Problems (§8D)

1. What is the degree measure of the angle with the given radian measure?
   (a) $\pi/4$ (f) $14\pi/3$
   (b) $\pi/6$ (g) $-\pi/2$
   (c) 2π (h) 0
   (d) $4\pi/3$ (i) -2π
   (e) $\pi/2$ (j) 30

2. What is the radian measure of the angle with the given degree measure?
   (a) $60°$ (e) $45°$
   (b) $150°$ (f) $30°$
   (c) $-400°$ (g) $-120°$
   (d) $0°$ (h) $\pi°$

3. See Figure 12. Prove the following.
   (a) The *Law of Sines*:

(7)
$$\frac{\sin A}{a} = \frac{\sin B}{b} = \frac{\sin C}{c}$$

   (b) $\sin^2 B = \dfrac{b^2 - p^2}{b^2 + q^2 - p^2}$

   (c) $c^2 = a^2 + b^2 - 2(h^2 - pq)$

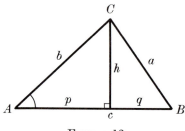

FIGURE 12

4. What is the maximum area of an isosceles triangle whose equal sides are each of length 10?

5. An observer is 1 mile from a spot where a rocket is launched vertically. When the rocket is 1 mile high, it is gaining altitude at the rate of 50 mph. How fast is the angle of elevation from the observer increasing at that instant?

6. In triangle ABC (with opposite sides a, b, c), a is increasing at 4 in./sec., while C stays fixed at $90°$. How fast is angle A increasing when $a = 10$ and $A = 60°$:

    (a) if the length of b remains constant (so that c varies)?

    (b) if the length of c remains constant (so that b varies)?

7. Two sides of a triangle are 12 and 20. The included angle is increasing at $10°$/min. When this angle is $30°$,

    (a) how fast is the length of the third side increasing?

    (b) how fast is the area increasing?

8. The equal sides of an isosceles triangle are increasing at 1 in./min. while their included angle increases at $30°$/min. How fast is the area increasing when the sides are each 10 and the angle is $30°$?

9. The angle of elevation of the sun is decreasing at a rate of $15°$/hr. When it is $30°$, what is the rate of increase of the length of the shadow cast by a 10-foot pole?

10. A light in a lighthouse 1 mile from a straight shoreline is rotating at 2 revolutions per minute. How fast is the beam moving along the shore when it passes the point a half mile from the point opposite the lighthouse?

11. We wish to hang a light bulb above a point Q on a table so as to maximize the intensity of the light at a point P on the table 6 feet from Q (Figure 13). How high should the bulb be hung if the intensity at P is inversely proportional to the square of the distance between P and the bulb, and directly proportional to $\sin \theta$?

12. For what angle θ does the trapezoid in Figure 14 have maximum area?

13. For what angle θ is the length of AB, Figure 15, a minimum?

14. A beam must be carried around a corner from a hall of width a to a hall of width b. What is the length of the longest beam that will turn the corner (Figure 16)?

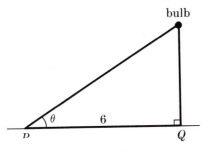

FIGURE 13

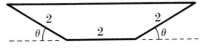

FIGURE 14

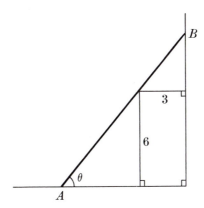

FIGURE 15

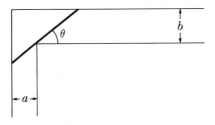

FIGURE 16

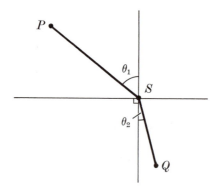

FIGURE 17

15. Light travels with velocity v_1 from a point P above the surface of a lake to a point S on the surface, and from there with velocity v_2 to a point Q below the surface (Figure 17). Show that the time required for the light to travel from P to Q will be a minimum if S is chosen so that

$$\frac{\sin \theta_1}{\sin \theta_2} = \frac{v_1}{v_2}.$$

16. Show that of all possible quadrilaterals with sides of given length, the one with largest area is such that each pair of opposite angles has sum π. [*Hint.* Express the area in terms of α and β (Figure 18). Obtain a relation between α and β by means of the Law of Cosines.]

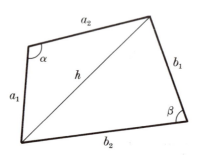

FIGURE 18

Answers to problems (§8D)

1. (a) $45°$ (b) $30°$ (c) $360°$ (d) $240°$ (e) $90°$ (f) $840°$

   (g) $-90°$ (h) $0°$ (i) $-360°$ (j) $\dfrac{5400°}{\pi}$

2. (a) $\pi/3$ (b) $5\pi/6$ (c) $-20\pi/9$ (d) 0   (e) $\pi/4$ (f) $\pi/6$
    (g) $-2\pi/3$ (h) $\pi^2/180$

4. 50

5. 25 radians/hour

6. (a) $\frac{1}{10}\sqrt{3}$ radians/sec.   (b) $\frac{2}{5}\sqrt{3}$ radians/sec.

7. (a) $\dfrac{5\pi}{3(34 - 15\sqrt{3})^{1/2}}$ units/min.   (b) $\frac{10}{3}\pi\sqrt{3}$ sq. units/min.

8. $\frac{25}{6}\pi\sqrt{3} + 5$

9. $\frac{10}{3}\pi$ ft./hr.

10. 5π mi./min.

11. $3\sqrt{2}$ feet

12. $\theta = 60°$

13. $\tan\theta = \sqrt[3]{2}$

14. $(a^{2/3} + b^{2/3})^{3/2}$

§8E. Integration

8E1. The basic formulas. As one might expect, integration of trigono-metric functions is more complicated and more interesting than differentia-tion. The methods of derivation are worth study, both for their own sake and because in some practical situation one may find himself lacking both a table of integrals and a lucky guess.

The following formulas are obvious:

$$(1) \qquad \int \sin x \, dx = -\cos x, \qquad \int \cos x \, dx = \sin x,$$

$$(2) \qquad \int \sec^2 x \, dx = \tan x, \qquad \int \csc^2 x \, dx = -\cot x,$$

$$(3) \qquad \int \sec x \tan x \, dx = \sec x, \qquad \int \csc x \cot x \, dx = -\csc x.$$

The antiderivatives of the tangent and the secant involve the logarithm. We state them only for the interval $(-\pi/2, \pi/2)$. Here $\cos x > 0$ and $\sec x > 0$. In fact, since $\sec^2 x > \tan^2 x$, we have $\sec x = |\sec x| > |\tan x|$ and therefore

$$\sec x + \tan x > 0 \qquad\qquad \left(-\frac{\pi}{2} < x < \frac{\pi}{2}\right).$$

The formulas are:

(4) $$\int \tan x \, dx = \log \sec x \qquad \left(-\frac{\pi}{2} < x < \frac{\pi}{2}\right),$$

(5) $$\int \sec x \, dx = \log(\sec x + \tan x) \qquad \left(-\frac{\pi}{2} < x < \frac{\pi}{2}\right).$$

They can be derived by fiddling with tangents and secants, as formulas (2) and (3) might suggest:

(6)
$$\int \tan x \, dx = \int \frac{\tan x \sec x}{\sec x} \, dx = \log \sec x,$$

$$\int \sec x \, dx = \int \frac{\sec x \, (\tan x + \sec x)}{\sec x + \tan x} \, dx = \log(\sec x + \tan x).$$

Formula (4) for the tangent can also be derived as follows:

(7) $$\int \tan x \, dx = \int \frac{\sin x}{\cos x} \, dx = -\log \cos x.$$

(This *is* formula (4), isn't it?)

Similarly, we obtain:

(8) $$\int \cot x \, dx = -\log \csc x \qquad (0 < x < \pi),$$

(9) $$\int \csc x \, dx = -\log(\csc x + \cot x) \qquad (0 < x < \pi).$$

Example 1.

(a) $$\int \sin 3x \, dx = -\tfrac{1}{3} \cos 3x$$

(b) $$\int \sin^2 x \cos x \, dx = \tfrac{1}{3} \sin^3 x$$

(c) $$\int e^{\cos x} \sin x \, dx = -e^{\cos x}$$

(d) $\displaystyle\int \tan x \, \sec^2 x \, dx = \int (\tan x)(\sec^2 x \, dx) = \frac{1}{2}\tan^2 x$

(e) $\displaystyle\int \tan x \, \sec^2 x \, dx = \int (\sec x)(\sec x \tan x) \, dx = \frac{1}{2}\sec^2 x$

8E2. $\int sin^m x \ cos^n x \ dx$**.** Here m and n represent integers ≥ 0. If one or the other is odd, the integral reduces to the form of a polynomial.

Example 2.

$$\int \sin^m x \, \cos^7 x \, dx = \int \sin^m x (\cos^2 x)^3 \cos x \, dx$$

$$= \int \sin^m x (1 - \sin^2 x)^3 \cos x \, dx.$$

This is in the form

$$\int u^m (1 - u^2)^3 \, du.$$

We expand and integrate the powers of u term by term. The work is tedious but straightforward.

When both m and n are even we are dealing with powers of $\sin^2 x$ and $\cos^2 x$. The procedure is to reduce the exponent by converting to $\cos 2x$. From formula (9) of §8B1 we get

(10) $\qquad \cos^2 x = \frac{1}{2}(1 + \cos 2x), \qquad \sin^2 x = \frac{1}{2}(1 - \cos 2x).$

Consequently,

(11) $\qquad \displaystyle\int \cos^2 x \, dx = \frac{1}{2} \int (1 + \cos 2x) \, dx = \frac{1}{2}x + \frac{1}{4}\sin 2x,$

and

(12) $\qquad \displaystyle\int \sin^2 x \, dx = \frac{1}{2} \int (1 - \cos 2x) \, dx = \frac{1}{2}x - \frac{1}{4}\sin 2x.$

For higher powers and products we apply (10) as many times as necessary.

Example 3.

$$\int \sin^4 x \, dx = \int (\sin^2 x)^2 \, dx$$

(13)
$$= \tfrac{1}{4} \int (1 - \cos 2x)^2 \, dx$$

$$= \tfrac{1}{4} \int (1 - 2\cos 2x + \cos^2 2x) \, dx$$

(14)
$$= \tfrac{1}{4} \int [1 - 2\cos 2x + \tfrac{1}{2}(1 + \cos 4x)] \, dx$$

$$= \tfrac{1}{4} \int (\tfrac{3}{2} - 2\cos 2x + \tfrac{1}{2}\cos 4x) \, dx$$

$$= \tfrac{3}{8}x - \tfrac{1}{4}\sin 2x + \tfrac{1}{32}\sin 4x.$$

8E3. $\int \sec^3 x \, dx$**.** This integral comes up often enough to warrant special attention. (Examples will appear in §8G.) Again, we fiddle with secants and tangents. Clearly, $\sec^3 x$ will arise when we differentiate $\sec x \tan x$; so we investigate:

(15)
$$\frac{d}{dx} (\sec x \tan x) = \sec^3 x + \tan^2 x \sec x$$

$$= \sec^3 x + (\sec^2 x - 1) \sec x = 2\sec^3 x - \sec x.$$

Hence

(16)
$$2\sec^3 x = \frac{d}{dx} (\sec x \tan x) + \sec x$$

and

(17)
$$\int \sec^3 x \, dx = \tfrac{1}{2} \sec x \tan x + \tfrac{1}{2} \log(\sec x + \tan x)$$

$$\left(-\frac{\pi}{2} < x < \frac{\pi}{2} \right).$$

This procedure amounts to integrating by parts; see Problem 9.

Problems (§8E)

1. Evaluate:

(a) $\int \sin^2 x \cos x \, dx$

(b) $\int \cos^2 x \sin x \, dx$

(c) $\int \sec^2 x \tan^3 x \, dx$

(d) $\int \tan^2 x \sec x \, dx$

$(-\pi/2 < x < \pi/2)$

(e) $\int \sin^3 x \, dx$

(f) $\int \cos^5 x \, dx$

(g) $\int \dfrac{\sin x}{1 + \cos x} \, dx$

(h) $\int \dfrac{1 - \sin x}{\cos x} \, dx$

(i) $\int \sin x \cos x \, dx$

(j) $\int \sin 2x \, dx$

(k) $\int \cos 4x \, dx$

(l) $\int (\sin x + \cos x)^2 \, dx$

2. Evaluate:

(a) $\int \tan^2 x \, dx$

(b) $\int \sec^3 x \tan x \, dx$

(c) $\int \tan^3 x \sec x \, dx$

(d) $\int \tan^3 x \, dx$

$(-\pi/2 < x < \pi/2)$

(e) $\int \tan^2 x \sec^2 x \, dx$

(f) $\int \sec^4 x \, dx$

(g) $\int \tan^4 x \, dx$

(h) $\int \sec x \sin x \, dx$

$(-\pi/2 < x < \pi/2)$

(i) $\int (\sec x - \cos x) \, dx$

$(-\pi/2 < x < \pi/2)$

(j) $\int \sin x \tan x \, dx$

$(-\pi/2 < x < \pi/2)$

(k) $\int \sec 2x \tan 2x \, dx$

(l) $\int \dfrac{\sin 3x}{\sec 3x} \, dx$

3. Evaluate:

(a) $\displaystyle\int \cos x \cot x\ dx$

$(0 < x < \pi)$

(b) $\displaystyle\int \csc^4 x\ dx$

(c) $\displaystyle\int \cot^2 x\ dx$

(d) $\displaystyle\int \cot^3 x\ dx$

$(0 < x < \pi)$

4. Evaluate:

(a) $\displaystyle\int \frac{\cos x - \sin x}{\cos x + \sin x}\ dx$

(b) $\displaystyle\int \frac{\cos x - \sin x}{(\cos x + \sin x)^2}\ dx$

(c) $\displaystyle\int \sin^3 x \cos^2 x\ dx$

(d) $\displaystyle\int \cos^4 x\ dx$

(e) $\displaystyle\int \sec^2 x \cot x\ dx$

$(0 < x < \pi/2)$

(f) $\displaystyle\int \frac{\log \tan x}{\sin 2x}\ dx$

(g) $\displaystyle\int \frac{1}{\sqrt{x}} \sin \sqrt{x}\ dx$

(h) $\displaystyle\int \sqrt{1 - \cos x}\ dx$

$(0 \le x \le 2\pi)$

(i) $\displaystyle\int \sin x \sin(\cos x)\ dx$

(j) $\displaystyle\int \frac{1}{x} \sin(\log x)\ dx$

5. Evaluate:

(a) $\displaystyle\int (\tan u + \cot u)^2\ du$

(b) $\displaystyle\int (\tan u + \cot u)^3\ du$

$(0 < u < \pi/2)$

(c) $\displaystyle\int (\tan u + \sec u)^2\ du$

(d) $\displaystyle\int (\tan u + \sec u)^3\ du$

$(-\pi/2 < u < \pi/2)$

(e) $\displaystyle\int (\sec u - \cos u)^2\ du$

(f) $\displaystyle\int (\sec u - \cos u)^3\ du$

$(-\pi/2 < u < \pi/2)$

6. Evaluate the following integrals. [*Hint.* Use formulas (14), (15), and (16) of Problem 9 in §8B.]

(a) $\displaystyle\int \sin 3x \sin x\ dx$

(b) $\displaystyle\int \sin 3x \sin 2x\ dx$

(c) $\displaystyle\int \cos 2x \cos 5x\ dx$

(d) $\displaystyle\int \cos 3x \cos 7x\ dx$

(e) $\displaystyle\int \sin 4x \cos x\ dx$

(f) $\displaystyle\int \sin 7x \cos 2x\ dx$

(g) $\displaystyle\int \sin x \sin 2x \sin 3x\ dx$

(h) $\displaystyle\int \cos x \sin 3x \cos 5x\ dx$

7. Evaluate:

(a) $\displaystyle\int x \sin x \, dx$

(d) $\displaystyle\int x^2 \cos x \, dx$

(b) $\displaystyle\int x \cos x \, dx$

(e) $\displaystyle\int x \sec^2 x \, dx \quad (-\pi/2 < x < \pi/2)$

(c) $\displaystyle\int x^2 \sin x \, dx$

(f) $\displaystyle\int x \sin^2 x \, dx$

8. Find:

(a) $\displaystyle\frac{d}{dx} (e^x \sin x)$

(c) $\displaystyle\frac{d}{dx} [e^x (\sin x - \cos x)]$

(b) $\displaystyle\frac{d}{dx} (e^x \cos x)$

(d) $\displaystyle\frac{d}{dx} [e^x (\sin x + \cos x)]$

Conclude that:

(18)   (e) $\displaystyle\int e^x \sin x \, dx = \tfrac{1}{2} e^x (\sin x - \cos x)$

(19)   (f) $\displaystyle\int e^x \cos x \, dx = \tfrac{1}{2} e^x (\sin x + \cos x)$

9. Integrate by parts until the original integral appears on both sides of the equation, but with opposite signs; then transpose and solve:

(a) $\displaystyle\int e^x \sin x \, dx$

(c) $\displaystyle\int \sec^3 x \, dx$

(b) $\displaystyle\int e^x \cos x \, dx$

10. Find the average value of the given function over the interval indicated.
(a) $\cos x$, $[0, \pi/2]$ (c) $\sin 2x$, $[0, \pi]$
(b) $\sin x$, $[0, \pi]$ (d) $\sec^2 x$, $[0, \pi/3]$

11. Find the length of the graph of $y = \log \cos x$ ($0 \le x \le \pi/4$).

12. Find the volume generated when the region bounded by the graph of $y = \sin x$ ($0 \le x \le \pi$) and the x-axis is revolved about
(a) the x-axis   (b) the y-axis

13. Find the volume generated when the region bounded by the graph of $y = \tan x$, the x-axis, and the line $x = \pi/4$, is revolved about the x-axis.

14. Find the volume generated when the region bounded by the graph of $y = \sec x$, the x-axis, the y-axis, and the line $x = \pi/4$, is revolved about the x-axis. .

15. Find the volume generated when the region bounded by the graph of $y = e^x \sin x$ and the x-axis, $0 \le x \le \pi$, is revolved about
(a) the x-axis   (b) the y-axis

Answers to problems (§8E)

1. (a) $\frac{1}{3}\sin^3 x$ (b) $-\frac{1}{3}\cos^3 x$ (c) $\frac{1}{4}\tan^4 x$
   (d) $\frac{1}{2}\sec x \tan x - \frac{1}{2}\log(\sec x + \tan x)$
   (e) $-\cos x + \frac{1}{3}\cos^3 x$ (f) $\sin x - \frac{2}{3}\sin^3 x + \frac{1}{5}\sin^5 x$
   (g) $-\log(1 + \cos x)$ (h) $\log(1 + \sin x)$
   (i) $-\frac{1}{4}\cos 2x$, or $\frac{1}{2}\sin^2 x$, or $-\frac{1}{2}\cos^2 x$
   (j) $-\frac{1}{2}\cos 2x$, or $\sin^2 x$, or $-\cos^2 x$ (k) $\frac{1}{4}\sin 4x$ (l) $x - \frac{1}{2}\cos 2x$

2. (a) $\tan x - x$ (b) $\frac{1}{3}\sec^3 x$ (c) $\frac{1}{3}\sec^3 x - \sec x$
   (d) $\frac{1}{2}\tan^2 x - \log \sec x$ (e) $\frac{1}{3}\tan^3 x$ (f) $\frac{1}{3}\tan^3 x + \tan x$
   (g) $\frac{1}{3}\tan^3 x - \tan x + x$ (h) $\log \sec x$ (i) $\log(\sec x + \tan x) - \sin x$
   (j) $\log(\sec x + \tan x) - \sin x$ (k) $\frac{1}{2}\sec 2x$ (l) $-\frac{1}{12}\cos 6x$

3. (a) $\cos x - \log(\csc x + \cot x)$ (b) $-\cot x - \frac{1}{3}\cot^3 x$ (c) $-\cot x - x$
   (d) $-\frac{1}{2}\cot^2 x + \log \csc x$

4. (a) $\log|\cos x + \sin x|$ (b) $-\dfrac{1}{\cos x + \sin x}$ (c) $-\frac{1}{3}\cos^3 x + \frac{1}{5}\cos^5 x$

   (d) $\frac{1}{32}(12x + 8\sin 2x + \sin 4x)$ (e) $\log \tan x$

   (f) $\frac{1}{4}(\log \tan x)^2$ (g) $-2\cos\sqrt{x}$ (h) $-2\sqrt{2}\cos\dfrac{x}{2}$ (i) $\cos\cos x$

   (j) $-\cos\log x$

5. (a) $\tan u - \cot u$ (b) $\frac{1}{2}(\tan^2 u - \cot^2 u) + 2\log \tan u$
   (c) $2(\sec u + \tan u) - u$
   (d) $2\tan u(\sec u + \tan u) + \log(1 - \sin u)$
   (e) $\tan u - \frac{3}{2}u + \frac{1}{4}\sin 2u$
   (f) $\frac{1}{2}\sec u \tan u - \frac{5}{2}\log(\sec u + \tan u) + 2\sin u + \frac{1}{3}\sin^3 u$

6. (a) $\frac{1}{2}(\frac{1}{2}\sin 2x - \frac{1}{4}\sin 4x)$ (b) $\frac{1}{2}(\sin x - \frac{1}{5}\sin 5x)$
   (c) $\frac{1}{2}(\frac{1}{3}\sin 3x + \frac{1}{7}\sin 7x)$ (d) $\frac{1}{2}(\frac{1}{4}\sin 4x + \frac{1}{10}\sin 10x)$
   (e) $-\frac{1}{2}(\frac{1}{3}\cos 3x + \frac{1}{5}\cos 5x)$ (f) $-\frac{1}{2}(\frac{1}{5}\cos 5x + \frac{1}{9}\cos 9x)$
   (g) $-\frac{1}{4}(\frac{1}{2}\cos 2x + \frac{1}{4}\cos 4x - \frac{1}{6}\cos 6x)$
   (h) $\frac{1}{4}(\cos x + \frac{1}{3}\cos 3x - \frac{1}{7}\cos 7x - \frac{1}{9}\cos 9x)$

7. (a) $\sin x - x\cos x$ (b) $\cos x + x\sin x$
   (c) $2x\sin x + 2\cos x - x^2\cos x$
   (d) $2x\cos x - 2\sin x + x^2\sin x$ (e) $x\tan x - \log \sec x$
   (f) $\frac{1}{8}(2x^2 - 2x\sin 2x - \cos 2x)$

8. (a) $e^x(\cos x + \sin x)$ (b) $e^x(\cos x - \sin x)$ (c) $2e^x \sin x$
   (d) $2e^x \cos x$

10. (a) $\dfrac{2}{\pi}$ (b) $\dfrac{2}{\pi}$ (c) 0 (d) $\dfrac{3}{\pi}\sqrt{3}$

11. $\log(1 + \sqrt{2})$

12. (a) $\pi^2/2$ (b) $2\pi^2$

13. $\pi(1 - \pi/4)$

14. π

15. (a) $\dfrac{\pi}{4}(e^{2\pi} - 1)$ (b) $\pi(\pi e^\pi - e^\pi - 1)$

§8F. Inverse trigonometric functions

8F1. The arcsine. The sine was defined as a function of arc length. It certainly has no inverse, as it repeats values every 2π. When restricted to the interval $-\pi/2 \le s \le \pi/2$, however, it is increasing (Figure 1) and we can

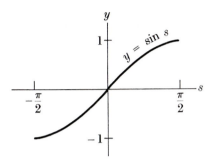

FIGURE 1

invert to express arc length in terms of it. This inverse function is called the arcsine, denoted *arcsin*. Thus,

$$s = \arcsin y \qquad (-1 \le y \le 1)$$

means

$$y = \sin s \qquad \left(-\frac{\pi}{2} \le \overset{s}{y} \le \frac{\pi}{2}\right).$$

For $-1 < y < 1$,

$$(1) \qquad \arcsin y = \int_0^y \frac{1}{\sqrt{1 - t^2}}\, dt \qquad (-1 < y < 1),$$

as was pointed out in §8A2, equation (6). For the endpoints,

$$\arcsin(-1) = -\frac{\pi}{2} \qquad \text{and} \qquad \arcsin 1 = \frac{\pi}{2}.$$

The arcsine is continuous on $-1 \le y \le 1$ (§6D3). On $-1 < y < 1$, it is differentiable, with

$$\frac{d}{dy} \arcsin y = \frac{1}{\sqrt{1 - y^2}} \qquad (-1 < y < 1).$$

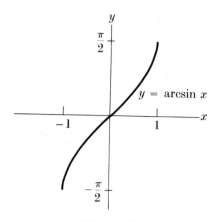

When speaking generally of the arcsine function one is apt to use x for the independent variable (Figure 2):

$$(2) \qquad\qquad y = \arcsin x \qquad\qquad (-1 \leq x \leq 1),$$

$$(3) \qquad\qquad \frac{d}{dx} \arcsin x = \frac{1}{\sqrt{1 - x^2}} \qquad\qquad (-1 < x < 1),$$

and

$$(4) \qquad\qquad \int \frac{1}{\sqrt{1 - x^2}} \, dx = \arcsin x.$$

Example 1. For $-1 < x < 1$:

(a) $\displaystyle\int \frac{1}{\sqrt{1 - x^2}} \, dx = \arcsin x$

(b) $\displaystyle\int \frac{x}{\sqrt{1 - x^2}} \, dx = -\sqrt{1 - x^2}$

(c) $\displaystyle\int \frac{x}{1 - x^2} \, dx = \log \frac{1}{\sqrt{1 - x^2}}$

(d) $\displaystyle\int \frac{1}{1 - x^2} \, dx = \frac{1}{2} \int \left(\frac{1}{1 + x} + \frac{1}{1 - x} \right) dx = \log \sqrt{\frac{1 + x}{1 - x}}$

8F2. The arctangent. The tangent has no inverse, as it repeats values every π. However, it is an increasing function on the interval $(-\pi/2,\ \pi/2)$ and so it has an inverse there. This inverse is called the arctangent, denoted *arctan*:

$$(5) \qquad y = \arctan x \qquad means \qquad x = \tan y \qquad \left(-\frac{\pi}{2} < y < \frac{\pi}{2}\right).$$

Since the tangent assumes all real values on $(-\pi/2,\ \pi/2)$ (§8C1), the domain of the arctangent is $\mathscr{R}$. The graph is shown in Figure 3.

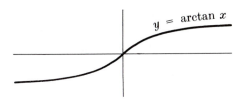

$y = \arctan x$

FIGURE 3

The derivative of $\tan y$, $\sec^2 y$, is never 0. By the Inverse-Function Theorem (§4G), the function $y = \arctan x$ is differentiable. Since $\tan y = x$, we have

$$(6) \qquad \sec^2 y\,\frac{dy}{dx} = 1, \qquad \frac{dy}{dx} = \frac{1}{\sec^2 y} = \frac{1}{1 + \tan^2 y}\,;$$

that is,

$$(7) \qquad \frac{d}{dx}\arctan x = \frac{1}{1 + x^2}.$$

The antiderivative form is

$$(8) \qquad \int \frac{1}{1 + x^2}\,dx = \arctan x.$$

Example 2.

(a) $\int \dfrac{1}{1 + x^2}\, dx = \arctan x$

(b) $\int \dfrac{x}{1 + x^2}\, dx = \log \sqrt{1 + x^2}$

(c) $\int \dfrac{x}{\sqrt{1 + x^2}}\, dx = \sqrt{1 + x^2}$

(d) $\int \dfrac{1}{\sqrt{1 + x^2}}\, dx = \log(x + \sqrt{1 + x^2})$

(Check by differentiating. The integral (d) is discussed in Example 1 of §8G5 below.)

8F3. The arcsecant. The secant is continuous and increasing on $[0, \pi/2)$, where it assumes all real values ≥ 1 (§8C1). Hence it has a continuous inverse defined on the interval $\{x: x \geq 1\}$ (§2E9). This inverse is called the arcsecant, denoted *arcsec*:

(9) $\qquad\qquad\qquad\qquad y = \text{arcsec } x \qquad\qquad\qquad\qquad (x \geq 1)$

means

(10) $\qquad\qquad\qquad\qquad x = \sec y \qquad\qquad\qquad\qquad (0 \leq y < \pi/2).$

The graph is shown in Figure 4. (The arcsecant can be defined in addition for $x \leq -1$, but to keep things simple we will stick with what we have.)

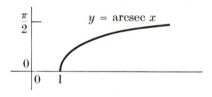

FIGURE 4

The derivative of $\sec y$, $\sec y \tan y$, is never 0 for $0 < y < \pi/2$, i.e., for $x = \sec y > 1$. By the Inverse-Function Theorem (§4G), the function $y = \text{arcsec } x$ is differentiable for $x > 1$. Since $\sec y = x$, we have

(11) $\quad \sec y \tan y \dfrac{dy}{dx} = 1, \qquad \dfrac{dy}{dx} = \dfrac{1}{\sec y \tan y} = \dfrac{1}{\sec y \sqrt{\sec^2 y - 1}}$;

$$(0 < y < \pi/2);$$

that is,

(12)
$$\frac{d}{dx} \operatorname{arcsec} x = \frac{1}{x\sqrt{x^2 - 1}}$$
$(x > 1)$.

The antiderivative form is

(13)
$$\int \frac{1}{x\sqrt{x^2 - 1}}\, dx = \operatorname{arcsec} x$$
$(x > 1)$.

Example 3. For $x > 1$:

(a) $\displaystyle\int \frac{1}{x\sqrt{x^2 - 1}}\, dx = \operatorname{arcsec} x$

(b) $\displaystyle\int \frac{x}{\sqrt{x^2 - 1}}\, dx = \sqrt{x^2 - 1}$

(c) $\displaystyle\int \frac{x}{x^2 - 1}\, dx = \log \sqrt{x^2 - 1}$

(d) $\displaystyle\int \frac{1}{x^2 - 1}\, dx = \frac{1}{2}\int \left(\frac{1}{x - 1} - \frac{1}{x + 1}\right) dx = \log \sqrt{\frac{x - 1}{x + 1}}.$

Problems (§8F)

1. Find the following:

(a) $\displaystyle\frac{d}{dx} \arcsin 2x$ (g) $\displaystyle\frac{d}{dy} \log \arctan y$

(b) $\displaystyle\frac{d}{dx} \arctan 3x$ (h) $\displaystyle\frac{d}{dy} \arctan \log y$

(c) $\displaystyle\frac{d}{dx} \arcsin \sin x$ (i) $\displaystyle\frac{d}{dx} \arcsin \sqrt{1 - x^2}$

(d) $\displaystyle\frac{d}{dx} \arctan \cot x$ (j) $\displaystyle\frac{d}{dx} \arctan \frac{1}{x}$

(e) $\displaystyle\frac{d}{du} \arctan e^u$ (k) $\displaystyle\frac{d}{dx} \operatorname{arcsec} x^2$

(f) $\displaystyle\frac{d}{dt} e^{\arctan t^2}$

2. Evaluate:

(a) $\displaystyle\int \frac{1}{\sqrt{1-2x^2}}\,dx$

(b) $\displaystyle\int \frac{1}{4+x^2}\,dx$

(c) $\displaystyle\int \frac{1}{\sqrt{9-x^2}}\,dx$

(d) $\displaystyle\int \frac{1}{1+8x^2}\,dx$

(e) $\displaystyle\int \frac{\sqrt{x}}{\sqrt{1-x^3}}\,dx$

(f) $\displaystyle\int \frac{1}{\sqrt{(1/x)-x^2}}\,dx$

(g) $\displaystyle\int \arcsin x\,dx$

(h) $\displaystyle\int \arctan x\,dx$

(i) $\displaystyle\int \frac{e^x}{\sqrt{1-e^{2x}}}\,dx$

(j) $\displaystyle\int \frac{1}{\sqrt{e^{2x}-1}}\,dx$

(k) $\displaystyle\int \frac{\sqrt{x}}{1+x^3}\,dx$

(l) $\displaystyle\int \frac{1}{e^x+e^{-x}}\,dx$

(m) $\displaystyle\int \frac{1}{1+x^2}\,e^{\arctan x}\,dx$

(n) $\displaystyle\int \frac{1}{\sqrt{1-x^2}}\,e^{\log\arcsin x}\,dx$

(o) $\displaystyle\int \frac{1}{x\sqrt{x-1}}\,dx$

(p) $\displaystyle\int \arctan \frac{1}{x}\,dx$

3. Evaluate:

(a) $\displaystyle\int x\arctan x\,dx$

(b) $\displaystyle\int x\arctan \frac{1}{x}\,dx$

(c) $\displaystyle\int x^2\arctan x\,dx$

4. Evaluate:

(a) $\displaystyle\int \log(a^2+x^2)\,dx$

(b) $\displaystyle\int \frac{1}{x^2}\log(a^2+x^2)\,dx$

Answers to problems (§8F)

1. (a) $\dfrac{2}{\sqrt{1-4x^2}}$　　(b) $\dfrac{3}{1+9x^2}$　　(c) $\dfrac{\cos x}{|\cos x|}$　　(d) -1

(e) $\dfrac{e^u}{1+e^{2u}}$　　(f) $\dfrac{2t}{1+t^4}\,e^{\arctan t^2}$

(g) $\dfrac{1}{(1+y^2)\arctan y}$　　(h) $\dfrac{1}{y[1+(\log y)^2]}$

(i) $-\dfrac{x}{|x|\sqrt{1-x^2}}$　　(j) $-\dfrac{1}{1+x^2}$　　(k) $\dfrac{2}{x\sqrt{x^4-1}}$

2. (a) $\dfrac{1}{\sqrt{2}}$ arcsin $\sqrt{2}\,x$ (b) $\frac{1}{2}$ arctan $\dfrac{x}{2}$ (c) arcsin $\dfrac{x}{3}$

(d) $\dfrac{1}{2\sqrt{2}}$ arctan $2\sqrt{2}\,x$ (e) $\frac{2}{3}$ arcsin $x^{3/2}$ (f) $\frac{2}{3}$ arcsin $x^{3/2}$

(g) x arcsin $x + \sqrt{1-x^2}$ (h) x arctan $x - \frac{1}{2}\log(1+x^2)$ (i) arcsin e^x

(j) arcsec e^x or $-$arcsin e^{-x} (k) $\frac{2}{3}$ arctan $x^{3/2}$ (l) arctan e^x (m) $e^{\text{arctan } x}$

(n) $\frac{1}{2}(\text{arcsin } x)^2$ (o) 2 arcsec $\sqrt{x}$ (p) x arctan $\dfrac{1}{x} + \frac{1}{2}\log(1+x^2)$

3. (a) $\frac{1}{2}(x^2$ arctan $x +$ arctan $x - x)$

(b) $\frac{1}{2}\left(x^2$ arctan $\dfrac{1}{x} -$ arctan $x + x\right)$

(c) $\frac{1}{6}[2x^3$ arctan $x + \log(1+x^2) - x^2]$

4. (a) $2a$ arctan $\dfrac{x}{a} + x\log(a^2+x^2) - 2x$ $(a \neq 0)$, $2(x\log|x| - x)$ $(a = 0)$

(b) $\dfrac{2}{a}$ arctan $\dfrac{x}{a} - \dfrac{1}{x}\log(a^2+x^2)$ $(a \neq 0)$, $-2\left(\dfrac{1}{x}\log|x| + \dfrac{1}{x}\right)$ $(a = 0)$

§8G. Trigonometric substitution

8G1. Discussion. In §5F we considered a change of variable in an integral by means of a substitution of the form $u = g(x)$, where x is the given variable of integration and u is the new one. Here we look at substitutions of the form $x = h(u)$. These are simply the others in reverse.

In the applications, h will be a trigonometric function. The method is used to convert algebraic expressions involving $a^2 - x^2$ or $a^2 + x^2$ to trigonometric expressions.

8G2. Integration by substitution. Let f be continuous on $[a, b]$. Then

(1)
$$\int_a^b f(x)\,dx = \int_\alpha^\beta f\big(h(u)\big)h'(u)\,du,$$

whenever h and h' are continuous on $[\alpha, \beta]$ (or $[\beta, \alpha]$) and

(2) $h(\alpha) = a$ and $h(\beta) = b$.

Briefly: The substitution

(3) $x = h(u)$, $dx = h'(u)\,du$

is valid under the integral sign.

Proof. This is simply equation (7) of §5F2 read from right to left. The letters have been changed to reflect the present viewpoint: the x, u, h, α, β here are the u, x, g, a, b there.

We remark that any choice of α and β will do in (2) so long as the hypotheses about continuity are satisfied.

Similarly, the antiderivative form of (1) is

(4)
$$\int f(x)\,dx = \int f(h(u))h'(u)\,du.$$

In applying the theorem we use the substitutions described below. In each case, the domain of the new variable is chosen so that a key trigonometric function will be positive, as noted. This simplifies keeping track of signs.

8G3. The expression $a^2 - x^2$, where $x^2 \leq a^2$. We substitute

(5)
$$x = a\sin u \qquad \left(-\frac{\pi}{2} \leq u \leq \frac{\pi}{2}\right).$$

Note that for $a > 0$,

(6)
$$a\cos u = \sqrt{a^2 - x^2} \geq 0 \qquad \left(-\frac{\pi}{2} \leq u \leq \frac{\pi}{2}\right).$$

8G4. The area of a sector. As an illustration of this substitution we show that *the area of a circular sector of radius a and central angle θ is*

(7)
$$S = \tfrac{1}{2}a^2\theta.$$

Note. The sector of central angle 2π is the entire disc. It follows from (7) that the area of a disc of radius a is πa^2. If we consider the area under the quarter-circle $x^2 + y^2 = a^2$ in the first quadrant, we obtain the following useful integration formula:

(8)
$$\int_0^a \sqrt{a^2 - x^2}\,dx = \tfrac{1}{4}\pi a^2 \qquad (a > 0).$$

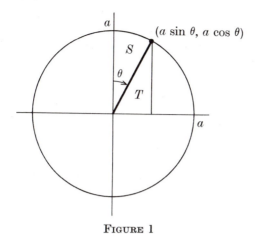

$(a \sin \theta, a \cos \theta)$

S

θ

T

a

a

FIGURE 1

Proof of (7). Figure 1 depicts the case $0 < \theta \le \pi/2$, with a convenient choice of coordinate axes. Larger values of θ are handled by a similar argument or by invoking additivity.

Let S and T be the areas of the sector and triangle, as marked. Their sum is A, the area under the curve. For the triangle we have, simply,

$$(9) \qquad T = \tfrac{1}{2}a \sin \theta \, a \cos \theta = \tfrac{1}{4}a^2 \sin 2\theta.$$

For A we have

$$(10) \qquad A = \int_0^{a \sin \theta} \sqrt{a^2 - x^2} \, dx.$$

Substitute $x = a \sin u$. Then

$$dx = a \cos u \, du \qquad \text{and} \qquad \sqrt{a^2 - x^2} = a \cos u.$$

Hence

$$(11) \qquad A = a^2 \int_0^{\theta} \cos^2 u \, du = \tfrac{1}{2}a^2 \int_0^{\theta} (1 + \cos 2u) \, du$$

$$= \tfrac{1}{2}a^2 (\theta + \tfrac{1}{2} \sin 2\theta).$$

Hence

$$S = A - T = \tfrac{1}{2}a^2 \, \theta.$$

8G5. The expression $a^2 + x^2$. We substitute

$$(12) \qquad x = a \tan u \qquad \left(-\frac{\pi}{2} < u < \frac{\pi}{2} \right).$$

Note that for $a > 0$,

(13) $$a \sec u = \sqrt{a^2 + x^2} > 0 \qquad \left(-\frac{\pi}{2} < u < \frac{\pi}{2}\right).$$

Example 1. Find

$$\int \frac{1}{\sqrt{1 + x^2}}\, dx.$$

Solution. Put $x = \tan u$ $(-\pi/2 < u < \pi/2)$. Then

$$dx = \sec^2 u\, du \qquad \text{and} \qquad \sqrt{1 + x^2} = \sec u.$$

Hence

$$\int \frac{1}{\sqrt{1 + x^2}}\, dx = \int \frac{1}{\sec u}\, \sec^2 u\, du = \int \sec u\, du$$

$$= \log(\sec u + \tan u) = \log(x + \sqrt{1 + x^2}).$$

8G6. The expression $x^2 - a^2$, where $x \geq a > 0$. We substitute

(14) $$x = a \sec u \qquad \left(0 \leq u < \frac{\pi}{2}\right).$$

Note that

(15) $$a \tan u = \sqrt{x^2 - a^2} \geq 0 \qquad \left(0 \leq u < \frac{\pi}{2}\right).$$

Example 2. Find

$$\int_1^2 \frac{1}{x} \sqrt{x^2 - 1}\, dx.$$

Solution. Put $x = \sec u$ $(0 \leq u \leq \pi/3)$. Then

$$dx = \sec u \tan u\, du \qquad \text{and} \qquad \sqrt{x^2 - 1} = \tan u.$$

Hence

$$\int_1^2 \frac{1}{x} \sqrt{x^2 - 1}\, dx = \int_0^{\pi/3} \frac{1}{\sec u} \tan u \sec u \tan u\, du$$

$$= \int_0^{\pi/3} \tan^2 u\, du = \int_0^{\pi/3} (\sec^2 u - 1)\, du$$

$$= \tan u - u \Big|_0^{\pi/3} = \sqrt{3} - \frac{\pi}{3}.$$

Problems (§8G)

1. Derive the following formulas. Then check by differentiating.

(a) $\int \sqrt{a^2 - x^2}\,dx = \tfrac{1}{2}(x\sqrt{a^2 - x^2} + a^2 \arcsin x/a) \quad (a > 0)$

(b) $\int \sqrt{a^2 + x^2}\,dx \doteq \tfrac{1}{2}[x\sqrt{a^2 + x^2} + a^2 \log(x + \sqrt{a^2 + x^2})]$

(c) $\int \sqrt{x^2 - a^2}\,dx = \tfrac{1}{2}[x\sqrt{x^2 - a^2} - a^2 \log(x + \sqrt{x^2 - a^2})]$

2. Evaluate:

(a) $\int x^2\sqrt{1 - x^2}\,dx$

(b) $\int x^3\sqrt{25 - x^2}\,dx$

(c) $\int \dfrac{1}{\sqrt{4 - x^2}}\,dx$

(d) $\int \dfrac{x}{\sqrt{1 - x^2}}$

(e) $\int (1 - 4x^2)^{3/2}\,dx$

(f) $\int \dfrac{1}{(1 - 4x^2)^{3/2}}\,dx$

(g) $\int \dfrac{x^2}{\sqrt{1 - x^2}}\,dx$

(h) $\int x \arcsin x\,dx$

3. Evaluate:

(a) $\int \sqrt{1 + x^2}\,dx$

(b) $\int \dfrac{1}{x^2\sqrt{x^2 + 9}}\,dx$

(c) $\int \dfrac{1}{x\sqrt{x^2 + 1}}\,dx \ (x > 0)$

(d) $\int \dfrac{x^3}{4 + x^2}\,dx$

(e) $\int \dfrac{1}{(1 + 36x^2)^{3/2}}\,dx$

(f) $\int \dfrac{1}{x^2 + x^4}\,dx$

4. Evaluate:

(a) $\int x\sqrt{x^2 - 16}\,dx$

(b) $\int \sqrt{x^2 - 1}\,dx \quad (x \geq 1)$

(c) $\int \dfrac{1}{\sqrt{x^2 - 1}}\,dx \quad (x \geq 1)$

(d) $\int \dfrac{x}{\sqrt{4x^2 - 1}}\,dx$

(e) $\int \dfrac{1}{x\sqrt{x^2 - 36}}\,dx \ (x \geq 6)$

(f) $\int \dfrac{1}{(9x^2 - 1)^{3/2}}\,dx$

5. Evaluate:

(a) $\displaystyle\int_0^1 x\sqrt{1-x^2}\,dx$

(e) $\displaystyle\int_0^1 \frac{1}{4-x^2}\,dx$

(b) $\displaystyle\int_0^{1/\sqrt{2}} \frac{x^2}{\sqrt{1-x^2}}\,dx$

(f) $\displaystyle\int_{2^{-1/2}}^{2^{-1/4}} \frac{x}{\sqrt{1-x^4}}\,dx$

(c) $\displaystyle\int_0^1 \frac{1}{2x^2+3}\,dx$

(g) $\displaystyle\int_{4/\sqrt{3}}^4 \frac{1}{x^2\sqrt{x^2-4}}\,dx$

(d) $\displaystyle\int_{1+3\sin3/5}^{1+3\sin4/5} \frac{1}{\sqrt{9-(x-1)^2}}\,dx$

(h) $\displaystyle\int_0^1 \frac{x+3}{x^2+9}\,dx$

(i) $\displaystyle\int_{2\sin(\pi/14)-1}^{2\sin(\pi/7)-1} \frac{1}{\sqrt{3-(x^2+2x)}}\,dx$

6. Find the area enclosed by the *ellipse*

$$\frac{x^2}{a^2}+\frac{y^2}{b^2}=1 \qquad\qquad (a>0,\,b>0).$$

7. Find the lengths of the following graphs.
(a) $y=x^2$ $(0\le x\le 2)$
(b) $y=\sqrt{x}$ $(0\le x\le 4)$
(c) $y=e^x$ $(0\le x\le 1)$
(d) $y=\log x$ $(1\le x\le \sqrt{3})$
(e) $y=\log(1-x^2)$ $(0\le x\le \frac{1}{2})$

8. Find the area of the surface generated when the given curve is revolved about the x-axis.

(a) $y=\sin x$ $(0\le x\le \pi)$ (c) $y=\dfrac{1}{x}$ $(1\le x\le 2)$

(b) $y=e^{-x}$ $(0\le x\le 1)$ (d) $y=e^x$ $(0\le x\le 1)$

9. Find the volume generated when the circle $x^2+(y-2)^2=1$ is revolved about the x-axis. Use two methods, disc and shell.

Answers to problems (§8G)

2. (a) $\frac{1}{8}[x(2x^2-1)\sqrt{1-x^2}+\arcsin x]$ (b) $-\frac{1}{15}(50+3x^2)(25-x^2)^{3/2}$

(c) $\arcsin\dfrac{x}{2}$ (d) $-\sqrt{1-x^2}$ (e) $\frac{1}{16}[2x(5-8x^2)\sqrt{1-4x^2}+3\arcsin 2x]$

(f) $\dfrac{x}{\sqrt{1-4x^2}}$ (g) $\frac{1}{2}(\arcsin x-x\sqrt{1-x^2})$

(h) $\frac{1}{4}(2x^2\arcsin x-\arcsin x+x\sqrt{1-x^2})$

3. (a) $\frac{1}{2}[x\sqrt{1 + x^2} + \log(x + \sqrt{1 + x^2})]$　　(b) $-\frac{1}{9}\frac{\sqrt{x^2 + 9}}{x}$

(c) $-\log\frac{\sqrt{1 + x^2} + 1}{x}$　　(d) $\frac{1}{2}x^2 - 2\log(4 + x^2)$

(e) $\frac{x}{\sqrt{1 + 36x^2}}$　　(f) $-\left(\frac{1}{x} + \arctan x\right)$

4. (a) $\frac{1}{3}(x^2 - 16)^{3/2}$　　(b) $\frac{1}{2}[x\sqrt{x^2 - 1} - \log(x + \sqrt{x^2 - 1})]$

(c) $\log(x + \sqrt{x^2 - 1})$　　(d) $\frac{1}{4}\sqrt{4x^2 - 1}$　　(e) $\frac{1}{6}\operatorname{arcsec}\frac{x}{6}$

(f) $-\frac{x}{\sqrt{9x^2 - 1}}$

5. (a) $\frac{1}{3}$　　(b) $\frac{1}{8}(\pi - 2)$　　(c) $\frac{1}{\sqrt{6}}\arctan\sqrt{\frac{2}{3}}$　　(d) $\frac{1}{5}$　　(e) $\frac{1}{4}\log 3$

(f) $\pi/24$　　(g) $\frac{1}{8}(\sqrt{3} - 1)$　　(h) $\frac{1}{2}\log\frac{10}{9} + \arctan\frac{1}{3}$　　(i) $\pi/14$

6. πab

7. (a) $\sqrt{17} + \frac{1}{4}\log(4 + \sqrt{17})$　　(b) $\sqrt{17} + \frac{1}{4}\log(4 + \sqrt{17})$

(c) $\log\frac{\sqrt{e^2 + 1} - 1}{\sqrt{2} - 1} + \sqrt{e^2 + 1} - (1 + \sqrt{2})$　　(d) $\log\frac{1 + \sqrt{2}}{\sqrt{3}} + 2 - \sqrt{2}$

(e) $\log 3 - \frac{1}{2}$

8. (a) $2\pi[\sqrt{2} + \log(1 + \sqrt{2})]$

(b) $\pi\left(1 + \sqrt{2} - \frac{\sqrt{1 + e^2}}{e^2} + \log\frac{1 + \sqrt{2}}{1 + \sqrt{1 + e^2}}\right)$

(c) $\pi\left(\sqrt{2} - \frac{1}{4}\sqrt{17} + \log\frac{4 + \sqrt{17}}{1 + \sqrt{2}}\right)$

(d) $\pi\left(e\sqrt{1 + e^2} - \sqrt{2} + \log\frac{e + \sqrt{1 + e^2}}{1 + \sqrt{2}}\right)$

9. $4\pi^2$

CHAPTER 9

Vectors in the Plane

In this chapter we probe a little deeper into properties of plane curves. Our new tool is vectors. §§9A–9B discuss the elementary algebra and geometry of vectors. The rest of the chapter (§§9C–9J) concerns vector-valued functions. Using vector methods in calculus, we analyze such things as tangents, motion, arc length, and curvature.

§9A. Algebraic operations in $\mathscr{R}^2$

9A1. The algebra of vectors. In $\mathscr{R}$, we are accustomed to identifying points with numbers. We speak of points when thinking geometrically, and we speak of numbers when thinking algebraically.

In somewhat similar vein, we shall now introduce algebraic operations into the plane, $\mathscr{R}^2$. The points of $\mathscr{R}^2$, furnished with these operations, are called *vectors*.

First we define *addition* of vectors. Let $P = (x, y)$ and $Q = (u, v)$ be vectors in $\mathscr{R}^2$. Their sum, $P + Q$, or $(x, y) + (u, v)$, is defined as follows:

(1)
$$P + Q = (x, y) + (u, v) = (x + u, y + v).$$

Clearly, addition of vectors is associative and commutative:

$$(P + Q) + R = P + (Q + R) \qquad \text{and} \qquad P + Q = Q + P.$$

The vector $O = (0, 0)$ is called the *zero* vector. For any vector $P = (x, y)$, we have

$$P + O = P.$$

Let us define $-P$ by: $-P = (-x, -y)$. Evidently,

$$P + (-P) = O.$$

We also define subtraction. If $P = (x, y)$ and $Q = (u, v)$ then, by definition,

$$P - Q = P + (-Q) = (x - u, y - v).$$

In particular, $P - P = O$.

Example 1. Let $P = (3, 4)$, $Q = (-2, 5)$, $R = (0, -6)$. Then

$$-P = (-3, -4), \ -Q = (2, -5), \ -R = (0, 6);$$
$$P + Q = (3, 4) + (-2, 5) = (3 - 2, 4 + 5) = (1, 9);$$
$$P - Q = (3, 4) - (-2, 5) = (3 + 2, 4 - 5) = (5, -1);$$
$$Q + R = (-2, 5) + (0, -6) = (-2, -1);$$
$$(P + Q) + R = (1, 9) + (0, -6) = (1, 3);$$
$$P + (Q + R) = (3, 4) + (-2, -1) = (1, 3).$$

In discussions about vectors, *numbers* are usually referred to as *scalars*.

If k is a scalar (number) and $P = (x, y)$ is a vector, then we define a product, kP, as follows:

(2)
$$kP = (kx, ky).$$

The vector kP is called a *scalar multiple* of P.

Example 2. If $P = (3, 4)$ and $k = 2$, then

$$kP = 2(3, 4) = (6, 8).$$

Observe that the sum of two vectors is a vector, and that a scalar multiple of a vector is a vector.

The following identities are clear from the definitions.

$$kO = O, \qquad\qquad OP = O,$$

(3) $$\qquad 1P = P, \qquad\qquad (-1)P = -P,$$

$$k(P + Q) = kP + kQ, \qquad (k + l)P = kP + lP.$$

In particular, $P + P = 2P$.

Vectors are frequently denoted by $\mathbf{P}$, $\vec{P}$, or other special symbols.

9A2. *The geometry of vectors; the parallelogram law.* The relation between P and kP $(P \neq O)$ is shown in Figure 1. The length of the segment from O to kP is $|k|$ times that of OP. The vectors on the same side of O as P are the scalar multiples kP for $k > 0$; they are called *parallel* to P. Those on the opposite side of O are the scalar multiples kP for $k < 0$; they are *opposite* to P.

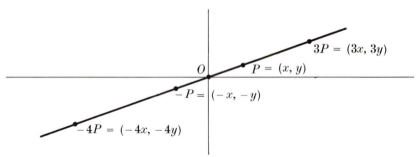

FIGURE 1

Figure 2 depicts the relations among vectors P, Q, and $R = P + Q$. (It is assumed that $P \neq O$, $Q \neq O$, and that P and Q are neither parallel nor opposite.) Notice that segments OP and QR are parallel (see (1)), as are OQ and PR. Thus, addition of vectors obeys the *parallelogram law*: if $R = P + Q$, then $OPRQ$ is a parallelogram.

The same figure shows the relations among P, R, and $Q = R - P$.

Example 3. The vectors $P = (5, 2)$, $Q = (1, 3)$, and $R = P + Q = (6, 5)$ are shown in Figure 3.

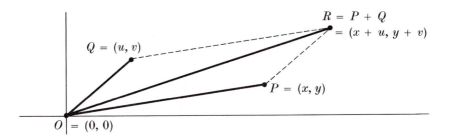

FIGURE 2

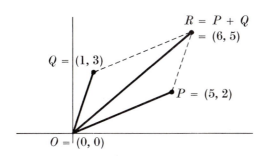

FIGURE 3

9A3. Pictorial representation of vectors. In many applications, the length and direction of a vector are more important than its coordinates. To call attention to them in a way that gives an immediate visual impression, we may draw the vector as an arrow with its foot at the origin and its head at the point (Figure 4).

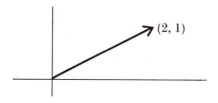

FIGURE 4

Figure 5 carries this pictorial device a step further, showing vectors A, B, and $A + B$. We find $A + B$ by sliding the arrow representing B so as to attach its foot to the head of A. The head of the displaced arrow is then at $A + B$. Figure 6 shows this "slide-and-attach" procedure for the sum of three vectors.

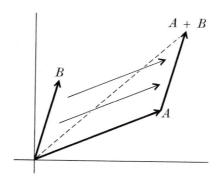

FIGURE 5

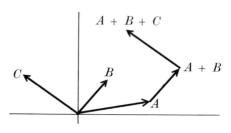

FIGURE 6

An important use of the device is in indicating a vector such as a force or a velocity associated with a given point. We will encounter these situations later on.

Problems (§9A)

1. Express each of the following as a single vector.

(a) $(2, 3) + (7, -2)$ (g) $(3, 1) + 6(0, 0)$

(b) $(3, 1) - (0, 0)$ (h) $2(1, 3) + 3(-2, 6)$

(c) $(1, 1) - (3, -2)$ (i) $(3 + 4)(6, 5)$

(d) $3(2, 4)$ (j) $3[(2, 6) + (-1, 1)]$

(e) $(-1)(0,2)$ (k) $(3 + 6)[(1, 6) - (6, -2)]$

(f) $(1, 2) - (3, 4) + (5, 2)$ (l) $(4 - 6)[(-1, 2) + (3, -3)]$

2. Solve for the vector A:

(a) $A + (1, 3) = (0, 4)$ (d) $A + 3A = (-4, 12)$

(b) $\frac{2}{3}A = (-2, 6)$ (e) $(6, -3) + A = (6, -3)$

(c) $2A + (2, 3) = A - (1, 2)$ (f) $(6, -3) + A = (6, 3)$

3. Solve for a and b:
    (a) $(a, 3) = (2, b)$
    (b) $(2a + 7, 3) = (-11, b - 2)$
    (c) $(a + b, a - b) = (4, 2)$
    (d) $3[(a, b) - (4, 2)] = (6, 8) - (a, b)$
    (e) $3[(a, b) - (4, 2)] = (6, 8) - (b, a)$
    (f) $(a^2 + 1, -a) = (17, |b| - b)$

4. Solve for P and Q:
    (a) $P + Q = (4, -6), \quad P - Q = (-5, 3)$
    (b) $2P + Q = (5, 3), \quad -P + 3Q = (8, 5)$
    (c) $P + Q = Q, \quad 2P + 3Q = (6, 3)$
    (d) $P + 3Q = 2P - 2Q = (4, 6)$
    (e) $3(P - 2Q) = 2(Q + 3P) = (7, -14)$

5. Find scalars s and t such that:
    (a) $s(1, 2) + t(3, 4) = (7, 8)$
    (b) $s(3, -1) + t(1, -1) = (-1, 3)$
    (c) $s(-2, 3) + t(-6, 5) = (4, -6)$

6. Two sides of a parallelogram are OP and OQ. What are the lengths of the diagonals if
    (a) $P = (3, 2)$ and $Q = (1, 4)$?
    (b) $P = (-4, 3)$ and $Q = (6, -1)$?
    (c) $P = (-2, -3)$ and $Q = (3, -3)$?

7. (a) Find three vectors:
        (i) none of which has any coordinate zero,
        (ii) no two of which are parallel or opposite, and
        (iii) whose sum is O.
    (b) Find four such vectors.

8. Let $V_1 = (a_1, b_1)$ and $V_2 = (a_2, b_2)$ be non-zero vectors.
    (a) Show that if $V_1 = kV_2$ for some scalar k, then $a_1b_2 = a_2b_1$.
    (b) Conversely, show that if $a_1b_2 = a_2b_1$, then there is a scalar k such that
    $V_1 = kV_2$. [*Hint.* If $a_2 \neq 0$, define $k = a_1/a_2$.]

9. Let V_1 and V_2 be non-zero vectors that are neither parallel nor opposite.
    (a) Show geometrically that for any vector A, there are scalars s and t such that $sV_1 + tV_2 = A$. [*Hint.* The line through A parallel to OV_2 intersects OV_1.]
    (b) Prove the same result algebraically by solving for s and t. [*Hint.* Use Problem 8(b).]

Answers to problems (§9A)

1. (a) $(9, 1)$ (b) $(3, 1)$ (c) $(-2, 3)$ (d) $(6, 12)$ (e) $(0, -2)$
   (f) $(3, 0)$ (g) $(3, 1)$ (h) $(-4, 24)$ (i) $(42, 35)$ (j) $(3, 21)$
   (k) $(-45, 72)$ (l) $(-4, 2)$

2. (a) $(-1, 1)$ (b) $(-3, 9)$ (c) $(-3, -5)$ (d) $(-1, 3)$ (e) $(0, 0)$
    (f) $(0, 6)$

3. (a) $a = 2, \ b = 3$ (b) $a = -9, \ b = 5$ (c) $a = 3, \ b = 1$
    (d) $a = 4\frac{1}{2}, \ b = 3\frac{1}{2}$ (e) $a = 5, \ b = 3$ (f) $a = -4, \ b = -2$

4. (a) $P = (-\frac{1}{2}, -\frac{3}{2}), \ Q = (\frac{9}{2}, -\frac{9}{2})$ (b) $P = (1, \frac{4}{7}), \ Q = (3, \frac{13}{7})$
    (c) $P = (0, 0), \ Q = (2, 1)$ (d) $P = (\frac{5}{2}, \frac{15}{4}), \ Q = (\frac{1}{2}, \frac{3}{4})$
    (e) $P = (\frac{4}{3}, -\frac{8}{3}), \ Q = (-\frac{1}{2}, 1)$

5. (a) $s = -2, \ t = 3$ (b) $s = 1, \ t = -4$ (c) $s = -2, \ t = 0$

6. (a) $2\sqrt{2}, \ 2\sqrt{13}$ (b) $2\sqrt{29}, \ 2\sqrt{2}$ (c) $5, \ \sqrt{37}$

§9B. Length and scalar product

9B1. Length of a vector. The *length* or *magnitude* of a vector P is defined to be the length of the segment OP. It is designated $|P|$. If $P = (x, y)$, then,

$$(1) \qquad |P| = \sqrt{x^2 + y^2}.$$

For any scalar k,

$$|kP| = |k||P|.$$

Here $|k|$ means the absolute value of the real number k, while $|P|$ and $|kP|$ denote the lengths of the vectors P and kP.

Note that $|P - Q|$ is equal to the distance between P and Q.

Example 1. If $P = (3, 4)$ and $Q = (2, -1)$, then

$$|P| = \sqrt{3^2 + 4^2} = 5;$$

$$-2P = (-6, -8), \qquad |-2P| = 2|P| = 10;$$

distance between P and $Q = |P - Q|$

$$= \sqrt{(3 - 2)^2 + (4 + 1)^2} = \sqrt{26}.$$

A useful observation is that if P is any non-zero vector, then

$$\frac{P}{|P|}$$

(i.e., $\frac{1}{|P|} P$) is a *unit* vector; that is, a vector of length 1.

9B2. Angle between two vectors. Let $P_1 = (x_1, y_1)$ and $P_2 = (x_2, y_2)$ be two non-zero vectors. By the angle between P_1 and P_2 we mean the angle θ

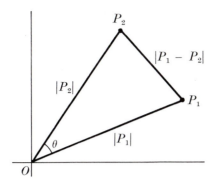

FIGURE 1

between the segments OP_1 and OP_2 (Figure 1). It turns out that $\cos \theta$ is readily calculated in terms of the coordinates of P_1 and P_2. In fact, the Law of Cosines (§8D3) states that

$$(2) \qquad |P_1 - P_2|^2 = |P_1|^2 + |P_2|^2 - 2|P_1||P_2| \cos \theta.$$

Now, $|P_1 - P_2|^2 = (x_1 - x_2)^2 + (y_1 - y_2)^2$. That is,

$$(3) \qquad |P_1 - P_2|^2 = |P_1|^2 + |P_2|^2 - 2(x_1 x_2 + y_1 y_2).$$

Equating (2) and (3), we get

$$(4) \qquad \cos \theta = \frac{x_1 x_2 + y_1 y_2}{|P_1||P_2|}.$$

9B3. Scalar product. The expression appearing in the numerator in (4) is called the *scalar product* (or *dot* product, or *inner* product) of P_1 and P_2. It is designated $P_1 \cdot P_2$. Thus, for $P_1 = (x_1, y_1)$, $P_2 = (x_2, y_2)$,

$$(5) \qquad P_1 \cdot P_2 = (x_1, y_1) \cdot (x_2, y_2) = x_1 x_2 + y_1 y_2.$$

The term *scalar product* emphasizes that this product of two vectors is a scalar.

The scalar product is defined by (5) for any two vectors, zero or not. The angle between two vectors, however, is defined only when both vectors are different from zero.

Example 2. The scalar product of the vectors $P_1 = (4, -5)$ and $P_2 = (3, 2)$ is

$$P_1 \cdot P_2 = (4)(3) + (-5)(2) = 2.$$

In terms of scalar products, if θ is the angle between the (non-zero) vectors P_1 and P_2, then, by (4),

$$(6) \qquad \cos \theta = \frac{P_1 \cdot P_2}{|P_1||P_2|}.$$

Example 3. Find the angle θ between the vectors $P_1 = (1, 4)$ and $P_2 = (3, -2)$.

Solution. We have

$$P_1 \cdot P_2 = (1)(3) + (4)(-2) = -5, \qquad |P_1| = \sqrt{17}, \qquad |P_2| = \sqrt{13}.$$

Therefore,

$$\cos \theta = \frac{-5}{\sqrt{17}\sqrt{13}} \approx -\frac{5}{14.8} \approx -0.34,$$

$$\theta \approx 110°.$$

Let P, Q, and R be vectors and let k be a scalar. The following identities about scalar products are direct consequences of the definition.

$$(7) \qquad \begin{gathered} P \cdot Q = Q \cdot P, \qquad P \cdot O = 0, \qquad P \cdot P = |P|^2, \\ P \cdot (Q + R) = P \cdot Q + P \cdot R, \qquad (kP) \cdot Q = k(P \cdot Q). \end{gathered}$$

In particular, $(-P) \cdot Q = -(P \cdot Q)$.

Example 4. If $P = (3, 1)$, $Q = (-1, -1)$, $R = (2, 5)$, then

$$\begin{aligned} P \cdot Q &= (3)(-1) + (1)(-1) = -4 = Q \cdot P, \\ P \cdot O &= (3)(0) + (1)(0) = 0, \\ |P|^2 &= P \cdot P = 3^2 + 1^2 = 10, \qquad |P| = \sqrt{10}, \\ P \cdot R &= (3)(2) + (1)(5) = 11, \qquad Q + R = (-1, -1) + (2, 5) = (1, 4), \\ P \cdot (Q + R) &= (3)(1) + (1)(4) = 7 = P \cdot Q + P \cdot R, \\ 4P &= (12, 4), \qquad (4P) \cdot Q = (12)(-1) + (4)(-1) = -16 = 4(P \cdot Q). \end{aligned}$$

9B4. Perpendicularity. Let P and Q be vectors.

Suppose first that neither one is O, and let θ be the angle between them. If θ is a right angle, we say, naturally enough, that P and Q are perpendicular. The same condition is expressed by $\cos \theta = 0$. By (6), this is the same as $P \cdot Q = 0$.

In case either P or Q is O, we will say by convention that they are perpendicular. Here, too, $P \cdot Q = 0$.

Thus, *the condition for perpendicularity of P and Q is*

(8) $$P \cdot Q = 0.$$

Perpendicular vectors are also called *orthogonal* vectors.

Example 5. $P = (1, 3)$ and $Q = (6, -2)$ are perpendicular, since

$$P \cdot Q = (1)(6) + (3)(-2) = 0.$$

But P is not perpendicular to $R = (5, -1)$, since

$$P \cdot R = (1)(5) + (3)(-1) \neq 0.$$

Problems (§9B)

1. Find the lengths of the following vectors.
   (a) $(1, 2)$ (b) $(3, -1)$ (c) $(-1, 4)$

2. Find the following scalar products.
   (a) $(1, 2) \cdot (3, -4)$ (f) $(-1, -1) \cdot (2, -3)$
   (b) $(2, 1) \cdot (3, -4)$ (g) $(-1, 4) \cdot (-1, 2)$
   (c) $(3, -4) \cdot (1, 2)$ (h) $(1, 3) \cdot [(2, 4) - (1, 2)]$
   (d) $(6, 0) \cdot (0, 6)$ (i) $[7(2, -1)] \cdot (1, -2)$
   (e) $(3, -1) \cdot (3, -1)$ (j) $(2, 7) \cdot (0, 0)$

3. Which of the following pairs of vectors are perpendicular?
   $(2, 6), (-1, 3)$ $(2, -8), (-4, 1)$
   $(2, 6), (3, -1)$ $(2, 8), (-4, 1)$
   $(2, 6), (1, 0)$ $(2, 3), (3, 2)$
   $(1, 4), (4, -1)$ $(2, 3), (4, 2)$
   $(1, 4), (-4, 1)$ $(2, -3), (3, 2)$
   $(-4, 1), (1, 4)$

4. Find a so that $(1, a)$ is perpendicular to:
   (a) $(2, 3)$ (c) $(0, -3)$ (e) $(a, 1)$
   (b) $(-1, 6)$ (d) $(1, 1)$ (f) $(-1, a)$

5. (a) Is there a vector P such that $P \cdot P = -3$? -2?
   (b) If $P \cdot P = 0$, can P be different from O?
   (c) If $P \cdot Q = 0$, can P and Q both be different from O?
   (d) Describe (geometrically) all those vectors P for which the scalar product of P with itself is equal to 3.

6. Let A and B be non-zero vectors. Prove that:
   (a) If A and B are perpendicular, then $|A + B|^2 = |A|^2 + |B|^2$.
   (b) Conversely, if $|A + B|^2 = |A|^2 + |B|^2$, then A and B are perpendicular.

7. Let $A = (a, b)$ be a fixed, non-zero vector.
   (a) A point P moves in such a way that $A \cdot P = A \cdot A$. Show that its path is a straight line.
   (b) A point P moves in such a way that $A \cdot P = P \cdot P$. Show that its path is a circle.

8. In each case, find the vector with positive x-coordinate that is perpendicular to the given vector and has the same length.
   (a) $(-5, 2)$ (c) $(-2, -1)$
   (b) $(0, 1)$ (d) $(3, -2)$

9. Prove by vector calculation:
   (a) The diagonals of a rhombus are perpendicular.
   (b) If the diagonals of a parallelogram are equal, the parallelogram is a rectangle.

10. Let A and B be non-zero vectors, and let P be the foot of the perpendicular from A onto OB. (The vector P is called the *projection* of A on B.) Prove that

$$P = \frac{A \cdot B}{|B|^2} B.$$

11. Find the maximum value of $(4, -2) \cdot Q$ as Q moves along the parabola $y = x^2$.

12. A vertical line cuts the parabolas $y = x^2$ and $y = 1 - x^2$ in points P and Q. What is the maximum value of $P \cdot Q$?

13. Let $Q = (1, 3)$ and $R = (3, 4)$. What are the vectors P such that $P \cdot Q = P \cdot R$?

14. Find the angle between the given vectors.
   (a) $(-2, 1)$, $(3, 2)$ (d) $(4, 0)$, $(2, 2)$
   (b) $(4, -2)$, $(4, 8)$ (e) $(3, 1)$, $(-3, 1)$
   (c) $(1, 0)$, $(2, 0)$ (f) $(3, 1)$, $(-3, -1)$

15. Show by vector calculation that the bisector of the angle between the two equal sides of an isosceles triangle bisects the third side and is perpendicular to it.

Answers to problems (§9B)

1. (a) $\sqrt{5}$ (b) $\sqrt{10}$ (c) $\sqrt{17}$
2. (a) -5 (b) 2 (c) -5 (d) 0 (e) 10 (f) 1 (g) 9 (h) 7
   (i) 28 (j) 0
3. $(2, 6)$, $(3, -1)$ $(1, 4)$, $(4, -1)$ $(1, 4)$, $(-4, 1)$ $(-4, 1)$, $(1, 4)$
   $(2, 8)$, $(-4, 1)$ $(2, -3)$, $(3, 2)$
4. (a) $-\frac{2}{3}$ (b) $\frac{1}{6}$ (c) 0 (d) -1 (e) 0 (f) 1 or -1
8. (a) $(2, 5)$ (b) $(1, 0)$ (c) $(1, -2)$ (d) $(2, 3)$
11. $\max (4, -2) \cdot Q = 2$ (at $x = 1$)
12. $\max P \cdot Q = 1$ (at $x = 1$ or -1)

13. all vectors of the form $P = (x, -2x)$

14. (a) $\cos \theta = -4/\sqrt{65} \approx -0.50$, $\theta \approx 120°$ (b) $90°$ (c) 0 (d) $45°$

    (e) $\cos \theta = -0.8$, $\theta \approx 145°$ (f) $180°$

§9C. Lines

9C1. Direction vectors of a line. Let l be a given line. Let V be any point in the plane such that the segment OV is parallel to l (Figure 1). We call the vector V a *direction vector* for l. If t is any real number $\neq 0$, then tV is also a direction vector for l. Conversely, all direction vectors for l are of this form. In particular, $-V$ is also a direction vector for l.

all direction vectors have this form

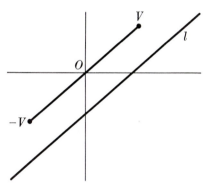

FIGURE 1

There is a simple relation between points of l and direction vectors for l (Figure 2).

    (a) *If P and Q are distinct points of l, then $V = P - Q$ is a direction vector for l.*

    (b) *If Q is a point of l and V is a direction vector for l, then $P = Q + V$ is a point of l.*

Both assertions are merely restatements of the parallelogram law for addition and subtraction of vectors.

    A direction vector V defines a direction or orientation on l: the direction from Q toward P (if V is parallel to $P - Q$), or the direction from P toward Q (if V is parallel to $Q - P$).

Example 1. Let $P = (3, 1)$ and $Q = (2, -2)$ be points of a line l. Then

$$V = P - Q = (1, 3)$$

is a direction vector for l (Figure 3).

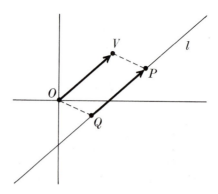

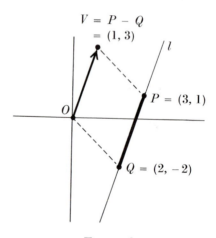

FIGURE 3

Example 2. Let $Q = (2, 1)$ be a point of l and let $V = (-1, 1)$ be a direction vector for l. Then

$$P = Q + V = (1, 2)$$

is a point of l (Figure 4).

9C2. Vector and parametric equations of a line. Let Q be a given point on a line l and let V be a direction vector for l. If P is a point of l then the line PQ is parallel to OV. Conversely, if PQ is parallel to OV then P lies on l. Under these conditions, $P - Q = tV$, i.e.,

(1) $$P = Q + tV,$$

[handwritten annotations: tV = P - Q direction; t is either + or is; tv is vector zero; Vector equation of line l]

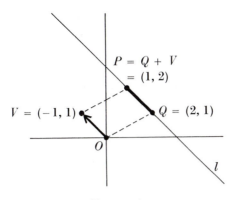

FIGURE 4

for some number t. We call (1) a vector equation of l. If $P = (x, y)$, $Q = (x_0, y_0)$, and $V = (a, b)$, this takes the form

$$(2) \qquad P = (x, y) = (x_0, y_0) + t(a, b).$$

The coordinate equations can be written separately:

$$(3) \qquad x = x_0 + at, \qquad y = y_0 + bt.$$

These are called *parametric equations* of the line, the variable t being the *parameter*.

Example 3. Find the vector equation of the line through the point $Q = (4, 1)$ with direction vector $V = (1, 3)$.

Solution. The equation is

$$P = (4, 1) + t(1, 3) = (4 + t, 1 + 3t).$$

The parametric form is

$$x = 4 + t, \qquad y = 1 + 3t.$$

Vector equations handle vertical lines as easily as nonvertical.

Example 4. Find the vector equation of the line through the point $(3, 0)$ with direction vector $(0, 1)$.

Solution. The equation is

$$P = (3, 0) + t(0, 1) = (3, t).$$

If we are given two points on a line, we subtract to get a direction vector and then proceed as before. If P_1 and P_2 are points of l, then $P_2 - P_1$ is a direction vector, and the equation of l is

(4)
$$P = P_1 + t(P_2 - P_1).$$

Example 5. Find a vector equation for the line l through the points $P_1 = (0, 1)$ and $P_2 = (2, 5)$.

Solution. Here $P_2 - P_1 = (2, 5) - (0, 1) = (2, 4)$. As a simpler direction vector, we take $(1, 2)$. The equation is

$$P = (0, 1) + t(1, 2) = (t, 1 + 2t).$$

Example 6. Find the point of intersection of the lines

$$P = (1, 2) + t(2, 3) \quad \text{and} \quad P = (1, 4) + t(4, -1).$$

Solution. The first thing to realize is that the parametric values in the two equations need not be the same at a common point. So we will call the parameters s and t:

$$P = (1, 2) + s(2, 3), \qquad P = (1, 4) + t(4, -1).$$

The condition of intersection is

$$x = 1 + 2s = 1 + 4t$$
$$y = 2 + 3s = 4 - t.$$

Solving, we get $s = \frac{4}{7}, t = \frac{2}{7}$. The common point, obtained from either of these parametric values, is $(\frac{15}{7}, \frac{26}{7})$.

It is easy to get from the rectangular equation of a line to the vector equation, or vice versa.

Example 7.
(a) What is the vector equation of the line $y = 2x + 1$?

Solution. Put $x = t$. Then $y = 2t + 1$. These are parametric equations. The vector form is
$$P = (t, 2t + 1) = (0, 1) + t(1, 2).$$

(b) What is the rectangular equation of the line

$$P = (3, -2) + t(1, 8)?$$

Solution. Since $x = 3 + t$ and $y = -2 + 8t$, we have $8x = 24 + 8t$ and therefore $8x - y = 26$.

If two lines are perpendicular, so are their direction vectors; and conversely. Hence if P_1 and Q_1 are distinct points on one line and P_2 and Q_2 are distinct points on another, then the condition that the lines be perpendicular is

(5) $$(P_1 - Q_1) \cdot (P_2 - Q_2) = 0.$$

Example 8. The line through $P_1 = (4, -1)$ and $Q_1 = (3, 4)$ has direction vector $P_1 - Q_1 = (1, -5)$. The line through $P_2 = (3, 3)$ and $Q_2 = (-2, 2)$ has direction vector $P_2 - Q_2 = (5, 1)$. Since

$$(P_1 - Q_1) \cdot (P_2 - Q_2) = (1, -5) \cdot (5, 1) = 0,$$

the lines are perpendicular.

Example 9. Find the vector equation of the line through $(1, 2)$ perpendicular to (i.e., with direction vector perpendicular to) the vector $(4, 1)$.

Solution. By inspection, a vector perpendicular to $(4, 1)$ is $(1, -4)$. Hence an equation is

$$P = (1, 2) + t(1, -4).$$

Problems (§9C)

1. Find a direction vector for the line through the two given points.
   (a) $(1, 2)$ and $(3, 2)$ (c) $(0, 0)$ and $(4, 1)$
   (b) $(-1, 2)$ and $(-3, 2)$ (d) $(0, 0)$ and $(-4, -1)$

2. A line with the given direction vector V passes through the given point Q. Find another point on the line.
   (a) $V = (1, 2)$, $Q = (2, 1)$ (c) $V = (3, -1)$, $Q = (-3, 2)$
   (b) $V = (-1, 1)$, $Q = (0, 0)$ (d) $V = (0, 1)$, $Q = (1, -2)$

3. Find vector and parametric equations of each line in Problem 2.

4. Find vector and parametric equations of the line through the two given points.
   (a) $(0, 0)$ and $(1, 1)$ (d) $(1, 2)$ and $(1, 3)$
   (b) $(2, -7)$ and $(3, 4)$ (e) $(3, 1)$ and $(-2, 1)$
   (c) $(-1, -2)$ and $(1, 2)$ (f) $(1, 0)$ and $(0, 1)$

5. Find the slopes of the following lines.
   (a) $P = (3, -1) + t(2, 1)$ (d) $P = (4 + t, 2 - 3t)$
   (b) $P = (2, 3) + t(-3, 2)$ (e) $P = (t, 3t)$
   (c) $P = (1, 0) + t(1, 4)$ (f) $P = (-1 + t, 3)$

6. Find a rectangular equation for each line in Problem 5.

7. Find a direction vector for each of the following lines.
   (a) $2x - 3y = 2$ (d) $2x + 3y = 1$
   (b) $x + 2y = 7$ (e) $5y = 2$
   (c) $-x + 7y = 0$ (f) $2x = -1$

8. Write vector and parametric equations for each line in Problem 7.

9. A line through a point Q has direction vector Q. What are two other points on the line?

10. Find the equation of the line through the given point Q and perpendicular to the given vector V.

(a) $Q = (1, 0)$, $V = (-2, 1)$ (c) $Q = (3, -2)$, $V = (1, 0)$
(b) $Q = (-1, -3)$, $V = (1, -2)$ (d) $Q = (-1, -2)$, $V = (0, 1)$

11. In each case, decide using vector methods whether the line through the first pair of points is perpendicular to the line through the second pair.

(a) $(7, 3)$ and $(3, 2)$; $(3, 1)$ and $(4, -3)$
(b) $(8, 3)$ and $(4, 2)$; $(3, 1)$ and $(2, -3)$
(c) $(6, 6)$ and $(3, 5)$; $(6, 6)$ and $(3, 4)$
(d) $(6, 6)$ and $(3, 5)$; $(6, 6)$ and $(9, 5)$
(e) $(6, 6)$ and $(3, 5)$; $(6, 6)$ and $(7, 3)$
(f) $(2, 1)$ and $(0, 0)$; $(4, 4)$ and $(5, 2)$

12. In each case, decide using vector methods whether the three given points are the vertices of a right triangle.

(a) $(1, 2)$, $(4, 11)$, and $(4, 1)$
(b) $(0, 1)$, $(10, -1)$, and $(1, 4)$
(c) $(4, 11)$, $(-1, 1)$, and $(9, -5)$
(d) $(4, 11)$, $(-1, 1)$, and $(1, 0)$
(e) $(4, 11)$, $(-1, 1)$, and $(8, 9)$

13. In the given triangle ABC, write the equation of the line through each vertex and perpendicular to the opposite side.

(a) $A = (1, 2)$, $B = (-2, 0)$, $C = (3, 5)$
(b) $A = (-3, 5)$, $B = (2, -1)$, $C = (-3, -1)$

14. The line through $(1, 2)$ and $(4, 3)$ has various equations of the form $P = (1, 2) + tV$. Find V so that $P = (4, 3)$ when

(a) $t = 1$ (b) $t = -1$ (c) $t = 2$

15. Find the point of intersection of the two given lines.

(a) $P = (5, 3) + t(1, 2)$, $P = (4, -9) + t(-1, 3)$
(b) $P = (0, 1) + t(1, 0)$, $P = (13, 5) + t(2, 1)$
(c) $P = (t, 3t)$, $P = (-1, 7) + t(3, -1)$
(d) $P = (2, 1) + t(6, 3)$, $P = (-2, 5) + t(4, -10)$

16. Let A and B be non-zero vectors. Show that

$$\frac{A}{|A|} + \frac{B}{|B|}$$

is a direction vector of the bisector of angle AOB.

17. Show that the midpoint of the segment AB (where $A \neq B$) is $\frac{1}{2}(A + B)$. [*Hint*. First show that $\frac{1}{2}(A + B)$ lies on the line AB.]

18. Use the midpoint formula (Problem 17) to show that the diagonals of a parallelogram bisect each other.

19. Use the midpoint formula (Problem 17) to show that the midpoints of the sides of any quadrilateral are vertices of a parallelogram.

20. Using the midpoint formula (Problem 17), show that the segment joining the midpoints of two sides of a triangle is parallel to the third side and half its length.

21. Let A_1, A_2, and A_3 be the vertices of a triangle and let M_1, M_2, and M_3 be the midpoints of their opposite sides. Using the midpoint formula (Problem 17), show that

$$(A_1 - M_1) + (A_2 - M_2) + (A_3 - M_3) = 0.$$

22. Let A_1, A_2, $\cdots$, A_n be the vertices of a regular polygon with center O. Use the midpoint formula (Problem 17) to show that

$$A_1 + A_2 + \cdots + A_n = O.$$

[*Hint.* Note that $(A_1 + A_3)/2 = kA_2$, $(A_2 + A_4)/2 = kA_3$, etc., where $0 < k < 1$. Hence the sum, S, satisfies $S = kS$.]

Answers to problems (§9C)

1. (a) (1, 0)   (b) (1, 0)   (c) (4, 1)   (d) (4, 1)

3. (a) $P = (2, 1) + t(1, 2)$; $x = 2 + t, y = 1 + 2t$
   (b) $P = t(-1, 1)$; $x = -t, y = t$
   (c) $P = (-3, 2) + t(3, -1)$; $x = -3 + 3t, y = 2 - t$
   (d) $P = (1, 0) + t(0, 1)$; $x = 1, y = t$

4. (a) $P = t(1, 1)$; $x = t, y = t$
   (b) $P = (3, 4) + t(1, 11)$; $x = 3 + t, y = 4 + 11t$
   (c) $P = t(1, 2)$; $x = t, y = 2t$
   (d) $P = (1, 0) + t(0, 1)$; $x = 1, y = t$
   (e) $P = (0, 1) + t(1, 0)$; $x = t, y = 1$
   (f) $P = (1, 0) + t(1, -1)$; $x = 1 + t, y = -t$

5. (a) $\frac{1}{2}$ (b) $-\frac{2}{3}$ (c) 4   (d) -3 (e) 3   (f) 0

6. (a) $x - 2y = 5$ (b) $2x + 3y = 13$ (c) $4x - y = 4$
   (d) $3x + y = 14$ (e) $y = 3x$ (f) $y = 3$

7. (a) (3, 2)   (b) (2, -1)   (c) (7, 1)   (d) (3, -2)   (e) (1, 0)
   (f) (0, 1)

8. (a) $P = (1, 0) + t(3, 2)$; $x = 1 + 3t, y = 2t$
   (b) $P = (7, 0) + t(2, -1)$; $x = 7 - 2t, y = t$
   (c) $P = t(7, 1)$; $x = 7t, y = t$
   (d) $P = (\frac{1}{2}, 0) + t(3, -2)$; $x = \frac{1}{2} - 3t, y = 2t$
   (e) $P = (0, \frac{2}{5}) + t(1, 0)$; $x = t, y = \frac{2}{5}$
   (f) $P = (-\frac{1}{2}, 0) + t(0, 1)$; $x = -\frac{1}{2}, y = t$

10. (a) $P = (1, 0) + t(1, 2)$ (b) $P = (-1, -3) + t(2, 1)$
    (c) $P = (3, 0) + t(0, 1)$ (d) $P = (0, -2) + t(1, 0)$

11. (a) yes   (b) no   (c) no   (d) no   (e) yes   (f) yes

12. (a) yes   (b) no   (c) no   (d) yes   (e) yes

13. (a) $P = (1, 2) + t(1, -1)$, $P = (-2, 0) + t(3, -2)$, $P = (3, 5) + t(2, -3)$
   (b) $P = (-3, 0) + t(0, 1)$, $P = (0, -1) + t(1, 0)$, $P = (-3, -1) + t(6, 5)$

14. (a) $(3, 1)$ (b) $(-3, -1)$ (c) $(\frac{3}{2}, \frac{1}{2})$

15. (a) $(2, -3)$ (b) $(5, 1)$ (c) $(2, 6)$ (d) $(0, 0)$

§9D. Vector-valued functions; parametric equations

The vector equation of a line, $P = Q + tV$, defines the vector P as a function of the parameter, t. This is an example of a *vector-valued* function.

In general, we define a vector-valued function whenever we associate to each number t (in some interval I) a vector (x, y) in the plane. Denoting the function by P, we have

$$(1) \qquad\qquad P(t) = (x, y) \qquad\qquad (t \in I).$$

Each coordinate is an ordinary (real-valued) function of t; say

$$(2) \qquad\qquad x = f(t), \quad y = g(t) \qquad\qquad (t \in I).$$

Thus,

$$(3) \qquad\qquad P(t) = \big(f(t), g(t)\big) \qquad\qquad (t \in I).$$

As before, the coordinate equations (2) are called *parametric* equations, the variable t being the *parameter*.

When we think of a vector-valued function geometrically rather than algebraically, we call it a *curve*. In this sense, a curve is more than just the set of values of the function, but the set of values along with the information as to how they were arrived at. (This information may be important, as when the equation represents motion, t denoting time.) By tradition, the word "curve" is used with either meaning, the context making clear which one is intended.

Example 1. The vector-valued function

$$P(t) = (1, 2) + t(3, 1) \qquad\qquad (t \in \mathscr{R})$$

defines a line (a special type of curve). The line is oriented, in order of increasing values of the parameter, in the direction from the point $P(0) = (1, 2)$ toward $P(1) = (4, 3)$; that is, parallel to the direction vector $(3, 1)$. Another equation of this line is

$$Q(t) = (4, 3) + t(6, 2).$$

It gives the same orientation, since the direction vector $(6, 2)$ is parallel to $(3, 1)$. A function that imparts the opposite orientation is

$$R(t) = (1, 2) - t(3, 1).$$

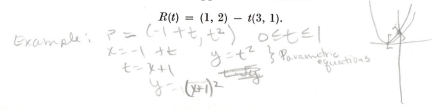

Example 2. The vector-valued function

$$P(t) = (\cos t, \sin t) \qquad (0 \le t \le 2\pi)$$

describes the unit circle, $x^2 + y^2 = 1$. Note that the point $(1, 0)$ appears twice: as $P(0)$ and as $P(2\pi)$. The parametric equations are

$$x = \cos t, \qquad y = \sin t \qquad (0 \le t \le 2\pi).$$

The circle is generated by starting at $(1, 0)$ and proceeding counterclockwise one revolution, ending with a repetition of the starting point.

The equations

$$x = \cos t, \qquad y = \sin t \qquad (t \ge 0)$$

describe the same circle starting at $(1, 0)$ and going around and around endlessly, also counterclockwise. The equations

$$x = \cos t, \qquad y = -\sin t \qquad (t \in \mathscr{R})$$

give the same circle traversed clockwise, and infinitely often, with neither beginning nor end.

Example 3. Consider particles whose motions in the plane are given by.

$$P(t) = (t, t^2) \qquad (0 \le t \le 1),$$

$$Q(t) = \left(\cos \frac{\pi}{2} t, \cos^2 \frac{\pi}{2} t\right) \qquad (0 \le t \le 1),$$

$$R(t) = (\sin \pi t, \sin^2 \pi t) \qquad (0 \le t \le 1).$$

Each particle traverses the curve $y = x^2$ $(0 \le x \le 1)$. However, their motions are different. As t increases from 0 to 1, $P(t)$ starts at $(0, 0)$ and runs up along the curve to $(1, 1)$; $Q(t)$ traverses the curve in the opposite direction, from $(1, 1)$ to $(0, 0)$; and $R(t)$ goes from $(0, 0)$ to $(1, 1)$ and then back again (Figure 1).

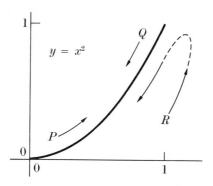

FIGURE 1

Vector or parametric equations are not used just for curves already familiar. They can also describe curves that would be complicated or impossible to express directly in rectangular coordinates.

Example 4. Find the path traced by a point P on a circle as the circle rolls along a straight line. (This curve is called a *cycloid*.)

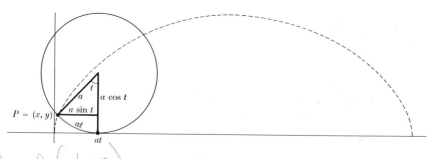

center at (at, a)

FIGURE 2

Solution. Let the circle roll along the positive x-axis, starting with P at the origin. Let the radius of the circle be a. When the circle has rotated through t radians, it has rolled a distance at. From Figure 2,

$$x = at - a \sin t, \qquad y = a - a \cos t.$$

Thus, the cycloid is given in terms of the angle of rotation as parameter by the equations

$$x = a(t - \sin t), \qquad y = a(1 - \cos t).$$

Algebraic notation is used with vector-valued functions in the obvious way, as in the expressions

$$t^3(\cos t, \sin t), \qquad P(t) \cdot Q(t),$$

and so on.

Problems (§9D)

1. Name three points on the given curve.
   (a) $P(t) = (t^3 - 1, t^2 - 1)$ (c) $P(t) = (t \sin t, t \cos t)$
   (b) $P(t) = (1, 1 - t)$ (d) $P(t) = (te^t, te^{-t})$

2. Name three points on the given curve.
   (a) $x = t + \sin t, \quad y = t - \cos t$
   (b) $x = 1 + t^2, \quad y = 1 - t^2$

   (c) $x = t + \log t, \quad y = 1 + \dfrac{1}{t}$

3. Express each of the following curves parametrically.

   (a) $y = x$ (d) $2x - 3y + 4 = 0$

   (b) $y = x^2$ (e) $x^2 + y^2 = 30$

   (c) $y = 3x + 1$ (f) $2x^2 + 3y^2 = 10$

4. Find a rectangular equation for the curve defined by the given pair of parametric equations.

   (a) $x = \cos t$, $y = 2 \sin t$ $(0 \le t \le 2\pi)$

   (b) $x = t - 1$, $y = t + 1$

   (c) $x = \tan t$, $y = \sec t$

   (d) $x = t^2$, $y = t^3$ $(-1 \le t \le 1)$

   (e) $x = e^t$, $y = e^{-t}$

   (f) $x = 2 \cos t$, $y = 3 \sin t$ $(0 \le t \le 2\pi)$

   (g) $x = \log t$, $y = e^t$

   (h) $x = t^2 + t$, $y = t^2 - t$

   (i) $x = (t + 1)^2$, $y = t - 1$

5. Given that $P(t) = (t + 1, 1/t)$, $Q(t) = (2t, g(t))$, and $P(t) \cdot Q(t) = 2t^2 + 2t + 1/t - 1$, find $g(t)$.

6. What are the maximum and minimum values of $(1, 1) \cdot Q$ as Q moves around the circle $x^2 + y^2 = 1$? [*Hint.* Express the circle parametrically.]

7. A circle of radius a rolls along a straight line. A point P is attached to the circle at a distance b from the center. Show that the equations of the path traced by P are

$$x = at - b \sin t, \qquad y = a - b \cos t,$$

where t is the angle of rotation of the circle.

Note. When $b = a$, we get the *cycloid* discussed in Example 4.

When $b < a$ (Figure 3(a)), the path is called a *curtate* cycloid. When $b > a$ (Figure 3(b)), as for a point on the flange of a railroad wheel, we get a *prolate* cycloid.

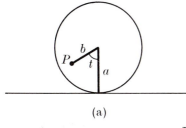 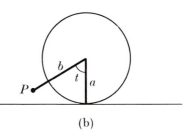

          (a)   (b)

FIGURE 3

8. The curve obtained by "unwinding a string wrapped around a circle" is called an *involute* of the circle. In Figure 4, $P = (x, y)$ is a point on the involute, PQ is tangent to the circle at Q, and the length of PQ is equal to the length of the arc $\overset{\frown}{Q_0 Q}$. Show that parametric equations of the involute are

$$x = a(\cos t + t \sin t), \qquad y = a(\sin t - t \cos t),$$

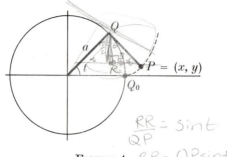

$$\frac{RR}{QP} = \sin t$$

$$RP = QP \sin t$$

FIGURE 4

where a is the radius of the circle. [*Hint.* Look at the triangle PQR, where $R = (a \cos t, y)$.]

9. Figure 5 shows circles centered at O with radii a and b. The point $P = (x, y)$ is determined as indicated in the figure. As the radial line rotates, P traces out an *ellipse*. Show that parametric equations of the ellipse are

$$x = a \cos t, \qquad y = b \sin t.$$

Then find the rectangular equation of the ellipse. Sketch the graph.

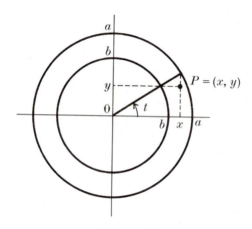

FIGURE 5

10. Figure 6 shows circles centered at O with radii a and b. The point $P = (x, y)$ is determined as indicated in the figure. As the radial line rotates, P traces out a *hyperbola*. Notice that P is not defined when $t = \pi/2$ nor when $t = -\pi/2$, so that the hyperbola consists of two branches, not connected to one another.

(a) Show that parametric equations of the hyperbola are

$$x = a \sec t, \qquad y = b \tan t.$$

Then find the rectangular equation of the hyperbola.

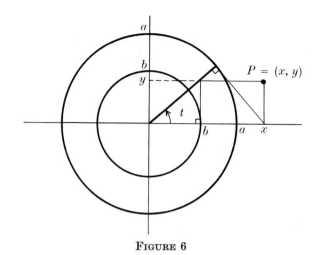

FIGURE 6

(b) Show that for $0 \leq t < \pi/2$, the hyperbola lies under the line

$$y = \frac{b}{a} x,$$

the vertical distance between them being

$$b \frac{\cos t}{1 + \sin t}.$$

Note that this distance approaches 0 as $t \to \pi/2$. (Because of this fact, this line is an *asymptote* of the hyperbola.) Corresponding relations hold in the other three quadrants. Sketch the graph.

Answers to problems (§9D)

3. (a) $x = t,\ y = t$ (b) $x = t,\ y = t^2$ (c) $x = t,\ y = 3t + 1$
    (d) $x = 3t + 1,\ y = 2t + 2$ (e) $x = \sqrt{30}\cos t,\ y = \sqrt{30}\sin t$
    (f) $x = \sqrt{5}\cos t,\ y = \sqrt{10/3}\sin t$

4. (a) $4x^2 + y^2 = 4$ (b) $y = x + 2$.   (c) $y^2 = 1 + x^2$
    (d) $y = x^{3/2}\ (0 \leq x \leq 1)$ (e) $y = 1/x\ (x > 0)$ (f) $9x^2 + 4y^2 = 36$
    (g) $y = e^{e^x}$ (h) $(x - y)^2 = 2(x + y)$ (i) $x = (y + 2)^2$

5. $g(t) = 1 - t$

6. $\sqrt{2},\ -\sqrt{2}$

§9E. Calculus with vectors

9E1. *Limits and continuity.* The underlying idea of limit and continuity is the same for vector-valued functions as for real-valued functions. Consider a vector-valued function

(1) $$P(t) = \big(f(t),\ g(t)\big).$$

The assertion

$$\text{(2)} \qquad\qquad \lim_{t \to t_0} P(t) = (a, b)$$

is to be defined so as to express in a precise way the rough notion that $P(t)$ is close to (a, b) when t is close to (but not equal to) t_0. Now, clearly, a point $\big(f(t), g(t)\big)$ is close to (a, b) provided that $f(t)$ is close to a and $g(t)$ is close to b. We accordingly define (2) to mean that

$$\text{(3)} \qquad\qquad \lim_{t \to t_0} f(t) = a \qquad \text{and} \qquad \lim_{t \to t_0} g(t) = b.$$

(In other words, we define limits coordinatewise.)

We say that P is *continuous* at t_0 if both f and g are continuous at t_0; equivalently, if

$$\text{(4)} \qquad\qquad \lim_{t \to t_0} P(t) = P(t_0). \qquad \text{or}$$

P is continuous at

In the main, the functions we deal with will be continuous—i.e., continuous at each point of their domains.

9E2. Derivative; tangent; velocity. Let

$$P(t) = \big(f(t), g(t)\big)$$

be a continuous vector-valued function. What shall we mean by a tangent vector to P at a point t_0? Suppose P represents the path of a particle, t denoting time; what shall we mean by the velocity of the particle at time t_0?

We have

$$P(t_0) = \big(f(t_0), g(t_0)\big).$$

A change in t by an amount Δt results in a new point $t_0 + \Delta t$ and functional values

$$\text{(5)} \qquad\qquad P(t_0 + \Delta t) = \big(f(t_0 + \Delta t), g(t_0 + \Delta t)\big).$$

To focus on the change in values let us write

$$\text{(6)} \qquad\qquad \Delta P = P(t_0 + \Delta t) - P(t_0).$$

Notice that ΔP is a *vector* (Figure 1). It tells us how far $P(t_0 + \Delta t)$ lies from $P(t_0)$, and in what direction (provided ΔP is not the zero vector).

In terms of f and g,

$$\text{(7)} \qquad\qquad \Delta P = \big(f(t_0 + \Delta t) - f(t_0), g(t_0 + \Delta t) - g(t_0)\big).$$

vector divided by scalar is vector

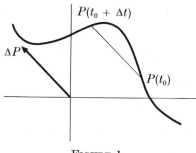

FIGURE 1

The vector ΔP is the change in P resulting from the change Δt. The vector

$$(8) \qquad \frac{\Delta P}{\Delta t} = \left(\frac{f(t_0 + \Delta t) - f(t_0)}{\Delta t} , \frac{g(t_0 + \Delta t) - g(t_0)}{\Delta t} \right)$$

is the average rate of change of P per unit change in t (between t_0 and $t_0 + \Delta t$).

In the application to motion, $\Delta P/\Delta t$ is the average velocity of the particle over the interval from t_0 to $t_0 + \Delta t$.

If $\Delta P \neq O$, $\Delta P/\Delta t$ is a direction vector for the segment from $P(t_0)$ to $P(t_0 + \Delta t)$.

When Δt is small, $\Delta P/\Delta t$ is an approximation to our intuitive notion of the "instantaneous" velocity "at" the point t_0, and an approximation to a tangent vector at t_0.

We now define the *derivative* of P at t_0 by:

$$(9) \qquad P'(t_0) = \lim_{\Delta t \to 0} \frac{\Delta P}{\Delta t} \; ;$$

that is $\big($see $(8)\big)$,

$$(10) \qquad P'(t_0) = \big(f'(t_0), g'(t_0) \big).$$

Thus, on any interval where f and g are differentiable, P' is defined as a vector-valued function. If $P'(t)$ is not zero we call it a *tangent vector* to the curve P at the point t. When P represents the path of a particle, the *velocity vector*, V, is also defined to be P'.

The tangent line at t is defined to be the line through the point $P(t)$ with direction vector $P'(t)$, the tangent vector. Hence the equation of the tangent line at t_0 is

$$(11) \qquad T(t) = P(t_0) + t P'(t_0).$$

If we wish to drop the subscript we need a different letter for the new parameter:

(12) $$T(s) = P(t) + sP'(t).$$

Example 1. Find the derivative of
$$P(t) = (t^2 + t, 1 - t^3).$$
What is the equation of the tangent line at any point t?

Solution. The derivative is
$$P'(t) = (2t + 1, -3t^2).$$
Hence the equation of the tangent line at any point t is
$$T(s) = (t^2 + t, 1 - t^3) + s(2t + 1, -3t^2).$$
At $t = 1$, for example,
$$T(s) = P(1) + sP'(1) = (2, 0) + s(3, -3).$$

Example 2. Find the tangent vector to the unit circle,
$$P(t) = (\cos t, \sin t).$$
Solution. We have
$$P'(t) = (-\sin t, \cos t).$$
Notice that $P(t) \cdot P'(t) = 0$, showing that the tangent is perpendicular to the radius.

Example 3. Where on the cycloid
$$x = a(t - \sin t), \qquad y = a(1 - \cos t) \qquad (0 \le t \le 2\pi)$$
(§9D, Example 4) does the tangent point up at an angle of $45°$?

Solution. See Figure 2. We are asked for the point at which P' is parallel to $(1, 1)$; that is, at which $x' = y'$. Since
$$x' = a(1 - \cos t), \qquad y' = a \sin t,$$
the condition is
$$1 - \cos t = \sin t.$$
Squaring and simplifying, we get
$$(\cos t)(1 - \cos t) = 0.$$
Then either $\cos t = 0$, so that $t = \pi/2$ or $3\pi/2$; or $\cos t = 1$, so that $t = 0$ or 2π.

If $P'(t_0) \neq 0$, it is a direction vector for the tangent line at $P(t_0)$

$Q = P(t_0) + t P'(t_0)$ gives tangent line.

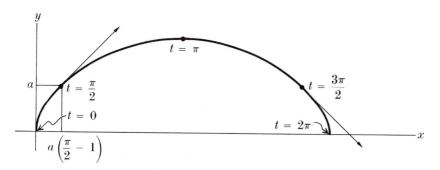

FIGURE 2

At $t = \pi/2$ we have $x' = a$ and $y' = a$, so this is a solution. Here,

$$x = a\left(\frac{\pi}{2} - 1\right), \qquad y = a.$$

At $t = 3\pi/2$, $x' = a$ and $y' = -a$ and the tangent vector is pointing $45°$ *down*. At $t = 0$ or 2π, both x' and y' are 0 and there is no tangent vector.

When P' exists we say that P is *differentiable*. Then f and g are differentiable real-valued functions, and are therefore continuous. But this means that P is continuous. In brief, *differentiable vector-valued functions are continuous*.

9E3. Properties of the derivative. Derivatives of vector-valued functions have algebraic properties reminiscent of derivatives of real-valued functions. This is to be expected, since if $P = (x, y)$, then $P' = (x', y')$. We see at once that:

   (i) If P is constant, then $P' = O$. Conversely, if $P' = O$, then P is constant.
   (ii) $(P + Q)' = P' + Q'$.
   (iii) $(kP)' = kP'$ if k is a constant scalar $(k \in \mathscr{R})$.
Next, let us show that:
   (iv) $(P \cdot Q)' = P \cdot Q' + Q \cdot P'$.

The calculation is straightforward. In terms of $P_1 = (x_1, y_1)$ and $P_2 = (x_2, y_2)$, we have $P_1' = (x_1', y_1')$ and $P_2' = (x_2', y_2')$, and

$$P_1 \cdot P_2 = x_1 x_2 + y_1 y_2.$$

Therefore,

$$
\begin{aligned}
(P_1 \cdot P_2)' &= (x_1 x_2' + x_2 x_1') + (y_1 y_2' + y_2 y_1') \\
&= (x_1 x_2' + y_1 y_2') + (x_2 x_1' + y_2 y_1') \\
&= P_1 \cdot P_2' + P_2 \cdot P_1'.
\end{aligned}
$$

A useful consequence of (iv) is

$$\text{(v)} \quad \frac{d}{dt}|P|^2 = (P \cdot P)' = 2(P \cdot P').$$

[margin handwriting: $P(t) = (\cos t, \sin t)$ — unit circle — $\leftarrow P(t)$ — $|P(t)| = 1$, so, $P(t) \cdot P'(t) = 0$ — $P'(t) = (-\sin t, \cos t)$ — $P(t) \cdot P'(t) = -\cos t \sin t + \sin t \cos t$ — $P(t) \cdot P'(t) = 0$]

From (i) and (v) we get

(vi) If $|P|$ is constant then $P \cdot P' = 0$.

Example 4. The equation of the unit circle is $|P(t)| = 1$. By (vi), $P(t) \cdot P'(t) = 0$. (Cf. Example 2.)

In addition, we have the following product and quotient formulas. If f is a differentiable scalar-valued function and P a differentiable vector-valued function, then

[margin handwriting: $\frac{d}{dt}(|P(t)|)$ — By chain rule — $2|P(t)|\frac{d}{dt}(|P(t)|) = \frac{d}{dt}(|P(t)|)^2 = 2P(t) \cdot P'(t)$ — $2P(t) \cdot P'(t) = 2|P(t)|\frac{d}{dt}(|P(t)|) = 2(P(t) \cdot P'$ — $\frac{d}{dt}(|P(t)|) = \frac{P(t) \cdot P'(t)}{|P(t)|}$]

(vii) $(fP)' = fP' + f'P$

and

(viii) $\left(\dfrac{P}{f}\right)' = \dfrac{fP' - f'P}{f^2}$ (wherever f is not zero).

These are obtained by straightforward calculation.

Example 5.

(a) If $P = t^2(e^t + t, \cos t)$, then

$$P' = t^2(e^t + 1, -\sin t) + 2t(e^t + t, \cos t).$$

(b) If

$$P = \frac{(e^t + t, \cos t)}{t^2}$$

then

$$P' = \frac{t^2(e^t + 1, -\sin t) - 2t(e^t + t, \cos t)}{t^4}.$$

Next, suppose that P is a differentiable function of t, where t is a differentiable function of another real variable, say r. Then the chain rule (§3F2) leads to the vector chain rule:

$$\text{(ix)} \quad \frac{dP}{dr} = \frac{dt}{dr}\frac{dP}{dt} \qquad \text{or} \qquad \frac{dP}{dr} = \frac{dP}{dt}\frac{dt}{dr}.$$

In the second equation, the scalar dt/dr is written on the right to make the chain formula look more familiar.

Example 6. If $P = (\cos t, \sin t)$, where $t = r^2 + 1$, then

$$\frac{dP}{dr} = \frac{dt}{dr}\frac{dP}{dt} = 2r(-\sin t, \cos t)$$

$$= 2r(-\sin(r^2 + 1), \cos(r^2 + 1)).$$

Problems (§9E)

1. Evaluate the following limits.

(a) $\displaystyle\lim_{t \to 2} (t^2, t + 1)$

(b) $\displaystyle\lim_{t \to 1} \left(\frac{t^2 - 1}{t - 1}, \frac{t^3 - 1}{t - 1}\right)$

(c) $\displaystyle\lim_{t \to 0} \left(\frac{\sin t}{t}, \frac{1 - \cos t}{t}\right)$

2. Find ΔP (defined as in (6)) in each of the following situations.

(a) $P = (t^2, t + 1)$, $t_0 = 2$, $\Delta t = 1$

(b) $P = \left(\frac{1}{t + 1}, \frac{1}{t - 1}\right)$, $t_0 = 3$, $\Delta t = 1$

(c) $P = \left(\sqrt{t}, \frac{1}{\sqrt{t}}\right)$, $t_0 = 4$, $\Delta t = 2$

3. Evaluate:

(a) $\dfrac{d}{dt}(t^2, t + 1)$

(c) $\dfrac{d}{dt}\left(\sqrt{t}, \frac{1}{\sqrt{t}}\right)$

(b) $\dfrac{d}{dt}\left(\frac{1}{t + 1}, \frac{1}{t - 1}\right)$

4. Find $P'(t)$:

(a) $P(t) = (t - \sin t, t + \cos t)$

(c) $P(t) = \left(\log t, \log \frac{1}{t}\right)$

(b) $P(t) = (te^t, te^{-t})$

5. Find $P'(t_0)$. Then write the equation of the tangent line to the curve at t_0.

(a) $P(t) = \left(\frac{1}{t}, t^2 + 1\right)$, $t_0 = 1$

(b) $P(t) = (e^t, e^{-t})$, $t_0 = 0$

(c) $P(t) = (t + 1, t^2 - 1)$, $t_0 = 3$

(d) $P(t) = (\sin t, \cos 2t), \quad t_0 = \pi$

(e) $P(t) = (\log t, e^t), \quad t_0 = 1$

(f) $P(t) = \left(\dfrac{t^2 + 1}{t^2 - 1}, \dfrac{1}{t^2}\right), \quad t_0 = 2$

6. Show that if P is differentiable, then

(x) $$\frac{d}{dt}|P| = \frac{P \cdot P'}{|P|} \qquad\qquad (P(t) \neq O).$$

7. Compute $\dfrac{d}{dt}|P|$ for each of the following. [You may use formula (x) of Problem 6.]

(a) $P = (\cos t^2, \sin t^2)$ (c) $P = (e^{-t} \cos t, e^{-t} \sin t)$

(b) $P = (t^2, 2t)$ (d) $P = (t \cos t, t \sin t)$

8. Find $\dfrac{d}{dt}(P \cdot P')$.

(a) $P = (2t + 1, t - 7)$ (c) $P = (1 - \cos t, t - \sin t)$

(b) $P = (e^t, e^{-t})$ (d) $P = (\cos t, \sin t)$

9. Find the values of t at which $|P|$ is a maximum or minimum.

(a) $P = (t^2, t + 3)$ (c) $P = (t^2, t)$

(b) $P = (t, t + 1)$ (d) $P = (e^t, e^{-t})$

10. Let $P = (t^2 - 1, t^3 - t)$. Show that the tangent lines to the curve, at the point where the curve crosses itself, are perpendicular.

11. Problem 9 of §9D derives the equation of an *ellipse*:

$$P = (a \cos t, b \sin t).$$

Show that for each point P of the ellipse, P' also lies on the ellipse.

12. The curve $P(t) = \big(f(t), g(t)\big)$ satisfies the stated condition for all t. Describe the curve geometrically.

(a) $f(t) = 1$ (d) $P'(t) = (a, b)$ (constant)

(b) $f(t) = g(t)$ (e) $f(t)^2 + g(t)^2 = 2$

(c) $f'(t) = g'(t)$

Answers to problems (§9E)

1. (a) $(4, 3)$ (b) $(2, 3)$ (c) $(1, 0)$

2. (a) $(5, 1)$ (b) $-(\tfrac{1}{20}, \tfrac{1}{6})$ (c) $(\sqrt{6} - 2, \tfrac{1}{6}\sqrt{6} - \tfrac{1}{2})$

3. (a) $(2t, 1)$ (b) $-\left(\dfrac{1}{(t + 1)^2}, \dfrac{1}{(t - 1)^2}\right)$ (c) $\tfrac{1}{2}(t^{-1/2}, -t^{-3/2})$

4. (a) $(1 - \cos t, 1 - \sin t)$ (b) $\big((1 + t)e^t, (1 - t)e^{-t}\big)$ (c) $(1/t, -1/t)$

5. (a) $P'(1) = (-1, 2)$, $T(s) = (1, 2) + s(-1, 2)$
   (b) $P'(0) = (1, -1)$, $T(s) = (1, 1) + s(1, -1)$
   (c) $P'(3) = (1, 6)$, $T(s) = (4, 8) + s(1, 6)$
   (d) $P'(\pi) = (-1, 0)$, $y = 1$
   (e) $P'(1) = (1, e)$, $T(s) = (0, e) + s(1, e)$
   (f) $P'(2) = -(\frac{8}{9}, \frac{1}{4})$, $T(s) = (\frac{5}{3}, \frac{1}{4}) + s(\frac{8}{9}, \frac{1}{4})$

7. (a) 0 (b) $2t(t^2 + 2)/|t|\sqrt{t^2 + 4}$ $(t \neq 0)$ (c) $-e^{-t}$
   (d) $t/|t|$ $(t \neq 0)$

8. (a) 5 (b) $2(e^{2t} + e^{-2t})$ (c) $1 - \cos t + t \sin t$ (d) 0

9. (a) minimum at $t = -1$ (b) minimum at $t = -\frac{1}{2}$
   (c) minimum at $t = 0$ (d) minimum at $t = 0$

12. (a) the line $x = 1$ (or some part of it)
    (b) the line $y = x$ (or some part of it)
    (c) the line $y = x + c$ (or some part of it), for suitable c
    (d) a single point if $a = b = 0$; otherwise, the line $ay = bx + c$ for suitable c
    (e) the circle center origin radius $\sqrt{2}$ (or some part of it)

§9F. Graphs of real-valued functions

9F1. Real-valued functions. Let

(1)
$$x = f(t), y = g(t) \qquad (t \in I)$$

be parametric equations, where f and g are differentiable. As we know, the curve need not be the graph of y as a function of x, nor of x as a function of y. For instance, the circle

$$x = \cos t, y = \sin t \qquad (0 \leq t \leq 2\pi)$$

is not the graph of a real-valued function. Of course, we know how to break up the circle into portions, each one of which is such a graph:

$$y = \sqrt{1 - x^2} \quad \text{and} \quad y = -\sqrt{1 - x^2} \quad (-1 \leq x \leq 1),$$

or

$$x = \sqrt{1 - y^2} \quad \text{and} \quad x = -\sqrt{1 - y^2} \quad (-1 \leq y \leq 1).$$

In each case, the function is differentiable on the interior of the interval.

Generally speaking, this kind of analysis can be performed on any parametric curve we will be dealing with, except at a few isolated points. The following discussion shows how. Let $x = f(t)$ and $y = g(t)$ be as in (1).

Suppose dx/dt is never zero in the interior of I. Then y is a continuous function of x for t ∈ I and is a differentiable function of x for t in the interior of I; and

(2)
$$\frac{dy}{dx} = \frac{dy/dt}{dx/dt} \qquad (t \in \text{interior of } I).$$

Proof. Since f' is never 0 in the interior of I, $x = f(t)$ has a continuous inverse on I (§4A4). Moreover, by the Inverse-Function Theorem (§4G), applied to the interior of I, this inverse is differentiable and

(3)
$$\frac{dt}{dx} = \frac{1}{dx/dt} \qquad (t \in \text{interior of } I).$$

Hence $y = g(t) = g(f^{-1}(x))$ is continuous for $t \in I$; and, by the chain rule, y is differentiable and

(4)
$$\frac{dy}{dx} = \left(\frac{dy}{dt}\right)\left(\frac{dt}{dx}\right) = \frac{dy/dt}{dx/dt} \qquad (t \in \text{interior of } I).$$

Note that this result agrees with §9E2, according to which the tangent line has direction vector $(dx/dt, dy/dt)$.

Similarly, *if dy/dt is never zero in the interior of I, then x is a continuous function of y for t ∈ I, and*

(5)
$$\frac{dx}{dy} = \frac{dx/dt}{dy/dt} \qquad (t \in \text{interior of } I).$$

Example 1. In the parametric equations of the cycloid,

$$x = a(t - \sin t), \qquad y = a(1 - \cos t)$$

(§9D, Example 4), x and y are differentiable (and hence continuous) everywhere:

$$\frac{dx}{dt} = a(1 - \cos t), \qquad \frac{dy}{dt} = a \sin t.$$

Evidently, dx/dt is never 0 for $0 < t < 2\pi$. Therefore, y is a continuous function of x for $0 \le t \le 2\pi$ and is differentiable for $0 < t < 2\pi$; and

$$\frac{dy}{dx} = \frac{a \sin t}{a(1 - \cos t)} = \frac{\sin t}{1 - \cos t} \qquad (0 < t < 2\pi).$$

(i) If $P'(t) = 0$ for all t
       then $P(t)$ is constant,
   or if P is constant $P' = 0$
(ii) $(P+Q)' = P' + Q'$

Since dy/dt is never 0 for $0 < t < \pi$, x is a continuous function of y for $0 \leq t \leq \pi$ and is differentiable for $0 < t < \pi$; and

$$\frac{dx}{dy} = \frac{a(1 - \cos t)}{a \sin t} = \frac{1 - \cos t}{\sin t} \qquad (0 < t < \pi).$$

(See §9E2, Figure 2.)

9F2. The second derivative. The formula

$$(6) \qquad\qquad y' = \frac{dy}{dx} = \frac{dy/dt}{dx/dt}$$

was obtained with the help of the chain rule. For information about the concavity of the graph, we may wish to find

$$(7) \qquad\qquad y'' = \frac{dy'}{dx}.$$

This task requires the chain rule again:

$$(8) \qquad\qquad y'' = \frac{dy'}{dx} = \left(\frac{dy'}{dt}\right)\left(\frac{dt}{dx}\right) = \frac{dy'/dt}{dx/dt}.$$

Similarly,

$$(9) \qquad\qquad x'' = \frac{dx'}{dy} = \left(\frac{dx'}{dt}\right)\left(\frac{dt}{dy}\right) = \frac{dx'/dt}{dy/dt}.$$

Example 2. For the cycloid in Example 1 we obtained

$$\frac{dx}{dt} = a(1 - \cos t) \qquad \text{and} \qquad y' = \frac{dy}{dx} = \frac{\sin t}{1 - \cos t} \qquad (0 < t < 2\pi).$$

From the latter,

$$\frac{dy'}{dt} = \frac{(1 - \cos t)(\cos t) - (\sin t)(\sin t)}{(1 - \cos t)^2} = -\frac{1}{1 - \cos t},$$

and consequently,

$$y'' = \frac{dy'}{dx} = \frac{dy'/dt}{dx/dt} = -\frac{1}{a(1 - \cos t)^2} \qquad (0 < t < 2\pi).$$

Since this is always negative, the graph is concave down for $0 \leq t \leq 2\pi$ (§4D5).

Problems (§9F)

1. Express dy/dx and d^2y/dx^2 as functions of t, and indicate concavity of the graph $\left(\text{viewed as } y = F(x)\right)$.
   - (a) $x = t^2 - t, y = t^3 + 1$
   - (b) $x = t^2 - t, y = t^3 + t^2 + 1$
   - (c) $x = t + \log t, y = e^t$
   - (d) $x = \tan t - t, y = \cot t + t$
   - (e) $x = \cos t + t \sin t, y = \sin t - t \cos t$
   - (f) $x = e^t \cos t, y = e^t \sin t$

2. Express dy/dx and d^2y/dx^2 as functions of t. Also, find a rectangular equation for the curve and sketch the graph.
   - (a) $x = \cos 2t, y = \cos t$
   - (b) $x = 1 + t, y = t^2$
   - (c) $x = t^2, y = t^3$
   - (d) $x = 4t, y = t^2$
   - (e) $x = \sin^4 t, y = \cos^4 t$
   - (f) $x = \tan t, y = \sec t$
   - (g) $x = \cot t, y = \sin^2 t$

Answers to problems (§9F)

1. (a) $y' = 3\dfrac{t^2}{2t - 1}$, $y'' = 6\dfrac{t(t - 1)}{(2t - 1)^3}$ $(t \neq \frac{1}{2})$. Concave down on $t \leq 0$ and on

$\frac{1}{2} < t \leq 1$, concave up on $0 \leq t < \frac{1}{2}$ and on $t \geq 1$.

(b) $y' = \dfrac{3t^2 + 2t}{2t - 1}$, $y'' = 2\dfrac{3t^2 - 3t - 1}{(2t - 1)^3}$ $(t \neq \frac{1}{2})$. Let $t_1 = (3 - \sqrt{21})/6$ and

$t_2 = (3 + \sqrt{21})/6$; concave down on $t \leq t_1$ and on $\frac{1}{2} < t \leq t_2$, concave up on $t_1 \leq t < \frac{1}{2}$ and on $t \geq t_2$.

(c) $y' = \dfrac{te^t}{1 + t}$, $y'' = \dfrac{t(1 + t + t^2)e^t}{(1 + t)^3}$ $(t > 0)$. Concave up on $t > 0$.

(d) $y' = -\cot^4 t$, $y'' = 4\cot^5 t \csc^2 t$ $(t \neq 0, \pi/2, \cdots)$. Concave up on $0 < t < \pi/2$, $\pi < t < 3\pi/2$, etc.; concave down on $\pi/2 < t < \pi$, $3\pi/2 < t < 2\pi$, etc.

(e) $y' = \tan t$, $y'' = (1/t)\sec^3 t$ $(t \neq 0; t \neq \pi/2, 3\pi/2, \cdots)$. Concave down on $0 < t < \pi/2$; then concave down on $\pi/2 < t < 3\pi/2$, concave up on $3\pi/2 < t < 5\pi/2$, etc. The opposite concavity holds on the negative of each interval.

(f) $y' = \dfrac{\cos t + \sin t}{\cos t - \sin t}$, $y'' = \dfrac{2e^{-t}}{(\cos t - \sin t)^3}$ $(t \neq \pi/4, 5\pi/4, \cdots)$. Concave down

on $\pi/4 < t < 5\pi/4$, concave up on $5\pi/4 < t < 9\pi/4$, etc.

2. (a) $y' = \frac{1}{4}\sec t,\ y'' = -\frac{1}{16}\sec^3 t\ (t \neq 0,\ \pi/2,\cdots);\ x = 2y^2 - 1\ (-1 \leq y \leq 1)$
  (b) $y' = 2t,\ y'' = 2;\ y = (x-1)^2$
  (c) $y' = 3t/2,\ y'' = 3/4t\ (t \neq 0);\ x = y^{2/3}$
  (d) $y' = t/2,\ y'' = \frac{1}{2};\ y = (x/4)^2$
  (e) $y' = -\cot^2 t,\ y'' = \frac{1}{2}\csc^6 t\ (t \neq 0,\ \pi/2,\cdots);\ \sqrt{x} + \sqrt{y} = 1$
  (f) $y' = \sin t,\ y'' = \cos^3 t\ (t \neq \pi/2,\ 3\pi/2,\cdots);\ y^2 = 1 + x^2$
  (g) $y' = -2\sin^3 t \cos t,\ y'' = 2\sin^4 t\ (4\cos^2 t - 1)\ (t \neq 0,\ \pi,\cdots);$
      $y = 1/(1+x^2)$

§9G. Motion in the plane

9G1. Velocity and speed. A vector-valued function,

$$P(t) = \big(f(t), g(t)\big),$$

may be used to describe the position of a particle at time t. In this application one often wishes to distinguish between the function (including information about t) and its set of values. The set of values itself is then called the *trajectory*. If one car drives through town north to south while another goes east to west, then their trajectories will intersect, though not necessarily for the same value of t.

We assume P differentiable. Then the velocity vector is defined as

(1)
$$V(t) = P'(t) = \big(f'(t), g'(t)\big).$$

Where this is not zero, it is parallel to the tangent vector (§9E2).

The magnitude of the velocity vector is called the *speed*:

(2)
$$|V(t)| = \sqrt{f'(t)^2 + g'(t)^2}.$$

Observe that the speed, $|V(t)|$, is a number, not a vector, and that $|V(t)| \geq 0$.

Example 1. The function

$$P(t) = (\cos t, \sin t)$$

describes motion around the unit circle. The velocity vector is

$$V(t) = P'(t) = (-\sin t, \cos t).$$

When $t = 0$, so that $P(t) = (1, 0)$, the velocity vector is $V(0) = (0, 1)$, which is directed vertically upward. At $P(\pi/2) = (0, 1)$, the velocity vector is $V(\pi/2) = (-1, 0)$, which is directed horizontally to the left.

In this motion, the particle moves at constant speed 1, since, for every t,

$$|V(t)|^2 = \sin^2 t + \cos^2 t = 1.$$

Example 2. Show that a particle with constant velocity is moving in a straight line.

Solution. If $V(t) = (a, b)$, where a and b are constants, then, taking anti-derivatives, we get

$$P(t) = (at + c, bt + d),$$

where c and d are constants. This is the equation of a straight line.

9G2. Acceleration. The intuitive notion of velocity as rate of change of position led to the precise definition of the velocity vector as the derivative of the position vector (§9E2). Similarly, we now define the acceleration vector, A, as the derivative of the velocity:

(3)
$$A(t) = V'(t).$$

Thus, for $P(t) = \big(f(t), g(t)\big)$,

(4)
$$A(t) = P''(t) = \big(f''(t), g''(t)\big).$$

Example 3. The equation

$$P(t) = (h + a \cos t, k + a \sin t) = (h, k) + (a \cos t, a \sin t),$$

where $a > 0$, describes motion about the circle

$$(x - h)^2 + (y - k)^2 = a^2$$

with center (h, k) and radius a. The velocity vector is

$$V(t) = (-a \sin t, a \cos t),$$

the speed is $|V(t)| = a$, and the acceleration vector is

$$A(t) = (-a \cos t, -a \sin t).$$

Thus, the acceleration at each point is directed "toward the center of the circle"— that is, the position vector plus the acceleration vector is a constant, the center of the circle.

9G3. Force; Newton's laws. According to *Newton's second law*, a force $F(t)$ acting on a particle of mass m imparts an acceleration equal to $F(t)/m$ (for properly chosen units):

(5)
$$A(t) = \frac{1}{m} F(t).$$

$F = mA$ *= force acting on particle*

mass of particle *acceleration vector*

Note that force is a vector. The effect of two forces acting on the same particle is the force equal to their vector sum (the "resultant"). These are laws of physics.

Example 4. A particle of mass $\frac{1}{2}$ is acted on by a force $F_1(t) = (1 - t, 1 + t)$ and a force $F_2(t) = (t, -1)$. At time $t = 0$ the particle is at the origin with velocity $(1, 0)$. Describe the motion.

Ex. 5:

Solution. The effective force is

suppose we know that the force acting on a particle of mass m.

$$F(t) = F_1(t) + F_2(t) = (1, t).$$

$F(t) = (t, \sin t)$

Then

At $t = 0$, suppose the particle is at the origin and moving with a velocity

$$A(t) = (1/\tfrac{1}{2})F(t) = (2, 2t).$$

$V(0) = (1, 0)$

Describe the subsequent motion.

Taking antiderivatives, we get

$$V(t) = (2t + c_1, t^2 + c_2),$$

where c_1 and c_2 are constants. Since $V(0) = (1, 0)$, we have $c_1 = 1$, $c_2 = 0$, and

$$V(t) = (2t + 1, t^2).$$

Taking antiderivatives again, we get

$$P(t) = (t^2 + t + k_1, \tfrac{1}{3}t^3 + k_2).$$

$P(t) = (f(t), g(t))$ *is our final answer*

Since $P(0) = 0$, the constants are both 0 and we have, finally,

$(t, \sin t) = m(f''(t), g''(t))$

$= (m f''(t), m g''(t))$

$$P(t) = (t^2 + t, \tfrac{1}{3}t^3).$$

$m f'' = t$

$f' = \frac{t^2}{2m} + c_1$

As a particular case, let us ask what happens if the force is zero over an interval. Then the acceleration is zero and therefore the velocity is constant. The motion is then along a straight line (Example 2). This is *Newton's first law*: a particle on which no force is acting moves along a straight line at constant speed.

$m g''(t) = \sin t$

$g'(t) = -\cos t + c_2$

$V(0) = (1, 0)$

$f'(0) = 1$

$g'(0) = 0$

Problems (§9G)

$\frac{0^2}{m} + c_1 = 1$ *or*

$c_1 = 1$

$\frac{-\cos 0}{m} = c_2$

1. A particle moves so that at time t it is at the point $P(t)$. Find its velocity, speed, and acceleration at time t.

$c_2 = -\frac{1}{m}$

(a) $P(t) = (2 \cos t, 3 \sin t)$ (d) $P(t) = (t \cos t, t \sin t)$

$f'(t) = \frac{t^2}{2m}$

(b) $P(t) = (t^2 - 1, t^2 + 1)$ (e) $P(t) = (3t + 1, 2t - 1)$

$g'(t) = -\frac{\cos t}{m} + 1$

(c) $P(t) = (e^t, e^{t-t})$ (f) $P(t) = (t^2, t^3)$

$f(t) = \frac{t^3}{6m} + t + \frac{1}{m}$

$g(t) = \frac{-\sin t}{m} + t + c_4$

$P(0) = (f(0), g(0)) = (0, 0)$

$f(0) = c_3 = 0$

$g(0) = c_4 = 0$

$f(t) = \frac{t^3}{6m} + c_3$

$\frac{-\sin t}{m} + t/m$

$g(t) = \sin t + \frac{t}{m}$

$P(t) = \left(\frac{t^3}{6m} + c_1, \frac{-\sin t}{6m} + t + t/m\right)$

2. Describe the motion of a particle given the following information about its position P, its velocity V, and its acceleration A.

   (a) $V = (\sin t, \cos t)$, $P(0) = (0, 1)$

   (b) $V = (t, t^2)$, $P(0) = (1, 0)$

   (c) $A = (0, 0)$, $V(0) = (1, 0)$, $P(0) = (1, 1)$

   (d) $A = (t, 3)$, $V(0) = (2, 1)$, $P(0) = (0, 0)$

3. This problem refers to the cycloid,

$$P(t) = a(t - \sin t, 1 - \cos t),$$

discussed in §9D, Example 4. Let C denote the center of the rolling circle. Show that the acceleration vector always has the same direction and length as the segment from P to C (i.e., $P + P'' = C$).

4. A particles moves so that at time t it is at $P(t)$. It does not pass through the origin, but is nearest the origin at time $t = 2$. Assuming that P is differentiable, show that $P(2) \cdot P'(2) = 0$.

5. A particle moves in the plane so that its velocity vector is always perpendicular to its position vector. Show that its trajectory lies on a circle.

6. A projectile is fired from the ground at an angle θ. The initial speed is k. Assume that the only force acting on the projectile is that due to gravity; this force is directed downward with magnitude mg, where m is the mass of the projectile and g is the gravitational constant. Thus, the force is the constant vector $F = (0, -mg)$. At time $t = 0$, the projectile is at the origin; its velocity then is $V(0) = (k \cos \theta, k \sin \theta)$ (Figure 1).

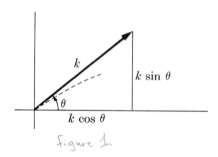

figure 1

   (a) Show that the position of the projectile at time t is given by

$$P(t) = (kt \cos \theta, -\tfrac{1}{2}gt^2 + kt \sin \theta).$$

   (b) How high does the projectile rise?

   (c) How far does it travel before hitting the ground?

   (d) At what speed will it strike the ground?

   (e) For what initial angle θ will it travel the farthest?

   (f) For given θ, what is the effect on the distance traveled if the initial speed is doubled?

Answers to problems (§9G)

1. (a) $V = (-2 \sin t, 3 \cos t)$, $|V| = \sqrt{4 + 5 \cos^2 t}$, $A = -(2 \cos t, 3 \sin t)$

   (b) $V = (2t, 2t)$, $|V| = 2\sqrt{2}\,|t|$, $A = (2, 2)$

   (c) $V = (e^t, -e^{-t})$, $|V| = \sqrt{e^{2t} + e^{-2t}}$, $A = (e^t, e^{-t})$

   (d) $V = (\cos t - t \sin t, \sin t + t \cos t)$, $|V| = \sqrt{1 + t^2}$,
       $A = (-2 \sin t - t \cos t, 2 \cos t - t \sin t)$

   (e) $V = (3, 2)$, $|V| = \sqrt{13}$, $A = (0, 0)$

   (f) $V = (2t, 3t^2)$, $|V| = |t|\sqrt{4 + 9t^2}$, $A = (2, 6t)$

2. (a) $P = (1 - \cos t, 1 + \sin t)$ (b) $P = (1 + t^2/2, t^3/3)$

   (c) $P = (1 + t, 1)$ (d) $P = (2t + t^3/6, t + 3t^2/2)$

6. (b) $(k^2/2g) \sin^2\theta$ (c) $(k^2/g) \sin 2\theta$ (d) k (e) $45°$

   (f) it is quadrupled

§9H. Length of arc

9H1. Properties of length. In this section we develop a formula for the length of a curve presented in parametric form. The result represents one of the great advantages of parametric equations over rectangular. In many instances, the parametric equations can be handled in one stroke, where the rectangular would have to be considered in several pieces, and only after a certain amount of fussing to decide what the pieces are.

We wish to find the length of a curve

$$P(t) = (f(t), g(t))$$

over a closed interval $a \le t \le b$. To do this we assume that P has a continuous derivative (as is usually the case). This is because our methods lead to an integral involving f' and g'.

We proceed in familiar fashion. Let L_u^v denote the length we are trying to define, between $t = u$ and $t = v$. Of course, arc length is to be additive:

(A) $$L_a^u + L_u^v = L_a^v.$$

Now we seek bounds for L_u^v. Imagine walking across a field, edging eastward and northward as you go. [This means that $dx/dt \ge 0$ and $dy/dt \ge 0$.] Your friend does the same, but at each instant edges faster than you in both directions. [His dx/dt and dy/dt are both greater than yours.] Clearly, he walks farther than you in any given length of time.

By considering absolute values we include south as well as north and west as well as east. Our formulation is:

(1)   If $|f_1'(t)| \ge |f_2'(t)|$ and $|g_1'(t)| \ge |g_2'(t)|$, for every t, then the arc $(f_1(t), g_1(t))$ is longer than $(f_2(t), g_2(t))$.

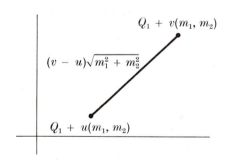

$$Q_1 + v(m_1, m_2)$$

$$(v - u)\sqrt{m_1^2 + m_2^2}$$

$$Q_1 + u(m_1, m_2)$$

FIGURE 1

We shall adopt this as a principle. Actually, we need it only when one or the other of the arcs is a straight line segment, say $Q(t) = Q_1 + t(m_1, m_2)$. Here $(dx/dt, dy/dt)$ is simply the direction vector, (m_1, m_2). See Figure 1. Notice that the length of the segment from $t = u$ to $t = v$ is

(2) $$(v - u)\sqrt{m_1^2 + m_2^2}.$$

9H2. Definition of arc length. Our assumption (1) leads easily to a definition and formula for length. By (1), the arc from u to v is shorter than a segment with direction vector

$$(\max_{[u,v]} |f'|, \max_{[u,v]} |g'|).$$

(Since f' and g' are continuous, we know these maxima exist.) Consequently $\big($see (2)$\big)$,

(B) $$L_u^v \leq (v - u)\sqrt{(\max_{[u,v]}|f'|)^2 + (\max_{[u,v]}|g'|)^2}.$$

Similarly, L_u^v is bounded from below by the corresponding expression in $\min_{[u,v]} f'$ and $\min_{[u,v]} g'$. The two together give us condition **(B)** (§6A) with respect to the function

$$\sqrt{f'^2 + g'^2}.$$

We were led to this from the principle (1). We did not prove this principle but only assumed it. No one is obliged to accept it. But, by the General Theorem for the Integral (§6A), anyone who does assume (1) $\big($and **(A)**$\big)$ is now forced to define the length of the curve

$$P(t) = \big(f(t), g(t)\big)$$

on $[a, b]$ as

(3)
$$L = \int_a^b \sqrt{f'^2 + g'^2}.$$

Briefly, $L = \int_a^b |P'|$.

In case a function is given in the form $y = \varphi(x)$, we may represent it parametrically as $x = t$, $y = \varphi(t)$. Then the definition given here is in agreement with the one in §6D2.

Let us denote the length of the curve between a and t by s. Then

(4)
$$s = \int_a^t |P'|.$$

[handwritten annotations:]
$P(t) = (\cos t, \sin t)$
$0 \le t \le 2\pi$
$P'(t) = (-\sin t, \cos t)$
$|P'(t)| = \sqrt{(-\sin t)^2 + (\cos t)^2}$
$|P'(t)| = 1$
Arc length $= \int_0^{2\pi} 1\,dt = 2\pi$

By the Fundamental Theorem of Calculus,

(5)
$$\frac{ds}{dt} = |P'(t)|.$$

Example 1. Find the length of the curve

$$P(t) = (t^3, t^2) \qquad\qquad (0 \le t \le 2).$$

Solution. This is the curve $y = x^{2/3}$ discussed in §6D2, Example 2. Here we have

$$x'^2 + y'^2 = (3t^2)^2 + (2t)^2 = t^2(9t^2 + 4)$$

and so

$$L = \int_0^2 t\sqrt{9t^2 + 4}\ dt = \tfrac{1}{27}(9t^2 + 4)^{3/2}\Big|_0^2 = \tfrac{8}{27}(10\sqrt{10} - 1).$$

Example 2. Find the length of the arch of the cycloid

$$x = a(t - \sin t), \qquad y = a(1 - \cos t) \qquad (0 \le t \le 2\pi)$$

(§9D, Example 4).

Solution. Here

$$x' = a(1 - \cos t) \qquad \text{and} \qquad y' = a \sin t \qquad P'(t) = (x', y')$$

so that

$$|P'(t)|^2 = x'^2 + y'^2 = 2a^2(1 - \cos t) = 4a^2 \sin^2 \frac{t}{2}$$

(see §8B(9)). Since $a > 0$, and since $\sin(t/2) \geq 0$ for $0 \leq t \leq 2\pi$, we have

$$\sqrt{4a^2 \sin^2 \frac{t}{2}} = 2a \sin \frac{t}{2} = |P'(t)| \quad (0 \leq t \leq 2\pi).$$

Consequently,

$$L = 2a \int_0^{2\pi} \sin \frac{t}{2}\, dt = 2a \left(-2 \cos \frac{t}{2}\right)\Big|_0^{2\pi} = 8a.$$

This answer seems reasonable, as it is a respectable amount greater than the horizontal distance, $2\pi a$.

9H3. Motion. If $P(t)$ represents position of a particle at time t, then $P'(t)$ is the velocity vector and $|P'(t)|$ is the speed at time t (§9G1). Thus, (4) states that the distance traveled is the integral of the speed, and (5) says that the speed is the derivative of the distance.

Example 3. A particle moves in such a way that the distance traveled up to time t is te^t $(t \geq 0)$. What is its speed?

Solution. Its speed at any instant t is

$$\frac{d}{dt}(te^t) = (1 + t)e^t.$$

Example 4. A particle travels according to the law

$$x = \cos 2t, \qquad y = \sin 2t \qquad\qquad (0 \leq t \leq 2\pi).$$

What distance does it cover?

Solution. We have

$$x'^2 + y'^2 = (-2 \sin 2t)^2 + (2 \cos 2t)^2 = 4.$$

Hence

$$L = \int_0^{2\pi} \sqrt{4} = 4\pi.$$

The particle winds around the circle twice.

Problems (§9H)

1. Find the lengths of the following curves.

   (a) $P = (2 \cos^3 t,\ 2 \sin^3 t),\quad 0 \leq t \leq \pi/2$
   (b) $P = (e^t \cos t,\ e^t \sin t),\quad 0 \leq t \leq \pi$
   (c) $P = 2(\cos t + t \sin t,\ \sin t - t \cos t),\quad 0 \leq t \leq 2$
   (d) $P = (2t - 1,\ t + 2),\quad 0 \leq t \leq 3$

   (e) $P = \left(t,\ \dfrac{t^3}{6} + \dfrac{1}{2t}\right),\quad 1 \leq t \leq 4$

2. $P(t)$ represents the position of a particle at time t. Find the distance traveled over the stated time interval.

(a) $P(t) = (t^2, 1 - t^2), \quad 0 \leq t \leq 1$

(b) $P(t) = (t, t^2), \quad 0 \leq t \leq 1$ *Hint* $t = \tan x \quad dt = \frac{1}{2}\sec^3 x \, dy$

(c) $P(t) = (\cos^2 t, \sin^2 t), \quad 0 \leq t \leq \pi/2$

(d) $P(t) = (\log\sqrt{1 + t^2}, \arctan t), \quad 0 \leq t \leq 1$

3. A particle moves according to the stated law, in which s represents the distance traveled up to time t. When is the speed a maximum or minimum?

(a) $s = (t - 2)^3 + t + 8 \; (t \geq 0)$ (c) $s = 1 - (2t + 1)e^{-2t} \; (t \geq 0)$

(b) $s = 1 + t - \cos t$

Answers to problems ($\S 9H$)

1. (a) 3   (b) $\sqrt{2}\,(e^\pi - 1)$ (c) 4   (d) $3\sqrt{5}$ (e) $\frac{8.7}{8}$

2. (a) $\sqrt{2}$ (b) $\frac{1}{2}\sqrt{5} + \frac{1}{4}\log(2 + \sqrt{5})$ (c) $\sqrt{2}$ (d) $\log(1 + \sqrt{2})$

3. (a) minimum at $t = 2$

(b) maximum at $t = \pi/2, 5\pi/2, \cdots$; minimum at $t = 3\pi/2, 7\pi/2, \cdots$

(c) maximum at $t = \frac{1}{2}$

§9I. Curvature

9I1. Introduction. In this section and the next we study the geometry of curves. Our object of study is a curve

$$(1) \qquad\qquad P(t) = \big(f(t), g(t)\big),$$

where t ranges over some interval I. We assume that P is twice differentiable on I and that P' is never zero in the interior.

The first assumption implies that P has a continuous derivative and hence that, between any two points, the curve has a length which we can compute. Pick a reference point $t = a$. According to §9H(4), the length, s, from a to any parametric value t, is

$$(2) \qquad\qquad s = \int_a^t \sqrt{f'^2 + g'^2}.$$

This is *directed* arc length. For $t < a$ we get $s < 0$.

The second assumption means that there is a tangent vector at each point in the interior of I. It also means that t is a differentiable function of s there. For, from (2),

$$(3) \qquad\qquad \frac{ds}{dt} = \sqrt{f'(t)^2 + g'(t)^2} > 0 \qquad (t \in \text{interior of } I).$$

By the Inverse-Function Theorem (§4G),

$$(4) \qquad \frac{dt}{ds} = \frac{1}{\sqrt{f'(t)^2 + g'(t)^2}} \qquad (t \in \text{interior of } I).$$

9I2. Definition of curvature. What are our intuitive feelings about the "curvature" of a curve? Somehow, it should express how fast you are turning compared with how far you are traveling. We all agree that a *circle* has *constant* curvature. And a small circle has greater curvature than a large one (Figure 1).

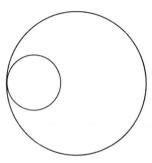

FIGURE 1

The curvature of a parabola (for example) is *changing* from point to point. It is greatest at the vertex and decreases steadily as we go out from the vertex (Figure 2). We are speaking of curvature *at a point*. Of course this will be defined as a derivative of some sort.

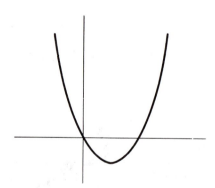

FIGURE 2

Back to our curve $P(t)$ (1). Let α denote the angle from the positive x-axis to the tangent vector, $P'(t)$. Then $d\alpha/ds$ is the rate at which the tangent

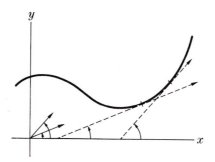

FIGURE 3

is turning per unit change in s, the length along the arc. (Figure 3). Hence $|d\alpha/ds|$ is the rate of turning without regard to left or right. We adopt this as our definition of curvature. Thus, we define the *curvature* κ (kappa) by:

(5)
$$\kappa = \left|\frac{d\alpha}{ds}\right|.$$

For some purposes it is more convenient to deal with the reciprocal. We define ρ (rho), the *radius of curvature*, by:

(6)
$$\rho = \frac{1}{\kappa} = \left|\frac{ds}{d\alpha}\right| \qquad\qquad (\kappa \neq 0).$$

The next example gives evidence that the definition is aptly chosen.

Example 1. Find the radius of curvature of a circle of radius a.

Solution. With the center at the origin, the parametric equations are

$$x = a\cos t, \qquad y = a\sin t \qquad\qquad (0 \leq t \leq 2\pi).$$

Here $s = at$ (Figure 4), so that $ds/dt = a$.

Since the tangent makes a constant angle (namely, 90°) with the radius, α changes at the same rate as the central angle, t. Hence $dt/d\alpha = 1$ and we have

$$\frac{ds}{d\alpha} = \frac{ds}{dt}\frac{dt}{d\alpha} = a.$$

The radius of curvature of the circle is the radius of the circle.

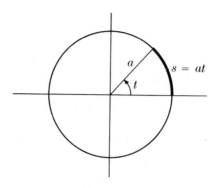

FIGURE 4

9I3. Formula for curvature. We cannot ordinarily compute ρ or κ as easily as for the circle. We want formulas directly in terms of the original data (1). We will prove:

(7)
$$\kappa = \frac{|f'g'' - g'f''|}{(f'^2 + g'^2)^{3/2}}$$
($t \in$ interior of I.)

Now, $\kappa = |d\alpha/ds|$ and $d\alpha/ds = (d\alpha/dt)(dt/ds)$. Hence to get κ, we evaluate $d\alpha/dt$ and dt/ds. We already know dt/ds from (4): $dt/ds = 1/\sqrt{f'^2 + g'^2}$. Hence (7) will follow if we can show that

(8)
$$\frac{d\alpha}{dt} = \frac{f'g'' - g'f''}{f'^2 + g'^2}$$
($t \in$ interior of I).

Proof of (8). First consider any point where $f'(t) > 0$ (Figure 5). Then $-\pi/2 < \alpha < \pi/2$ and

(9)
$$\alpha = \arctan \frac{g'}{f'}.$$

Hence

(10)
$$\frac{d\alpha}{dt} = \frac{1}{1 + (g'/f')^2} \frac{d}{dt}\left(\frac{g'}{f'}\right),$$

which simplifies to (8).

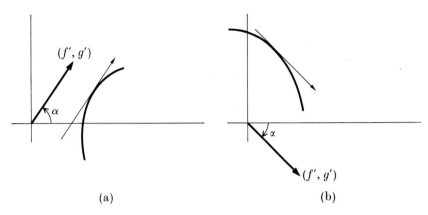

FIGURE 5

In case $f'(t) < 0$, we replace α by $\beta = \alpha - \pi$ (Figure 6(a)) or $\beta = \alpha + \pi$ (Figure 6(b)), so that $-\pi/2 < \beta < \pi/2$. Since

(11) $$\frac{d\alpha}{d\beta} = 1 \quad \text{and} \quad \beta = \arctan \frac{g'}{f'},$$

we again get (8).

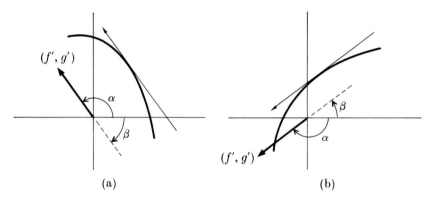

FIGURE 6

Finally, if $f'(t) = 0$, then, by assumption, $g'(t) \neq 0$. We replace α by $\beta = \pi/2 - \alpha$ if $g'(t) > 0$ (Figure 7), and by $\beta = -\pi/2 - \alpha$ if $g'(t) < 0$ (Figure 8). Then for nearby values of t, $-\pi/2 < \beta < \pi/2$. In these cases we have

(12) $$\frac{d\alpha}{d\beta} = -1 \quad \text{and} \quad \beta = \arctan \frac{f'}{g'}.$$

Again we get (8).

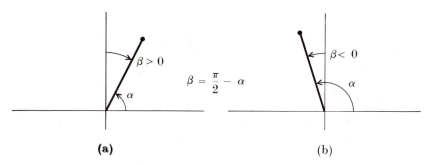

(a) (b)

FIGURE 7

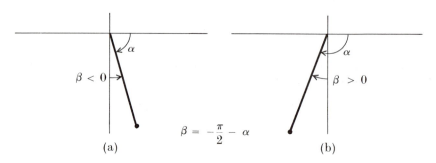

(a) (b)

FIGURE 8

This establishes the formula for κ (7). The corresponding formula for ρ is, of course,

(13)
$$\rho = \frac{(f'^2 + g'^2)^{3/2}}{|f'g'' - g'f''|} \qquad (f'g'' - g'f'' \neq 0)$$

Example 2. Analyze the curvature along the cycloid

$$x = a(t - \sin t), \qquad y = a(1 - \cos t) \qquad (0 \leq t \leq 2\pi)$$

(§9D, Example 4).

Discussion. We have

$$x' = a(1 - \cos t), \qquad y' = a \sin t.$$

We note that $P' = (x', y') = (0, 0)$ at $t = 0$ and $t = 2\pi$ but that $|P'| \neq 0$ for $0 < t < 2\pi$. Next,

$$x'' = a \sin t, \qquad y'' = a \cos t.$$

Therefore,

$$x'^2 + y'^2 = 2a^2(1 - \cos t), \qquad x'y'' - y'x'' = a^2(\cos t - 1).$$

Consequently,

$$\rho = \frac{(x'^2 + y'^2)^{3/2}}{|x'y'' - y'x''|} = 2\sqrt{2}\, a\sqrt{1 - \cos t} \qquad (0 < t < 2\pi).$$

We see that ρ is an increasing function on $(0, \pi]$, with the value $4a$ at $t = \pi$. Thus, at the flattest part, $t = \pi$, the cycloid is bending at the same rate as a circle of radius $4a$, 4 times that of the rolling circle. As $t \to 0$, $\rho \to 0$. In other words, as we go backwards along the curve toward the starting point, it winds more and more tightly, becoming tighter than any given circle, no matter how small.

The behavior on $[\pi, 2\pi)$ is symmetric to that on $(0, \pi]$.

914. y as a function of x. For a curve given in the form $y = \varphi(x)$, formulas (7) and (13) reduce to

(14)
$$\kappa = \frac{|y''|}{(1 + y'^2)^{3/2}}, \qquad \rho = \frac{(1 + y'^2)^{3/2}}{|y''|}.$$

To derive these we simply take $t = x$. We then have $x = f(t) = t$, $y = g(t) = \varphi(t)$. For a curve given in the form $x = \psi(y)$, one can write down corresponding formulas.

Example 3. Discuss the curvature of the parabola $y = x^2$.

Discussion. Here $y' = 2x$, $y'' = 2$, and

$$\rho = \tfrac{1}{2}(1 + 4x^2)^{3/2}.$$

This shows analytically (i.e., by mathematical analysis) that ρ is a minimum at the vertex and increases as we move away from it in either direction.

The minimum value, at $x = 0$, is $\rho = \tfrac{1}{2}$.

The point $(2, 4)$ doesn't seem very far out along the curve. Would you care to guess the value of ρ there?

For $x = 2$,

$$\rho = \tfrac{1}{2}(1 + 16)^{3/2} > \tfrac{1}{2} \times 16^{3/2} = 32.$$

The curve at that point is flatter than a circle of radius 32(!).

Problems (§9I)

1. Find the radius of curvature of the given curve at the given value of t.
   (a) $P = (t^2, t^3), \quad t = 2$
   (b) $P = (e^t, e^{-t}), \quad t = 0$
   (c) $P = (2 \cos t, \sin t), \quad t = 0$
   (d) $P = (\tan t, \cot t), \quad t = \pi/4$

2. Find the curvature (as a function of t).
   (a) $P = (e^t \sin t, e^t \cos t)$ (c) $P = (t^2, \log t)$
   (b) $P = (2 \cos^3 t, 2 \sin^3 t)$ (d) $P = (4 \cos t, 2 \sin t)$

3. Find the curvature (as a function of x).
   (a) $y = \sin x$ (d) $y = \log \sec x$

   (b) $y = e^x$ (e) $y = x + \dfrac{1}{x}$

   (c) $y = \sqrt{x}$ (f) $y = \sqrt{4 - x^2}$

4. Find the curvature at the given point.
   (a) $y^2 = x^3$, $\quad$ (1, 1)   (c) $x^{1/2} + y^{1/2} = 2$, $\quad$ (1, 1)
   (b) $y^2 = x^3 + 8$, $\quad$ (1, 3)   (d) $(y - 1)^2 = 4x$, $\quad$ (4, 5)

5. Find the maximum and minimum values (if any) of the curvature.
   (a) $y = e^x$
   (b) $y = \sin x$ $\quad (0 \le x \le 2\pi)$

   (c) $y = \dfrac{1}{x}$

   (d) $y = \log x$
   (e) $P = (a \cos t, b \sin t)$ $\quad (0 \le t \le 2\pi)$, $\quad$ where $a > b > 0$
   (f) $y = \frac{1}{3}x^3$

6. What can be said about the curvature of $y = f(x)$
   (a) at an interior maximum or minimum of f?
   (b) at a point of inflection?

7. Find the curvature at the local maxima and minima of the given functions.

   (a) $y = \sin x$ (c) $y = \dfrac{1}{\sqrt{1 - x^2}}$

   (b) $y = 1 - x^2$ (d) $y = x + \dfrac{1}{x}$

8. Show that the maximum value of the curvature of $y = x^3 - 3x$ does not occur at a local maximum or minimum of the function.

Answers to problems (§91)

1. (a) $\frac{8}{3}10^{3/2}$ $\quad$ (b) $\sqrt{2}$ $\quad$ (c) $\frac{1}{2}$ $\quad$ (d) $\sqrt{2}$

2. (a) $\frac{1}{2}\sqrt{2}\, e^{-t}$ $\quad$ (b) $\frac{1}{3}|\csc 2t|$ $\quad$ (c) $\dfrac{4t^2}{(1 + 4t^4)^{3/2}}$

   (d) $\dfrac{1}{(1 + 3 \sin^2 t)^{3/2}}$

3. (a) $\dfrac{|\sin x|}{(1 + \cos^2 x)^{3/2}}$ $\quad$ (b) $\dfrac{e^x}{(1 + e^{2x})^{3/2}}$ $\quad$ (c) $\dfrac{2}{(1 + 4x)^{3/2}}$

   (d) $\cos x$ $\quad$ (e) $\dfrac{2|x^3|}{[x^4 + (x^2 - 1)^2]^{3/2}}$ $\quad$ (f) $\frac{1}{2}$

4. (a) $\frac{6}{169}\sqrt{13}$ (b) $\frac{22}{75}\sqrt{5}$ (c) $\frac{1}{4}\sqrt{2}$ (d) $\frac{1}{50}\sqrt{5}$

5. (a) max $\kappa = \frac{2}{9}\sqrt{3}$ at $x = -\frac{1}{2}\log 2$
   (b) max $\kappa = 1$ at $x = \pi/2, 3\pi/2$;   min $\kappa = 0$ at $x = 0, \pi, 2\pi$
   (c) max $\kappa = \frac{1}{2}\sqrt{2}$ at $x = 1, -1$
   (d) max $\kappa = \frac{2}{9}\sqrt{3}$ at $x = 1/\sqrt{2}$
   (e) max $\kappa = a/b^2$ at $t = 0, \pi, 2\pi$;   min $\kappa = b/a^2$ at $t = \pi/2, 3\pi/2$
   (f) max $\kappa = \frac{1}{18}5^{5/4}6^{1/2}$ at $x = 5^{-1/4}$;   min $\kappa = 0$ at $x = 0$

7. (a) 1   (b) 2   (c) 1   (d) 2

§9J. Unit vectors and circle of curvature

9J1. *The unit tangent vector.* Let

$$(1) \qquad P(t) = \big(f(t), g(t)\big)$$

be as in §9I. For a finer analysis of curvature and related ideas we introduce certain unit vectors. The *unit tangent vector* is the tangent vector of length 1. It is denoted by T. Thus,

$$(2) \qquad T = \frac{P'(t)}{|P'(t)|} = \frac{1}{|P'|}(f', g').$$

Let α denote, as heretofore, the angle from the positive x-axis to T. Then, evidently,

$$(3) \qquad T = (\cos\alpha, \sin\alpha)$$

and $\big($see $(2)\big)$

$$(4) \qquad \cos\alpha = \frac{f'}{|P'|}, \qquad \sin\alpha = \frac{g'}{|P'|}.$$

Example 1. Find the unit tangent vector to the parabola $P(t) = (t, t^2)$ at any point.

Solution. Here $P' = (1, 2t)$ and $|P'| = \sqrt{1 + 4t^2}$. Therefore,

$$T = \frac{P'}{|P'|} = \frac{1}{\sqrt{1 + 4t^2}}(1, 2t).$$

9J2. *The unit normal vector.* From (3) we get

$$(5) \qquad \frac{dT}{dt} = \frac{dT}{d\alpha}\frac{d\alpha}{dt} = \frac{d\alpha}{dt}(-\sin\alpha, \cos\alpha).$$

Where $d\alpha/dt \neq 0$, this is a non-zero vector. Clearly, it is perpendicular to T. The unit vector in the direction of dT/dt is called the *unit normal vector*, denoted by N. ("Normal" is another word for perpendicular.) Therefore,

(6)
$$N = \frac{dT/dt}{|dT/dt|} = \frac{d\alpha/dt}{|d\alpha/dt|} (-\sin \alpha, \cos \alpha).$$

Thus,

(7)
$$N = (-\sin \alpha, \cos \alpha) \quad \text{if} \quad \frac{d\alpha}{dt} > 0,$$

$$N = (\sin \alpha, -\cos \alpha) \quad \text{if} \quad \frac{d\alpha}{dt} < 0.$$

We do not define N at a point where $d\alpha/dt = 0$.

It is desirable to express N in terms of the original data (1). Formula (8) of §9I shows that $d\alpha/dt$ has the same sign as $f'g'' - g'f''$. Therefore (see (4) and (7)),

(8)
$$N = \frac{1}{|P'|} (-g', f') \quad \text{if} \quad f'g'' - g'f'' > 0,$$

$$N = \frac{1}{|P'|} (g', -f') \quad \text{if} \quad f'g'' - g'f'' < 0.$$

Finally, N is not defined at any point where $f'g'' - g'f'' = 0$.

Example 2. Find the unit normal vector to the parabola $P(t) = (t, t^2)$ at any point.

Solution. Here $P' = (f', g') = (1, 2t)$ and $P'' = (0, 2)$. Then $f'g'' - g'f'' = 2 > 0$. By (8),

$$N = \frac{1}{|P'|} (-g', f') = \frac{1}{\sqrt{1 + 4t^2}} (-2t, 1).$$

Example 3. Find the unit tangent and normal vectors to the cubic $P(t) = (t, t^3)$ at any point.

Solution. We have $P' = (f', g') = (1, 3t^2)$ and $P'' = (0, 6t)$. Then

$$T = \frac{1}{\sqrt{1 + 9t^4}} (1, 3t^2).$$

Next, $f'g'' - g'f'' = 6t$. By (8),

$$N = \frac{1}{\sqrt{1 + 9t^4}}(-3t^2, 1) \qquad \text{when} \qquad t > 0$$

and

$$N = \frac{1}{\sqrt{1 + 9t^4}}(3t^2, -1) \qquad \text{when} \qquad t < 0.$$

When $t = 0$, N is not defined.

9J3. Direction of the normal. We now show that the unit normal vector N is always directed *toward the concave side* of the curve.

When $d\alpha/dt > 0$, α is increasing with t. As we move in the direction of increasing t, the curve is bending counterclockwise (Figure 1). In this case, by (7),

$$(9) \qquad N = (-\sin\alpha, \cos\alpha) = \big(\cos(\alpha + \pi/2), \sin(\alpha + \pi/2)\big).$$

So N is directed 90° counterclockwise from T. Therefore N is directed toward the concave side of the curve (Figure 1).

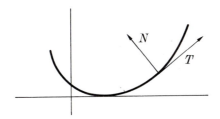

FIGURE 1

If $d\alpha/dt < 0$ then α is decreasing with increasing t, the curve bends clockwise,

$$(10) \qquad N = (\sin\alpha, -\cos\alpha) = \big(\cos(\alpha - \pi/2), \sin(\alpha - \pi/2)\big),$$

and N is directed 90° clockwise from T (Figure 2). Again, N is directed toward the concave side.

If the curve has no concave side then $d\alpha/dt = 0$ (and hence $\kappa = 0$) and N is not defined. This is the situation at an inflection point or along a straight-line segment.

Example 4. In Example 2 we saw that the normal vector to the parabola $P(t) = (t, t^2)$ is in the direction $(-2t, 1)$. Hence for $t > 0$ it is directed to the

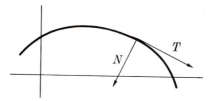

FIGURE 2

left and for $t < 0$ it is directed to the right (and, for $t = 0$, straight up). In all cases it points toward the concave side of the parabola (Figure 3).

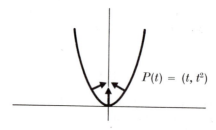

FIGURE 3

Example 5. In Example 3 we saw that the normal vector to the cubic $P(t) = (t, t^3)$ is in the direction $(-3t^2, 1)$ for $t > 0$ and in the direction $(3t^2, -1)$ for $t < 0$. In both cases this is toward the concave side of the curve (Figure 4). At $t = 0$ there is an inflection point and N is not defined.

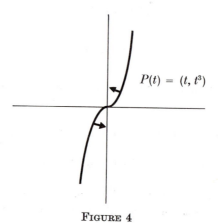

FIGURE 4

9J4. Circle of curvature. By the *circle of curvature* to the curve P at a point t is meant that circle that is tangent to the curve at t, has the same radius of curvature, and lies on the concave side. See Figure 5. Just as the tangent approximates the curve by matching its direction at a point, the circle of curvature approximates the curve by matching both its direction and its rate of turning.

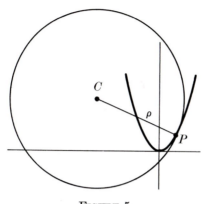

FIGURE 5

The circle of curvature is easily located. Its radius is ρ. To get its center, start at P and move in the direction of N a distance ρ. Thus, the center is the point

$$(11) \qquad\qquad C = P + \rho N.$$

It is called the *center of curvature* (Figure 5).

Example 6. Find the circle of curvature to the parabola $P(t) = (t, t^2)$ at the points $(0, 0)$ and $(2, 4)$.

Solution. Recapitulating several earlier examples, we have:

$$P' = (f', g') = (1, 2t), \qquad P'' = (0, 2), \qquad f'g'' - g'f'' = 2,$$

$$\rho = \tfrac{1}{2}(1 + 4t^2)^{3/2}, \qquad N = \frac{1}{\sqrt{1 + 4t^2}}(-2t, 1).$$

The center of curvature is the point

$$C = P + \rho N = (t, t^2) + (1 + 4t^2)(-t, \tfrac{1}{2}) = (-4t^3, \tfrac{1}{2} + 3t^2).$$

At $t = 0$ we have $\rho = \tfrac{1}{2}$ and $C = (0, \tfrac{1}{2})$. The circle of curvature is $x^2 + (y - \tfrac{1}{2})^2 = \tfrac{1}{4}$. At $t = 2$ we have $\rho = \tfrac{1}{2} \cdot 17^{3/2}$ and $C = (-32, 12\tfrac{1}{2})$. The circle is

$$(x + 32)^2 + (y - \tfrac{25}{2})^2 = \tfrac{1}{4} \cdot 17^3.$$

Example 7. Find the circle of curvature to the cubic $P(t) = (t, t^3)$ at the point $(1, 1)$.

Solution. We have

$$P' = (f', g') = (1, 3t^2), \qquad P'' + (0, 6t), \qquad f'g'' - g'f'' = 6t.$$

At $t = 1$,

$$P' = (1, 3), \qquad f'g'' - g'f'' = 6, \qquad \rho = \tfrac{1}{6} \times 10^{3/2},$$

and

$$N = \frac{1}{\sqrt{10}}\,(-g', f') = \frac{1}{\sqrt{10}}\,(-3, 1).$$

The center of curvature is

$$C = P + \rho N = (1, 1) + \tfrac{1}{6} \times 10^{3/2}\,\frac{1}{\sqrt{10}}\,(-3, 1) = (-4, \tfrac{8}{3}).$$

The circle is

$$(x + 4)^2 + (y - \tfrac{8}{3})^2 = \tfrac{1000}{36}.$$

9J5. Remark on the definition of curvature. Since

$$\frac{dT}{ds} = \frac{dT}{d\alpha}\frac{d\alpha}{ds}$$

and $dT/d\alpha$ is a unit vector $\big($namely, $(-\sin \alpha, \cos \alpha)\big)$, we have $|dT/ds| = |d\alpha/ds|$. Therefore,

$$(12) \qquad\qquad \kappa = \left|\frac{dT}{ds}\right|.$$

One could take (12) as a definition of curvature. Since T has constant length, only its direction changes with s; $|dT/ds|$ tells us how fast. One can then proceed from (12) directly to the formula for κ in terms of f and g (formula (7) of §9I), without ever mentioning the angle α. This is the method used in three dimensions (§11G below).

Problems (§9J)

1. Establish the formula:

$$\rho N = \frac{f'^2 + g'^2}{f'g'' - g'f''}\,(-g', f').$$

2. Find the unit tangent, unit normal, and center of curvature as functions of t.
Write the equation of the circle of curvature corresponding to the given value of t.

   (a) $P(t) = (t, \sin t)$ $(0 < t < \pi)$; $t = \pi/2$

   (b) $P(t) = (\cos t, \sin t)$; $t = \pi/4$

   (c) $P(t) = (2 \cos^3 t, 2 \sin^3 t)$ $(0 < t < \pi/2)$; $t = \pi/4$

   (d) $P(t) = (e^t, e^{-t})$; $t = 0$

   (e) $P(t) = \left(\dfrac{1}{1+t}, \dfrac{1}{1-t} \right)$ $(-1 < t < 1)$; $t = 0$

3. Express the center of curvature as a function of x.

   (a) $y = \sqrt{x}$ (b) $y = e^x$ (c) $y = \sqrt{1 - x^2}$

4. Write a vector equation for the center of curvature of the cycloid $P(\theta) = (\theta - \sin \theta, 1 - \cos \theta)$.

Answers to problems (§9J)

2. (a) with the abbreviation $a = \sqrt{1 + \cos^2 t}$: $T = (1/a)(1, \cos t)$,
   $N = (1/a)(\cos t, -1)$, $C = (t, \sin t) + a^2(\cot t, -\csc t)$;
   $(x - \pi/2)^2 + y^2 = 1$

   (b) $T = (-\sin t, \cos t)$, $N = -(\cos t, \sin t)$, $C = (0, 0)$; $x^2 + y^2 = 1$

   (c) $T = (-\cos t, \sin t)$, $N = (\sin t, \cos t)$,
   $C = 2(\cos^3 t, \sin^3 t) + 3 \sin 2t(\sin t, \cos t)$;
   $(x - 2\sqrt{2})^2 + (y - 2\sqrt{2})^2 = 9$

   (d) with the abbreviation $a = \sqrt{e^{2t} + e^{-2t}}$: $T = (1/a)(e^t, -e^{-t})$,
   $N = (1/a)(e^{-t}, e^t)$, $C = (e^t, e^{-t}) + \tfrac{1}{2}a^2(e^{-t}, e^t)$;
   $(x - 2)^2 + (y - 2)^2 = 2$

   (e) with the abbreviation $a = \sqrt{2(1 + 6t^2 + t^4)}$:

   $$T = \frac{1}{a}(-(1 - t)^2, (1 + t)^2),$$

   $$N = \frac{1}{a}((1 + t)^2, (1 - t)^2),$$

   $$C = \left(\frac{1}{1+t}, \frac{1}{1-t} \right) + \frac{a^2}{4(1 - t^2)^3}((1 + t)^2, (1 - t)^2);$$

   $(x - \tfrac{3}{2})^2 + (y - \tfrac{3}{2})^2 = \tfrac{1}{2}$

3. (a) $(3x + \tfrac{1}{2}, -4x\sqrt{x})$ (b) $(x - 1 - e^{2x}, 2e^x + e^{-x})$ (c) $(0, 0)$

4. $C(\theta) = (\theta + \sin \theta, \cos \theta - 1)$

CHAPTER 10

Polar Coordinates

In this chapter we apply calculus to study tangents, area, and arc length in polar coordinates.

§10A. Polar coordinates

10A1. Polar coordinates. It is often convenient to locate a point directly in terms of its distance and direction from the origin. Direction is specified by an angle, θ, measured counterclockwise from the positive x-axis. Distance is given as the directed distance, r, measured along the terminal side of the angle. The numbers r and θ—which are written as a pair, in that order:

$$(r, \theta)$$

—are called the *polar* coordinates of the point.

A point has many pairs of polar coordinates. The point Q in Figure 1 has polar coordinates $(2, \pi/6)$. It also has polar coordinates $(2, \pi/6 + 2\pi)$, $(2, \pi/6 - 4\pi)$, etc. Any multiple of 2π added to the θ-coordinate of a point yields another θ-coordinate of the same point.

Still another pair of polar coordinates for Q is $(-2, \pi/6 + \pi)$. Negative r signifies directed distance measured in the opposite direction.

The value $r = 0$ specifies the origin, irrespective of the value of θ. For instance, $(0, 0)$, $(0, \pi/3)$, $(0, -\pi/2)$ are all polar coordinates of the origin.

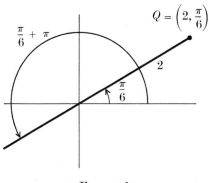

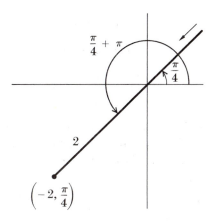

Figure 1

Example 1. Locate the point with polar coordinates $(-2, \pi/4)$.

Solution. We mark off an angle $\pi/4$, then move 2 units the "wrong" way. In other words, we mark off an angle $\pi/4$, then an additional π, and then measure off 2 units (Figure 2).

Figure 2

10A2. Relation to rectangular coordinates. There are simple relations between polar and rectangular coordinates (Figure 3).

If the polar coordinates of a point are (r, θ), then its rectangular coordinates are

(1) $x = r \cos \theta, \quad y = r \sin \theta.$

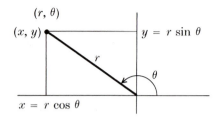

FIGURE 3

To go from rectangular to polar coordinates is more involved. Given the rectangular coordinates (x, y), we find all polar coordinates (r, θ) from the relations

(2) $$r^2 = x^2 + y^2, \qquad \cos \theta = \frac{x}{r}, \qquad \sin \theta = \frac{y}{r}.$$

When x and y are not both zero, this is done as follows. First find

(3) $$|r| = \sqrt{x^2 + y^2}.$$

One collection of pairs (r, θ) is then determined by

(4) $$r = |r|, \qquad \cos \theta = \frac{x}{|r|}, \qquad \sin \theta = \frac{y}{|r|}.$$

The values of $\cos \theta$ and $\sin \theta$ thus obtained determine a unique θ in the range $0 \le \theta < 2\pi$. For the other solutions, add or subtract 2π, 4π, 6π, etc.

The remaining pairs (r, θ) are those satisfying

(5) $$r = -|r|, \qquad \cos \theta = -\frac{x}{|r|}, \qquad \sin \theta = -\frac{y}{|r|}.$$

To find these values of θ, simply add or subtract π, 3π, 5π, etc. to any one of the values from (4).

Finally, if x and y are both 0, then $r = 0$ and θ may be anything at all.

Remark. The conversion to polar coordinates can be expressed more concisely by:

(6) $$r^2 = x^2 + y^2,$$

(7) $$\theta = \arctan y/x \quad \text{(if } x \ne 0\text{)}, \qquad \theta = \pi/2 \quad \text{(if } x = 0\text{)}.$$

But then there remains another step: to choose the sign of r consistent with the choice of θ.

Example 2. Find polar coordinates for the point with rectangular coordinates $(\frac{1}{2}, \frac{1}{2}\sqrt{3})$.

Solution. Here $r^2 = \frac{1}{4} + \frac{3}{4} = 1$. Choosing $r = 1$ we get

$$\cos\theta = \tfrac{1}{2}, \qquad \sin\theta = \tfrac{1}{2}\sqrt{3},$$

so that $\theta = \pi/3$ plus any multiple of 2π $\big($Figure 4(a)$\big)$. Choosing $r = -1$, we get

$$\cos\theta = -\tfrac{1}{2}, \qquad \sin\theta = -\tfrac{1}{2}\sqrt{3},$$

so that $\theta = \pi/3 + \pi, \pi/3 + 3\pi$, etc. $\big($Figure 4(b)$\big)$.

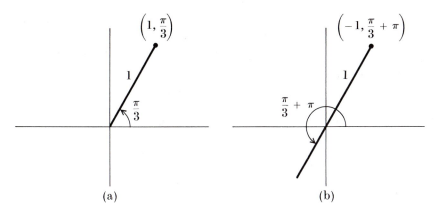

(a)   (b)

FIGURE 4

Hence polar coordinates for the point are $(1, \pi/3)$, $(-1, \pi/3 + \pi)$, or any pair obtained from either of these by adding a multiple of 2π to the θ-coordinate.

Example 3. Find polar coordinates for the point with rectangular coordinates $(-3, 4)$.

Solution. Here $r^2 = 25$. Choosing $r = 5$, we get

$$\cos\theta = -\tfrac{3}{5}, \qquad \sin\theta = \tfrac{4}{5}.$$

Let θ_1 denote the angle between $\pi/2$ and π determined by these equations. The other solutions are θ_1 plus any multiple of 2π. Note that $\theta_1 = \pi - \arctan\frac{4}{3}$ $\big($Figure 5(a)$\big)$. Choosing $r = -5$, we get

$$\cos\theta = \tfrac{3}{5}, \qquad \sin\theta = -\tfrac{4}{5},$$

so that $\theta = \theta_1 + \pi, \theta_1 + 3\pi$, etc. $\big($Figure 5(b)$\big)$.

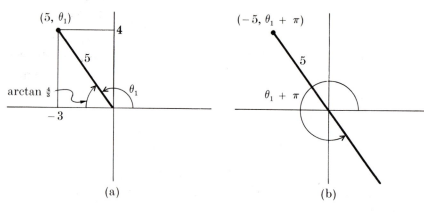

Figure 5

Problems (§10A)

1. Find the rectangular coordinates of the points with the given polar coordinates. Plot the points.

(a) $(3, 2\pi/3)$

(b) $(4, \pi/6)$

(c) $(-2, \pi/4)$

(d) $(-1, 11\pi/6)$

(e) $(1, 0)$

(f) $(0, 1)$

(g) $(2, -5\pi/3)$

(h) $(-1, -7\pi/6)$

(i) $(3, 3\pi/4)$

(j) $(13, \arctan \frac{5}{12})$

(k) $(3, 3\pi/2)$

(l) $(2, -\pi/2)$

(m) $(-1, -\pi)$

(n) $(2, \pi/2)$

(o) $(-3, -\pi/2)$

2. Find two pairs of polar coordinates for the points with the following rectangular coordinates. Choose them so that the two r-coordinates have opposite signs.

(a) $(-\sqrt{3}, 1)$

(b) $(2, -2)$

(c) $(3, \sqrt{3})$

(d) $(-1, 0)$

(e) $(0, 1)$

(f) $(-2, -2\sqrt{3})$

(g) $(\sqrt{6}, -\sqrt{2})$

(h) $(0, -3)$

(i) $(\sqrt{2}, \sqrt{2})$

(j) $(1, 2)$

(k) $(5, -12)$

(l) $(-3, -4)$

3. Find all pairs (x_0, y_0) that are both polar and rectangular coordinates for the same point.

4. A regular pentagon is inscribed in the circle $r = 3$ with one vertex on the positive x-axis. Write the polar coordinates of all the vertices.

5. (a) Show that the distance between the points with polar coordinates (r_1, θ_1) and (r_2, θ_2) is

$$\sqrt{r_1^2 + r_2^2 - 2r_1r_2 \cos(\theta_1 - \theta_2)}.$$

(b) Using (a), find the polar equation of the circle with center $(2, \pi/3)$ (polar coordinates) and radius 3.

Answers to problems (§10A)

1. (a) $(-3/2, 3\sqrt{3}/2)$ (b) $(2\sqrt{3}, 2)$ (c) $(-\sqrt{2}, -\sqrt{2})$
   (d) $(-\sqrt{3}/2, 1/2)$ (e) $(1, 0)$ (f) $(0, 0)$
   (g) $(1, \sqrt{3})$ (h) $(\sqrt{3}/2, -1/2)$ (i) $(-3\sqrt{2}/2, 3\sqrt{2}/2)$
   (j) $(12, 5)$ (k) $(0, -3)$ (l) $(0, -2)$
   (m) $(1, 0)$ (n) $(0, 2)$ (o) $(0, 3)$
2. (a) $(2, 5\pi/6), (-2, -\pi/6)$ (b) $(2\sqrt{2}, -\pi/4), (-2\sqrt{2}, 3\pi/4)$
   (c) $(2\sqrt{3}, \pi/6), (-2\sqrt{3}, 7\pi/6)$ (d) $(-1, 0), (1, \pi)$
   (e) $(1, \pi/2), (-1, 3\pi/2)$ (f) $(4, 4\pi/3), (-4, \pi/3)$
   (g) $(2\sqrt{2}, -\pi/6), (-2\sqrt{2}, 5\pi/6)$ (h) $(3, 3\pi/2), (-3, \pi/2)$
   (i) $(2, \pi/4), (-2, 5\pi/4)$ (j) $(\sqrt{5}, \arctan 2), (-\sqrt{5}, \pi + \arctan 2)$
   (k) $(13, -\arctan \frac{12}{5}), (-13, \pi - \arctan \frac{12}{5})$
   (l) $(5, \pi + \arctan \frac{4}{3}), (-5, \arctan \frac{4}{3})$
3. all points $(t, 0)$
4. $(3, 0), (3, 2\pi/5), (3, 4\pi/5), (3, 6\pi/5), (3, 8\pi/5)$
5. (b) $r^2 - 4r \cos(\theta - \pi/3) = 5$

§10B. Graphs of polar equations

10B1. Graphs. The graph of a polar equation is the set of all points for which *at least* one pair of polar coordinates satisfies the equation. To determine when a point (r_0, θ_0) lies on the graph of an equation $r = f(\theta)$ requires care, since some coordinates of the point may satisfy the equation even while others do not.

The principle is clear from the preceding discussion. For $r_0 \neq 0$, the point will be on the graph provided there is at least one integer n (positive, negative, or zero) such that either

(1) $(r_0, \theta_0 + 2n\pi)$ or $(-r_0, \theta_0 + \pi + 2n\pi)$

satisfies the equation. For $r_0 = 0$, the issue is simply whether the equation $f(\theta) = 0$ has a solution.

Example 1. Show that $(2, 3\pi/2)$ lies on the graph of $r^2 = 4 \sin \theta$.

Solution. We note that the given coordinates do not satisfy the equation:

$$2^2 \neq 4 \sin \frac{3\pi}{2} ;$$

and neither will $(2, \theta)$ for any θ differing from $3\pi/2$ by a multiple of 2π. However, $(-2, \pi/2)$ are coordinates of the same point and do satisfy the equation:

$$(-2)^2 = 4 \sin \frac{\pi}{2}.$$

Example 2. Show that the points $(2, \pi/2)$ and $(0, \pi/2)$ lie on the graph of $r = 2 \sin^2\theta$.

Solution. The point $(2, \pi/2)$ lies on the graph, because the given coordinates satisfy the equation:

$$2 = 2 \sin^2 \frac{\pi}{2}.$$

The point $(0, \pi/2)$ lies on the graph even though $0 \neq 2 \sin^2(\pi/2)$. This is because $(0, 0)$ are also coordinates for the same point and $0 = 2 \sin^2 0$.

10B2. Some examples of graphs. In this section we study the graphs of some simple polar equations. When confronted with a polar equation, one may be tempted to convert immediately to rectangular coordinates, returning to familiar ground. Sometimes this is helpful, but not always, as a simple polar equation may lead to an impossibly complicated rectangular one.

We begin with the two very simplest equations.

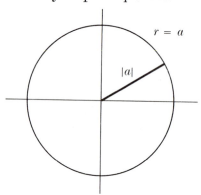

FIGURE 1

Example 3.

(a) *The circle $r = a$ (where a is a constant $\neq 0$).*

Evidently, the graph is a circle with center at the origin and radius $|a|$ (Figure 1).

(b) *The line $\theta = \beta$ (where β is a constant).*

Evidently, the graph is a line through the origin making an angle β with the x-axis (Figure 2).

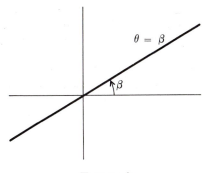

FIGURE 2

Notice that the graphs of $r = a$ and $\theta = \beta$ intersect in two points, (a, β) and $(-a, \beta)$ (Figure 3).

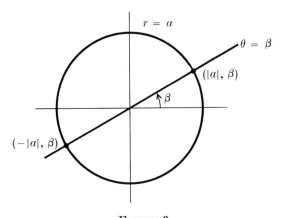

FIGURE 3

Next, we look at several examples involving trigonometric functions. Typically, $f(\theta)$ is a simple expression in terms of $\sin \theta$, $\cos \theta$, etc. In these circumstances it is sufficient to compute values of r for $0 \le \theta < 2\pi$, as all other values are repetitions of these.

Example 4. *The circle* $r = 2 \cos \theta$.

We can convert quickly to rectangular coordinates by multiplying through by r:

$$r^2 = 2r \cos \theta = 2x, \quad \text{i.e.,} \quad x^2 + y^2 = 2x,$$

which we recognize as a circle. In standard form, this is

$$(x - 1)^2 + y^2 = 1,$$

which exhibits the fact that the center is $(1, 0)$ and the radius 1. (*Caution.* Whenever we multiply an equation through by r we introduce the root $r = 0$ and thus add the origin to the graph. In the present instance, the origin was on the graph to start with.)

Now let us analyze the graph directly from the polar equation. When $\theta = 0$, $r = 2 \cos 0 = 2$. As θ increases from 0 to $\pi/2$, $2 \cos \theta$ decreases from 2 to 0, and we obtain the portion of the graph shown in Figure 4(a). As θ increases from $\pi/2$ to π, $2 \cos \theta$ decreases from 0 to -2; the added points are included in Figure 4(b).

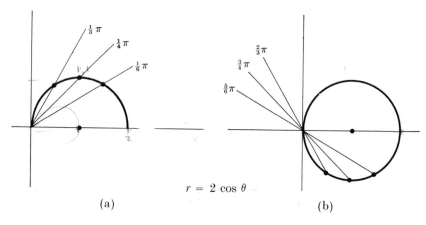

$$r = 2 \cos \theta$$

(a)   (b)

FIGURE 4

Now as θ increases from π to $3\pi/2$, the points in Figure 4(a) are retraced, and as θ increases from $3\pi/2$ to 2π, the lower half of the graph is retraced.

Example 5. *The cardioid* $r = 1 + \cos \theta$.

As θ increases from 0 to $\pi/2$, $1 + \cos \theta$ decreases from 2 to 1. Between $\pi/2$ and π, $1 + \cos \theta$ decreases further, from 1 to 0. Since $\cos(\pi + \alpha) = \cos(\pi - \alpha)$, the graph between π and 2π is a reflection in the x-axis of the portion already described. The heart-shaped curve thus obtained is shown in Figure 5.

Example 6. *The limaçon* $r = 1 + 2 \cos \theta$.

When $\theta = 0$, $r = 3$. At first, as θ increases, r decreases, becoming 0 when $2 \cos \theta = -1$, i.e., $\theta = 2\pi/3$; as θ continues to π, $1 + 2 \cos \theta$ continues to decrease, from 0 to -1. As in Example 5, the graph between π and 2π is a reflection in the x-axis of the portion already described. The graph is shown in Figure 6.

Example 7. *The four-leaved rose* $r = \sin 2\theta$.

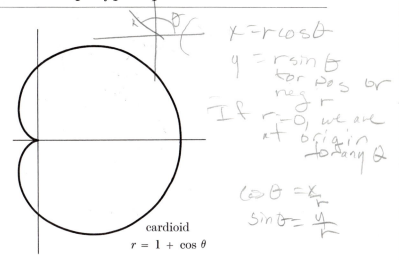

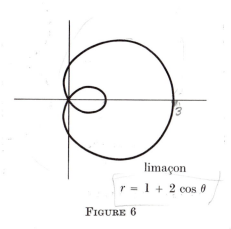

cardioid
$r = 1 + \cos \theta$

FIGURE 5

limaçon
$r = 1 + 2 \cos \theta$

FIGURE 6

The leaf in the first quadrant is traced out as θ goes from 0 to $\pi/2$, so that 2θ goes from 0 to π (Figure 7). Notice that we *calculate* $\sin 2\theta$ but *plot* at θ. Values of θ between $\pi/2$ and π (2θ between π and 2π) generate the leaf in the *fourth* quadrant; those between π and $3\pi/2$ (2θ between 2π and 3π) the leaf in the third quadrant; and those between $3\pi/2$ and 2π (2θ between 3π and 4π) the leaf in the second quadrant.

Example 8. *The lemniscate* $r^2 = \sin \theta$.

If (r, θ) is on the graph, so is $(-r, \theta)$. As θ increases from 0 to $\pi/2$, r^2 increases from 0 to 1. This gives us the graph in the first and third quadrants shown in Figure 8(a). Since $\sin(\pi - \alpha) = \sin \alpha$, the graph between $\pi/2$ and π is a reflection in the y-axis of the portion just described. This gives us Figure 8(b). When

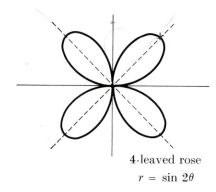

4-leaved rose

$r = \sin 2\theta$

FIGURE 7

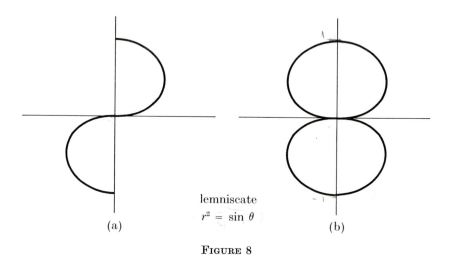

lemniscate

$r^2 = \sin \theta$

(a)   (b)

FIGURE 8

$\pi < \theta < 2\pi$, $\sin \theta < 0$ and $r^2 = \sin \theta$ is impossible; hence these values of θ do not generate any new points.

Finally, we consider a non-trigonometric example.

Example 9. *The spiral* $r = e^{\theta/2\pi}$.
Here, r always increases with θ. The graph is shown in Figure 9.

10B3. Intersection of two graphs. Sometimes we want to know when two polar graphs intersect. We cannot always find out simply by solving their two equations simultaneously. For instance, the equations

$$r = \theta, \qquad r = \theta + 2\pi$$

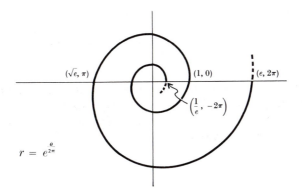

$$r = e^{\frac{\theta}{2\pi}}$$

FIGURE 9

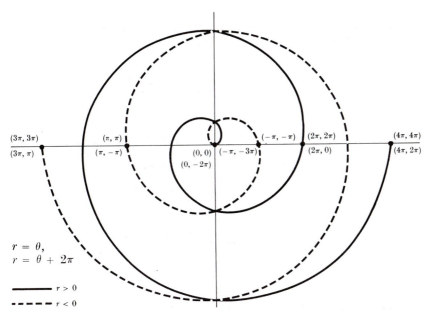

$$r = \theta,$$
$$r = \theta + 2\pi$$

———— $r > 0$

‐‐‐‐ $r < 0$

FIGURE 10

have no common solution at all, yet their graphs are identical (Figure 10).

To determine whether two graphs intersect at the origin, we test each separately to see whether it does pass through the origin.

To find where $r = f(\theta)$ and $r = g(\theta)$ intersect otherwise, we are faced, in principle, with infinitely many equations:

(2) $$f(\theta + 2n\pi) = g(\theta + 2m\pi)$$

for each integer n and each integer m, and

(3) $$f(\theta + 2n\pi) = -g(\theta + \pi + 2m\pi)$$

for all integers n and m.

In practice we can usually narrow down the system of equations to just a few, since f and g will usually appear as simple expressions in $\sin \theta$, $\cos \theta$, etc. In this case their values repeat every 2π and the entire system (2) and (3) is reduced to just two equations:

(4) $$f(\theta) = g(\theta)$$

and

(5) $$f(\theta) = -g(\theta + \pi) \qquad (0 \le \theta < 2\pi).$$

Example 10. Find the points of intersection of the circle $r = f(\theta) = \cos \theta$ with the cardioid $r = g(\theta) = 1 - \cos \theta$ (Figure 11).

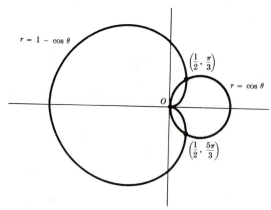

FIGURE 11

Solution. The origin is a point of intersection. All others are determined by equations (4) and (5); here, these take the form

(6) $$\cos \theta = 1 - \cos \theta$$

and

(7) $$\cos \theta = -1 + \cos(\theta + \pi) \qquad (0 \le \theta < 2\pi).$$

Equation (6) gives $\cos \theta = \frac{1}{2}$, $\theta = \pi/3$ or $5\pi/3$. The points are

$$\left(\frac{1}{2}, \frac{\pi}{3}\right) \qquad \text{and} \qquad \left(\frac{1}{2}, \frac{5\pi}{3}\right).$$

Equation (7) reduces to $\cos \theta = -\frac{1}{2}$; hence $\theta = \pi/3 + \pi$ or $5\pi/3 + \pi$. The points are

$$\left(-\frac{1}{2}, \frac{\pi}{3} + \pi\right) \quad \text{and} \quad \left(-\frac{1}{2}, \frac{5\pi}{3} + \pi\right),$$

the same points as before.

Problems (§10B)

1. Which of the following points lie on the graph of $r^2 = \sin \theta$:

   $(-1, \pi/2)$, $(\sqrt{3}/2, \pi/3)$, $(1, 3\pi/2)$, $(2^{-1/4}, -\pi/4)$, $(0, \pi/6)$?

2. Find a polar equation having the same graph as the given rectangular equation.

   (a) $y = x^2$

   (b) $xy = 1$

   (c) $x^2 + y^2 + 2x - 3y = 0$

   (d) $3x - 2y = 7$

   (e) $y = 3$

   (f) $3x - 2y = 0$

3. Find a rectangular equation having the same graph as the given polar equation.

   (a) $r = 3 \sin \theta$

   (b) $r^2 = \sin 2\theta$

   (c) $r = \cos 2\theta$

   (d) $\theta = -\pi/4$

   (e) $r = -2$

   (f) $r = 1 + \cos \theta$

4. Sketch the graphs of the following polar equations.

   (a) $r = \sin 3\theta$

   (b) $r = \cos (\theta/2)$

   (c) $r = \tan \theta$

   (d) $r = 2/(\cos \theta)$

   (e) $r = \theta$

   (f) $r^2 = \cos \theta$

   (g) $r = 2 \cos 2\theta$

   (h) $r = 1 + \sin 2\theta$

   (i) $r = 2 + \cos \theta$

   (j) $r = 3 \csc \theta$

   (k) $r = \cot \theta$

   (l) $r^2 = \cos 2\theta$

   (m) $r = 1 + 2 \sin \theta$

   (n) $r = 2 \sin^2(\theta/2)$

   (o) $r = \cos 4\theta$

5. In each case, find all points of intersection of the graphs of the two polar equations.

   (a) $r = 1 + \cos \theta$,　$r^2 = \frac{1}{2} \cos \theta$

   (b) $r = \sin^2 \theta$,　$r = -1$

   (c) $r = 4 \cos \theta$,　$\theta = \pi/3$

   (d) $r = 1 + \sin \theta$,　$r = 4 - 2 \sin \theta$

Answers to problems (§10B)

1. $(-1, \pi/2)$, $(1, 3\pi/2)$, $(2^{-1/4}, -\pi/4)$, $(0, \pi/6)$

2. (a) $r = \tan \theta \sec \theta$　　(b) $r^2 = 2 \csc 2\theta$

   (c) $r = 3 \sin \theta - 2 \cos \theta$　　(d) $3r \cos \theta - 2r \sin \theta = 7$

   (e) $r = 3 \csc \theta$　　(f) $\tan \theta = \frac{3}{2}$

3. (a) $x^2 + y^2 = 3y$ (b) $(x^2 + y^2)^2 = 2xy$ (c) $(x^2 + y^2)^3 = (x^2 - y^2)^2$
   (d) $x + y = 0$ (e) $x^2 + y^2 = 4$ (f) $x^2 + y^2 = x + \sqrt{x^2 + y^2}$
5. (a) $(0, 0)$, $(\frac{1}{2}, 2\pi/3)$, $(\frac{1}{2}, 4\pi/3)$ (b) $(1, \pi/2)$, $(1, 3\pi/2)$
   (c) $(0, 0)$, $(2, \pi/3)$ (d) $(2, \pi/2)$

§10C. Parametric equations; tangent

10C1. Parametric equations of a polar curve. In this section we discuss the tangent to a curve in terms of its polar equation, say

(1) $$r = f(\theta).$$

Since we know about tangents in terms of rectangular coordinates, we attack the problem that way. The formulas for converting to rectangular coordinates are

(2) $$x = r \cos \theta, \qquad y = r \sin \theta.$$

With $r = f(\theta)$, they become

(3) $$x = f(\theta) \cos \theta, \qquad y = f(\theta) \sin \theta.$$

These are the parametric equations of the curve in terms of the parameter θ. We will continue to write them in the simpler form (2), remembering that r there stands for $f(\theta)$.

In vector form, the equation of the curve is

(4) $$P(\theta) = (x, y) = (r \cos \theta, r \sin \theta).$$

10C2. Tangent vector and tangent line. If r is a differentiable function of θ, then x and y are also differentiable. We will assume that x' and y' are not both 0. The tangent vector is then given by

(5) $$P'(\theta) = (x', y')$$

(see §9E2). By (2),

(6) $$x' = -r \sin \theta + r' \cos \theta, \qquad y' = r \cos \theta + r' \sin \theta.$$

One may now write down the equation of the tangent line at (r, θ):

(7) $$T(t) = P(\theta) + tP'(\theta),$$

where $P(\theta)$ and $P'(\theta)$ are given by (4), (5), and (6). Instead, one usually considers the angle between the radial line and the tangent line, for which there is a simple formula. We now look at that angle.

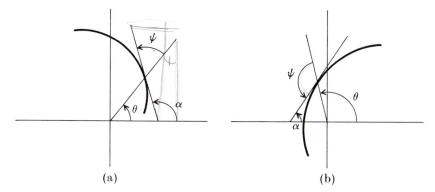

(a)   (b)

FIGURE 1

10C3. The angle ψ from the radial line to the tangent line. To avoid complications we will assume $r \neq 0$. Then, as we shall show,

$$(8) \qquad \tan \psi = \frac{r}{r'} \quad (if\ r' \neq 0), \qquad \psi = \frac{\pi}{2} \quad (if\ r' = 0).$$

Proof. Let α denote the inclination of the tangent line. Then, as pictures of various cases will show (Figure 1),

$$(9) \qquad \psi = \alpha - \theta \quad \text{plus a multiple of} \quad \pi.$$

Therefore,

$$(10) \qquad \tan \psi = \tan(\alpha - \theta).$$

We will compute $\sin(\alpha - \theta)$ and $\cos(\alpha - \theta)$. Since $(\cos \alpha, \sin \alpha)$ is the unit tangent vector,

$$(11) \qquad \cos \alpha = \frac{x'}{|P'|} \quad \text{and} \quad \sin \alpha = \frac{y'}{|P'|}.$$

Consequently,

$$(12) \qquad \sin(\alpha - \theta) = \frac{1}{|P'|} (y' \cos \theta - x' \sin \theta)$$

and

(13) $$\cos(\alpha - \theta) = \frac{1}{|P'|}\,(x'\cos\theta + y'\sin\theta).$$

When we substitute from (6) and multiply out, we find in each case that the terms in $\sin\theta\cos\theta$ cancel; we obtain:

(14) $$\sin(\alpha - \theta) = \frac{r}{|P'|}, \qquad \cos(\alpha - \theta) = \frac{r'}{|P'|}.$$

If $r' \neq 0$, we may divide to get $\tan\psi = \tan(\alpha - \theta) = r/r'$. If $r' = 0$ then $\cos(\alpha - \theta) = 0$ and so $\alpha - \theta = \pi/2$ and hence $\psi = \pi/2$.

Example 1. See Figure 2. Find ψ at the point $(1 + 1/\sqrt{2}, \pi/4)$ on the cardioid $r = 1 + \cos\theta$. (This curve was discussed in §10B2, Example 5.)

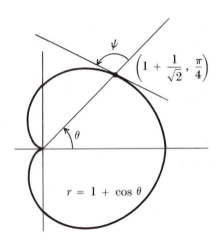

FIGURE 2

Solution. Here, $r' = -\sin\theta$. At $(1 + 1/\sqrt{2}, \pi/4)$,

$$r = \frac{\sqrt{2} + 1}{\sqrt{2}}, \qquad r' = -\frac{1}{\sqrt{2}},$$

and

$$\tan\psi = \frac{r}{r'} = -(\sqrt{2} + 1) \approx -2.4, \quad \psi \approx 115°.$$

Problems (§10C)

1. Find the angle ψ from the radial line to the tangent line; express it in terms of θ. Sketch the situation at the given values of θ.

    (a) $r = e^\theta$; $\theta = \pi/4, \pi/3$

    (b) $r = 2 + \cos\theta$; $\theta = \pi/6$

    (c) $r = \theta^2$; $\theta = \pi$

    (d) $r = \sin\theta$; $\theta = \pi/4$

    (e) $r = 1 - \cos\theta$; $\theta = \pi/2$

2. Show that the two curves have perpendicular tangent lines at the given point of intersection. [*Hint.* Show that $\tan\psi_1 \tan\psi_2 = -1$.]

    (a) $r = 1 + \cos\theta$, $r = 1 - \cos\theta$; $\theta = \pi/2$

    (b) $r = \theta$, $r = 1/\theta$; $\theta = 1$

    (c) $r = 2\sin\theta$, $r = 2\cos\theta$; $\theta = \pi/4$

    (d) $r = 1 - \cos\theta$, $r = \dfrac{1}{1 - \cos\theta}$; $\theta = \pi/2$

3. Find the slope of the tangent line to the given curve at the given point.

    (a) $r = 1 + \cos\theta$; $\theta = \pi/2$ (c) $r = e^{2\theta}$; $\theta = \pi/4$

    (b) $r = \sec^2\theta$; $\theta = \pi/4$ (d) $r = 2 + \sin\theta$; $\theta = \pi/6$

4. Consider the graph of $r = f(\theta)$ (polar coordinates). Using formula (7) of §9I3, show that the curvature at any point (r, θ) is given by

$$(15) \qquad\qquad \kappa = \frac{|r^2 + 2r'^2 - rr''|}{(r^2 + r'^2)^{3/2}}.$$

(It is assumed that (r, θ) is an interior point of an interval on which f is twice differentiable and f' is never 0.)

5. Use formula (15) of Problem 4 to find the curvature of the given curve at the given point.

    (a) $r = 2\cos\theta$, $\theta = 2$

    (b) $r = \dfrac{1}{1 - \cos\theta}$, $\theta = \pi/3$

    (c) $r = 1 + 2\cos\theta$, $\theta = 0$ and $\theta = \pi$

Answers to problems (§10C)

1. (a) $\pi/4$ (b) $-\arctan((2 + \cos\theta)/(\sin\theta))$ (c) $\arctan\theta/2$
    (d) θ (e) $\arctan((1 - \cos\theta)/(\sin\theta))$

3. (a) 1   (b) 3   (c) 3   (d) $-3\sqrt{3}$

5. (a) 1   (b) $\frac{1}{8}$ (c) $\frac{5}{9}$, 3

§10D. Integration in polar coordinates

10D1. Area. Let $r = f(\theta)$ be continuous on an interval $\alpha \le \theta \le \beta$. We wish to define and compute the area of the region bounded by the graph of f and the lines $\theta = \alpha$ and $\theta = \beta$. We assume $\beta - \alpha \le 2\pi$. For convenience, we assume $f \ge 0$. See Figure 1.

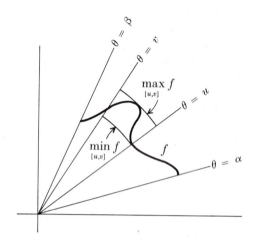

FIGURE 1

Denote by A_u^v the area between the lines $\theta = u$ and $\theta = v$ ($u \le v$). Of course, we will require that areas add up properly:

(A) $$A_\alpha^u + A_u^v = A_\alpha^v.$$

Next, since f is continuous, it assumes a maximum value and a minimum value on $[u, v]$. Consider the circular sectors with central angle $v - u$ and radii $\min_{[u,v]} f$ and $\max_{[u,v]} f$ (Figure 1). Obviously, A_u^v must lie between the area of the smaller and the area of the larger. Now, the area of a sector of radius a and angle θ is $\frac{1}{2}a^2\theta$ (§8G4). Thus we have:

(B) $$\tfrac{1}{2}(\min_{[u,v]} f)^2(v - u) \le A_u^v \le \tfrac{1}{2}(\max_{[u,v]} f)^2(v - u).$$

Because of **(A)** and **(B)**, we are forced to define

(1) $$A = \tfrac{1}{2} \int_\alpha^\beta f(\theta)^2 \, d\theta.$$

Briefly,

(2)
$$A = \tfrac{1}{2} \int_{\alpha}^{\beta} r^2 \, d\theta.$$

Example 1. Find the area of the region enclosed by the cardioid $r = 1 + \cos \theta$.

Solution. The graph is pictured in §10B2, Figure 5. The region is swept out as θ goes from 0 to 2π. Hence the area is:

$$A = \tfrac{1}{2} \int_{0}^{2\pi} (1 + 2 \cos \theta + \cos^2\theta) \, d\theta = \tfrac{3}{2}\pi.$$

Example 2. Find the area of the region between the graphs of $r = \cos \theta$ and $r = 1 - \cos \theta$ (Figure 2).

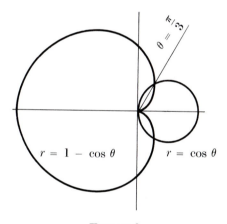

FIGURE 2

Solution. Because of symmetry, we compute twice the upper half. The points of intersection are $(0, 0)$ and $(\tfrac{1}{2}, \pi/3)$ (§10B3, Example 10). The region in question is divided into two parts by the line segment joining them, and we can find these areas separately and add. The total area is, thus,

$$A = 2 \cdot \tfrac{1}{2} \left[\int_{0}^{\pi/3} (1 - 2 \cos \theta + \cos^2\theta) \, d\theta + \int_{\pi/3}^{\pi/2} \cos^2\theta \, d\theta \right]$$

$$= \int_{0}^{\pi/3} (1 - 2 \cos \theta) \, d\theta + \int_{0}^{\pi/2} \cos^2\theta \, d\theta$$

$$= \frac{\pi}{3} - \sqrt{3} + \frac{\pi}{4} = \tfrac{7}{12}\pi - \sqrt{3} \approx 0.1.$$

10D2. Length of arc. It is easy to express arc length in polar form. Let $r = f(\theta)$ be defined for $\alpha \le \theta \le \beta$. We assume that this function has a continuous derivative. Then the length of the curve is

$$(3) \qquad L = \int_\alpha^\beta \sqrt{f(\theta)^2 + f'(\theta)^2}\, d\theta.$$

Briefly,

$$(4) \qquad L = \int_\alpha^\beta \sqrt{r^2 + r'^2}\, d\theta.$$

Proof. We derive this by considering the parametric equations,

$$(5) \qquad x = r\cos\theta, \qquad y = r\sin\theta.$$

In these terms we have, according to §9H(3),

$$(6) \qquad L = \int_\alpha^\beta \sqrt{x'^2 + y'^2}\, d\theta.$$

Here,

$$(7) \qquad x' = -r\sin\theta + r'\cos\theta, \qquad y' = r\cos\theta + r'\sin\theta,$$

and a straightforward calculation shows that

$$(8) \qquad x'^2 + y'^2 = r^2 + r'^2.$$

This gives us (4).

Example 3. Find the length of the graph of $r = 1 - \cos\theta$.

Solution. The graph appears as part of Figure 2. Here $r' = \sin\theta$,

$$r^2 + r'^2 = 2 - 2\cos\theta,$$

and

$$L = \sqrt{2} \int_0^{2\pi} \sqrt{1 - \cos\theta}\, d\theta.$$

To evaluate this we use the identity

$$1 - \cos\theta = 2\sin^2\frac{\theta}{2}.$$

Note that $\sin \theta/2 \geq 0$ for $0 \leq \theta \leq 2\pi$, so that

$$\sqrt{\sin^2 \frac{\theta}{2}} = \sin \frac{\theta}{2} \qquad (0 \leq \theta \leq 2\pi).$$

Hence,

$$L = 2 \int_0^{2\pi} \sin \frac{\theta}{2} \, d\theta = 2 \left[-2 \cos \frac{\theta}{2} \right]_0^{2\pi} = 8.$$

Problems (§10D)

1. Find the area of the region bounded by the given curve and radial lines.
   (a) $r = e^\theta$, $\quad 0 \leq \theta \leq 1$; $\quad \theta = 0, \theta = 1$
   (b) $r = 4 \sec \theta$; $\quad \theta = 0, \theta = \pi/4$
   (c) $r = 3 + \cos \theta$; $\quad \theta = 0, \theta = \pi$
   (d) $r = \theta^2$; $\quad \theta = 0, \theta = \pi/2$

2. Find the area of the region bounded by the given curves.
   (a) The smaller loop of $r = 1 + 2 \cos \theta$ (§10B, Example 6)
   (b) One leaf of the rose $r = \sin 2\theta$ (§10B, Example 7)
   (c) One loop of the lemniscate $r^2 = \sin \theta$ (§10B, Example 8)
   (d) The circle $r = 6$
   (e) Inside $r = 4 \sin \theta$ and outside $r = 2$
   (f) One leaf of the three-leaved rose $r = 2 \sin 3\theta$

3. Find the length of the given curve.
   (a) $r = e^\theta$ $\quad (0 \leq \theta \leq 1)$
   (b) $r = \sin^3(\theta/3)$ $\quad (0 \leq \theta \leq \pi/2)$
   (c) $r = 2 \sec \theta$ $\quad (0 \leq \theta \leq \pi/4)$
   (d) $r = \theta^2$ $\quad (0 \leq \theta \leq 1)$
   (e) $r = \theta$ $\quad (0 \leq \theta \leq 1)$

4. Write an integral for the length of the given curve.
   (a) The limaçon $r = 1 + 2 \cos \theta$ (§10B, Example 6)
   (b) One leaf of $r = \sin 2\theta$ (§10B, Example 7)

   (c) $r = \dfrac{1}{1 + \cos \theta}$ $\quad (0 \leq \theta \leq \pi/2)$

5. Find the area enclosed between the two loops of the limaçon $r = 1 + 2 \cos \theta$ (§10B, Example 6). [*Caution. r* changes sign.]

Answers to problems (§10D)

1. (a) $\frac{1}{4}(e^2 - 1)$ (b) 8   (c) $19\pi/4$ (d) $\pi^5/320$
2. (a) $\pi - \frac{3}{2}\sqrt{3}$ (b) $\pi/8$ (c) 1   (d) 36π (e) $\frac{4}{3}\pi + 2\sqrt{3}$ (f) $\pi/3$

3. (a) $\sqrt{2}(e - 1)$ (b) $\frac{1}{4}\pi - \frac{3}{8}\sqrt{3}$ (c) 2
   (d) $\frac{1}{3}(5^{3/2} - 8)$ (e) $\frac{1}{2}[\sqrt{2} + \log(1 + \sqrt{2})]$

4. (a) $2\int_0^\pi \sqrt{4\cos\theta + 5}\, d\theta$ (b) $\int_0^{\pi/2} \sqrt{3\cos^2 2\theta + 1}\, d\theta$

   (c) $\sqrt{2}\int_0^{\pi/2} (1 + \cos\theta)^{-3/2}\, d\theta$

5. $\pi + 3\sqrt{3}$

CHAPTER 11

Space Coordinates, Vectors, and Curves

This chapter introduces the geometry of space. §§11A–11E treat the analytic geometry of three dimensions. Lines and planes are treated systematically, both through their vector equations and their coordinate equations. No formal knowledge of solid geometry is assumed, but we will make use of a few easily visualized facts. In particular, we will take the following facts for granted: (i) three points not on a line determine a plane; (ii) a line through two points in a plane lies wholly in the plane; (iii) if a line is perpendicular to each of two lines in a plane, it is perpendicular to every line in that plane.

The rest of the chapter (§§11F–11H) makes use of calculus. Much, though not all, of the discussion is a straightforward generalization of the results for the plane in Chapter 9.

§11A. Coordinates

11A1. Coordinates. We extend the *xy*-coordinate system for $\mathscr{R}^2$ to an *xyz*-system for three-dimensional space, $\mathscr{R}^3$, by introducing a *z*-axis through the origin perpendicular to both the *x*-axis and the *y*-axis. With these two axes both horizontal, the *z*-axis is vertical, with the upward direction chosen as positive. The origin is the point $O = (0, 0, 0)$.

The coordinate system divides space into eight octants. In drawings, the axes are positioned so as to present the first octant: $x \geq 0$, $y \geq 0$, $z \geq 0$ (Figure 1).

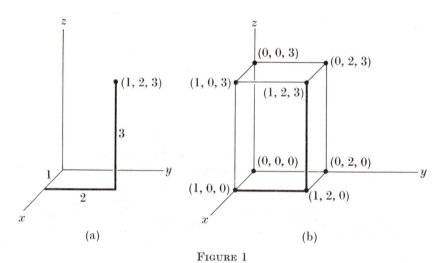

(a)   (b)

FIGURE 1

Even before introducing a formal study of the equations of lines and planes, we can observe a few evident relations. The *xy*-plane is the set of all points $(x, y, 0)$. A point lies in this plane if its *z*-coordinate is 0; and conversely. Hence the equation of the *xy*-plane is

$$z = 0.$$

Neither *x* nor *y* appears, indicating that *x* and *y* are totally unrestricted. Similarly, the equation of the *xz*-plane is $y = 0$ and the equation of the *yz*-plane is $x = 0$.

The *z*-axis is the set of all points $(0, 0, z)$. Hence it is represented by the pair of equations

(1) $x = 0, \qquad y = 0.$

The first of these is the *yz*-plane, the second the *xz*-plane. A point satisfies

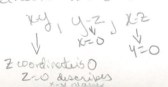

both equations if it lies in both planes; and conversely. Thus, the equations (1) characterize the z-axis as the intersection of the yz-plane with the xz-plane. Similarly, the equations of the y-axis are $x = 0$, $z = 0$; and the equations of the x-axis are $y = 0$, $z = 0$.

Considerations like these are not limited to the number 0. For instance, the equation of the plane parallel to the xy-plane and 3 units above it is $z = 3$.

11A2. Distance formula. The distance between $O = (0, 0, 0)$ and $P = (x, y, z)$ is

(2) $$\sqrt{x^2 + y^2 + z^2}.$$

The distance between $P_1 = (x_1, y_1, z_1)$ and $P_2 = (x_2, y_2, z_2)$ is

(3) $$d = \sqrt{(x_1 - x_2)^2 + (y_1 - y_2)^2 + (z_1 - z_2)^2}.$$

The proof consists of two applications of the Pythagorean Theorem. Figure 2(a) shows the special case, Figure 2(b) the general case.

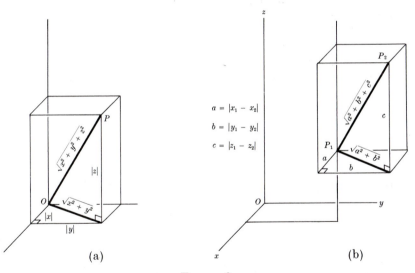

$$a = |x_1 - x_2|$$
$$b = |y_1 - y_2|$$
$$c = |z_1 - z_2|$$

(a)   (b)

FIGURE 2

Example 1. The distance from O to $(4, 1, -3)$ is

$$\sqrt{4^2 + 1^2 + (-3)^2} = \sqrt{16 + 1 + 9} = \sqrt{26}.$$

The distance between $P_1 = (1, -1, 2)$ and $P_2 = (2, 1, -3)$ is

$$\sqrt{(1 - 2)^2 + (-1 - 1)^2 + (2 + 3)^2} = \sqrt{1 + 4 + 25} = \sqrt{30}.$$

Special case
distance from $P(x,y,z)$ to $O(000$
$$D = \sqrt{x^2 + y^2 + Z^2}$$

11A3. Equation of a sphere. The sphere of radius a with center (h, k, l) is the set of all points whose distance from (h, k, l) is a. Hence a point (x, y, z) lies on the sphere if

$$(x - h)^2 + (y - k)^2 + (z - l)^2 = a^2, \tag{4}$$

and not otherwise. Therefore (4) is the equation of the sphere.

Example 2. The equation of the sphere of radius 7 with center $(-4, 4, 1)$ is

$$(x + 4)^2 + (y - 4)^2 + (z - 1)^2 = 49.$$

Problems (§11A)

1. Plot the following points.
   (a) $(2, 0, 0)$
   (b) $(1, 1, 1)$
   (c) $(1, 3, 2)$
   (d) $(1, 3, -2)$
   (e) $(0, 2, 2)$
   (f) $(0, 0, 1)$
   (g) $(1, -1, 1)$
   (h) $(-2, 1, 3)$
   (i) $(0, 3, 0)$

2. Find the distances between the following pairs of points.
   (a) $(1, 1, -2)$ and $(2, -1, 1)$
   (b) $(0, 1, 0)$ and $(0, 0, 1)$
   (c) $(-1, -2, -3)$ and $(1, 2, 3)$
   (d) $(0, 0, 0)$ and $(1, 2, -2)$
   (e) $(1, 2, 3)$ and $(1, 2, -3)$
   (f) $(3, 4, 6)$ and $(2, -1, 6)$

3. Show that the triangles with the given vertices are isosceles.
   (a) $(1, 0, -1)$, $(4, -2, 0)$, and $(-2, 1, 1)$
   (b) $(2, -1, 1)$, $(2, 1, 3)$, and $(4, 3, 3)$
   (c) $(1, 2, 2)$, $(2, 1, 0)$, and $(4, 0, -1)$
   (d) $(-3, -3, -4)$, $(0, 0, 0)$, and $(-4, 1, -2)$

4. Find the equation of the sphere with the given center and radius. Name three points on each sphere.
   (a) $(1, -1, 2)$, $r = 1$
   (b) $(3, 0, -1)$, $r = 3$
   (c) $(0, 0, 0)$, $r = 2$
   (d) $(-1, -1, 5)$, $r = 4$

5. Find the center and radius of the following spheres.
   (a) $x^2 + y^2 + z^2 + 4x - 2y - 6z = 50$
   (b) $x^2 + y^2 + z^2 + 4y = 12$
   (c) $x^2 + y^2 + z^2 + 2x + 2y + 2z = 1$
   (d) $x^2 + y^2 + z^2 - 10x + 6y - 4z = -37$

6. Describe the graph of the equation
   $$x^2 + y^2 + z^2 - 2x + 2y - 8z + 18 = 0.$$

7. The set of all points equidistant from $(1, -2, 2)$ and $(3, 2, 4)$ is the graph of what equation? List three of the points.

8. Find the equation of the sphere passing through the four points $(1, 3, 3)$, $(2, -1, 2)$, $(5, 0, 6)$, and $(4, 3, 2)$.

9. A bridge is 15 feet above the level of a river. A man crossing at 5 feet/second passes directly over a raft being carried by the current at 4 feet/second. How fast is the distance between them increasing one second later?

Answers to problems (§11A)

2. (a) $\sqrt{14}$ (b) $\sqrt{2}$ (c) $2\sqrt{14}$ (d) 3   (e) 6   (f) $\sqrt{26}$

4. (a) $(x-1)^2 + (y+1)^2 + (z-2)^2 = 1$
   (b) $(x-3)^2 + y^2 + (z+1)^2 = 9$
   (c) $x^2 + y^2 + z^2 = 4$
   (d) $(x+1)^2 + (y+1)^2 + (z-5)^2 = 16$

5. (a) center $(-2, 1, 3)$, radius 8
   (b) center $(0, -2, 0)$, radius 4
   (c) center $(-1, -1, -1)$, radius 2
   (d) center $(5, -3, 2)$, radius 1

6. the point $(1, -1, 4)$

7. $x + 2y + z = 5$

8. $(x-3)^2 + (y-1)^2 + (z-4)^2 = 9$

9. $41/\sqrt{266}$ ft./sec.

§11B. Vectors

11B1. Vectors in $\mathscr{R}^3$. Vectors are introduced in $\mathscr{R}^3$ as in $\mathscr{R}^2$. For $P_1 = (x_1, y_1, z_1)$ and $P_2 = (x_2, y_2, z_2)$, we define

$$P_1 + P_2 = (x_1 + x_2, y_1 + y_2, z_1 + z_2),$$

$$-P_1 = (-x_1, -y_1, -z_1),$$

(1) $$P_1 - P_2 = P_1 + (-P_2) = (x_1 - x_2, y_1 - y_2, z_1 - z_2),$$

$$kP_1 = (kx_1, ky_1, kz_1),$$

where k is a scalar (number). The *zero* vector is $O = (0, 0, 0)$. The familiar properties, such as $P + O = P$, and $P_1 + P_2 = P_2 + P_1$, hold as before.

Example 1. If $P_1 = (2, 4, -3)$ and $P_2 = (4, -2, 1)$, then
$$P_1 + P_2 = (2+4, 4-2, -3+1) = (6, 2, -2),$$
$$-P_1 = (-2, -4, 3),$$
$$P_1 - P_2 = (2-4, 4+2, -3-1) = (-2, 6, -4),$$
$$5P_1 = (10, 20, -15).$$

Let $P \neq O$. As before, the vectors on the line OP and on the same side of O as P are the vectors kP for $k > 0$ and are called parallel to P; while those on the opposite side of O are the vectors kP for $k < 0$ and are called opposite to P.

11B2. Midpoint formula. *The midpoint of a segment P_1P_2 is $(P_1 + P_2)/2$.*
The proof of this is just a matter of similar triangles. Sufficient details are indicated in Figure 1.

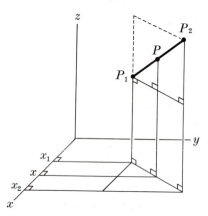

FIGURE 1

Example 2. If $P_1 = (4, 3, 1)$ and $P_2 = (-1, -1, 2)$, then the midpoint of P_1P_2 is $(\frac{3}{2}, 1, \frac{3}{2})$.

11B3. The parallelogram law. *The vectors $O, P, Q,$ and $P + Q$ lie in a plane and form the vertices of a parallelogram (if $P \neq O, Q \neq O,$ and P and Q are neither parallel nor opposite).*

Proof. (See Figure 2.) By the midpoint formula, the segment from O to $P + Q$ and the segment PQ have the same midpoint, $(P + Q)/2$. The plane of triangle OPQ contains the line through O and $(P + Q)/2$ and hence contains $P + Q$.

Since opposite sides of this plane quadrilateral are of equal length (§11A2), the figure is a parallelogram.

11B4. Length, angle, and scalar product. The *length* or *magnitude* of $P = (x, y, z)$, denoted $|P|$, is defined to be the length of the segment OP. Thus,

(2) $$|P| = \sqrt{x^2 + y^2 + z^2}.$$

Note that $|P_1 - P_2|$ is equal to the distance between P_1 and P_2.

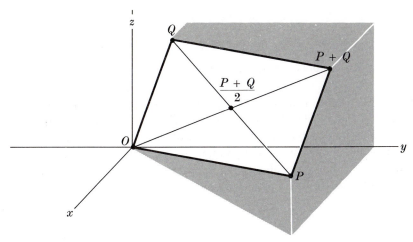

FIGURE 2

The *scalar product* of $P_1 = (x_1, y_1, z_1)$ and $P_2 = (x_2, y_2, z_2)$ is

(3)
$$P_1 \cdot P_2 = x_1 x_2 + y_1 y_2 + z_1 z_2.$$

It obeys the familiar laws: $P_1 \cdot P_2 = P_2 \cdot P_1$, etc. As in §9B we find that if θ is the angle between P_1 and P_2 (i.e., the angle between the segments OP_1 and OP_2), then

(4)
$$\cos \theta = \frac{P_1 \cdot P_2}{|P_1|\,|P_2|}.$$

Again, *the condition for perpendicularity is*

(5)
$$P_1 \cdot P_2 = 0.$$

Example 3. If $P_1 = (2, 3, 2)$ and $P_2 = (-1, 1, 5)$, then

$$|P_1| = \sqrt{4 + 9 + 4} = \sqrt{17}, \qquad |P_2| = \sqrt{1 + 1 + 25} = \sqrt{27}.$$

Next,

$$P_1 \cdot P_2 = (2)(-1) + (3)(1) + (2)(5) = 11.$$

Therefore,

$$\cos \theta = \frac{11}{\sqrt{27}\sqrt{17}} \approx \frac{11}{21.4} \approx 0.51, \qquad \theta \approx 60°.$$

Example 4. If $P_1 = (2, 3, 2)$ and $P_2 = (4, -6, 5)$, then

$$P_1 \cdot P_2 = (2)(4) + (3)(-6) + (2)(5) = 0.$$

Therefore these vectors are perpendicular.

11B5. Direction cosines of a vector. Directions in space are more complicated to describe than in the plane. The angle between a vector and the x-axis does not determine the direction of the vector; there are infinitely many possible directions, corresponding to the rays in a conical surface (Figure 3). The same considerations shows that there is no clear concept of positive or negative direction for an angle.

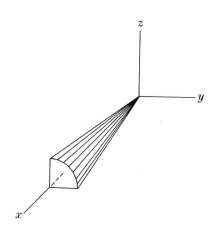

FIGURE 3

Given a non-zero vector V, we will consider *three* angles: the angles α, β, and γ that it makes with the positive x-, y-, and z-axes. We choose their measures to be between 0 and π (inclusive). The angles are called the *direction angles* of V. If $V = (x, y, z)$, then (see Figure 4)

(6)
$$\cos \alpha = \frac{x}{|V|}, \qquad \cos \beta = \frac{y}{|V|}, \qquad \cos \gamma = \frac{z}{|V|}.$$

These numbers are called the *direction cosines* of V.

If we are given $V = (x, y, z)$, we can find its direction cosines from (6), and then its direction angles. On the other hand, given the direction angles of a vector V, we know its direction cosines, and these are the coordinates of a vector parallel to V.

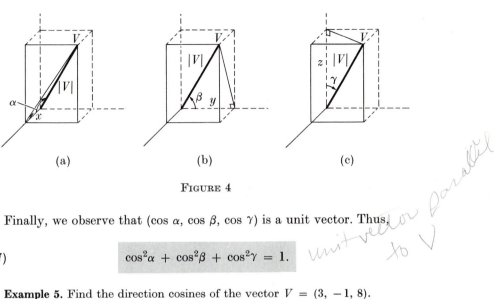

FIGURE 4

Finally, we observe that $(\cos \alpha, \cos \beta, \cos \gamma)$ is a unit vector. Thus,

unit vector parallel to V

(7)
$$\cos^2\alpha + \cos^2\beta + \cos^2\gamma = 1.$$

Example 5. Find the direction cosines of the vector $V = (3, -1, 8)$.

Solution. $|V| = \sqrt{3^2 + (-1)^2 + 8^2} = \sqrt{74}$. Therefore the direction cosines are

$$\cos \alpha = \frac{3}{\sqrt{74}}, \qquad \cos \beta = -\frac{1}{\sqrt{74}}, \qquad \cos \gamma = \frac{8}{\sqrt{74}}.$$

Example 6. If two of the direction angles of a vector are $\alpha = 60°$ and $\beta = 60°$, what is the third?

Solution. $\cos^2\alpha = \frac{1}{4}$ and $\cos^2\beta = \frac{1}{4}$, so

$$\cos^2\gamma = 1 - \tfrac{1}{4} - \tfrac{1}{4} = \tfrac{1}{2}.$$

Therefore $\cos \gamma = 1/\sqrt{2}$ or $-1/\sqrt{2}$, and $\gamma = 45°$ or $135°$.

Problems (§11B)

1. Write as a single vector:
   (a) $(1, 2, 3) + (-3, 1, 4)$
   (b) $3(1, -2, 2) + 2(0, 1, 4)$
   (c) $(-1, 3, 5) - (2, -1, 4)$
   (d) $-2(2, 1, 2) + 3(1, 4, 0)$

2. Find the vector A:
   (a) $A - (2, 3, 0) = (1, 4, 1)$
   (b) $\tfrac{3}{4}A = (0, 9, 5)$
   (c) $A + (1, 0, 4) = (3, 1, 1) - 2A$
   (d) $A + A = (0, 0, 0)$

3. Solve for a, b, and c:
   (a) $(a + 1, b - 1, 2c) = (2a, 1 - b, 1 - c)$
   (b) $a(a, b, c) = (3a, 3a, 0)$
   (c) $(a - b + c, a + b - c, 1) = (\tfrac{1}{3}, a - b, b + c)$

4. Solve for P, Q, and R:

    (a) $P + Q + R = (1, 0, 5)$, $P + Q = (0, 0, 6)$, $Q + R = (0, -2, 2)$

    (b) $P + Q + R = (2, 0, 2)$, $2P - Q = (2, 2, 3)$, $2Q - R = (-1, 1, 3)$

5. Find scalars r, s, and t such that:

    (a) $r(1, 2, 3) + s(1, 3, 2) + t(-1, -1, 0) = (0, 5, 7)$

    (b) $r(3, 1, 1) + s(2, 0, 2) + t(1, -1, -1) = (5, 1, 7)$

6. Two sides of a parallelogram are OP and OQ. What are the lengths of the diagonals if

    (a) $P = (3, 1, 2)$ and $Q = (1, 3, 4)$?

    (b) $P = (3, -1, -1)$ and $Q = (1, -3, 4)$?

7. Let A, B, and C be the vertices of a triangle. There are three points X such that A, B, C, and X are the vertices of a parallelogram. Find these three points and show that their sum is $A + B + C$.

8. Compute $P \cdot Q$.

    (a) $P = (3, -1, -1)$, $Q = (3, 1, 1)$

    (b) $P = (3, -1, -1)$, $Q = (3, 0, 4)$

    (c) $P = (2, -2, 3)$, $Q = (2, -2, 3)$

    (d) $P = (2, -2, 3)$, $Q = (3, 6, 2)$

9. Show that the three given vectors are the vertices of a right triangle. Which is the vertex of the right angle?

    (a) $(1, 7, 2)$, $(0, 7, -2)$,   and $(-1, 6, 1)$

    (b) $(0, -5, -2)$, $(-3, -2, 2)$,   and $(2, 1, -5)$

    (c) $(2, 4, -2)$, $(0, 1, -1)$,   and $(1, 1, 1)$

    (d) $(0, 0, 0)$, $(-2, 4, 1)$,   and $(1, 2, -6)$

    (e) $(1, 0, 0)$, $(0, 1, 1)$,   and $(0, 1, 0)$

10. Find three non-zero vectors perpendicular to the given vector: one with x-coordinate 0, one with y-coordinate 0, and one with z-coordinate 0. If possible, also find one with *no* coordinate 0.

    (a) $(1, 2, 3)$ (e) $(12, 3, -2)$

    (b) $(1, 0, 0)$ (f) $(5, -1, 6)$

    (c) $(-3, -2, 7)$ (g) $(-1, -2, 5)$

    (d) $(-1, 2, 4)$ (h) $(2, 1, -2)$

11. (a) Find three vectors:

        (i) none of which has any coordinate zero,

        (ii) no two of which are parallel or opposite, and

        (iii) whose sum is O.

    (b) Find four such vectors.

12. Find the direction cosines and direction angles of the following vectors.

    (a) $(1, 1, 1)$ (c) $(1, -1, 1)$ (e) $(3, 0, 4)$

    (b) $(2, 2, 2)$ (d) $(1, 2, 3)$

13. Find the angle between the two given vectors.

    (a) $(1, 1, 0)$ and $(1, 0, 0)$

    (b) $(3, 4, 0)$ and $(4, -3, 7)$

(c) (1, 2, 1)   and   (−1, 1, 2)
(d) (2, 2, 1)   and   (−1, 3, 1)
(e) (−1, −1, 2)   and   (1, −1, 4)
(f) (2, 0, 1)   and   (1, 2, −2)

14. Let A and B be non-zero vectors, and let P be the foot of the perpendicular from A onto OB. (The vector P is called the *projection* of A on B.) Prove that

$$P = \frac{A \cdot B}{|B|^2} B.$$

Answers to problems (§11B)

1. (a) (−2, 3, 7)   (b) (3, −4, 14)   (c) (−3, 4, 1)   (d) (−1, 10, −4)
2. (a) (3, 7, 1)   (b) (0, 12, $\frac{20}{3}$)   (c) ($\frac{2}{3}, \frac{1}{3}, -1$)   (d) (0, 0, 0)
3. (a) $a = b = 1, c = \frac{1}{3}$
   (b) $a = b = 3, c = 0;$ or $a = 0$, and b and c arbitrary
   (c) $a = 0, b = \frac{1}{3}, c = \frac{2}{3}$
4. (a) $P = (1, 2, 3),\quad Q = (-1, -2, 3),\quad R = (1, 0, -1)$
   (b) $P = (1, 1, 2),\quad Q = (0, 0, 1),\quad R = (1, -1, -1)$
5. (a) $r = 1, s = 2, t = 3$ (b) $r = 0, s = 3, t = -1$
6. (a) $2\sqrt{17}$ and $2\sqrt{3}$ (b) $\sqrt{41}$ and $\sqrt{33}$
8. (a) 7   (b) 5   (c) 17   (d) 0
9. (a) (−1, 6, 1)   (b) (0, −5, −2)   (c) (0, 1, −1)   (d) (0, 0, 0)
   (e) (0, 1, 0)
12. (a) $\cos \alpha = \cos \beta = \cos \gamma = 1/\sqrt{3} \approx 0.58,\quad \alpha = \beta = \gamma \approx 55°$
    (b) $\cos \alpha = \cos \beta = \cos \gamma = 1/\sqrt{3} \approx 0.58,\quad \alpha = \beta = \gamma \approx 55°$
    (c) $\cos \alpha = -\cos \beta = \cos \gamma = 1/\sqrt{3} \approx 0.58,\quad \alpha = \gamma \approx 55°, \beta \approx 125°$
    (d) $\cos \alpha = 1/\sqrt{14} \approx 0.27, \cos \beta = 2/\sqrt{14} \approx 0.54, \cos \gamma = 3/\sqrt{14} \approx 0.80;$
    $\alpha \approx 75°, \beta \approx 55°, \gamma \approx 35°$
    (e) $\cos \alpha = 0.60, \cos \beta = 0, \cos \gamma = 0.80;\quad \alpha \approx 55°, \beta \approx 90°, \gamma \approx 35°$
13. (a) 45°   (b) 90°   (c) 60°   (d) $\cos \theta = 5/3\sqrt{11} \approx 0.50,\quad \theta \approx 60°$
    (e) $\cos \theta = 4/3\sqrt{3} \approx 0.77, \theta \approx 40°$ (f) 90°

§11C. Lines

11C1. Direction vectors and direction numbers of a line. In $\mathscr{R}^3$, as in $\mathscr{R}^2$, we call a vector $V \neq O$ a *direction vector* for a line l if OV is parallel to l. Then the vectors tV ($t \neq 0$) are also direction vectors for l; and there are no others. The coordinates of a direction vector are called *direction numbers* of the line.

Let Q be a point of l. As before, the parallelogram law shows that if P is another point of l, then $V = P - Q$ is a direction vector for l; and, conversely, if V is a direction vector for l, then $P = Q + V$ is a point of l.

11C2. Vector and parametric equations of a line. As in $\mathscr{R}^2$, we find that if Q is a given point on a line l and V is a direction vector for l, then an equation for l is

$$(1) \qquad P = Q + tV.$$

We may put this into parametric form. Say $P = (x, y, z)$, $Q = (x_0, y_0, z_0)$, and $V = (a, b, c)$. Then we have

$$(2) \qquad x = x_0 + at, \qquad y = y_0 + bt, \qquad z = z_0 + ct.$$

Example 1. The vector equation of the line through $(1, -3, -3)$ with direction vector $(4, 5, -3)$ is

$$P = (1, -3, -3) + t(4, 5, -3).$$

The parametric equations are

$$x = 1 + 4t, \qquad y = -3 + 5t, \qquad z = -3 - 3t.$$

If we are given two points, P_1 and P_2, on l, then $P_1 - P_2$ is a direction vector for l and an equation is

$$(3) \qquad P = P_1 + t(P_2 - P_1).$$

Example 2. The vector equation of the line through $(1, 0, 4)$ and $(-3, -4, 2)$ is

$$P = (1, 0, 4) + t((-3, -4, 2) - (1, 0, 4))$$
$$= (1, 0, 4) - t(4, 4, 2),$$

or, more simply,

$$P = (1, 0, 4) + t(2, 2, 1).$$

The parametric equations are

$$x = 1 + 2t, \qquad y = 2t, \qquad z = 4 + t.$$

Example 3. Find the point of intersection, if any, of the lines

$$P = (5, -1, 3) + t(1, 3, -1) \qquad \text{and} \qquad P = (6, 2, 2) + t(1, 2, 1).$$

Solution. Since the parametric values in the two equations need not be the same at a common point, we will call the parameters s and t:

$$P = (5, -1, 3) + s(1, 3, -1), \qquad P = (6, 2, 2) + t(1, 2, 1).$$

The condition of intersection is

$$x = 5 + s = 6 + t$$
$$y = -1 + 3s = 2 + 2t$$
$$z = 3 - s = 2 + t,$$

that is,

$$s = 1 + t$$
$$3s = 3 + 2t$$
$$-s = -1 + t.$$

From the first two equations we get $s = 1, t = 0$, and this checks with the third equation. Therefore the lines do intersect. The common point, obtained from either of these parametric values, is $(6, 2, 2)$.

11C3. Coordinate equations of a line. If we solve for t in each of the parametric equations (2) and then equate the results, we get the "coordinate" equations

(4)
$$\frac{x - x_0}{a} = \frac{y - y_0}{b} = \frac{z - z_0}{c}.$$

Note that these are two equations, not three.

Example 4. Find the coordinate equations of the line

$$x = 1 + 4t, \qquad y = -3 + 5t, \qquad z = -3 - 3t.$$

Solution. Solving each equation for t and equating, we get

$$\frac{x - 1}{4} = \frac{y + 3}{5} = \frac{z + 3}{-3}.$$

If t is absent from one or two of the parametric equations, the coordinate equations assume a modified form.

Example 5. Find the coordinate equations of the line

$$x = 1 + t, \qquad y = 3, \qquad z = -2 + 2t.$$

Solution. Here we cannot express t in terms of y. On the contrary, we are told that y is a constant, 3. In this case, the two coordinate equations are

$$\frac{x - 1}{1} = \frac{z + 2}{2} \qquad \text{and} \qquad y = 3.$$

11C4. Angle between two lines. We speak of the angle between two lines, whether they intersect or not, by referring to direction vectors for the lines. If direction vectors V_1 and V_2 form an obtuse angle, then V_1 and $-V_2$, which are direction vectors for the same lines, form an acute angle. Hence we may define the angle θ between two lines to be the acute angle between direction vectors for the lines. Thus, $0 \le \theta \le \pi/2$. Consequently, $\cos \theta \ge 0$. Therefore, the angle θ between lines with direction vectors V_1 and V_2 is given by

$$\cos \theta = \frac{|V_1 \cdot V_2|}{|V_1||V_2|}, \qquad 0 \le \theta \le \pi/2.$$

Example 6. Find the angle between two lines with direction vectors $V_1 = (2, 3, 2)$ and $V_2 = (1, -1, -5)$.

Solution. $V_1 \cdot V_2 < 0$, so we work with V_1 and $-V_2$. The angle between these vectors was found in §11B4, Example 3 to be about $60°$.

Problems (§11C)

1. Write vector and parametric equations for the line through Q with direction vector V. Name three points on the line.
   - (a) $Q = (1, 2, 1)$, $V = (1, 2, 0)$
   - (b) $Q = (0, 1, 1)$, $V = (2, 3, -1)$
   - (c) $Q = (-2, 1, -3)$, $V = (1, 0, 0)$
   - (d) $Q = (0, 0, 0)$, $V = (5, 1, -3)$
   - (e) $Q = (5, 2, -1)$, $V = (1, 1, 1)$
   - (f) $Q = (1, 2, 3)$, $V = (1, 2, 3)$

2. Write coordinate equations for the lines of Problem 1.

3. Write vector and parametric equations for the line through R and Q. Name a third point on the line.
   - (a) $R = (1, -1, 2)$, $Q = (2, 1, -1)$
   - (b) $R = (2, -3, -1)$, $Q = (4, 5, 0)$
   - (c) $R = (1, -3, 5)$, $Q = (4, -3, 2)$
   - (d) $R = (1, 0, 0)$, $Q = (0, 1, 0)$
   - (e) $R = (2, 7, -1)$, $Q = (2, 4, -1)$
   - (f) $R = (-1, -1, 2)$, $Q = (4, 3, 1)$

4. Write coordinate equations for the lines of Problem 3.

5. Write vector and parametric equations for the lines with the given coordinate equations.

   - (a) $\dfrac{x-2}{3} = \dfrac{y+1}{7} = \dfrac{z-4}{2}$
   - (b) $x + 1 = y + 2 = z - 1$
   - (c) $\dfrac{x-3}{2} = \dfrac{y-1}{3}$, $z = 4$
   - (d) $\dfrac{x+1}{3} = y + 2 = \dfrac{z-1}{2}$
   - (e) $x = 1$, $y = 2$
   - (f) $x = 1$, $\dfrac{y-1}{3} = \dfrac{z+1}{2}$

6. Find the point of intersection of the two lines.
   - (a) $P = (1, 2, -2) + t(-1, 3, 1)$; $P = (0, 1, 1) + t(-1, 7, -1)$
   - (b) $P = (3, -1, 2) + t(1, -1, 2)$; $P = (5, 1, 5) + t(-1, 5, -3)$
   - (c) $x = 1 + t, y = 2 - t, z = t$; $x = -2 + t, y = 2 - 4t, z = 1 + 3t$
   - (d) $\dfrac{x+2}{3} = \dfrac{y-1}{-2} = z - 3$; $x - 3 = \dfrac{y-5}{3}, z = 4$

7. Find the angle between the two lines.
   - (a) $P = (1, 3, -1) + t(2, -2, 1)$; $P = (2, 0, 0) + t(1, 1, 3)$
   - (b) $P = (2, 0, 1) + t(4, -3, 6)$; $P = (1, 7, 5) + t(6, 0, -4)$

(c) $\dfrac{x-1}{2} = \dfrac{y+1}{3} = \dfrac{z-2}{-1}$; $\dfrac{x+3}{1} = \dfrac{y-2}{2}$, $z = 5$

(d) $x = 1, y = 2$; $P = (3, 1, 1) + t(0, 0, 1)$

8. Find the equation of a line through the given point Q, parallel to the *xy*-plane, and having a direction vector *perpendicular* to the given vector V.

    (a) $Q = (1, 2, 4), \quad V = (-1, 2, 2)$
    (b) $Q = (3, 1, -1), \quad V = (1, 3, 2)$
    (c) $Q = (4, -2, -3), \quad V = (1, 0, 0)$
    (d) $Q = (-2, 1, 5), \quad V = (5, -4, -2)$

9. Show that if A and B are non-zero vectors then

$$\frac{A}{|A|} + \frac{B}{|B|}$$

is a direction vector of the bisector of angle AOB.

10. Find the angle that a diagonal of a face of a cube makes with a diagonal of an adjoining face.

11. Show that the three medians of triangle ABC intersect at $\frac{1}{3}(A + B + C)$. [*Hint.* Check direction vectors.]

12. Let A, B, C, and D be any four points in $\mathscr{R}^3$. Quadrilateral $ABCD$ need not lie in a plane. Show that, nevertheless, the midpoints of its sides are the vertices of a parallelogram (or lie on a line) and hence lie in a common plane.

Answers to problems (§11C)

1. (a) $P = (1, 2, 1) + t(1, 2, 0)$; $x = t + 1, y = 2t + 2, z = 1$
  (b) $P = (0, 1, 1) + t(2, 3, -1)$; $x = 2t, y = 3t + 1, z = -t + 1$
  (c) $P = (-2, 1, -3) + t(1, 0, 0)$; $x = t, y = 1, z = -3$
  (d) $P = t(5, 1, -3)$; $x = 5t, y = t, z = -3t$
  (e) $P = (5, 2, -1) + t(1, 1, 1)$; $x = t + 5, y = t + 2, z = t - 1$
  (f) $P = t(1, 2, 3)$; $x = t, y = 2t, z = 3t$

2. (a) $\dfrac{x-1}{1} = \dfrac{y-2}{2}$, $z = 1$ (b) $\dfrac{x}{2} = \dfrac{y-1}{3} = \dfrac{z-1}{-1}$

  (c) $y = 1, \quad z = -3$ (d) $\dfrac{x}{5} = \dfrac{y}{1} = \dfrac{z}{-3}$

  (e) $\dfrac{x-5}{1} = \dfrac{y-2}{1} = \dfrac{z+1}{1}$ (f) $\dfrac{x}{1} = \dfrac{y}{2} = \dfrac{z}{3}$

3. (a) $P = (1, -1, 2) + t(1, 2, -3)$; $x = t + 1, y = 2t - 1, z = -3t + 2$
  (b) $P = (2, -3, -1) + t(2, 8, 1)$; $x = 2t + 2, y = 8t - 3, z = t - 1$
  (c) $P = (1, -3, 5) + t(1, 0, -1)$; $x = t + 1, y = -3, z = -t + 5$
  (d) $P = (1, 0, 0) + t(-1, 1, 0)$; $x = -t + 1, y = t, z = 0$
  (e) $P = (2, 7, -1) + t(0, 1, 0)$; $x = 2, y = t, z = -1$
  (f) $P = (-1, -1, 2) + t(5, 4, -1)$; $x = 5t - 1, y = 4t - 1, z = -t + 2$

4. (a) $\dfrac{x-1}{1} = \dfrac{y+1}{2} = \dfrac{z-2}{-3}$ (b) $\dfrac{x-2}{2} = \dfrac{y+3}{8} = \dfrac{z+1}{1}$

(c) $\dfrac{x-1}{1} = \dfrac{z-5}{-1}$, $\quad y = -3$ (d) $\dfrac{x-1}{-1} = \dfrac{y}{1}$, $\quad z = 0$

(e) $x = 2$, $\quad z = -1$ (f) $\dfrac{x+1}{5} = \dfrac{y+1}{4} = \dfrac{z-2}{-1}$

5. (a) $P = (2, -1, 4) + t(3, 7, 2)$; $\quad x = 3t + 2, y = 7t - 1, z = 2t + 4$
 (b) $P = (-1, -2, 1) + t(1, 1, 1)$; $\quad x = t - 1, y = t - 2, z = t + 1$
 (c) $P = (3, 1, 4) + t(2, 3, 0)$; $\quad x = 2t + 3, y = 3t + 1, z = 4$
 (d) $P = (-1, -2, 1) + t(3, 1, 2)$; $\quad x = 3t - 1, y = t - 2, z = 2t + 1$
 (e) $P = (1, 2, 0) + t(0, 0, 1)$; $\quad x = 1, y = 2, z = t$
 (f) $P = (1, 1, -1) + t(0, 3, 2)$; $\quad x = 1, y = 3t + 1, z = 2t - 1$

6. (a) $(-1, 8, 0)$ (b) $(6, -4, 8)$ (c) no intersection   (d) $(1, -1, 4)$

7. (a) $\cos \theta = 1/\sqrt{6} \approx 0.42$, $\quad \theta \approx 65°$ (b) $90°$
 (c) $\cos \theta = 8/\sqrt{70} \approx 0.95$, $\quad \theta \approx 20°$ (d) 0

8. (a) $P = (1, 2, 4) + t(2, 1, 0)$ (b) $P = (3, 1, -1) + t(3, -1, 0)$
 (c) $P = (4, -2, -3) + t(0, 1, 0)$ (d) $P = (-2, 1, 5) + t(4, 5, 0)$

10. $60°$

§11D. Planes: vector equations

11D1. Vector equation of a plane; parametric equations. Let Q be a point in space, and let V_1 and V_2 be two non-zero vectors that are neither parallel to each other nor opposite. The lines $P = Q + tV_1$ and $P = Q + tV_2$ determine a plane, p. For all numbers t_1 and t_2, the point $Q + t_1 V_1 + t_2 V_2$ lies in p (parallelogram law). Conversely, every point P of p is of this form, as is indicated in Figure 1. Hence the equation

(1) $$P = Q + t_1 V_1 + t_2 V_2,$$

in which Q, V_1, and V_2 are fixed, while t_1 and t_2 are two real variables, is a vector equation of the plane p.

The vectors V_1 and V_2 are called *direction vectors* for p. Thus, (1) is the equation of the plane through the given point Q with direction vectors V_1 and V_2.

We may put (1) into parametric form. Say $Q = (x_0, y_0, z_0)$, $V_1 = (a_1, b_1, c_1)$, and $V_2 = (a_2, b_2, c_2)$. Then (1) becomes:

(2) $$x = x_0 + a_1 t_1 + a_2 t_2, \quad y = y_0 + b_1 t_1 + b_2 t_2, \quad z = z_0 + c_1 t_1 + c_2 t_2.$$

These are parametric equations in two parameters, t_1 and t_2.

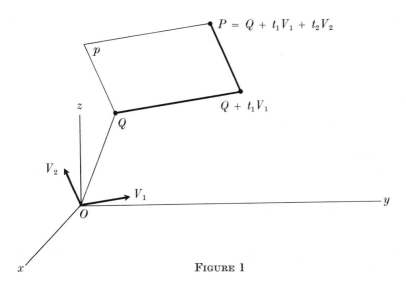

$$P = Q + t_1 V_1 + t_2 V_2$$

$$Q + t_1 V_1$$

FIGURE 1

Example 1. Find the vector equation of the plane through $Q = (5, -1, 3)$ with direction vectors $V_1 = (1, 2, 1)$ and $V_2 = (-1, -1, 3)$.

Solution. The equation is

$$P = (5, -1, 3) + t_1(1, 2, 1) + t_2(-1, -1, 3).$$

The parametric form is

$$x = 5 + t_1 - t_2, \qquad y = -1 + 2t_1 - t_2, \qquad z = 3 + t_1 + 3t_2.$$

To find the equation of the plane through a given line $P = Q + t_1 V_1$ and a given point R (not on the line), take for V_2 the direction vector of QR.

Example 2. Find the vector equation of the plane through the line $P = (0, 1, 2) + t(4, 4, 3)$ and the point $R = (1, 2, 3)$.

Solution. Direction vectors are $V_1 = (4, 4, 3)$, from the equation of the line, and $V_2 = (1, 1, 1)$, from the two points. The equation of the plane is

$$P = (0, 1, 2) + t_1(4, 4, 3) + t_2(1, 1, 1).$$

To find the equation of the plane through three given points, find V_1 and V_2 as direction vectors of two of the lines they determine.

Example 3. Find the vector equation of the plane through the three points $Q = (5, -1, 3)$, $R = (6, 2, 2)$, and $S = (4, -2, 0)$.

Solution. The plane is determined by the lines QR and QS, for which direction vectors are $V_1 = (1, 3, -1)$ and $V_2 = (1, 1, 3)$. Hence the equation is

$$P = (5, -1, 3) + t_1(1, 3, -1) + t_2(1, 1, 3).$$

11D2. Normal to a plane. A line or vector perpendicular to a plane (i.e., to every line in the plane) is also called a *normal* to the plane. Given the vector equation of a plane,

$$P = Q + t_1 V_1 + t_2 V_2,$$

we can easily find a normal vector, N. For, V_1 and V_2 are direction vectors of lines in the plane, and we need only choose N perpendicular to them:

(3) $$N \cdot V_1 = 0, \qquad N \cdot V_2 = 0.$$

These are only two equations, and N has three coordinates, but we determine the coordinates only up to a factor of proportionality.

Example 4. Find a normal to the plane

$$P = (1, 2, 3) + t_1(2, -1, 0) + t_2(-1, 1, 2).$$

Solution. Let $N = (a, b, c)$ be the normal vector sought. The conditions of perpendicularity, (3), are

$$2a - b = 0$$
$$-a + b + 2c = 0.$$

Picking $a = 1$, we get $b = 2$ and $c = -\frac{1}{2}$. Thus we may take $N = (1, 2, -\frac{1}{2})$; or, more conveniently,

$$N = (2, 4, -1).$$

Problems (§11D)

1. Find a vector equation of the plane through the given point Q and with vectors V_1 and V_2.
   (a) $Q = (5, 2, -7)$, $V_1 = (1, 2, 1)$, $V_2 = (3, -2, 0)$
   (b) $Q = (-1, 2, 1,)$, $V_1 = (-1, 1, 1)$, $V_2 = (5, 3, -2)$
   (c) $Q = (6, 3, 1)$, $V_1 = (1, 0, 0)$, $V_2 = (0, 1, 0)$
   (d) $Q = (4, -1, 3)$, $V_1 = (0, 0, 1)$, $V_2 = (1, 1, 1)$
   (e) $Q = (5, 0, 4)$, $V_1 = (1, 0, 2)$, $V_2 = (3, 0, 7)$

2. Find a vector equation of the plane containing the given point and line.
   (a) $(1, 2, 3)$; $P = (-2, 1, 4) + t(1, -1, 2)$
   (b) $(0, 0, 0)$; $P = (1, -5, 2) + t(4, 0, -3)$
   (c) $(1, 1, 1)$; the z-axis
   (d) $(0, 0, 1)$; the line through $(1, 0, 0)$ and $(0, 1, 0)$
   (e) $(4, 3, -5)$; $P = (7, 2, -3) + t(1, 12, 5)$
   (f) $(3, 1, 4)$; $\dfrac{x - 7}{6} = \dfrac{y + 2}{5} = \dfrac{z + 5}{4}$

3. Show that the lines

$$P = (3, 1, 4) + t(3, -2, 4) \qquad \text{and} \qquad P = (6, -7, 2) + t(-9, 6, -12)$$

are parallel, and find a vector equation of the plane containing both lines.

4. Find a vector equation of the plane through the three given points.

   (a) $(1, 0, 0)$, $(1, 2, 3)$,   and $(1, 1, 1)$

   (b) $(2, -5, 7)$, $(-1, 3, 6)$,   and $(4, 7, -5)$

   (c) $(6, 7, 0)$, $(-2, 3, 0)$,   and $(7, 9, 0)$

   (d) $(1, 1, -1)$, $(1, -1, 1)$,   and $(-1, 1, 1)$

   (e) $(0, 0, 0)$, $(1, 0, 1)$,   and $(0, 1, 1)$

   (f) $(5, 4, 3)$, $(-1, 1, 4)$,   and $(2, -2, 5)$

5. Find a vector equation of the plane through the given point and perpendicular to the given line.

   (a) $(6, 7, 3)$; $P = (1, 13, -6) + t(5, -3, 4)$

   (b) $(1, 5, 2)$; $P = (4, 1, 7) + t(3, -1, 0)$

   (c) $(2, 1, 1)$; $\dfrac{x + 3}{4} = \dfrac{y - 1}{3} = \dfrac{z + 5}{7}$

   (d) $(0, 0, 0)$; $P = (0, 0, 0) + t(1, 1, 1)$

6. Find a vector equation for the line through the given point and normal to the given plane.

   (a) $(1, 6, 2)$; $P = (1, 2, 5) + t_1(2, 3, 1) + t_2(4, -3, -1)$

   (b) $(-1, 4, 3)$; $P = (2, -3, 7) + t_1(1, 2, -1) + t_2(5, 7, 4)$

   (c) $(1, 2, 3)$;   the yz-plane

   (d) $(0, 0, 0)$; $P = (1, 0, 0) + t_1(1, -1, 0) + t_2(1, 0, -1)$

   (e) $(1, 2, 0)$; $P = (6, 3, 2) + t_1(1, 6, 0) + t_2(3, -5, 0)$

   (f) $(3, 5, -7)$; $P = t_1(1, 2, -3) + t_2(-1, 5, 4)$

7. Find a vector equation of the plane containing the given point and parallel to the given plane.

   (a) $(1, 2, -7)$; $P = (4, 3, 2) + t_1(1, 2, 7) + t_2(3, -1, 4)$

   (b) $(4, 6, 0)$; $P = (6, 7, -2) + t_1(2, 3, -2) + t_2(3, 2, -2)$

   (c) $(0, 0, 0)$; $P = (1, -4, 2) + t_1(1, 1, 0) + t_2(0, 0, 1)$

   (d) $(1, 1, 1)$; $P = (4, -17, 3) + t_1(2, 3, 2) + t_2(7, 2, -4)$

8. Which of the points

$$(5, 1, 1), \qquad (4, 3, 2), \qquad (9, 0, 3), \qquad (7, -1, -4)$$

lie in the plane

$$P = (1, 3, -2) + t_1(2, 1, 1) + t_2(-1, 3, 2)?$$

9. Find the distance from the origin to the plane

$$P = (5, 1, 3) + t_1(1, -1, 0) + t_2(1, 0, -2).$$

(By definition, this means the distance along the normal.)

10. Show that the planes

$$P = (7, 5, 2) + t_1(-1, 1, 2) + t_2(3, 1, 0)$$

and

$$P = (3, 11, -1) + t_1(0, 2, 3) + t_2(1, 1, 1)$$

are parallel. [*Hint.* Find a vector normal to one and show that it is also normal to the other.]

Answers to problems (§11D)

1. (a) $P = (5, 2, -7) + t_1(1, 2, 1) + t_2(3, -2, 0)$

2. (a) $P = (-2, 1, 4) + t_1(1, -1, 2) + t_2(3, 1, -1)$
   (b) $P = t_1(4, 0, 3) + t_2(1, -5, 2)$
   (c) $P = t_1(0, 0, 1) + t_2(1, 1, 1)$
   (d) $P = (1, 0, 0) + t_1(1, -1, 0) + t_2(1, 0, -1)$
   (e) $P = (7, 2, -3) + t_1(1, 12, 5) + t_2(3, -1, 2)$
   (f) $P = (7, -2, -5) + t_1(6, 5, 4) + t_2(4, -3, -9)$

3. $P = (3, 1, 4) + t_1(3, -8, -2) + t_2(3, -2, 4)$

4. (a) $P = (1, 0, 0) + t_1(0, 2, 3) + t_2(0, 1, 1)$
   (b) $P = (2, -5, 7) + t_1(3, -8, 1) + t_2(1, 6, -6)$
   (c) $P = (6, 7, 0) + t_1(2, 1, 0) + t_2(1, 2, 0)$
   (d) $P = (1, 1, -1) + t_1(0, 1, -1) + t_2(1, 0, -1)$
   (e) $P = t_1(1, 0, 1) + t_2(0, 1, 1)$
   (f) $P = (5, 4, 3) + t_1(6, 3, -1) + t_2(3, 6, -2)$

5. (a) $P = (6, 7, 3) + t_1(0, 4, 3) + t_2(3, 5, 0)$
   (b) $P = (1, 5, 2) + t_1(1, 3, 0) + t_2(0, 0, 1)$
   (c) $P = (2, 1, 1) + t_1(-3, 4, 0) + t_2(0, 7, -3)$
   (d) $P = t_1(1, -1, 0) + t_2(0, 1, -1)$

6. (a) $P = (1, 6, 2) + t(0, 1, -3)$
   (b) $P = (-1, 4, 3) + t(5, -3, -1)$
   (c) $P = (1, 2, 3) + t(1, 0, 0)$
   (d) $P = t(1, 1, 1)$
   (e) $P = (1, 2, 0) + t(0, 0, 1)$
   (f) $P = (3, 5, -7) + t(23, -1, 7)$

7. (a) $P = (1, 2, -7) + t_1(1, 2, 7) + t_2(3, -1, 4)$
   (b) $P = (4, 6, 0) + t_1(2, 3, -2) + t_2(3, 2, -2)$
   (c) $P = t_1(1, 1, 0) + t_2(0, 0, 1)$
   (d) $P = (1, 1, 1) + t_1(2, 3, 2) + t_2(7, 2, -4)$

8. only $(7, -1, -4)$

9. 5

§11E. Planes: coordinate equations

We now discuss equations of planes in coordinate form.

11E1. "Point-normal" equation of a plane. Let l be a line with direction vector (a, b, c) and let $P_0 = (x_0, y_0, z_0)$ be a given point.

The equation of the plane through P_0 and perpendicular to l is

(1)
$$a(x - x_0) + b(y - y_0) + c(z - z_0) = 0.$$

Proof. Let $P = (x, y, z)$ be any point in space. In terms of P, (1) states that

(2)
$$(a, b, c) \cdot (P - P_0) = 0.$$

If $P = P_0$ then P lies in the plane and satisfies the equation.

If $P \neq P_0$ then PP_0 is a line, with direction vector $P - P_0$, and (2) asserts that this line is perpendicular to l. This is the same as saying that P lies in the plane. So if P satisfies (2) it lies in the plane, and conversely.

Therefore (2) $\bigl($or (1)$\bigr)$ is the equation of the plane.

Example 1. Find the point-normal equation of the plane through $(5, -1, 3)$ perpendicular to the line

$$P = (7t, 1 - 4t, 3 + t).$$

Solution. The line has direction vector $(7, -4, 1)$. The required equation is

$$7(x - 5) - 4(y + 1) + (z - 3) = 0.$$

Example 2. Find the equation of the line through $(2, 1, -1)$ perpendicular to the plane

$$x + 3(y + 4) - 2(z + 2) = 5.$$

Solution. The required line has direction vector $(1, 3, -2)$ (from the coefficients in the equation of the plane). Hence its equation is

$$P = (2, 1, -1) + t(1, 3, -2).$$

A vector perpendicular to the normal to a plane is the direction vector for some line lying in the plane. Hence, to go from the point-normal equation of a plane to the vector equation, pick two such vectors. This can be done by inspection. Then write down the equation.

Example 3. Find the vector equation of the plane

$$3(x - 1) + 4(y - 2) - 5(z + 3) = 0.$$

Solution. The normal vector is $N = (3, 4, -5)$. By inspection, two vectors perpendicular to N are

$$V_1 = (4, -3, 0) \qquad \text{and} \qquad V_2 = (0, 5, 4).$$

A point in the plane is $(1, 2, -3)$ (from the given equation). Hence the vector equation is

$$P = (1, 2, -3) + t_1(4, -3, 0) + t_2(0, 5, 4).$$

To go from the vector form to the point-normal form, find a normal vector as explained in §11D2 and write down the equation.

Example 4. Find the point-normal equation of the plane

$$P = (1, 2, 3) + t_1(2, -1, 0) + t_2(-1, 1, 2).$$

Solution. The plane contains the point (1, 2, 3). In §11D2, Example 4, we found a normal vector to be $N = (2, 4, -1)$. Hence its point-normal equation is

$$2(x - 1) + 4(y - 2) - (z - 3) = 0.$$

11E2. General equation of a plane. We now show that every plane has an equation of the form

(3) $$ax + by + cz + d = 0 \qquad (a, b, c \text{ not all } 0),$$

and, conversely, that every equation of this form is the equation of a plane. In fact, if in (1) we put

(4) $$d = -(ax_0 + by_0 + cz_0),$$

we get (3). Since, in (1), (a, b, c) is a direction vector, the numbers a, b, c are not all 0. Conversely, suppose (3) is given, with, say, $c \neq 0$. Set $z_0 = -d/c$. Then (3) becomes

(5) $$ax + by + c(z - z_0) = 0,$$

which is in the form (1).

Equation (3) is called the *general equation* of the plane.

Note that in (3), (a, b, c) is still a direction vector normal to the plane.

Example 5.

(a) The general equation of the plane $7(x - 5) - 4(y + 1) + (z - 3) = 0$ is

$$7x - 4y + z - 42 = 0.$$

(b) A point-normal equation of the plane $3x - 2y + 5z = 6$ is

$$3x - 2y + 5(z - \tfrac{6}{5}) = 0.$$

Example 6. Find the point of intersection of the line

$$P(t) = (2 + t, -1 + 2t, 1 - t)$$

with the plane $3x - y + 2z = 7$.

Solution. We must find the value of t for which $P(t)$ lies in the plane. This will happen when

$$3(2 + t) - (-1 + 2t) + 2(1 - t) = 7,$$

that is, when $t = 2$. The required point is $P(2) = (4, 3, -1)$.

Example 7. Show that the line

$$P = (1 + 3t, 4 + 7t, -1 + 2t)$$

lies in the plane $4x - 2y + z + 5 = 0$.

Solution. We verify in one step that all points of the line satisfy the equation of the plane: whatever the value of t,

$$4(1 + 3t) - 2(4 + 7t) + (-1 + 2t) + 5 = 0.$$

To find the equation of the plane through three given points one can substitute their coordinates in the general equation, (3). This results in three simultaneous equations in the unknowns a, b, c, d.

A method that avoids solving three simultaneous equations is first to note a pair of direction vectors for the plane, then compute the normal, then write down the point-normal equation.

Example 8. Find the equation of the plane through the points $Q = (2, 3, 2)$, $R = (-1, 3, 5)$, and $S = (1, 6, 5)$.

Solution. Direction vectors for QR and QS are

$$V_1 = (-3, 0, 3) \quad \text{or} \quad (-1, 0, 1), \quad \text{and} \quad V_2 = (-1, 3, 3).$$

For the normal vector, $N = (a, b, c)$, we have $N \cdot V_1 = 0$ and $N \cdot V_2 = 0$; that is,

$$-a + c = 0$$
$$-a + 3b + 3c = 0.$$

Picking $a = 3$ we get $b = -2$ and $c = 3$. Then $N = (3, -2, 3)$. The equation of the plane is

$$3(x - 2) - 2(y - 3) + 3(z - 2) = 0,$$

or, in the general form,

$$3x - 2y + 3z - 6 = 0.$$

To find the line of intersection of two planes, proceed as in the following examples.

Example 9. Find the line of intersection of the planes $x = 3 - z$ and $y = 2z + 2$.

Solution. We observe that each of these equations is missing a variable. The common variable is z.

Let $P_1 = (x_1, y_1, z_1)$ be any point on the line. Its z-coordinate, z_1, will be our parameter t. Since P_1 lies on the plane $x = 3 - z$, we have $x_1 = 3 - t$. Similarly, $y_1 = 2t + 2$. Hence, parametric equations of the line are

$$x = 3 - t, \quad y = 2t + 2, \quad z = t.$$

In general, to find the intersection of two planes use their equations to express two of the variables x, y, and z in terms of the third. (Eliminate one of the variables between the two equations. Then start over and eliminate a different one.) Then proceed as in Example 9.

Example 10. Find the line of intersection of the planes

$$2x - 4y + 3z = 8,$$
$$x + 2y - z = 6.$$

Solution. Eliminating x (by subtracting twice the second equation from the first), we get

$$8y - 5z = 4.$$

Similarly, eliminating y, we have

$$4x + z = 20.$$

Then

$$x = \tfrac{1}{4}(20 - z) \qquad \text{and} \qquad y = \tfrac{1}{8}(4 + 5z).$$

Hence, parametric equations of the line sought are

$$x = 5 - \tfrac{1}{4}t, \qquad y = \tfrac{1}{2} + \tfrac{5}{8}t, \qquad z = t.$$

The direction vector of this line is $(-\tfrac{1}{4}, \tfrac{5}{8}, 1)$. A parallel vector is $(-2, 5, 8)$. Hence, more simply, we have

$$x = 5 - 2t, \qquad y = \tfrac{1}{2} + 5t, \qquad z = 8t.$$

11E3. Angle between two planes. The angle between two planes is defined to be the acute angle between their normals.

Example 11. Find the angle θ between the planes

$$3x - y + 3z = 1 \qquad \text{and} \qquad x + 2y - 4z = 2.$$

Solution. Normal vectors are $N_1 = (3, -1, 3)$ and $N_2 = (1, 2, -4)$.

Hence

$$\cos \theta = \frac{|N_1 \cdot N_2|}{|N_1||N_2|} = \frac{11}{\sqrt{19}\sqrt{21}} \approx \frac{11}{20} = 0.55, \quad \theta \approx 55°.$$

Problems (§11E)

1. Write the point-normal equation of the plane through the given point and perpendicular to the given line.

(a) $(3, 5, 14)$; $P = (1, -4, 2) + t(7, 2, 9)$

(b) $(1, -2, 3)$; $P = (0, 1, 2) + t(2, 5, 1)$

(c) $(-2, 7, 6)$;   the y-axis

(d) $(0, 0, 0)$; $\dfrac{x + 5}{-2} = \dfrac{y + 2}{3}$, $z = 6$

(e) $(1, 1, 1)$; $P = t(1, 1, 1)$

2. Write a vector equation of the line through the given point and perpendicular to the given plane.
   (a) $(1, 2, -1)$; $2x + 3y - z = 6$
   (b) $(2, -1, 7)$; $(x - 4) + 2(y - 3) - (z + 2) = 0$
   (c) $(1, 0, 0)$; $z = 6$
   (d) $(3, -2, 12)$; $2(x + 4) - 7(y + 1) + 2(z - 3) = 0$
   (e) $(0, 0, 0)$; $x + y + z = 0$

3. Write a vector equation of the given plane.
   (a) $2x - 3y + z = 15$
   (b) $(x - 2) - 2(y - 3) + 3(z + 4) = 0$
   (c) $2x - 3y = 10$
   (d) $5(x + 1) + 2(y - 1) + 3z = 6$
   (e) $y = z$

4. Write the point-normal equation of the given plane.
   (a) $P = (1, 7, 4) + t_1(2, 1, -3) + t_2(-1, 5, 7)$
   (b) $P = (2, 2, 1) + t_1(-1, 1, 0) + t_2(4, 11, -1)$
   (c) $P = t_1(1, 8, -3) + t_2(-1, 3, 2)$
   (d) $P = (0, 5, 0) + t_1(1, 0, 0) + t_2(0, 0, 1)$

5. Write the general equation of the given plane.
   (a) $3(x - 1) - 2(y + 1) - 3(z + 2) = 0$
   (b) $-3(x - 1) + 2(y + 2) - 2(z - 6) = 0$
   (c) $4(x - 3) + 5y - 3(z - 2) = 0$

6. Write a point-normal equation of the given plane.
   (a) $x - 3y + 4z = 15$ (c) $3x + 4y - 5z = 6$
   (b) $x + 2y - 3z + 4 = 0$ (d) $x + y + z = 0$

7. Find the point of intersection of the given line and plane.
   (a) $P = (3, -1, 2) + t(1, 3, 1)$; $2x - 3y + z + 13 = 0$
   (b) $P = (1, -3, 7) + t(4, -3, 5)$; $3(x - 4) + 2(y - 2) - 5(z - 7) = 0$
   (c) $P = t(3, 17, -5)$; $4x - 3y - 5z + 14 = 0$
   (d) $P = (4, -1, 3) + t(0, 0, 1)$; $x + 4y + 3z = 18$

8. Which of the planes

$$x + 2y + z = 2,   x - 3y - 2z = 4,$$

$$3x - 4y - 3z = 10,   -4(x - 2) + 7(y + 1) + 5z = 0$$

contains the line

$$P = (3, 1, -2) + t(1, -3, 5) ?$$

9. Write the general equation of the plane through the three given points.
   (a) $(1, 1, 5)$, $(-3, 2, 10)$,   and $(5, 3, -7)$
   (b) $(0, 7, 4)$, $(3, 4, 2)$,   and $(5, 3, 1)$
   (c) $(8, 1, 3)$, $(0, 3, 1)$,   and $(1, 4, 2)$
   (d) $(1, 0, 0)$, $(0, 0, 1)$,   and $(0, 1, 0)$

10. Write a vector equation of the line of intersection of the given planes.
    (a) $4x + 2y + z = 9,$ $7x - y - 2z = 8$
    (b) $2x + 2y + 5z = 9,$ $2x + y - 3z = 10$
    (c) $4x + 3y + z = 14,$ $3x + y - 3z = -4$
    (d) $x - 2y + z = -2,$ $2x - 3y - 2z = -29$

11. Find the angle between the given planes.
    (a) $x - 2y + 5z = 7,$ $4x + y - 7z = 4$
    (b) $3x + 12y - 8z = 1,$ $11x - 4y + 5z = 21$
    (c) $5x - y - 3z = 14,$ $x + y - z = 3$
    (d) $-4x + 3y + 3z = 6,$ $y = x$

12. Find the angle between two faces of a regular tetrahedron. (The four faces are congruent equilateral triangles.)

Answers to problems (§11E)

1. (a) $7(x - 3) + 2(y - 5) + 9(z - 14) = 0$
   (b) $2(x - 1) + 5(y + 2) + (z - 3) = 0$
   (c) $y - 7 = 0$
   (d) $2x - 3y = 0$
   (e) $(x - 1) + (y - 1) + (z - 1) = 0$

2. (a) $P = (1, 2, -1) + t(2, 3, -1)$
   (b) $P = (2, -1, 7) + t(1, 2, -1)$
   (c) $P = (1, 0, 0) + t(0, 0, 1)$
   (d) $P = (3, -2, 12) + t(2, -7, 2)$
   (e) $P = t(1, 1, 1)$

3. (a) $P = (5, 0, 5) + t_1(3, 2, 0) + t_2(0, 1, 3)$
   (b) $P = (2, 3, -4) + t_1(2, 1, 0) + t_2(0, 3, 2)$
   (c) $P = (5, 0, 0) + t_1(0, 0, 1) + t_2(3, 2, 0)$
   (d) $P = (-1, 1, 2) + t_1(2, -5, 0) + t_2(0, 3, -2)$
   (e) $P = t_1(1, 0, 0) + t_2(0, 1, 1)$

4. (a) $2(x - 1) - (y - 7) + (z - 4) = 0$
   (b) $(x - 2) + (y - 2) + 15(z - 1) = 0$
   (c) $25x + y + 11z = 0$
   (d) $y - 5 = 0$

5. (a) $3x - 2y - 3z - 11 = 0$
   (b) $3x - 2y + 2z - 19 = 0$
   (c) $4x + 5y - 3z - 6 = 0$

6. (a) $x - 3(y + 5) + 4z = 0$
   (b) $(x + 4) + 2y - 3z = 0$
   (c) $3(x - 2) + 4y - 5z = 0$
   (d) $x + y + z = 0$

7. (a) $(7, 11, 6)$ (b) $(-3, 0, 2)$ (c) $(3, 17, -5)$ (d) $(4, -1, 6)$

8. $x - 3y - 2z = 4$ and $-4(x - 2) + 7(y + 1) + 5z = 0$

9. (a) $11x + 14y + 6z - 55 = 0$ (b) $x - y + 3z - 5 = 0$
   (c) $2x + 3y - 5z - 4 = 0$ (d) $x + y + z - 1 = 0$

10. (a) $P = (1, \frac{11}{3}, -\frac{7}{3}) + t(1, -5, 6)$
    (b) $P = (0, 7, -1) + t(11, -16, 2)$
    (c) $P = (0, \frac{19}{5}, \frac{13}{5}) + t(2, -3, 1)$
    (d) $P = (-10, -1, 6) + t(7, 4, 1)$

11. (a) $\cos \theta = 33/\sqrt{30 \times 66} \approx 0.73, \quad \theta \approx 45°$
    (b) $\cos \theta = 55/\sqrt{162 \times 217} \approx 0.29, \quad \theta \approx 75°$
    (c) $\cos \theta = 7/\sqrt{3 \times 35} \approx 0.69, \quad \theta \approx 45°$
    (d) $\cos \theta = 7/\sqrt{34 \times 2} \approx 0.85, \quad \theta \approx 30°$

12. $\cos \theta = \frac{1}{3}, \quad \theta \approx 70°$

§11F. Vector-valued functions

The concept of vector-valued function and the results developed in §9D and §9E for two dimensions extend immediately to three dimensions.

11F1. Vector-valued functions. A function P that associates to each number t in some interval a point or vector,

(1) $$P(t) = \big(f(t), g(t), h(t)\big),$$

in $\mathcal{R}^3$ is called a *vector-valued function*. The parametric form of (1) is

(2) $$x = f(t), \qquad y = g(t), \qquad z = h(t).$$

When we think of a vector-valued function geometrically, we call it a *curve*.

Example 1. Lines are curves. They are the curves for which the coordinate functions (2) are all linear (but not all constant, which would yield only a point).

Example 2. Describe the curve

$$P(t) = \left(\cos t, \sin t, \frac{t}{2}\right).$$

Solution. The curve is a circular *helix*. As (x, y) winds around the unit circle, z pulls it up (or down). With each revolution (a change in t by 2π), the level changes by the amount π (Figure 1).

11F2. Limits and continuity; derivative and tangent. Limits and continuity are defined coordinatewise. For $P(t)$ as in (1),

(3) $$\lim_{t \to t_0} P(t) = (a, b, c)$$

means

(4) $$\lim_{t \to t_0} f(t) = a, \qquad \lim_{t \to t_0} g(t) = b, \qquad and \qquad \lim_{t \to t_0} h(t) = c.$$

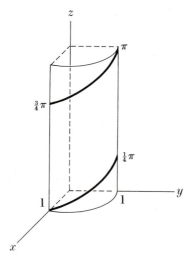

FIGURE 1

We say that P is *continuous* at t_0 provided that

$$(5) \qquad\qquad \lim_{t \to t_0} P(t) = P(t_0).$$

The *derivative* at t_0 is defined by

$$(6) \qquad\qquad P'(t_0) = \lim_{\Delta t \to 0} \frac{P(t_0 + \Delta t) - P(t_0)}{\Delta t}.$$

We find that

$$(7) \qquad\qquad P'(t_0) = \big(f'(t_0), g'(t_0), h'(t_0)\big).$$

The derivative has all the properties discussed in §9E3:

$$(8) \qquad (P + Q)' = P' + Q', \qquad (P \cdot Q)' = P \cdot Q' + Q \cdot P',$$

$$(9) \qquad (fP)' = fP' + f'P, \qquad \left(\frac{P}{f}\right)' = \frac{fP' - f'P}{f^2}, \qquad \frac{dP}{ds} = \frac{dP}{dt}\frac{dt}{ds}.$$

The proofs are the same as in §9E3, with the additional coordinate taken into account.

When $P'(t_0)$ is not zero, we may interpret it as a *tangent* vector at t_0. The tangent line at t_0 is the line through $P(t_0)$ with direction vector $P'(t_0)$.

If P is differentiable, i.e., if its derivative exists, then it is continuous.

Example 3. Find the equation of the tangent line to the circular helix of Example 2,

$$P(t) = \left(\cos t,\, \sin t,\, \frac{t}{2}\right)$$

at $t = \pi/2$.

Solution. The derivative is

$$P'(t) = (-\sin t,\, \cos t,\, \tfrac{1}{2}).$$

The equation of the tangent line is

$$T(s) = P(\pi/2) + sP'(\pi/2) = (0,\, 1,\, \pi/4) + s(-1,\, 0,\, \tfrac{1}{2}).$$

11F3. Motion in space. The discussion of motion given in §9G extends at once to three dimensions. Let

$$P(t) = \big(f(t),\, g(t),\, h(t)\big)$$

represent the position of a particle at time t. The set of values of P is called the *trajectory* of the particle. The derivative of P is the *velocity*, V, and the derivative of V is the *acceleration*, A. The magnitude of the velocity vector is called the *speed*:

$$|V| = \sqrt{f'(t)^2 + g'(t)^2 + h'(t)^2}.$$

Example 4. The position of a particle is given by

$$P(t) = (1 + 2t,\, t^2,\, -1 + t^3).$$

Describe its situation at time $t = 2$. When is the magnitude of the acceleration a minimum?

Solution. The velocity and acceleration vectors are

$$V(t) = P'(t) = (2,\, 2t,\, 3t^2)$$

and

$$A(t) = V'(t) = (0,\, 2,\, 6t).$$

We have

$$P(2) = (5,\, 4,\, 7), \qquad V(2) = (2,\, 4,\, 12),$$

and

$$|V(2)| = \sqrt{2^2 + 4^2 + 12^2} = 2\sqrt{41} \approx 12.8.$$

At $t = 2$, the particle is at the point $P(2) = (5,\, 4,\, 7)$ and is headed toward the point

$$P(2) + V(2) = (7,\, 8,\, 19)$$

at a speed of $|V(2)| \approx 12.8$.

For the magnitude of the acceleration vector, we have

$$|A(t)|^2 = 4 + 36t^2,$$

which achieves a minimum value of 4 at $t = 0$.

11F4. Length of arc. We wish to define the length of a curve

$$P(t) = \big(f(t), g(t), h(t)\big)$$

on a closed interval $a \le t \le b$. We assume that P has a continuous derivative on this interval. Then, reasoning just as in §9H, we find ourselves forced to define the length by:

$$(10) \qquad L = \int_a^b \sqrt{f'^2 + g'^2 + h'^2}.$$

Briefly, $L = \int_a^b |P'|$.

If s denotes the length of arc between a and t, then, as before,

$$(11) \qquad s = \int_a^t |P'|.$$

Again, by the Fundamental Theorem of Calculus,

$$(12) \qquad \frac{ds}{dt} = |P'| = \sqrt{f'(t)^2 + g'(t)^2 + h'(t)^2}.$$

Example 5. Find the length of one turn of the circular helix of §11F1, Example 2:

$$P(t) = \left(\cos t, \sin t, \frac{t}{2}\right).$$

Solution. Here $P'(t) = (-\sin t, \cos t, \tfrac{1}{2})$, so that

$$|P'(t)| = \sqrt{\sin^2 t + \cos^2 t + \tfrac{1}{4}} = \tfrac{1}{2}\sqrt{5}.$$

The length requested is

$$L = \int_0^{2\pi} \tfrac{1}{2}\sqrt{5} = \sqrt{5}\,\pi \approx 2.2\pi,$$

about 10 percent longer than the circle in the plane.

As before, equation (11), applied to motion, states that the distance traveled is the integral of the speed; and (12) says that the speed is the derivative of the distance.

Example 6. A particle moves in space in such a way that the distance traveled up to time t is

$$t + (1 + t) \log(1 + t) \qquad\qquad (t \geq 0).$$

What is its speed?

Solution. Its speed at any instant t is

$$\frac{d}{dt}[t + (1 + t)\log(1 + t)] = 2 + \log(1 + t).$$

Problems (§11F)

1. Compute P' and P''.
   (a) $P = (t^2, 2t + 1, 1 - t^2)$ (d) $P = (3t - 2, t + 1, 2t + 5)$
   (b) $P = (e^t, e^{-t}, t)$ (e) $P = (t, t^2, t^3)$
   (c) $P = (\cos 2t, \sin 2t, t^2)$

2. Write the equation of the tangent line to the curve at the given point.
   (a) $P = (t, t^2, t^3)$, $t_0 = 1$
   (b) $P = (\cos t, \sin t, t^2)$, $t_0 = \pi/2$
   (c) $P = (\log t, t, e^t)$, $t_0 = 1$
   (d) $P = (3t + 2, t^2, 5 - t)$, $t_0 = 0$
   (e) $P = (t \cos t, t \sin t, t)$, $t_0 = 0$

3. The plane perpendicular to the tangent line at t_0 is called the *normal* plane to the curve at t_0. In Problem 2, find the normal plane to the given curve at the point t_0.

4. Find the velocity, speed, and acceleration at time t_0 of a particle whose position at time t is $P(t)$.
   (a) $P(t) = (t \cos t, t \sin t, t)$, $t_0 = 0$
   (b) $P(t) = (t, t^2, \frac{2}{3}t^3)$, $t_0 = 3$
   (c) $P(t) = (\sin t, \cos t, \log \cos t)$, $t_0 = \pi/4$
   (d) $P(t) = (1/t, t, t^2/2)$, $t_0 = 1$
   (e) $P(t) = (e^t, e^{-t}, t^2/\sqrt{2})$, $t_0 = 1$
   (f) $P(t) = (1 - t, 2t, 3t - 2)$, $t_0 = \pi$

5. Compute $(P \cdot Q)'$.
   (a) $P = (t^2 - 1, 3t, t^3)$, $Q = (t, 1/t, 1 + t^2)$
   (b) $P = (\sin t, \cos t, e^t)$, $Q = (\sin t, \cos t, -e^{-t})$
   (c) $P = (\log t, e^t, t)$, $Q = (t, \cos t, t^2)$
   (d) $P = (2t - 1, 3t + 1, 1 - t)$, $Q = (t + 3, 1 - 2t, 3t - 1)$

6. Establish the following formulas.
    (a) $(P \cdot Q)' = P \cdot Q' + Q \cdot P'$
    (b) $(fP)' = fP' + f'P$

    (c) $\left(\dfrac{P}{f}\right)' = \dfrac{fP' - f'P}{f^2}$

    (d) $|P|' = \dfrac{P \cdot P'}{|P|}$

7. Verify that P' is perpendicular to

$$\left(\frac{P'}{|P'|}\right)'.$$

(You may use Problem 6.)

8. Compute $|P|'$. (See Problem 6.)
    (a) $P = (\cos t, \sin t, t/2)$
    (b) $P = (t, t^2, 1 + t^2)$
    (c) $P = (5 \sin t, 12 \sin t, 13 \cos t)$
    (d) $P = (e^t, e^{-t}, te^t)$

9. Show that the tangent lines to the circular helix

$$P = (\cos t, \sin t, t/2)$$

(Examples 2 and 3) make a constant angle with the z-axis.

10. Find the length of the given curve.
    (a) $P(t) = (t^3/3, t^2/\sqrt{2}, t), \quad 0 \le t \le 3$
    (b) $P(t) = (\sin t, \cos t, \log \cos t), \quad 0 \le t \le \pi/4$
    (c) $P(t) = (6 \cos t, 8 \cos t, 10 \sin t), \quad 0 \le t \le 2$
    (d) $P(t) = (3t^2, 4\sqrt{2}\, t^{3/2}, 6t), \quad 0 \le t \le 2$

Answers to problems (§11F)

1. (a) $P' = (2t, 2, -2t), \quad P'' = (2, 0, -2)$
   (b) $P' = (e^t, -e^{-t}, 1), \quad P'' = (e^t, e^{-t}, 0)$
   (c) $P' = (-2 \sin 2t, 2 \cos 2t, 2t), \quad P'' = (-4 \cos 2t, -4 \sin 2t, 2)$
   (d) $P' = (3, 1, 2), \quad P'' = (0, 0, 0)$
   (e) $P' = (1, 2t, 3t^2), \quad P'' = (0, 2, 6t)$
2. (a) $T(s) = (1, 1, 1) + s(1, 2, 3)$
   (b) $T(s) = (0, 1, \pi^2/4) + s(1, 0, -\pi)$
   (c) $T(s) = (0, 1, e) + s(1, 1, e)$
   (d) $T(s) = (2, 0, 5) + s(3, 0, -1)$
   (e) $T(s) = s(1, 0, 1)$
3. (a) $x + 2y + 3z = 6$ (b) $x - \pi z + \pi^3/4 = 0$
   (c) $x + y + ez = e^2 + 1$ (d) $3x - z = 1$ (e) $x + z = 0$

4. (a) $V = (1, 0, 1)$, $|V| = \sqrt{2}$, $A = (0, 2, 0)$
    (b) $V = (1, 6, 18)$, $|V| = 19$, $A = (0, 2, 12)$
    (c) $V = \frac{1}{2}(\sqrt{2}, -\sqrt{2}, -2)$, $|V| = \sqrt{2}$, $A = -\frac{1}{2}(\sqrt{2}, \sqrt{2}, 4)$
    (d) $V = (-1, 1, 1)$, $|V| = \sqrt{3}$, $A = (2, 0, 1)$
    (e) $V = (e, -1/e, 2)$, $|V| = e + 1/e$, $A = (e, 1/e, 2)$
    (f) $V = (-1, 2, 3)$, $|V| = \sqrt{14}$, $A = (0, 0, 0)$
5. (a) $5t^4 + 6t^2 - 1$ (b) 0
    (c) $e^t(\cos t - \sin t) + \log t + 3t^2 + 1$ (d) $10 - 14t$

8. (a) $\dfrac{1}{2} \dfrac{t}{\sqrt{t^2 + 4}}$ (b) $\dfrac{t(3 + 4t^2)}{\sqrt{(1 + t^2)(1 + 2t^2)}}$ (c) 0

    (d) $\dfrac{(1 + t + t^2)e^{2t} - e^{-2t}}{\sqrt{(1 + t^2)e^{2t} + e^{-2t}}}$

10. (a) 12 (b) $\log(1 + \sqrt{2})$ (c) 20 (d) 24

§11G. The geometry of curves

11G1. Introduction. Our study of curves in space is modeled on the discussion of plane curves in §9I and §9J. Consider a curve

(1) $$P(t) = \big(f(t), g(t), h(t)\big),$$

where t ranges over some interval I. We assume that P is twice differentiable on I and that P' is never zero in the interior. If s denotes directed arc length from a reference point $t = a$, then $\big($see §11F(11)$\big)$,

(2) $$s = \int_a^t |P'|, \qquad \frac{ds}{dt} = |P'(t)|, \qquad \frac{dt}{ds} = \frac{1}{|P'(t)|} \, .$$

In contrast to the two-dimensional case, we have no angle of inclination of the tangent line to refer to. Consequently, the discussion here will rely on manipulations with vectors.

11G2. Unit vectors and curvature. The *unit tangent* vector is

(3) $$T = \frac{P'(t)}{|P'(t)|} = \frac{dP/dt}{ds/dt} = \frac{dP}{ds} \, .$$

Next, we define the normal. In the plane there were two directions normal to T and we picked one of the two for N. This time there are a whole planeful of directions normal to T. We single out one of them.

Since $T \cdot T = 1$, we have

$$2T \cdot \frac{dT}{ds} = 0.$$

Hence dT/ds is perpendicular to T. The *unit normal* (or *principal normal*) vector is defined by

(4) $$N = \frac{dT/ds}{|dT/ds|} \qquad (|dT/ds| \neq 0).$$

Equivalently,

(5) $$N = \frac{dT/dt}{|dT/dt|} \qquad (|dT/dt| \neq 0).$$

As suggested in §9J5, we define the *curvature* by

(6) $$\kappa = \left| \frac{dT}{ds} \right|.$$

Since T has constant length, only its direction changes with s; $|dT/ds|$ tells us how fast. Finally, the *radius* of curvature is

(7) $$\rho = \frac{1}{\kappa} \qquad (\kappa \neq 0).$$

11G3. Formulas for curvature. We now develop formulas for the curvature and the normal vector that are convenient for computation. They involve the derivatives,

$$P' = (f', g', h') \qquad \text{and} \qquad P'' = (f'', g'', h'').$$

We use the following abbreviations:

(8) $$p_{11} = P' \cdot P' = |P'|^2, \qquad p_{12} = P' \cdot P'', \qquad p_{22} = P'' \cdot P''.$$

In these terms, the formulas are as follows.

For computing both N and κ:

$$(9) \qquad N = \frac{p_{11}P'' - p_{12}P'}{|p_{11}P'' - p_{12}P'|} \qquad \text{and} \qquad \kappa = \frac{|p_{11}P'' - p_{12}P'|}{p_{11}^2}.$$

If all we want is κ, not N, we use:

$$(10) \qquad \kappa^2 = \frac{p_{11}p_{22} - p_{12}^2}{p_{11}^3}.$$

Then we get κ by taking the square root.

Before establishing these formulas we present an example.

Example 1. Find the unit tangent and normal vectors and the curvature of the curve

$$P(t) = (\tfrac{1}{3}t^3, t^2, 2t)$$

at $t = 1$.

Solution. We have

$$P'(t) = (t^2, 2t, 2), \qquad P''(t) = (2t, 2, 0).$$

The values at $t = 1$ are

$$P' = (1, 2, 2), \qquad P'' = (2, 2, 0),$$

and

$$p_{11} = 9, \qquad p_{12} = 6, \qquad p_{22} = 8.$$

The unit tangent is $T = (\tfrac{1}{3}, \tfrac{2}{3}, \tfrac{2}{3})$. Next,

$$p_{11}P'' - p_{12}P' = 9(2, 2, 0) - 6(1, 2, 2) = 6(2, 1, -2)$$

and

$$|p_{11}P'' - p_{12}P'| = 6 \times 3 = 18.$$

Hence from (9),

$$N = \frac{6(2, 1, -2)}{18} = \frac{1}{3}(2, 1, -2) \qquad \text{and} \qquad \kappa = \frac{18}{9^2} = \frac{2}{9}.$$

If we wanted only κ, not N, we would use (10) to get

$$p_{11}p_{22} - p_{12}^2 = 9 \times 8 - 6^2 = 36, \qquad \kappa^2 = \frac{36}{9^3}, \qquad \kappa = \frac{2}{9}.$$

11G4. Derivation of the formulas. By direct computation,

$$
(11) \qquad \frac{d}{dt}\,|P'| = \frac{d}{dt}\,\sqrt{f'^2 + g'^2 + h'^2} = \frac{p_{12}}{|P'|}.
$$

Therefore,

$$
(12) \qquad \begin{aligned}
\frac{dT}{dt} &= \frac{d}{dt}\frac{P'}{|P'|} = \frac{1}{|P'|^2}\left(|P'|P'' - \frac{p_{12}}{|P'|}\,P'\right)\\[2mm]
&= \frac{1}{|P'|^3}\,(p_{11}P'' - p_{12}P').
\end{aligned}
$$

Hence

$$
(13) \qquad N = \frac{dT/dt}{|dT/dt|} = \frac{p_{11}P'' - p_{12}P'}{|p_{11}P'' - p_{12}P'|}.
$$

To get κ we combine the expression for dT/dt in (12) with the equation $dt/ds = 1/|P'|$. Since $|P'|^4 = p_{11}^2$ we obtain

$$
(14) \qquad \kappa = \left|\frac{dT}{ds}\right| = \left|\frac{dT}{dt}\frac{dt}{ds}\right| = \frac{|p_{11}P'' - p_{12}P'|}{p_{11}^2}.
$$

We now have both parts of formula (9).

Finally, to get formula (10) for κ^2, we square in (14). This gives

$$
(15) \qquad \begin{aligned}
\kappa^2 &= \frac{1}{p_{11}^4}\,[p_{11}^2(P'' \cdot P'') - 2p_{11}p_{12}(P' \cdot P'') + p_{12}^2(P' \cdot P')]\\[2mm]
&= \frac{1}{p_{11}^3}\,(p_{11}p_{22} - p_{12}^2),
\end{aligned}
$$

as desired.

Problems (§11G)

1. Find the radius of curvature at the given point. (Additional information about these curves will be asked for in §11H, Problem 1.)
  (a) $P(t) = (t, t^2, t^3)$, $t = 1$
  (b) $P(t) = (t\cos t, t\sin t, t)$, $t = 0$
  (c) $P(t) = (\sin t, \cos t, \log\cos t)$, $t = 0$
  (d) $P(t) = (e^t + e^{-t}, e^t - e^{-t}, 4t)$, $t = 0$

2. Find the unit tangent, the unit normal, and the radius of curvature at the given point. (Additional information about these curves will be asked for in §11H, Problem 2.)
  (a) $P(t) = (t + 1, 2t - 1, 1 - t^2)$, $t = 0$
  (b) $P(t) = (t, 2t, \tfrac{2}{3}t^{3/2})$, $t = 9$

(c) $P(t) = (t, t^2, 1 + t^2), \quad t = 1$

(d) $P(t) = (e^t \cos t, e^t \sin t, e^t), \quad t = 0$

3. Show that if the curvature of a curve is 0 at every point, then the curve is a straight line. [*Hint.* Write $P = (F(s), G(s), H(s))$ and apply (6).]

4. Consider the circular helix $P(t) = (\cos t, \sin t, t/2)$. Show that the unit normal vector $N(t)$ lies in the xy-plane and that $P(t) + N(t)$ lies on the z-axis.

5. Show that for a particle moving at constant speed,

$$N = \frac{P''}{|P''|}.$$

Answers to problems (§11G)

1. (a) $\frac{7}{19}\sqrt{266}$ (b) 1   (c) $\frac{1}{2}\sqrt{2}$ (d) 10

2. (a) $T = \dfrac{1}{\sqrt{5}}(1, 2, 0), \quad N = (0, 0, -1), \quad \rho = \frac{5}{2}$

(b) $T = \dfrac{1}{\sqrt{14}}(1, 2, 3), \quad N = \dfrac{1}{\sqrt{70}}(-3, -6, 5), \quad \rho = \frac{84}{5}\sqrt{70}$

(c) $T = (\frac{1}{3}, \frac{2}{3}, \frac{2}{3}), \quad N = \frac{1}{6}\sqrt{2}\,(-4, 1, 1), \quad \rho = \frac{27}{4}\sqrt{2}$

(d) $T = \dfrac{1}{\sqrt{3}}(1, 1, 1), \quad N = \dfrac{1}{\sqrt{2}}(-1, 1, 0), \quad \rho = \frac{3}{2}\sqrt{2}$

§11H. The osculating plane

11H1. The osculating plane. Let

(1) $$P(t) = (f(t), g(t), h(t))$$

be as in §11G. Consider a fixed point

$$P(t_0) = (x_0, y_0, z_0).$$

We assume that $|dT/dt| \neq 0$ at t_0, so that $N(t_0)$ is defined. This is equivalent to saying that $\kappa \neq 0$. (See §11G2.)

The plane through $P(t_0)$ with direction vectors $T(t_0)$ and $N(t_0)$ (see §11D) is called the *osculating plane* at $t = t_0$. A vector equation of the osculating plane is, then,

(2) $$Q = P(t_0) + r_0 T(t_0) + r_2 N(t_0),$$

with parameters r_1 and r_2 $\big(\S11D(1)\big)$. Now, T has the same direction as P'; and N, as we saw in §11G(9), has the same direction as

$$(3) \qquad p_{11}P'' - p_{12}P'.$$

Hence any expression of the form $r_1T + r_2N$ can be written in the form $t_1P' + t_2P''$. Conversely, every expression of the latter form can be written the first way: given t_1 and t_2, simply solve for r_1 and r_2, using the assumption (§11G1) that $p_{11} \neq 0$. Therefore,

$$(4) \qquad Q = P(t_0) + t_1P'(t_0) + t_2P''(t_0)$$

is also a vector equation of the osculating plane.

To obtain the point-normal equation $\big(\S11E(1)\big)$, we need a normal to the plane as coefficient vector. This will be a non-zero vector perpendicular to both

$$P' = (f', g', h')$$

and

$$P'' = (f'', g'', h'').$$

We verify by inspection that the vector

$$(5) \qquad B = (g'h'' - h'g'', h'f'' - f'h'', f'g'' - g'f'')$$

is perpendicular to P' and P''. (To find it in the first place, one can proceed as described in §11D2.) It will always turn out that $B \neq O$, so that B does qualify as a coefficient vector.

To prove this we show that $B \cdot B > 0$. After some lengthy but straightforward algebraic manipulations, we find that

$$(6) \qquad B \cdot B = p_{11}p_{22} - p_{12}^2.$$

Thus, by the formula for κ^2 in §11G(10),

$$(7) \qquad B \cdot B = p_{11}^3 \kappa^2 > 0.$$

The point-normal equation of the osculating plane at t_0 is, therefore,

$$(8) \qquad \begin{aligned}(g'h'' - h'g'')(x - x_0) + (h'f'' - f'h'')(y - y_0) \\ + (f'g'' - g'f'')(z - z_0) = 0,\end{aligned}$$

where the derivatives are evaluated at t_0.

Example 1. Find the equation of the osculating plane to the curve

$$P(t) = (\tfrac{1}{3}t^3, t^2, 2t)$$

at $t_0 = 1$.

Solution. We have

$$P' = (t^2, 2t, 2) \quad \text{and} \quad P'' = (2t, 2, 0).$$

At $t_0 = 1$,

$$P = (\tfrac{1}{3}, 1, 2), \quad P' = (1, 2, 2), \quad P'' = (2, 2, 0).$$

Then

$$B = \big((2)(0) - (2)(2), (2)(2) - (1)(0), (1)(2) - (2)(2)\big) = -2(2, -2, 1).$$

Hence the equation of the osculating plane is

$$2(x - \tfrac{1}{3}) - 2(y - 1) + (z + 2) = 0.$$

Remark. It can be shown (with the help of additional hypotheses about derivatives) that in a certain precise sense, the osculating plane at t_0 is the plane having closest contact with the curve there. Thus, it is aptly named.

11H2. The osculating circle. The geometry of curves is an interesting and difficult subject all its own. We have only touched on it. Before leaving it we make a few comments to tie in with the results developed for plane curves.

We begin with a remark about the unit normal vector. In the plane, we saw that N is directed toward the concave side of the curve (§9J3). For a space curve, the notion of "the concave side" is not so clear cut. So we reason as follows.

In general, it makes sense to say that a function is moving in the direction of its derivative. For example, a curve is moving in the direction of its tangent. Applying this principle to the unit tangent vector, T, we conclude that T is moving in the direction of N. Thus, N is pointed in the direction in which the curve is turning.

We now turn to the *osculating circle* or *circle of curvature* at $t = t_0$. This is defined to be the circle of radius ρ that lies in the osculating plane, is tangent to the curve at t_0, and whose center lies in the direction of the normal vector. Hence its center, the *center of curvature*, is the point

(9) $$C = P + \rho N.$$

Evidently, the osculating circle is the intersection with the osculating plane of the sphere with center C and radius ρ.

Example 2. Consider the curve $P = (\frac{1}{3}t^3, t^2, 2t)$ at the point $t_0 = 1$. In §11G, Example 1 we found that

$$\kappa = \tfrac{2}{9} \qquad \text{and} \qquad N = \tfrac{1}{3}(2, 1, -2).$$

Hence $\rho = \frac{9}{2}$ and the center of curvature is

$$C = (\tfrac{1}{3}, 1, 2) + \tfrac{9}{2} \times \tfrac{1}{3}(2, 1, -2) = (\tfrac{10}{3}, \tfrac{5}{2}, -1).$$

The sphere with center C and radius ρ is

$$(x - \tfrac{10}{3})^2 + (y - \tfrac{5}{2})^2 + (z + 1)^2 = \tfrac{81}{4}.$$

The osculating circle is the intersection of this sphere with the osculating plane,

$$2(x - \tfrac{1}{3}) - 2(y - 1) + (z - 2) = 0$$

(obtained in Example 1 above).

Problems (§11H)

1. In §11G, Problem 1, find a vector equation and a point-normal equation of the osculating plane and find the center of curvature of each curve at the given point.

2. In §11G, Problem 2, find a vector equation and a point-normal equation of the osculating plane and find the center of curvature of each curve at the given point.

Answers to problems (§11H)

1. (a) $Q = (1, 1, 1) + t_1(1, 2, 3) + t_2(0, 1, 3),$
   $3x - 3y + (z - 1) = 0, \quad C = (-3\frac{1}{19}, -1\frac{18}{19}, 4\frac{6}{19})$
   (b) $Q = t_1(1, 0, 1) + t_2(0, 1, 0), \quad x - z = 0, \quad C = (0, 1, 0)$
   (c) $Q = (0, 1, 0) + t_1(1, 0, 0) + t_2(0, 1, 1), \quad (y - 1) - z = 0, \quad C = (0, \frac{1}{2}, -\frac{1}{2})$
   (d) $Q = (2, 0, 0) + t_1(0, 1, 2) + t_2(1, 0, 0), \quad 2y - z = 0, \quad C = (12, 0, 0)$

2. (a) $Q = (1, -1, 1) + t_1(1, 2, 0) + t_2(0, 0, 1),$
   $2(x - 1) - (y + 1) = 0, \quad C = (1, -1, -\frac{3}{2})$
   (b) $Q = t_1(1, 2, 3) + t_2(0, 0, 1), \quad 2x - y = 0, \quad C = (-41\frac{2}{5}, -82\frac{4}{5}, 102)$
   (c) $Q = (1, 0, 1) + t_1(1, 2, 2) + t_2(0, 1, 1),$
   $y - (z - 1) = 0, \quad C = (-8, 3\frac{1}{4}, 4\frac{1}{4})$
   (d) $Q = (1, 0, 1) + t_1(1, 1, 1) + t_2(0, 2, 1),$
   $(x - 1) + y - 2(z - 1) = 0, \quad C = (-\frac{1}{2}, 1\frac{1}{2}, 1)$

CHAPTER 12

Functions of Several Variables

Functions of two or more variables arise in all areas of mathematics and its applications. The volume enclosed by a right circular cylinder is a function of the radius of the base and the height. For a given f, the integral $\int_u^v f$ is a function of the two variables u and v. The distance an object falls through the atmosphere depends on (is a function of) the duration of the fall, the force of gravity, and the resistance of the air; air resistance, in turn, depends upon the shape of the object and other factors. A merchant's profits depend on his cost outlay, fixed expenses, selling prices, number of items sold, etc. The more carefully one analyzes, the more variables one is likely to think of. One must then strike a balance between the accuracy of the description and the difficulty of the resulting mathematical problem.

The present chapter presents some basic definitions (§12A) and gives a brief account of the graphs of some standard functions (§12B). §12C is devoted to technical results about continuous functions that will not be needed until Chapter 14. Most of this material is set in small type. One way to treat this section is to read the part that isn't.

§12A. Functions of two variables

12A1. Functions. We consider *real-valued* functions of two variables. Let A be a set in $\mathscr{R}^2$. The members of A are pairs of real numbers (or points in the plane). A function f with domain A associates to each pair (x, y) in A a real number, $z = f(x, y)$.

Example 1. The volume, V, enclosed by a right circular cylinder is a function of its radius, r, and its height, h; say

$$V = f(r, h).$$

Explicitly,

$$f(r, h) = \pi r^2 h.$$

This function is defined for all $r > 0$ and all $h > 0$; that is, the domain of f is the first quadrant in the rh-plane.

12A2. Domains. Domains of functions can in principle be any sets whatever in the plane. Those we actually encounter will be unexciting: either the entire plane, or else portions of the plane delineated in uncomplicated ways by points or continuous curves.

We now give some examples. The purpose is not to show that you have to worry about domains but that you don't have to worry.

Example 2. *The entire plane minus the single point* $(0, 0)$.
This is the domain of the function

$$z = g(x, y) = \frac{1}{x^2 + y^2}.$$

Example 3. *The unit disc,* $x^2 + y^2 \leq 1$.
This is the domain of the function

$$z = h(x, y) = \sqrt{1 - x^2 - y^2}.$$

Example 4. *The quadrant* $\{(x, y) : x > 0, y > 0\}$.
This appeared as a domain in Example 1.

Other typical possibilities are shown in Figure 1. The first set is bounded by two vertical lines and the graphs of two continuous functions. The next is similar but with x and y interchanged. The disc $x^2 + y^2 \leq 1$ appears as a special case of either type. The third set is unbounded to the right. The fourth is unbounded above and below. In all cases the set may include some or all of its boundary.

12A3. Regions; boxes. The sets discussed above are examples of *regions*. The characteristic property of regions is somewhat technical to describe. We shall state it in terms of *boxes*. Boxes are the two-dimensional analogue of intervals. If I and J are intervals, then the set of all points (x, y), for $x \in I$ and $y \in J$, is a box (Figure 2). If the intervals are both open, it is an *open* box; if they are both closed, a *closed* box. In any case, the box is denoted $I \times J$; this is read I "cross" J. Thus,

(1) $$I \times J = \{(x, y) : x \in I, y \in J\}.$$

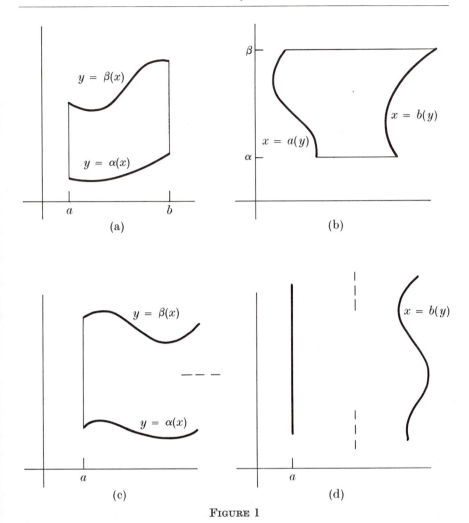

FIGURE 1

The property in question is this: if (x_0, y_0) is any point of a region not on a boundary curve, then there exists an open box containing (x_0, y_0) and lying wholly in the region.

All sets of the forms discussed above have this property. Henceforth, the expression, "Let f be a function defined on a region," is a signal not to brood but to relax.

Remark. To see why the above sets are regions, consider the typical case in which the set is bounded above by the graph of a continuous function $y = \beta(x)$ (Figure 3). Then $\beta(x_0) > y_0$. Pick a number r between them: $\beta(x_0) > r > y_0$. Since β is continuous, there is an open interval I about x_0 throughout which $\beta(x) > r$ (§2C2). It should now be clear how to complete the proof.

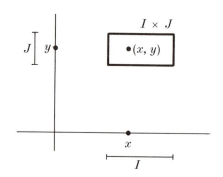

FIGURE 2

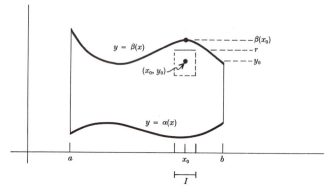

FIGURE 3

A region that includes none of its boundary points may be called *open*. Thus an open region contains an open box about each of its points.

In three dimensions, a typical region is bounded above and below by *surfaces* $z = A(x, y)$ and $z = B(x, y)$, where A and B are continuous (defined in §12A4 following) on some two-dimensional region in the xy-plane. In more than three variables the analogy continues, though the geometry becomes less concrete.

12A4. Limits and continuity. The underlying idea of limit and continuity is as before. Let f be defined on an open region, R, and let (x_0, y_0) be a point of R. The assertion

(2) $$\lim_{(x,y)\to(x_0,y_0)} f(x, y) = c$$

is to express in a precise way the rough idea that $f(x, y)$ is close to c when (x, y) is close to (but not equal to) (x_0, y_0). This last means that x is close to

x_0 and y is close to y_0. So we define (2) to mean that, given any open interval E about c, there is an open box $I \times J$ about (x_0, y_0) that f takes, except for the point (x_0, y_0) itself, into E.

We say that f is continuous at (x_0, y_0) if

$$(3) \qquad\qquad \lim_{(x,y)\to(x_0,y_0)} f(x, y) = f(x_0, y_0).$$

This may also be written

$$(4) \qquad\qquad \lim_{(\Delta x,\Delta y)\to(0,0)} f(x_0 + \Delta x, y_0 + \Delta y) = f(x_0, y_0).$$

The definition of continuity of f at (x_0, y_0) requires that f be defined on some open box containing (x_0, y_0). For this reason, open regions (§12A3) are appropriate sets on which to consider continuity.

We may extend the definition of continuity to boundary points of a region (corresponding to the idea in one dimension of continuity at an endpoint of an interval). To do this we interpret the phrase "open box about (x_0, y_0)" in a relative way. That is, we do not require that f take the entire box into K, but only the part of the box that meets the region. With this understanding, all regions become appropriate sets on which to consider continuity.

In the main, the functions we deal with will be continuous, that is, continuous at each point of their domains.

12A5. Recognizing continuous functions. It is easy to recognize continuous functions by observing the way in which complicated functions are built up from simple ones.

Some functions of two variables depend on fewer. First, there are the constant functions; these, of course, are continuous. Secondly, a function may depend on just one variable, as

$$h(x, y) = 3x, \qquad \text{or} \qquad k(x, y) = 4y^2 + 1.$$

These are continuous functions of one variable, and it is very easy to see that all such functions are continuous as functions of two variables.

Theorems on combining continuous functions are proved by the same methods as for one variable. If $f(x, y)$ and $g(x, y)$ are continuous, then so are their sum, difference, product, and quotient (where $g(x, y) \neq 0$). Likewise, compositions of continuous functions are continuous, so that, for instance, $\sin f(x, y)$, $\exp f(x, y)$, and $\log f(x, y)$ (where $f(x, y) > 0$) are continuous.

By combining these possibilities we can recognize all sorts of complicated functions as continuous. For example,

$$\cos (x + y)e^{\sin(xy)} \quad \text{and} \quad \tfrac{1}{2} \arctan(x - y) + \log(x^2 + y^2)$$

are continuous.

The definitions and results about limits and continuity extend to functions of any number of variables.

Problems (§12A)

1. Determine the natural domain of the given function, i.e., the set of all values for which the defining formula is meaningful.

(a) $z = x^2 + y^2 + 1$ (g) $z = \dfrac{1}{x} + \dfrac{1}{y}$

(b) $z = \sqrt{x} + \sqrt{y}$ (h) $z = \dfrac{1}{x + y}$

(c) $z = \sqrt{x + y}$ (i) $z = \tan x + \tan y$
(d) $z = \sqrt{x}\sqrt{y}$ (j) $z = \tan(x + y)$
(e) $z = \sqrt{xy}$ (k) $z = \arctan(x + y)$
(f) $z = \log(x^2 + y^2)$

2. State which of the points (x, y) is in the box $I \times J$.
   (a) $I = (0, 2)$, $J = (0, 2)$; $(x, y) = (1, 1), (1, 2), (2, 1), (\tfrac{3}{4}, 1), (\tfrac{3}{4}, \tfrac{3}{4}), (0, 0)$
   (b) $I = [0, 2]$, $J = [3, 4]$; $(x, y) = (1, 1), (1, 2), (1, 4), (2, 3), (2, 4), (3, 1)$
   (c) $I = (-1, 1)$, $J = [-2, 0]$; $(x, y) = (-1, 0), (0, 0), (0, 1), (1, 0),$
   $(-\tfrac{1}{2}, -\tfrac{1}{2}), (\tfrac{1}{2}, \tfrac{1}{2})$

3. Specify an open box containing the point z_0 and lying wholly in the region R. Choose the open box so that the corresponding closed box also lies in R.
   (a) $z_0 = (2, 1)$; R is the open first quadrant,

   $$\{(x, y): x > 0, y > 0\}.$$

   (b) $z_0 = (2, 1)$; R is the open disc

   $$\{(x, y): (x - 1)^2 + (y - 1)^2 < \tfrac{9}{4}\}.$$

   (c) $z_0 = (1, 2)$; R is the open region bounded by the parabola $y = 4 - x^2$ and the x-axis.
   (d) $z_0 = (\tfrac{1}{2}, \tfrac{1}{2})$; R is the open region bounded by the parabolic curves $y = x^2$ and $y = \sqrt{x}$.

Answers to problems (§12A)

1. (a) the entire plane   (b) $\{(x, y): x \geq 0 \text{ and } y \geq 0\}$
   (c) the set of all points on or above the line $x + y = 0$
   (d) $\{(x, y): x \geq 0 \text{ and } y \geq 0\}$
   (e) the set of all points in the first quadrant or the third quadrant, or on either of the coordinate axes
   (f) the plane minus the origin
   (g) the plane minus the coordinate axes
   (h) the plane minus the line $x + y = 0$
   (i) the plane minus the horizontal lines $y = (n + \frac{1}{2})\pi$ and the vertical lines $x = (n + \frac{1}{2})\pi$, n an integer (positive, negative, or 0)
   (j) the plane minus the lines $x + y = (n + \frac{1}{2})\pi$, n an integer
   (k) the entire plane

2. (a) $(1, 1)$, $(\frac{3}{4}, 1)$, $(\frac{3}{4}, \frac{3}{4})$ (b) $(1, 4)$, $(2, 3)$, $(2, 4)$ (c) $(0, 0)$, $(-\frac{1}{2}, -\frac{1}{2})$

§12B. Graphs

The graph of a function $z = f(x, y)$ (or of an equation $F(x, y, z) = 0$) is a *surface*. The ability to make a rough sketch of a surface can be an important aid in analyzing a problem. The main device is to intersect the surface with various planes and study the resulting curves. Such a curve is called the *trace* of the surface on the plane. The helpful choices are planes parallel to the coordinate planes. The equation of the trace can then be written down at once, and its graph sketched. The trace of $z = f(x, y)$ on the plane $x = x_0$, for instance, is the curve $z = f(x_0, y)$ (in that plane), a function of y alone (Figure 1).

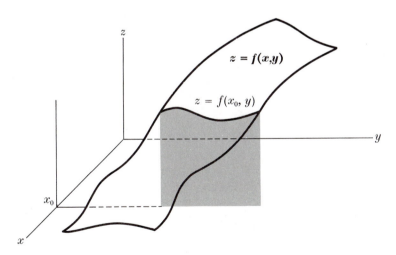

FIGURE 1

The simplest surfaces are those whose equations involve only one or only two of the three variables x, y, z. Absence of a variable means that there is no restriction on its values. Hence these surfaces are *cylindrical* surfaces.

Example 1. *The plane $z = 1$.* (Figure 2.)

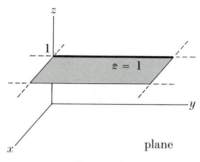

plane

FIGURE 2

As we know, this is the horizontal plane at height 1.

Example 2. *The plane $z = x$.* (Figure 3.)

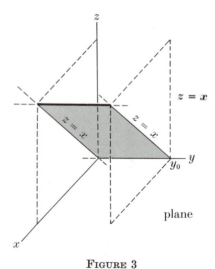

plane

FIGURE 3

The trace in each plane $y = y_0$ is the line $z = x$ in that plane. The surface is the set of all points (x, y, x), i.e., z is equal to x, and y is anything at all.

Example 3. *The circular cylinder* $x^2 + y^2 = 1$. (Figure 4.)

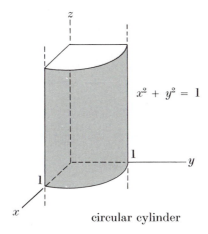

circular cylinder

FIGURE 4

The trace on each horizontal plane is the circle $x^2 + y^2 = 1$ in that plane.

Example 4. *The parabolic cylinder* $z = y^2$. (Figure 5.)

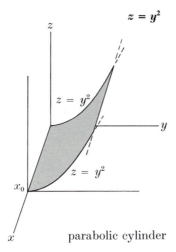

parabolic cylinder

FIGURE 5

The trace on each plane $x = x_0$ is the parabola $z = y^2$ in that plane.

Next we give a sketchy account of some equations containing all three variables.

Example 5. *The plane $ax + by + cz = 1$ $(a \neq 0, b \neq 0, c \neq 0)$. (Figure 6.)*

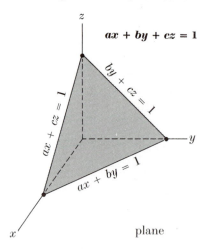

FIGURE 6

This is the plane through the three points $(1/a, 0, 0)$, $(0, 1/b, 0)$, and $(0, 0, 1/c)$. The traces in the coordinate planes are the lines $ax + by = 1$, $ax + cz = 1$, and $by + cz = 1$.

Example 6. *The sphere $x^2 + y^2 + z^2 = a^2$ $(a > 0)$. (Figure 7.)*

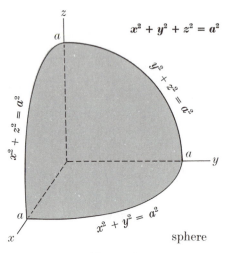

FIGURE 7

This is the set of all points whose distance from the origin is a. The traces on the coordinate planes are circles with the same center and radius.

Example 7. *The ellipsoid $x^2/a^2 + y^2/b^2 + z^2/c^2 = 1$ $(a > 0, b > 0, c > 0)$.* (Figure 8.)

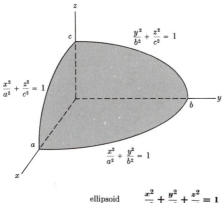

ellipsoid $\dfrac{x^2}{a^2} + \dfrac{y^2}{b^2} + \dfrac{z^2}{c^2} = 1$

FIGURE 8

The traces on the coordinate planes are the *ellipses*

$$\frac{x^2}{a^2} + \frac{y^2}{b^2} = 1, \qquad \frac{x^2}{a^2} + \frac{z^2}{c^2} = 1, \qquad \text{and} \qquad \frac{y^2}{b^2} + \frac{z^2}{c^2} = 1.$$

The traces on planes parallel to the coordinate planes are also ellipses.

Example 8. *The paraboloid of revolution $z = x^2 + y^2$.* (Figure 9.)

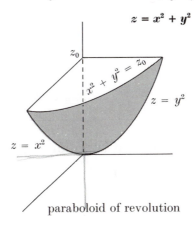

paraboloid of revolution

FIGURE 9

This intersects the xy-plane in the origin only. At each higher level, $z = z_0$, the trace is a circle, of radius $\sqrt{z_0}$.

The traces in the xz- and yz-planes are parabolas, $z = x^2$ and $z = y^2$. The surface is generated by revolving these about the z-axis.

Example 9. *The hyperbolic paraboloid* $z = y^2 - x^2$. (Figure 10.)

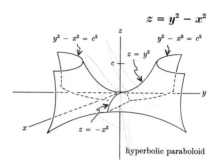

$$z = y^2 - x^2$$

$$y^2 - x^2 = c^2$$ $$y^2 - x^2 = c^2$$

$$z = y^2$$

$$z = -x^2$$

hyperbolic paraboloid

FIGURE 10

The trace on each plane $x = a$ is a parabola, $z = y^2 - a^2$, opening upward. The trace on each plane $y = b$ is the parabola $z = b^2 - x^2$, opening downward. The trace on each plane $z = c$ is a *hyperbola*, $y^2 - x^2 = c^2$. The result is the saddle-shaped surface shown in Figure 10.

Problems (§12B)

1. Sketch the following surfaces.

(a) $x + y + z = 1$

(b) $2x - 3z = 1$

(c) $y = \sin x$

(d) $z = \dfrac{x^2}{4} + \dfrac{y^2}{9}$

(e) $z^2 = \dfrac{x^2}{4} + \dfrac{y^2}{9}$

(f) $z^2 = x^2 + y^2$

(g) $x = z^2$

(h) $y = x^2 - z^2$

§12C. Properties of continuous functions

The results developed in this section will not be needed until Chapter 14 on integration.

12C1. Fundamental Lemma on Closed Boxes. This is the result underlying the basic facts about continuous functions of two variables. It is the analogue of the Fundamental Lemma on Closed Intervals for continuous functions of one variable (§1H4).

The present lemma refers to points and boxes within a given closed box $[a, b] \times [\alpha, \beta]$. Let I and J be intervals in $[a, b]$ and $[\alpha, \beta]$ of more than one point and let $(x, y) \in I \times J$. We will say for short that $I \times J$ is a box "about" (x, y) if I is an interval about x and J an interval about y—that is to say, if x is in the interior of I unless $x = a$ or $x = b$, and y is in the interior of J unless $y = \alpha$ or $y = \beta$. Figure 1(a) shows the typical case, where x is in the interior of I and y in the interior of J. Figure 1(b) indicates the eight special cases in which $x = a$ or $x = b$ or $y = \alpha$ or $y = \beta$.

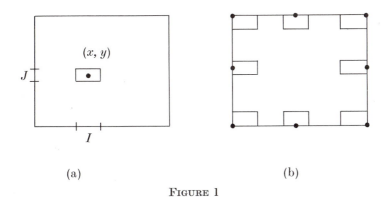

(a)   (b)

FIGURE 1

Another abbreviation is useful. Let $I_1 \times J_1$ and $I_2 \times J_2$ be intersecting boxes. By their "rectangular unions," we will mean the boxes

$$(1) \qquad (I_1 \cup I_2) \times (J_1 \cap J_2) \qquad \text{and} \qquad (I_1 \cap I_2) \times (J_1 \cup J_2).$$

See Figure 2.

We may now state the Fundamental Lemma on Closed Boxes as follows. *Let $\mathscr{K}$ be a set of closed boxes in $[a, b] \times [\alpha, \beta]$. Suppose that:*

(i) *Whenever two members of $\mathscr{K}$ intersect, their two rectangular unions belong to $\mathscr{K}$; and*

(ii) *Each point of $[a, b] \times [\alpha, \beta]$ has a box about it that belongs to $\mathscr{K}$.*

Then $[a, b] \times [\alpha, \beta]$ itself belongs to $\mathscr{K}$.

Proof. We apply the Fundamental Lemma on Closed Intervals (§1H4). Let $\mathscr{I}$ be the set of those intervals in $[a, b]$ whose product with $[\alpha, \beta]$ belongs to $\mathscr{K}$:

$$(2) \qquad\qquad\qquad I \in \mathscr{I} \qquad \text{if} \qquad I \times [\alpha, \beta] \in \mathscr{K}.$$

We are to prove that $[a, b] \in \mathscr{I}$.

By (i), the union of two intersecting members of $\mathscr{I}$ belongs to $\mathscr{I}$.

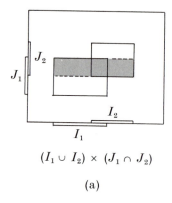

$(I_1 \cup I_2) \times (J_1 \cap J_2)$

(a)

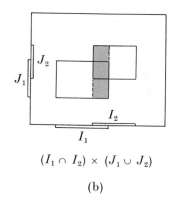

$(I_1 \cap I_2) \times (J_1 \cup J_2)$

(b)

FIGURE 2

Now we must show that each point $x \in [a, b]$ has an interval about it that belongs to $\mathscr{I}$. To do this, we apply the Fundamental Lemma on Closed Intervals (!). Let $\mathscr{J}$ be the set of those intervals in $[\alpha, \beta]$ whose product with some interval about x belongs to $\mathscr{K}$; that is,

(3) $J \in \mathscr{J}$ if there is an interval I about x such that $I \times J \in \mathscr{K}$.

We shall show that $[\alpha, \beta] \in \mathscr{J}$.

It follows from (i) that the union of two intersecting members of $\mathscr{J}$ belongs to $\mathscr{J}$.

By (ii), each point y of $[\alpha, \beta]$ has an interval about it that belongs to $\mathscr{J}$.

Hence, by the Fundamental Lemma on Closed Intervals, $[\alpha, \beta] \in \mathscr{J}$. This means that there is an interval I about x such that $I \times [\alpha, \beta] \in \mathscr{K}$. This, in turn, means that there is an interval about x (namely, I) that belongs to $\mathscr{I}$.

We now conclude, from the Fundamental Lemma on Closed Intervals, that $[a, b] \in \mathscr{I}$. This completes the proof.

12C2. Maximum-Value Theorem. *A function f continuous on a closed box* $K = [a, b] \times [\alpha, \beta]$ *is bounded there, and assumes a maximum and a minimum there.*

Proof. To show that f is bounded we apply the Fundamental Lemma on Closed Boxes (§12C1). The set $\mathscr{K}$ of the lemma will be the set of all closed boxes in K on which f is bounded.

(i) *If f is bounded on each of two intersecting boxes, it is bounded on both their rectangular unions.* This is clear.

(ii) *Each point of K has a closed box about it on which f is bounded.* This is a consequence of continuity.

By the Fundamental Lemma, f is bounded on K.

The argument for a maximum and a minimum is the same as in one dimension (§2E7).

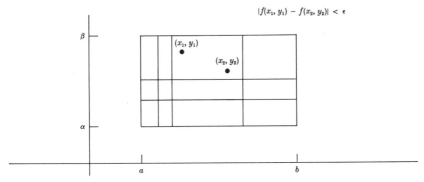

FIGURE 3

12C3. Partitions. Let $K = [a, b] \times [\alpha, \beta]$ be a closed box in $\mathscr{R}^2$. A gridwork of horizontal and vertical lines running across K (a finite number of each) determines a collection of smaller closed boxes whose union is K (adjacent boxes having just an edge in common). See Figure 3. We call such a collection a *partition* of K. We will refer to the sub-boxes making up the partition as the *segments* of the partition.

Let f be a function defined on K, and let ϵ be a positive number. As before, we will call a partition of K an "ϵ-partition" if, within any one segment, f varies by less than ϵ.

Suppose I_1 and I_2 are intersecting intervals in $[a, b]$, and J_1 and J_2 are intersecting intervals in $[\alpha, \beta]$. Suppose, further, that both $I_1 \times J_1$ and $I_2 \times J_2$ admit ϵ-partitions, say P_1 and P_2. Consider the "rectangular union,"

$$(I_1 \cup I_2) \times (J_1 \cap J_2)$$

(Figure 2). We form a partition, P, of this box by taking as points of subdivision all those that occur in P_1 or P_2 (within the confines indicated). The segments of P are then subsets of segments of P_1 or P_2. Consequently, f varies within each segment of P by less than ϵ. Therefore, P is an ϵ-partition.

Similarly, $(I_1 \cap I_2) \times (J_1 \cup J_2)$ admits an ϵ-partition.

12C4. Existence of ϵ-partitions. *Let f be continuous on a closed box $K = [a, b] \times [\alpha, \beta]$, and let ϵ be any positive number.*

Then there is a partition of K such that, within any one segment, f varies by less than ϵ. Briefly: K admits an ϵ-partition.

Proof. We apply the Fundamental Lemma on Closed Boxes (§12C1). The set $\mathscr{K}$ of the lemma will consist of all closed boxes in K that admit ϵ-partitions.

(i) *If each of two intersecting intervals admits an ϵ-partition, then so do both their rectangular unions.* This was noted above.

(ii) *Each point of K has a closed box about it that admits an ϵ-partition.* For, let (x_0, y_0) be any point of K. Since f is continuous, there are closed intervals I and J about x_0 and y_0 such that

$$|f(x, y) - f(x_0, y_0)| < \frac{\epsilon}{2}$$

whenever $x \in I$ and $y \in J$. Then f varies by less than ϵ on $I \times J$.

By the Fundamental Lemma, K admits an ϵ-partition.

Problems (§12C)

1. Let $I_1 = [0, 7]$, $I_2 = [2, 8]$, $J_1 = [3, 6]$, and $J_2 = [4, 8]$. Consider the following points in the plane:

$$(0, 3), \quad (1, 4), \quad (2, 5), \quad (3, 3), \quad (8, 5), \quad (6, 6),$$
$$(6, 7), \quad (1, 7), \quad (8, 8), \quad (7, 7)$$

(a) Which of them belong to the rectangular union

$$(I_1 \cup I_2) \times (J_1 \cap J_2)?$$

(b) Which of them belong to the rectangular union

$$(I_1 \cap I_2) \times (J_1 \cup J_2)?$$

2. Let $I_1 = [-3, 3]$, $I_2 = [-1, 6]$, $J_1 = [-3, 6]$, and $J_2 = [-1, 3]$. Consider the following points in the plane:

$$(-3, -3), \quad (-2, -2), \quad (-1, -1), \quad (-2, 0), \quad (3, 4),$$
$$(4, 5), \quad (5, 6), \quad (3, 0), \quad (4, 0), \quad (3, -3)$$

(a) Which of them belong to $(I_1 \cup I_2) \times (J_1 \cap J_2)$?

(b) Which of them belong to $(I_1 \cap I_2) \times (J_1 \cup J_2)$?

3. Let $K = [0, 1] \times [0, 1]$ and let $\epsilon = 0.1$. Specify an ϵ-partition for the given function f on the box K.
   (a) $f(x, y) = x$
   (b) $f(x, y) = x + y$
   (c) $f(x, y) = xy$
   (d) $f(x, y)$ is the constant function 2

Answers to problems (§12C)

1. (a) (1, 4), (2, 5), (8, 5), (6, 6)
   (b) (2, 5), (3, 3), (6, 6), (6, 7), (7, 7)
2. (a) (−1, −1), (−2, 0), (3, 0), (4, 0)
   (b) (−1, −1), (3, 4), (3, 0), (3, −3)
3. (a) any partition with vertical lines less than 0.1 apart
   (b) any partition with horizontal as well as vertical lines less than 0.05 apart
   (c) any partition with either horizontal or vertical lines less than 0.1 apart
   (d) any partition

CHAPTER 13

Partial Derivatives

This chapter is an introduction to the differential calculus of functions of two or more variables.

The basic material is presented in §§13A–13F. After that (§§13G–13J) the topics are more specialized.

§13A. Partial derivatives

13A1. *How do variables vary?* How does the volume in a cylinder vary with the radius and height? In

$$V = \pi r^2 h,$$

how does V vary as r and h vary? The question is not precise, because we have not agreed on how r and h are to vary. At the same rate? One increasing and the other decreasing?

If we stick to cylinders of radius 3 (Figure 1(a)), then we have $V = 9\pi h$, a function of h alone, and we can differentiate it: $dV/dh = 9\pi$. We can do the same with any other fixed radius. Writing $V = \pi r^2 h$, with the understanding that r is a constant, we get $dV/dh = \pi r^2$.

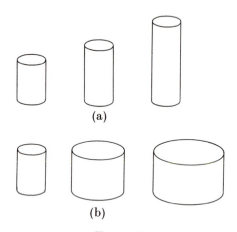

(a)

(b)

Figure 1

On the other hand, if we stick to cylinders of height 8 (Figure 1(b)), then $V = 8\pi r^2$, a function of r alone, and we can differentiate *that*: $dV/dr = 16\pi r$. If, in $V = \pi r^2 h$, h is understood to be a constant, then $dV/dr = 2\pi r h$.

This suggests studying variables one at a time. We vary one while holding the other fixed. Certainly in experiments this makes good sense. For functions of more than two variables, we vary one while holding *all* the others fixed. Explicitly, we differentiate with respect to one variable, regarding all the others as constant.

This procedure has two important advantages, neither of which should be underestimated. One is that the meaning and interpretation of what we are doing is perfectly clear. The other is that it is something we know how to do.

13A2. Partial derivatives. Consider a function of two variables,

$$z = f(x, y).$$

The derivative of f with respect to x, holding y constant, is called the *partial derivative of f with respect to x*. It is denoted

(1) $\dfrac{\partial z}{\partial x},\qquad z_x,\qquad \dfrac{\partial f}{\partial x},\qquad f_x,\qquad f_x(x,\,y),\qquad f_1(x,\,y),$

and other ways. The curly ∂ is used to emphasize that the derivative is different from an ordinary derivative.

The formal definition is

(2) $$f_x(x,\,y) \;=\; \lim_{\Delta x \to 0} \frac{f(x + \Delta x,\, y) - f(x,\, y)}{\Delta x}.$$

Similarly, the partial derivative of f with respect to y,

(3) $\dfrac{\partial z}{dy},\qquad z_y,\qquad \dfrac{\partial f}{dy},\qquad f_y,\qquad f_y(x,\,y),\qquad f_2(x,\,y),$

etc., is defined by

(4) $$f_y(x,\,y) \;=\; \lim_{\Delta y \to 0} \frac{f(x,\, y + \Delta y) - f(x,\, y)}{\Delta y}.$$

From our knowledge of ordinary derivatives, it is easy to recognize all sorts of functions of two variables for which the partial derivatives exist.

Example 1.
(a) If $f(x, y) = x^2 y + x$, then

$$f_x(x,\,y) = 2xy + 1,\qquad f_y(x,\,y) = x^2,$$
$$f_x(2,\,3) = 13,\qquad\qquad f_y(2,\,3) = 4.$$

(b) If $g(x, y) = ye^{xy}$, then

$$g_x(x,\,y) = y^2 e^{xy},\qquad g_y(x,\,y) = xye^{xy} + e^{xy}.$$

(c) If $h(x, y) = \sin x \cos y$, then

$$h_x(x,\,y) = \cos x \cos y,\qquad h_y(x,\,y) = -\sin x \sin y.$$

It is clear that partial derivatives are themselves functions of two variables.

The ideas and notation extend to functions of any number of variables.

Example 2. If $f(x, y, u, v) = xy + x^2 u + yuv$, then

$$f_x = y + 2xu, \qquad f_y = x + uv,$$
$$f_u = x^2 + yv, \qquad f_v = yu.$$

13A3. Geometrical significance of partial derivatives. Consider a function $z = f(x, y)$. Its graph is a surface. Consider a point (x_0, y_0, z_0). To hold y fixed at the value y_0 means to intersect the surface with the plane $y = y_0$ (Figure 2(a)). The intersection is the curve

(5) $$z = f(x, y_0)$$

in that plane. For each x, $f_x(x, y_0)$ is the slope of the tangent line to the curve (5). In particular, $f_x(x_0, y_0)$ is the slope of the curve at the point (x_0, y_0, z_0).

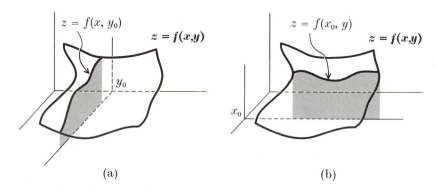

(a)   (b)

FIGURE 2

Similarly, to hold x fixed at x_0 means to intersect the surface with the plane $x = x_0$ (Figure 2(b)). The intersection is the curve

(6) $$z = f(x_0, y)$$

in that plane. For each y, $f_y(x_0, y)$ is the slope of the curve (6). In particular, $f_y(x_0, y_0)$ is the slope at (x_0, y_0, z_0).

Problems (§13A)

1. Find the partial derivatives with respect to x and y.

(a) $e^x \cos(xy)$

(l) $\log \dfrac{xy}{x^2 + y^2}$

(b) $\dfrac{x + y}{x^2 + y^2 + 1}$

(m) x^y

(c) $x^3 + 5x^2 y + 3xy^2 + 2y^3$

(n) $\log \sqrt{2x^2 + y^2}$

(d) $x \log y$

(o) $\dfrac{xy}{\sqrt{x^2 + y^2}}$

(e) $e^{(x+y)^2}$

(p) $\dfrac{x}{x - y}$

(f) $\sin(x + y) \cos(x - y)$

(q) xye^{xy}

(g) $(x^2 y + y)^4$

(r) $\dfrac{x + y}{x - y}$

(h) $x^3 + 3x + 1$

(s) $\sqrt{x^2 + y^2}$

(i) $y \tan(xy)$

(t) $\log(x^2 + y^2)$

(j) $\sqrt{1 - x^2 - y^2}$

(u) $\dfrac{xy}{x^2 + y^2}$

(k) $\arctan \dfrac{x}{y}$

(v) $\log \tan \dfrac{x}{y}$

2. Find the partial derivatives with respect to x, y, and z.

(a) $x^3 y z^2$

(e) $\dfrac{xy}{x^2 + y^2 + z^2}$

(b) $\arcsin \dfrac{xy}{z}$

(f) $x^3 + y^2 - z^2 - 2x + 3y + z$

(c) $e^{x^2 + y^2 + z^2}$

(g) $z \log \dfrac{y}{x}$

(d) $\cos(xy) \arctan(xz)$

(h) $e^{xyz} \sin x \cos y$

3. Consider the surface $z = x^2 + 2y^2$.
   (a) The plane $y = 2$ cuts the surface in a curve. Write the equations of the tangent line to this curve at $x = 3$.
   (b) The plane $x = 3$ cuts the surface in a curve. Write the equations of the tangent line to this curve at $y = 2$.

4. Consider the surface $z = (x^2 - 3y^2)^2$.
   - (a) The plane $x = 2$ cuts the surface in a curve. Write the equations of the tangent line to this curve at $y = 1$.
   - (b) The plane $y = 1$ cuts the surface in a curve. Write the equations of the tangent line to this curve at $x = 2$.

5. If $u = x \sin(x/y)$, show that

$$xu_x + yu_y = u.$$

6. If $u = \log(e^x + e^y)$, show that $u_x + u_y = 1$.

7. If $z = e^{1/(x+y)}$, show that $z_x - z_y = 0$.

8. If $z = (x - y)^3$, show that $z_x + z_y = 0$.

9. Show that the following functions satisfy the equation

$$x \frac{\partial u}{\partial x} + y \frac{\partial u}{\partial y} = 0..$$

   - (a) $u = \dfrac{y}{x}$
   - (c) $u = \log y - \log x$
   - (b) $u = \arctan \dfrac{y}{x}$
   - (d) $u = \dfrac{2xy}{x^2 + y^2}$

10. Exhibit a function $f(x, y)$ such that

$$f_x(x, y) = 2x \cos y + y \quad \text{and} \quad f_y(x, y) = x - x^2 \sin y + 1.$$

Answers to problems (§13A)

1. (We give $\partial/\partial x$ followed by $\partial/\partial y$.)
   - (a) $e^x \cos xy - ye^x \sin xy$; $-xe^x \sin xy$
   - (b) $(y^2 - 2xy - x^2 + 1)/(x^2 + y^2 + 1)^2$; $(x^2 - 2xy - y^2 + 1)/(x^2 + y^2 + 1)^2$
   - (c) $3x^2 + 10xy + 3y^2$; $5x^2 + 6xy + 6y^2$ (d) $\log y$; x/y
   - (e) $2(x + y)e^{(x+y)^2}$; $2(x + y)e^{(x+y)^2}$ (f) $\cos 2x$; $\cos 2y$
   - (g) $8xy^4(x^2 + 1)^3$; $4y^3(x^2 + 1)^4$ (h) $3(x^2 + 1)$; 0
   - (i) $y^2 \sec^2(xy)$; $\tan(xy) + xy \sec^2(xy)$
   - (j) $-x/(1 - x^2 - y^2)^{1/2}$; $-y/(1 - x^2 - y^2)^{1/2}$
   - (k) $y/(x^2 + y^2)$; $-x/(x^2 + y^2)$
   - (l) $(y^2 - x^2)/x(y^2 + x^2)$; $(x^2 - y^2)/y(x^2 + y^2)$
   - (m) yx^{y-1}; $x^y \log x$ (n) $2x/(2x^2 + y^2)$; $y/(2x^2 + y^2)$
   - (o) $y^3/(x^2 + y^2)^{3/2}$; $x^3/(x^2 + y^2)^{3/2}$
   - (p) $-y/(x - y)^2$; $x/(x - y)^2$ (q) $ye^{xy}(1 + xy)$; $xe^{xy}(1 + xy)$
   - (r) $-2y/(x - y)^2$; $2x/(x - y)^2$
   - (s) $x/(x^2 + y^2)^{1/2}$; $y/(x^2 + y^2)^{1/2}$
   - (t) $2x/(x^2 + y^2)$; $2y/(x^2 + y^2)$
   - (u) $y(y^2 - x^2)/(x^2 + y^2)^2$; $x(x^2 - y^2)/(x^2 + y^2)^2$
   - (v) $(1/y) \sec(x/y) \csc(x/y)$; $-(x/y^2) \sec(x/y) \csc(x/y)$

2. (We give $\partial/\partial x$; $\partial/\partial y$; $\partial/\partial z$.)
   - (a) $3x^2yz^2$; x^3z^2; $2x^3yz$
   - (b) with $a = (z^2 - x^2y^2)^{-1/2}$: $ay|z|/z$; $ax|z|/z$; $-axy/|z|$

(c) with $a = \exp(x^2 + y^2 + z^2)$: $2xa$; $2ya$; $2za$

(d) $z(\cos xy)/(1 + x^2z^2) - y \sin xy \arctan xz$;
    $-x \sin xy \arctan xz$; $x(\cos xy)/(1 + x^2z^2)$

(e) with $a = (x^2 + y^2 + z^2)^{-2}$: $ay(y^2 + z^2 - x^2)$; $ax(z^2 + x^2 - y^2)$; $-2axyz$

(f) $3x^2 - 2$; $2y + 3$; $-2z + 1$ (g) $-z/x$; z/y; $\log(y/x)$

(h) $(yz \sin x + \cos x)e^{xyz} \cos y$; $(xz \cos y - \sin y)e^{xyz} \sin x$;
    $xye^{xyz} \sin x \cos y$

3. (a) $y = 2, z = 6x - 1$ (b) $x = 3, z = 8y + 1$

4. (a) $x = 2, z = 13 - 12y$ (b) $y = 1, z = 8x - 15$

10. $x^2 \cos y + xy + y$

§13B. The Fundamental Lemma

13B1. Smooth functions. In developing facts about partial derivatives it is convenient to fix on a single set of hypotheses that handle a variety of results and are simple to state. An appropriate set turns out to be that the function under study be defined on a region, that its partial derivatives be defined on that region, and that they be continuous. To keep statements of theorems smooth, we abbreviate this to *smooth*. By definition, then, a *smooth* function is a function defined on a region and having continuous partial derivatives there. A *smooth surface* is the graph of a smooth function.

13B2. The Fundamental Lemma. This is the crucial result in the study of functions of two variables. It relates changes in the value of a function to changes in the x and y directions alone.

Let $z = f(x, y)$ be a smooth function. Consider the change in z from a point (x_0, y_0) to a new point, $(x_0 + \Delta x, y_0 + \Delta y)$. Call this change Δz:

$$(1) \qquad \Delta z = f(x_0 + \Delta x, y_0 + \Delta y) - f(x_0, y_0).$$

The **Fundamental Lemma** states that

$$(2) \qquad \Delta z = f_x(x_0, y_0)\, \Delta x + f_y(x_0, y_0)\, \Delta y + A\, \Delta x + B\, \Delta y;$$

or, more compactly,

$$(3) \qquad \Delta z = \frac{\partial z}{\partial x}\, \Delta x + \frac{\partial z}{\partial y}\, \Delta y + A\, \Delta x + B\, \Delta y;$$

where A and B are certain quantities that approach 0 as Δx and Δy approach 0, i.e., as $(\Delta x, \Delta y) \to (0, 0)$.

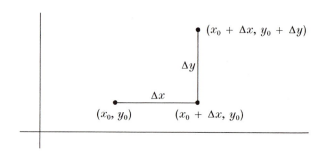

FIGURE 1

Proof. We analyze the change Δz in two steps, first changing x alone, then y alone (Figure 1).

As we hold y fixed at y_0 and go from (x_0, y_0) to $(x_0 + \Delta x, y_0)$, z changes by an amount we will call $\Delta_x z$:

$$(4) \qquad \Delta_x z = f(x_0 + \Delta x, y_0) - f(x_0, y_0).$$

Now, this is a familiar expression. If we divide by Δx and then let $\Delta x \to 0$, we get $f_x(x_0, y_0)$. Therefore, the quantity

$$(5) \qquad A = \frac{\Delta_x z}{\Delta x} - f_x(x_0, y_0) \qquad (\Delta x \neq 0)$$

approaches 0 as $\Delta x \to 0$. Let us also define $A = 0$ for $\Delta x = 0$. Then in either case,

$$(6) \qquad \Delta_x z = f_x(x_0, y_0)\, \Delta x + A\, \Delta x.$$

For the rest of the change in z we hold x fixed at $x_0 + \Delta x$ and go from $(x_0 + \Delta x, y_0)$ to $(x_0 + \Delta x, y_0 + \Delta y)$. Here z changes by an amount we will call $\Delta_y z$:

$$(7) \qquad \Delta_y z = f(x_0 + \Delta x, y_0 + \Delta y) - f(x_0 + \Delta x, y_0).$$

If we divide by Δy and then let $\Delta y \to 0$, we get $f_y(x_0 + \Delta x, y_0)$. For a neater expression, we note that for small Δx, this is approximately $f_y(x_0, y_0)$ (since f_y is continuous). Precisely, we define

$$(8) \qquad B = \frac{\Delta_y z}{\Delta y} - f_y(x_0, y_0) \quad \text{(if } \Delta y \neq 0), \qquad B = 0 \quad \text{(if } \Delta y = 0).$$

Then $B \to 0$ as $(\Delta x, \Delta y) \to (0, 0)$. Evidently,

$$(9) \qquad\qquad \Delta_y z = f_y(x_0, y_0)\, \Delta y + B\, \Delta y.$$

Since $\Delta z = \Delta_x z + \Delta_y z$, the equations (6) and (9), added together, give us the Fundamental Lemma, (2).

13B3. The Fundamental Lemma in several variables. In more than two variables the above ideas carry through in similar fashion. For example, suppose $u = f(x, y, z)$ is smooth, and let

$$(10) \qquad \Delta u = f(x_0 + \Delta x, y_0 + \Delta y, z_0 + \Delta z) - f(x_0, y_0, z_0).$$

Then

$$(11) \qquad \Delta u = \frac{\partial u}{\partial x}\,\Delta x + \frac{\partial u}{\partial y}\,\Delta y + \frac{\partial u}{\partial z}\,\Delta z + A\,\Delta x + B\,\Delta y + C\,\Delta z,$$

where A, B, and C all approach 0 as $(\Delta x, \Delta y, \Delta z) \to (0, 0, 0)$. (It is understood here that the partials are evaluated at (x_0, y_0, z_0).)

13B4. Smooth functions are continuous. If a function of one variable has a derivative, it is continuous. In more than one variable, however, the analogy does not carry through. Even though f_x and f_y exist, f need not be continuous. This is because partial differentiation refers to a limit taken only along a line (parallel to one of the axes), whereas continuity considers limits from all directions.

A *smooth* function, however, i.e., with *continuous* partials, is continuous. This can be seen directly from the Fundamental Lemma. Let $z = f(x, y)$ be smooth. Then the Fundamental Lemma, (2), holds. From (2) it is obvious that $\Delta z \to 0$ as $(\Delta x, \Delta y) \to (0, 0)$. This says that f is continuous. The same argument works in any number of variables.

§13C. The total differential

13C1. The total differential. If $z = f(x, y)$ is smooth, then, according to the Fundamental Lemma $\big(\S13B(3)\big)$,

$$(1) \qquad\qquad \Delta z = \frac{\partial z}{\partial x}\,\Delta x + \frac{\partial z}{\partial y}\,\Delta y + A\,\Delta x + B\,\Delta y,$$

where A and B approach 0 as $(\Delta x, \Delta y) \to (0, 0)$. Therefore, when Δx and Δy are "small," the products $A\,\Delta x$ and $B\,\Delta y$ are "very small." Hence for small Δx and Δy, we get a good approximation to Δz by discarding these last two terms. This is all somewhat vague, as we have said nothing about how good an approximation we get. But there are several advantages to outweigh this defect. One is that the common sense of the experimenter is a defense against gross misinterpretations. Another is that the approximation is relatively easy to compute, while the exact value may be a nuisance. A third is that an exact value often represents illusory accuracy in any case.

The approximation to Δz is denoted by dz or df and is called the **total differential**:

$$(2) \qquad dz = \frac{\partial z}{\partial x}\,\Delta x + \frac{\partial z}{\partial y}\,\Delta y, \qquad \text{or} \qquad df = f_x\,\Delta x + f_y\,\Delta y.$$

In this context, Δx and Δy are usually denoted by dx and dy. To summarize: if x changes from x_0 by a small amount dx, and y changes from y_0 by a small amount dy, then the resulting change in z is, approximately, the *total differential*

$$(3) \qquad dz = \frac{\partial z}{\partial x}\,dx + \frac{\partial z}{\partial y}\,dy.$$

The quantities dx and dy are also called differentials.

Example 1. A hotel manager is estimating the cost of carpeting a meeting room 10 yd. by 15 yd. at the rate of $10 per sq. yd. This comes to a total of $1500, but he wonders what to allow for errors in measurement, which he estimates at 2 percent in each direction.

If x and y are the dimensions of the room and S is the area, then

$$S = xy, \qquad \frac{\partial S}{\partial x} = y, \qquad \frac{\partial S}{\partial y} = x,$$

and

$$dS = \frac{\partial S}{\partial x}\,dx + \frac{\partial S}{\partial y}\,dy = y\,dx + x\,dy.$$

With $x = 15$, $y = 10$, $dx = 0.3$, and $dy = 0.2$, we get

$$dS = (10)(0.3) + (15)(0.2) = 6 \text{ sq. yd.}$$

The possible extra cost is $60.

In this example, as it happens, the calculation of the "exact" value is not forbidding. It comes out to be $60.60, clearly an instance of spurious accuracy.

Example 2. Find the approximate change in the volume of a right circular cone of radius 3 and height 6 when the radius is decreased by 0.2 and the height is increased by 0.1.

Solution. Here

$$V = \tfrac{1}{3}\pi r^2 h, \qquad \frac{\partial V}{\partial r} = \tfrac{2}{3}\pi r h, \qquad \frac{\partial V}{\partial h} = \tfrac{1}{3}\pi r^2,$$

and

$$dV = \frac{\partial V}{\partial r}\, dr + \frac{\partial V}{\partial h}\, dh = \tfrac{1}{3}\pi r(2h\, dr + r\, dh).$$

With $r = 3$, $h = 6$, $dr = -0.2$, and $dh = 0.1$, we get

$$dV = (\tfrac{1}{3}\pi)(3)[(12)(-0.2) + (3)(0.1)] = -2.1\pi.$$

Note that the original volume is

$$V = \tfrac{1}{3}\pi (3^2)(6) = 18\pi.$$

13C2. The total differential in more than two variables. In three or more variables the ideas are the same. If $u = f(x, y, z)$ is smooth, then

(4)
$$du = \frac{\partial u}{\partial x}\, dx + \frac{\partial u}{\partial y}\, dy + \frac{\partial u}{\partial z}\, dz$$

is an approximation to Δu.

Example 3. How is the volume of a rectangular box $4 \times 3 \times 2$ affected by a 1 percent error in the sides?

Solution. With $V = xyz$, we have

$$dV = yz\, dx + zx\, dy + xy\, dz.$$

With $dx = 0.01x$, $dy = 0.01y$, and $dz = 0.01z$, this is

$$dV = 0.03xyz = 0.03V.$$

The 3 errors of 1 percent in a side add to an error of 3 percent in their product. For $x = 4$, $y = 3$, $z = 2$, we have $V = 24$ and $dV = 0.72$.

13C3. The (total) differential in one variable. The total differential for functions of one variable is called, simply, the *differential*.
Let $y = f(x)$ be differentiable. Then

(5)
$$A = \frac{\Delta y}{\Delta x} - f'(x) \to 0 \qquad \text{as} \qquad \Delta x \to 0,$$

(6)
$$\Delta y = f'(x)\, \Delta x + A\, \Delta x,$$

and the approximating differential is

(7)
$$dy = f'(x)\, dx.$$

In the d-notation, this assumes the neat form

(8)
$$dy = \frac{dy}{dx}\, dx.$$

The relations among dy, Δy, $dx = \Delta x$, and $f'(x)$ are clarified in Figure 1. The point $(x + dx, y + dy)$ lies on the tangent to the graph at (x, y). Thus, $\Delta y - dy$ is the (directed) distance from the tangent line to the curve corresponding to the change dx. The accuracy of dy as an approximation to Δy is a measure of how close the tangent line is to the curve. Generally, the smaller dx, the better the approximation.

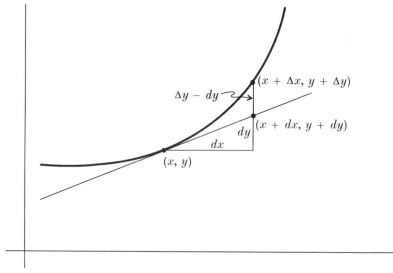

FIGURE 1

Example 4. Use the differential to approximate $\cos 46°$.

Solution. We let $y = \cos \theta$, so that

$$dy = -\sin \theta\, d\theta.$$

For given θ and $d\theta$, $\cos (\theta + d\theta)$ is approximately

$$y + dy = \cos \theta - \sin \theta\, d\theta.$$

We now look for a value of θ that is near 46°, so that $d\theta$ will be small, and for which we know both the sine and the cosine. Clearly, $\theta = 45°$ is what inspired this example in the first place. Since differentials result from differentiation, we must measure them in radians. We have

$$d\theta = 1° = \frac{\pi}{180} \text{ radians.}$$

Hence

$$\cos 46° \approx \cos 45° - (\sin 45°) \frac{\pi}{180}$$

$$\approx 0.71 - \frac{(0.71)(3.14)}{180} \approx 0.70.$$

Problems (§13C)

1. Compute the (total) differential.
   (a) $y = x^2$
   (b) $y = x \log x$
   (c) $u = \log \tan x$
   (d) $u = e^x(\sin x + \cos x)$
   (e) $z = \sqrt{x^2 + y^2}$
   (f) $V = \pi r^2 h$
   (g) $z = x \cos(xy)$
   (h) $z = y^x$
   (i) $u = \log \sqrt{x^2 + y^2}$

   (j) $z = \dfrac{xw}{x^2 + y^2 + w^2}$

   (k) $u = xy \sin z + xz \cos y$
   (l) $u = x \log z + z \log y$
   (m) $u = x \sin y + y \cos z + zw \sin x + wx \cos y$
2. Find the approximate change in u corresponding to the given values.
   (a) $u = x^3$; $x_0 = 2$; $dx = 0.1$
   (b) $u = y^4 - y^2$; $y_0 = 1$; $dy = 0.2$
   (c) $u = x^2 - 2xy + 2y^2$; $x_0 = 2, y_0 = 3$; $dx = 0.1, dy = -0.2$
   (d) $u = xy \sin(x + y)$; $x_0 = \pi/2, y_0 = \pi/3$; $dx = \pi/180, dy = \pi/90$
   (e) $u = (xy^2 + x)^4$; $x_0 = 2, y_0 = 1$; $dx = -\frac{1}{5}, dy = 1$
   (f) $u = \log \sqrt{x^2 + y^2}$; $x_0 = 3, y_0 = 4$; $dx = 1, dy = -1$
   (g) $u = (x + y)^2 + (y - z)^2$; $x_0 = 0, y_0 = 1, z_0 = 2$; $dx = 0.05, dy = 0.05, dz = 0.05$
   (h) $u = e^x \sin y + e^y \cos z$; $x_0 = 0, y_0 = 0, z_0 = 0$; $dx = 0.1, dy = -0.1, dz = 0.2$
   (i) $u = w^2 - 2x^2 + 3y^2 - 4z^2$; $w_0 = 3, x_0 = 2, y_0 = 1, z_0 = 0$; $dw = -0.1, dx = -0.2, dy = 0.1, dz = 0.2$

3. Approximate by means of the differential.

(a) $\sqrt[4]{17}$
(b) $\log 0.99$
(c) $\sin 30.5°$
(d) $1/997$
(e) $\sqrt{0.0037}$
(f) 2.1^5

(g) $\arcsin 0.49$
(h) $63^{-1/3}$
(i) $e^{0.01}$
(j) $3.97^{-1/2} \cos 62°$
(k) $\sqrt{4.03^2 + 2.97^2}$
(l) $\sqrt{2.02^2 + 1.98^2 + 1.03^2}$

4. What is the approximate change in
  (a) the volume in a cylinder of height 7 and radius 5 when the radius is increased by 0.2?
  (b) the area of a square of side 3 when the side is decreased by 0.1?
  (c) the surface area of a cube when the length of a side is increased by 2 percent?

5. What is the approximate volume of paint required to cover a sphere of radius 10 inches with a coat $\frac{1}{10}$ inch thick?

6. In a right triangle with legs 5 inches and 12 inches, the shorter leg is increased by $\frac{1}{5}$ inch and the longer is decreased by the same amount. By approximately how much is
  (a) the hypotenuse changed?
  (b) the area changed?

7. The heat given off by an electric heater is determined from the formula

$$H = \frac{kE^2}{R},$$

where E is the voltage, R the resistance, and k a constant. At a certain instant, E is 110 volts and R is 12 ohms. If the voltage drops to 100 volts, by about how much must the resistance be decreased to maintain the same heat?

8. If two resistors of resistance R_1 and R_2 are connected in parallel, then

$$\frac{1}{R} = \frac{1}{R_1} + \frac{1}{R_2},$$

where R is the resistance in the resulting system. Find the approximate change in R when R_1 is increased from 5 to 5.3 and R_2 is decreased from 4 to 3.8.

9. Two sides of a triangle are 100 feet and 120 feet. The angle between them is 30°. Approximate the change in area if each side is increased by 3 feet and the angle between them is decreased by 2°. [*Hint.* Area $= \frac{1}{2}ab \sin C$.]

Answers to problems (§13C)

1. (a) $dy = 2x\, dx$ (b) $dy = (1 + \log x)\, dx$ (c) $du = \sec x \csc x\, dx$

(d) $du = 2e^x \cos x\, dx$ (e) $dz = \dfrac{x}{\sqrt{x^2 + y^2}}\, dx + \dfrac{y}{\sqrt{x^2 + y^2}}\, dy$

(f) $dV = \pi r^2 \, dh + 2\pi rh \, dr$

(g) $dz = (\cos xy - xy \sin xy) \, dx - x^2 \sin xy \, dy$

(h) $dz = y^x \log y \, dx + xy^{x-1} \, dy$

(i) $du = \dfrac{x}{x^2 + y^2} \, dx + \dfrac{y}{x^2 + y^2} \, dy$

(j) $dz = w \dfrac{y^2 + w^2 - x^2}{(x^2 + y^2 + w^2)^2} \, dx - \dfrac{2xyw}{(x^2 + y^2 + w^2)^2} \, dy + x \dfrac{x^2 + y^2 - w^2}{(x^2 + y^2 + w^2)^2} \, dw$

(k) $du = (y \sin z + z \cos y) \, dx + (x \sin z - xz \sin y) \, dy$
$\qquad + (x \cos y + xy \cos z) \, dz$

(l) $du = \log z \, dx + (z/y) \, dy + (x/z + \log y) \, dz$

(m) $du = (\sin y + zw \cos x + w \cos y) \, dx + (x \cos y + \cos z - wx \sin y) \, dy$
$\qquad + (z \sin x + x \cos y) \, dw + (w \sin x - y \sin z) \, dz$

2. (a) 1.2   (b) 0.4   (c) -1.8 (d) $\dfrac{\pi^2}{270} - \dfrac{\pi^3\sqrt{3}}{720}$ (e) 921.6

(f) -0.04 (g) 0.2   (h) -0.2 (i) 1.6

3. (a) 2.03   (b) -0.01 (c) 0.508   (d) 0.001003   (e) 0.0608
(f) 40   (g) 0.51   (h) 0.2513   (i) 1.01   (j) 0.236
(k) 5.006   (l) 3.01

4. (a) an increase of $14\pi \approx 44$ (b) a decrease of 0.6
(c) an increase of 4 percent

5. $40\pi \approx 126$ cu. in.

6. (a) $-\frac{7}{65} \approx -0.11$ in.   (b) 0.7 sq. in.

7. $2\frac{2}{11}$ ohms

8. decreases by $\frac{1}{405}$

9. -16 sq. ft.

§13D. Directional derivatives

13D1. Definition of directional derivative. Let $z = f(x, y)$ be a smooth function. Consider any point (x_0, y_0). As we know, the partial derivatives $f_x(x_0, y_0)$ and $f_y(x_0, y_0)$ measure the rate of change of f along the positive directions of the coordinate axes. *Directional derivatives* generalize this notion to measure the rate of change along *any* direction α. See Figure 1. The point t units from (x_0, y_0) in the direction α is

(1) $\qquad (x_0, y_0) + t(\cos \alpha, \sin \alpha) = (x_0 + t \cos \alpha, y_0 + t \sin \alpha).$

The ratio

(2) $\qquad \dfrac{f(x_0 + t \cos \alpha, y_0 + t \sin \alpha) - f(x_0, y_0)}{t}$

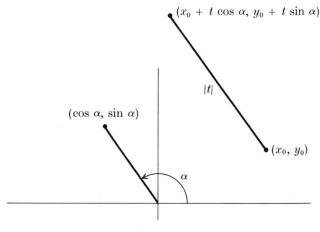

FIGURE 1

therefore represents the average change in z per unit change in t in the direction α. By definition, the *directional derivative* of f at (x_0, y_0) *in the direction* α is the limit of this quotient as $t \to 0$. We call it $f_{(\alpha)}(x_0, y_0)$. Thus,

$$(3) \qquad f_{(\alpha)}(x_0, y_0) = \lim_{t \to 0} \frac{f(x_0 + t \cos \alpha, y_0 + t \sin \alpha) - f(x_0, y_0)}{t}.$$

In the Δ-notation we have

$$(4) \qquad \Delta x = t \cos \alpha \qquad \text{and} \qquad \Delta y = t \sin \alpha.$$

As usual, we let

$$(5) \qquad \Delta z = f(x_0 + \Delta x, y_0 + \Delta y) - f(x_0, y_0).$$

Then

$$(6) \qquad f_{(\alpha)}(x_0, y_0) = \lim_{t \to 0} \frac{\Delta z}{t}.$$

It is clear that the directional derivative in the direction $\alpha = 0$ is simply the partial derivative with respect to x, while for $\alpha = \pi/2$ it is the partial with respect to y. (In symbols, $f_{(0)} = f_x$, $f_{(\pi/2)} = f_y$.)

13D2. Directional derivatives in terms of the partial derivatives.
Remarkably enough, every directional derivative can be expressed in terms
of the partials. The relation is

$$(7) \qquad\qquad f_{(\alpha)} = f_x \cos \alpha + f_y \sin \alpha.$$

Proof. By definition,

$$f_{(\alpha)} = \lim_{t \to 0} \frac{\Delta z}{t}.$$

Now, according to the Fundamental Lemma $\big(\S 13B(2)\big)$,

$$(8) \qquad\qquad \Delta z = f_x \, \Delta x + f_y \, \Delta y + A \, \Delta x + B \, \Delta y,$$

where $A \to 0$ and $B \to 0$ as $(\Delta x, \Delta y) \to (0, 0)$. Because

$$(9) \qquad\qquad \frac{\Delta x}{t} = \cos \alpha \qquad \text{and} \qquad \frac{\Delta y}{t} = \sin \alpha,$$

we have

$$(10) \qquad \frac{\Delta z}{t} = f_x \cos \alpha + f_y \sin \alpha + A \cos \alpha + B \sin \alpha.$$

Now, when t approaches 0, so do Δx and Δy, and hence so do A and B.
Taking limits in (10) therefore gives us (7).

Example 1. In certain circumstances, the temperature u at a point (x, y) is
given by

$$u = f(x, y) = e^{-x} \sin y.$$

Find its directional derivatives.

Solution. We have

$$f_x = -e^{-x} \sin y, \qquad f_y = e^{-x} \cos y.$$

The directional derivative in the direction α is

$$f_{(\alpha)} = e^{-x}(-\sin y \cos \alpha + \cos y \sin \alpha) = e^{-x} \sin(\alpha - y).$$

13D3. Gradient. The vector (f_x, f_y) is known as the *gradient* vector.
(Standard symbols for it are grad f or ∇f ("del" f).) Formula (7) expresses
the directional derivative as the scalar product of the gradient vector with
the unit vector $(\cos \alpha, \sin \alpha)$:

$$(11) \qquad\qquad f_{(\alpha)} = (f_x, f_y) \cdot (\cos \alpha, \sin \alpha).$$

We transform the right side by recalling the formula

$$(12) \qquad P_1 \cdot P_2 = |P_1||P_2| \cos \theta,$$

where θ is the angle between the two vectors ($\S 9B(6)$). Since

$$|(\cos \alpha, \sin \alpha)| = 1,$$

we get

$$(13) \qquad f_{(\alpha)} = |(f_x, f_y)||(\cos \alpha, \sin \alpha)| \cos \theta = |(f_x, f_y)| \cos \theta,$$

where θ is the angle between the gradient vector (f_x, f_y) and the direction vector $(\cos \alpha, \sin \alpha)$. From (13), it is evident that $f_{(\alpha)}$ attains a maximum value of $|(f_x, f_y)|$ when $\cos \theta = 1$, i.e., $\theta = 0$. Thus, the maximum of $f_{(\alpha)}$ occurs in the direction of the gradient vector and is equal to its magnitude. The minimum value, $-|(f_x, f_y)|$, occurs in the opposite direction ($\cos \theta = -1$, $\theta = \pi$). The minimum *absolute* value is 0, at $\theta = \pi/2$ or $3\pi/2$ (where $\cos \theta = 0$). Thus, the directional derivative is zero in the direction perpendicular to the gradient.

Of course, if both f_x and f_y are 0 at a point, then all directional derivatives are 0 at that point.

Example 2. Let $f(x, y) = x^2 + y^2$. In what direction is the directional derivative a maximum?

Solution. The maximum is in the direction of the gradient, $(2x, 2y)$, or (x, y). Its value is $|(2x, 2y)| = 2\sqrt{x^2 + y^2}$. The maximum directional derivative at any point is in the direction of the point, with magnitude twice the radius to the point.

13D4. Directional derivatives of functions of three variables. In three variables the ideas and results are analogous to those above. We consider a smooth function $u = f(x, y, z)$. A unit direction vector is specified as

$$(14) \qquad (\cos \alpha, \cos \beta, \cos \gamma).$$

The definition of directional derivative follows the previous pattern, and the directional derivative turns out to be equal to

$$(15) \qquad f_x \cos \alpha + f_y \cos \beta + f_z \cos \gamma.$$

It assumes a maximum in the direction of the gradient vector, (f_x, f_y, f_z), and it is 0 in all directions perpendicular to the gradient.

Example 3. Let $u = xy + yz$. Then the gradient of u is $(y, x + z, y)$ and the maximum directional derivative is

$$\sqrt{2y^2 + (x + z)^2}.$$

Problems (§13D)

1. Find the gradient of the given function

(a) $f(x, y) = x^3 + y^3 + x - y^2$ (c) $g(x, y) = \dfrac{xy}{x^2 + y^2}$

(b) $f(x, y) = x \cos y$ (d) $h(x, y) = 3x + 2y$

2. Express the directional derivative, $f_{(\alpha)}(x, y)$, in terms of the partial derivatives and α.

(a) $f(x, y) = x^3 - 3xy + 2y^2$ (c) $f(x, y) = \sqrt{x^2 + y^2}$
(b) $f(x, y) = (x + y)e^{xy}$ (d) $f(x, y) = \arctan(y/x)$

3. Find the directional derivative of f at P in the direction α.
(a) $f(x, y) = x^2 - 2y^2,\quad P = (1, 2),\quad \alpha = \pi/4$
(b) $f(x, y) = x \cos xy,\quad P = (1, 0),\quad \alpha = \pi/3$
(c) $f(x, y) = xe^y,\quad P = (2, 0),\quad \alpha = 150°$
(d) $f(x, y) = \arctan(y/x),\quad P = (1, 1),\quad \alpha = \arctan(\tfrac{2}{3})$

4. Find the directional derivative of f at P in the direction from P to Q.
(a) $f(x, y) = 3x^3 + x^2y + y^2,\quad P = (1, 3),\quad Q = (0, 4)$
(b) $f(x, y) = e^x \cos y,\quad P = (1, 2),\quad Q = (-2, 1)$
(c) $f(x, y) = \cos xy,\quad P = (0, 3),\quad Q = (2, -1)$
(d) $f(x, y) = \log(x^2 + y^2),\quad P = (5, -2),\quad Q = (0, 0)$

5. $f_{(\alpha)}$ achieves a maximum at P for $\alpha = \alpha_1$ and a minimum for $\alpha = \alpha_2$. Find $(\cos \alpha_1, \sin \alpha_1)$ and $(\cos \alpha_2, \sin \alpha_2)$.
(a) $f(x, y) = \sqrt{x^2 + y^2},\quad P = (1, 2)$
(b) $f(x, y) = \arctan(y/x),\quad P = (1, 1)$
(c) $f(x, y) = ye^x,\quad P = (0, 1)$

(d) $f(x, y) = \dfrac{1}{x^2 + y^2},\quad P = (3, 4)$

6. Let f and g be smooth functions. Show that:
(a) $\operatorname{grad}(f + g) = \operatorname{grad} f + \operatorname{grad} g$
(b) $\operatorname{grad}(fg) = f \operatorname{grad} g + g \operatorname{grad} f$

7. Find the gradient of f at P.
(a) $f(x, y, z) = xy + yz + xz,\quad P = (1, 2, 5)$
(b) $f(x, y, z) = \log(x^2 + y^2 + z^2),\quad P = (-2, 1, 2)$
(c) $f(x, y, z) = e^x \cos yz,\quad P = (0, 3, 0)$
(d) $f(x, y, z) = x/yz,\quad P = (3, 5, 1)$

8. Find the directional derivative of f at P in the given direction.
(a) $f(x, y, z) = x^2y + xz + yz^2,\quad P = (1, 2, 1)$, direction $(1/\sqrt{6}, -1/\sqrt{6}, 2/\sqrt{6})$
(b) $f(x, y, z) = \log(x^2 + y^2 + z^2),\quad P = (1, 0, 0)$, direction from P to $(2, 3, 1)$

(c) $f(x, y, z) = x \cos y + y \cos z + z \cos x$, $\quad P = (0, 1, 2)$, direction given
by the direction vector $(3, 5, -1)$

(d) $f(x, y, z) = \dfrac{1}{\sqrt{x^2 + y^2 + z^2}}$, $\quad P = (1, 2, 3)$, direction from P to the
origin

9. Find the maximum value of the directional derivative of f at P.
   (a) $f(x, y, z) = x^2 + y^2 + 2xyz$, $\quad P = (1, 3, 2)$
   (b) $f(x, y, z) = e^y \cos x + e^x \cos z$, $\quad P = (0, 0, 0)$
   (c) $f(x, y, z) = xe^{yz} + ye^{zx} + ze^{xy}$, $\quad P = (0, 1, 2)$
   (d) $f(x, y, z) = \sin xz + \cos yz$, $\quad P = (-1, 2, \pi/3)$

Answers to problems (§13D)

1. (a) $(3x^2 + 1, 3y^2 - 2y)$ (b) $(\cos y, -x \sin y)$
   (c) $(y(y^2 - x^2)/(x^2 + y^2)^2, x(x^2 - y^2)/(x^2 + y^2)^2)$ (d) $(3, 2)$
2. (a) $(3x^2 - 3y) \cos \alpha + (4y - 3x) \sin \alpha$
   (b) $e^{xy}(y^2 + xy + 1) \cos \alpha + e^{xy}(x^2 + xy + 1) \sin \alpha$
   (c) $[x/(x^2 + y^2)^{1/2}] \cos \alpha + [y/(x^2 + y^2)^{1/2}] \sin \alpha$
   (d) $[-y/(x^2 + y^2)] \cos \alpha + [x/(x^2 + y^2)] \sin \alpha$
3. (a) $-3\sqrt{2}$ (b) $\frac{1}{2}$ (c) $1 - \frac{1}{2}\sqrt{3}$ (d) $-\frac{1}{26}\sqrt{13}$
4. (a) $-4\sqrt{2}$ (b) $(\sin 2 - 3 \cos 2) e/\sqrt{10}$ (c) 0 (d) $-2/\sqrt{29}$
5. (a) $(1/\sqrt{5}, 2/\sqrt{5})$; $(-1/\sqrt{5}, -2/\sqrt{5})$
   (b) $(-1/\sqrt{2}, 1/\sqrt{2})$; $(1/\sqrt{2}, -1/\sqrt{2})$
   (c) $(1/\sqrt{2}, 1/\sqrt{2})$; $(-1/\sqrt{2}, -1/\sqrt{2})$
   (d) $(\frac{3}{5}, \frac{4}{5})$; $(-\frac{3}{5}, -\frac{4}{5})$
7. (a) $(7, 6, 3)$ (b) $(-\frac{4}{9}, \frac{2}{9}, \frac{4}{9})$ (c) $(1, 0, 0)$
   (d) $(\frac{1}{5}, -\frac{3}{25}, -\frac{3}{5})$
8. (a) $13/\sqrt{6}$ (b) $2/\sqrt{11}$ (c) $(3 \cos 1 + 5 \cos 2 + \sin 2 - 1)/\sqrt{35}$
   (d) $\frac{1}{14}$
9. (a) $2\sqrt{83}$ (b) $\sqrt{2}$ (c) $\sqrt{e^4 + 8e^2 + 18}$ (d) $(\frac{1}{9}\pi^2 + \frac{13}{4} + \sqrt{3})^{1/2}$

§13E. The chain rule

13E1. The chain rule. One often finds sets of variables related to other
sets of variables in intricate ways. The chain rule for functions of several
variables is a versatile tool enabling us to move efficiently from one set of
variables to another.

We state the chain rule in terms of two original variables and two new
variables. The notation is then sufficiently illustrative.

Let u be a smooth function of x and y, and let x and y, in turn, be smooth functions of s and t. The chain rule states that then u is a smooth function of s and t, with

(1)
$$\frac{\partial u}{\partial s} = \frac{\partial u}{\partial x}\frac{\partial x}{\partial s} + \frac{\partial u}{\partial y}\frac{\partial y}{\partial s}$$

and

(2)
$$\frac{\partial u}{\partial t} = \frac{\partial u}{\partial x}\frac{\partial x}{\partial t} + \frac{\partial u}{\partial y}\frac{\partial y}{\partial t}.$$

Proof. A change Δs results in changes $\Delta_s x$ and $\Delta_s y$, which in turn induce a change $\Delta_s u$. These last three are related by the Fundamental Lemma (§13B(3)):

(3)
$$\Delta_s u = \frac{\partial u}{\partial x}\Delta_s x + \frac{\partial u}{\partial y}\Delta_s y + A\,\Delta_s x + B\,\Delta_s y,$$

where $A \to 0$ and $B \to 0$ as $(\Delta_s x, \Delta_s y) \to (0, 0)$. Now, as $\Delta s \to 0$,

(4)
$$\frac{\Delta_s x}{\Delta s} \to \frac{\partial x}{\partial s} \quad \text{and} \quad \frac{\Delta_s y}{\Delta s} \to \frac{\partial y}{\partial s},$$

by hypothesis. Also, $\Delta_s x \to 0$ and $\Delta_s y \to 0$, so that $A \to 0$ and $B \to 0$. Putting these facts together, we see that

$$\frac{\Delta_s u}{\Delta s} \to \frac{\partial u}{\partial x}\frac{\partial x}{\partial s} + \frac{\partial u}{\partial y}\frac{\partial y}{\partial s},$$

which gives us (1). Similarly, we get (2).

If x and y are functions of t alone, the derivatives with respect to t are ordinary derivatives, and the chain rule becomes

(5)
$$\frac{du}{dt} = \frac{\partial u}{\partial x}\frac{dx}{dt} + \frac{\partial u}{\partial y}\frac{dy}{dt}.$$

Similarly, if u is a function of x alone, (1) and (2) reduce to

(6)
$$\frac{\partial u}{\partial s} = \frac{du}{dx}\frac{\partial x}{\partial s}, \quad \frac{\partial u}{\partial t} = \frac{du}{dx}\frac{\partial x}{\partial t}.$$

(Formula (6) also follows from the chain rule for functions of one variable.)

Example 1. If

$$u = x^2 - y^2, \qquad x = s + 2t, \qquad y = 3s - t,$$

then

$$\frac{\partial u}{\partial s} = (2x)(1) + (-2y)(3) = 2x - 6y$$

and

$$\frac{\partial u}{\partial t} = (2x)(2) + (-2y)(-1) = 4x + 2y.$$

Example 2. If

$$u = \frac{x}{y}, \qquad x = s^2 + t, \qquad y = t^2 + s,$$

then

$$\frac{\partial u}{\partial x} = \frac{1}{y}, \qquad \frac{\partial u}{\partial y} = -\frac{x}{y^2},$$

and we get

$$\frac{\partial u}{\partial s} = \left(\frac{1}{y}\right)(2s) - \left(\frac{x}{y^2}\right)(1) = \frac{2sy - x}{y^2},$$

and

$$\frac{\partial u}{\partial t} = \left(\frac{1}{y}\right)(1) - \left(\frac{x}{y^2}\right)(2t) = \frac{y - 2tx}{y^2}.$$

Example 3. Let $u = xy$, where x and y are differentiable functions of t. Then $\partial u/\partial x = y$, $\partial u/\partial y = x$, and formula (5) gives

$$\frac{du}{dt} = y\frac{dx}{dt} + x\frac{dy}{dt},$$

the product formula for derivatives.

Example 4. In a right triangle with sides x and y, x is increasing at 3 inches per second and y is decreasing at 2 inches per second. How is the area changing when $x = 10$ and $y = 6$?

Solution. Here

$$A = \tfrac{1}{2}xy, \qquad \frac{\partial A}{\partial x} = \tfrac{1}{2}y, \qquad \frac{\partial A}{\partial y} = \tfrac{1}{2}x, \qquad \frac{dx}{dt} = 3, \qquad \frac{dy}{dt} = -2,$$

and formula (5) gives

$$\frac{dA}{dt} = (\tfrac{1}{2}y)(3) + (\tfrac{1}{2}x)(-2) = \tfrac{3}{2}y - x.$$

When $x = 10$ and $y = 6$,

$$\frac{dA}{dt} = 9 - 10 = -1.$$

That is, the area is decreasing at 1 square inch per second.

Of course, we could have proceeded instead as in §3H, writing $A' = \frac{1}{2}(xy' + yx')$, etc.

13E2. Relations among variables. In the setting of the chain rule, the relations among the variables can be quite arbitrary. Intricate relations can arise in a natural way, and indeed the power of the chain rule resides in its ability to handle them.

Suppose that (x, y) are the rectangular coordinates of a point in the first quadrant, and let (r, θ) be its polar coordinates (Figure 1). How does y change as θ changes? What is $\partial y / \partial \theta$?

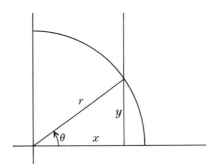

FIGURE 1

Clearly, the question has been incompletely formulated. If the point moves along a circle, y changes a certain way as θ changes. If the point moves along a vertical, y changes another way. In the first case, r was held fixed; in the second case, x was held fixed.

In other words, y is a function of r and θ, while r is a function of x and θ. Then y is also a function of x and θ, and the concept $\partial y / \partial \theta$ is ambiguous. One way to handle situations like this is to attach a subscript indicating which variable is being held fixed. Another is to stick to functional notation.

Example 5. In the situation just described we may write

$$y = f(r, \theta) = r \sin \theta, \qquad y = g(x, \theta) = x \tan \theta.$$

Then

$$\left. \frac{\partial y}{\partial \theta} \right|_r = f_\theta = r \cos \theta = x,$$

while

$$\frac{\partial y}{\partial \theta}\bigg|_x = g_\theta = x \sec^2\theta.$$

Example 6. A rectangular box is of height y and has a square base of side x. If A is the area of a wall and V is the volume, then

$$A = xy, \qquad V = x^2y = Ax = \frac{A^2}{y}.$$

Then

$$\frac{\partial V}{\partial x}\bigg|_y = 2xy = 2A, \qquad \text{while} \qquad \frac{\partial V}{\partial x}\bigg|_A = A.$$

Also,

$$\frac{\partial V}{\partial y}\bigg|_x = x^2, \qquad \text{while} \qquad \frac{\partial V}{\partial y}\bigg|_A = \frac{A^2}{-y^2} = -x^2.$$

In case u is a function of x and y where y is a differentiable function of x, formula (5) reduces to

(7)
$$\frac{du}{dx} = \frac{\partial u}{\partial x} + \frac{\partial u}{\partial y}\frac{dy}{dx}.$$

This exhibits a nice distinction between du/dx and $\partial u/\partial x$. Say $y = f(x)$ and

(8)
$$u = F(x, y) = F(x, f(x)) = G(x).$$

Then

(9)
$$\frac{du}{dx} = G'(x) \qquad \text{and} \qquad \frac{\partial u}{\partial x} = F_x(x, y).$$

Formula (7) states that

(10)
$$G'(x) = F_x(x, y) + F_y(x, y)f'(x).$$

Example 7. Let $u = x \log y + y \log x$, where y is a differentiable function of x. Then

$$\frac{\partial u}{\partial x} = \log y + \frac{y}{x} \qquad \text{and} \qquad \frac{\partial u}{\partial y} = \frac{x}{y} + \log x.$$

By (7),

$$\frac{du}{dx} = \left(\log y + \frac{y}{x}\right) + \left(\frac{x}{y} + \log x\right)\frac{dy}{dx}.$$

We could also find du/dx by differentiating through with respect to x, using the chain rule for one variable.

Example 8. With u as in Example 7, we have

$$\frac{du}{dx} = \left(\frac{x}{y}\frac{dy}{dx} + \log y\right) + \left(\frac{y}{x} + \log x \frac{dy}{dx}\right)$$

$$= \left(\log y + \frac{y}{x}\right) + \left(\frac{x}{y} + \log x\right)\frac{dy}{dx},$$

as before.

While the order of the steps is different, the two processes are really the same thing, as can be checked by tracing them back to the Fundamental Lemma.

In case u is a function of x, y, and z, and y is a function of x, the ∂-notation can lead to confusion. We may find ourselves writing

$$(11) \qquad\qquad \frac{\partial u}{\partial x} = \frac{\partial u}{\partial x} + \frac{\partial u}{\partial y}\frac{dy}{dx} \quad (?)$$

At this point, we should realize that we are using "$\partial u/\partial x$" to mean two different things. On the left side of (11), we were thinking of u as a function of x and z alone, and on the right side, as a function of x, y, and z.

Let us redo the problem using functional notation. We have, say,

$$(12) \qquad\qquad u = F(x, y, z) \qquad \text{and} \qquad y = g(x),$$

so that

$$(13) \qquad\qquad u = F\big(x, g(x), z\big) = G(x, z).$$

By the chain rule,

$$(14) \qquad\qquad G_x(x, z) = F_x(x, y, z) + F_y(x, y, z)g'(x)$$

—and the mystery has evaporated.

Example 9. Let $u = x^2 y + y^2 z$, where $y = e^x$. Then u may be viewed as a function of x and z alone. Find the partial derivative of this function with respect to x.

Solution. We have

$$u = F(x, y, z) = x^2 y + y^2 z \qquad \text{and} \qquad y = g(x) = e^x.$$

Then $u = F\big(x, g(x), z\big) = G(x, z)$ and

$$G_x(x, z) = F_x + F_y g' = 2xy + (x^2 + 2yz)e^x.$$

In terms of x and z alone,

$$G_x(x, z) = 2xe^x + (x^2 + 2e^x z)e^x.$$

We could also express u in terms of x and z in the first place:

$$u = G(x, z) = F(x, g(x), z) = x^2 e^x + e^{2x} z,$$

and compute $G_x(x, z)$ directly from there.

Problems (§13E)

1. Find $\partial u/\partial s$ and $\partial u/\partial t$.
   (a) $u = x^2 - 3xy + y^2$, $\quad x = 2s + 3t$, $\quad y = s - 2t$
   (b) $u = \arcsin(x + y)$, $\quad x = s \sin t$, $\quad y = s \cos t$
   (c) $u = e^{x/y}$, $\quad x = s + t$, $\quad y = s - t$

   (d) $u = \dfrac{1}{\sqrt{x^2 + y^2}}$, $\quad x = s \cos t$, $\quad y = s \sin t$ $\quad (s > 0)$

   (e) $u = (2x + y)^7$, $\quad x = e^{st}$, $\quad y = e^{-st}$
   (f) $u = x \log y$, $\quad x = s^2 + t^2$, $\quad y = s^2 - t^2$
   (g) $u = e^x \sin x$, $\quad x = s^2 + st + t^2$

2. Find the indicated partial derivatives.
   (a) $u = x^3 + x^2 y + xyz^2 + z^3$, $x = s^2 + t^2$, $y = 2st$, $z = s^2 - t^2$; $\quad \partial u/\partial s$
   (b) $w = \log(x^2 + y^2 + z^2)$, $x = t + s$, $y = t^2 + s^2$, $z = t^3 + s^3$; $\quad \partial w/\partial t$
   (c) $w = x^2 + 2y + z^2$, $x = t \sin s \cos r$, $y = t \sin r \cos s$, $z = t \cos s$; $\quad \partial w/\partial r$, $\partial w/\partial s$, $\partial w/\partial t$

   (d) $u = \dfrac{x + 1}{x^2 + y^2}$, $x = r^2 + s + t$, $y = rst$; $\quad \partial u/\partial r$, $\partial u/\partial s$, $\partial u/\partial t$

3. Compute the indicated partial derivatives at the given values of s and t.

   (a) $z = \dfrac{xy}{x^2 + y^2}$, $x = 2t + s$, $y = t - 2s$; $\quad \partial z/\partial x$ at $t = 1$, $s = -1$

   (b) $z = \sin(xy + y^2)$, $x = t^2 + s^2$, $y = t^2 - s^2$; $\quad \partial z/\partial y$ at $s = 1, t = 1$
   (c) $u = x^2 + y^2$, $x = s \sin t$, $y = s \cos t$; $\quad \partial u/\partial s$ and $\partial u/\partial t$ at $s = 2$, $t = \pi/2$
   (d) $u = \log(x + y)$, $x = \sin(s + t)$, $y = \sin(s - t)$; $\quad \partial u/\partial s$ and $\partial u/\partial t$ at $s = \pi/3, t = \pi/6$

4. Find du/dx.
   (a) $u = \log(x^2 + y^5)$, $\quad y = x^3 - 3x + 7$
   (b) $u = ze^{wz}$, $\quad z = x^2 + x - 1$, $\quad w = \sin x$
   (c) $u = x + \cos xy$, $\quad y = e^x$
   (d) $u = xyz$, $\quad z = \cos y$, $\quad y = \sin x$

5. A right circular cone has height 10 inches and radius 4 inches. How is the volume changing if the radius is increasing at a rate of 1 inch per second and the height is decreasing at 2 inches per second?

6. The pressure p, volume V, and absolute temperature T of a perfect gas in a closed system are related by the equation $pV = kT$, where k is a constant. At a certain instant, p is 20 pounds per square inch, V is 30 cubic feet, and T is 300°. The gas is being compressed so that V is decreasing by 1 cubic foot per minute, and the pressure is rising at a rate of 3 pounds per square inch per minute. How is the temperature changing at that instant?

7. Let $z = f(y + ax)$, where f is differentiable. Show that

$$a \frac{\partial z}{\partial y} = \frac{\partial z}{\partial x}.$$

8. Let $u = f(x + 2t, y + 3t)$, where f is smooth. Show that

$$\frac{\partial u}{\partial t} = 2 \frac{\partial u}{\partial x} + 3 \frac{\partial u}{\partial y}.$$

9. Suppose u and v are smooth functions of x and y, and u is a differentiable function of v. Show that

$$\frac{\partial u}{\partial x} \frac{\partial v}{\partial y} = \frac{\partial u}{\partial y} \frac{\partial v}{\partial x}.$$

[There are two cases: $du/dv \neq 0$ and $du/dv = 0$.]

10. A smooth function $f(x, y)$ is said to be *homogeneous of degree n* if it satisfies the the formula

(15) $$f(tx, ty) = t^n f(x, y).$$

Show that if f is homogeneous of degree n, then it satisfies *Euler's formula*:

(16) $$xf_x(x, y) + yf_y(x, y) = nf(x, y).$$

[*Hint.* Start with the definition of homogeneity. Differentiate both sides with respect to t. Then set $t = 1$.]

11. The following functions are homogeneous. In each case, find the value of n for which (15) holds (Problem 10). Then verify Euler's formula (16).

(a) $f(x, y) = \dfrac{x^2}{x + y}$ (c) $f(x, y) = \arctan \dfrac{y}{x}$

(b) $f(x, y) = \sqrt{x^2 + y^2}$ (d) $f(x, y) = (x + y)^3$

12. Is every smooth function homogeneous (Problem 10)?

Answers to problems (§13E)

1. (We give $\partial u/\partial s$, then $\partial u/\partial t$.)
   (a) $x - 4y$; $12x - 13y$
   (b) $(\sin t + \cos t)/[1 - (x + y)^2]^{1/2}$; $s(\cos t - \sin t)/[1 - (x + y)^2]^{1/2}$
   (c) $[(y - x)/y^2]e^{x/y}$; $[(y + x)/y^2]e^{x/y}$
   (d) $-1/s^2$; 0

(e) $7t(2x + y)^6(2e^{st} - e^{-st})$; $7s(2x + y)^6(2e^{st} - e^{-st})$
(f) $2s \log y + 2sx/y$; $2t \log y - 2tx/y$
(g) $e^x(\sin x + \cos x)(2s + t)$; $e^x(\sin x + \cos x)(2t + s)$

2. (a) $2s(3x^2 + 2xy + yz^2 + 2xyz + 3z^2) + 2t(x^2 + xz^2)$
   (b) $(2x + 4yt + 6zt^2)/(x^2 + y^2 + z^2)$
   (c) $\partial w/\partial r = 2t \cos r \cos s - 2xt \sin r \sin s$;
        $\partial w/\partial s = 2xt \cos r \cos s - 2t \sin r \sin s - 2zt \sin s$;
        $\partial w/\partial t = 2x \cos r \sin s + 2 \sin r \cos s + 2z \cos s$
   (d) with $a = (x^2 + y^2)^{-2}$:
        $\partial u/\partial r = 2a[r(y^2 - 2x - x^2) - yst(x + 1)]$;
        $\partial u/\partial s = a[y^2 - 2x - x^2 - 2yrt(x + 1)]$;
        $\partial u/\partial t = a[y^2 - 2x - x^2 - 2yrs(x + 1)]$

3. (a) 0.24   (b) 2   (c) $\partial u/\partial s = 4$; $\partial u/\partial t = 0$
   (d) $\partial u/\partial s = \sqrt{3}/3$; $\partial u/\partial t = -\sqrt{3}/3$

4. (a) $[2x + 15y^4(x^2 - 1)]/(x^2 + y^5)$
   (b) $e^{wz}[(1 + wz)(2x + 1) + z^2 \cos x]$
   (c) $1 - (1 + x)e^x \sin(xe^x)$
   (d) $yz + (x \cos x)(z - y \sin y)$

5. increasing at 16π cu. in./sec.

6. increasing at $35°$/min.

11. (a) 1   (b) 1   (c) 0   (d) 3

12. no

§13F. Tangent plane

13F1. Definition of the tangent plane. Corresponding to the notion of tangent line to a curve is the notion of tangent plane to a surface. Let $z = f(x, y)$ be a smooth surface, and consider any point $Q_0 = (x_0, y_0, z_0)$ on the surface. The plane $y = y_0$ intersects the surface in the curve

(1) $$P_1(x) = (x, y_0, f(x, y_0)),$$

and the plane $x = x_0$ intersects it in the curve

(2) $$P_2(y) = (x_0, y, f(x_0, y)).$$

Their tangent vectors at the point Q_0 are

(3) $P_1'(x_0) = (1, 0, f_x(x_0, y_0))$ and $P_2'(y_0) = (0, 1, f_y(x_0, y_0))$.

These tangent vectors, together with the point Q_0, determine a plane, p (Figure 1). Clearly, any tangent plane worthy of the name will contain the two tangent lines $Q_0 + tP_1'(x_0)$ and $Q_0 + tP_2'(y_0)$, and hence can only be p.

We now derive the point-normal equation of p (§11E(1)):

(4) $$a(x - x_0) + b(y - y_0) + c(z - z_0) = 0.$$

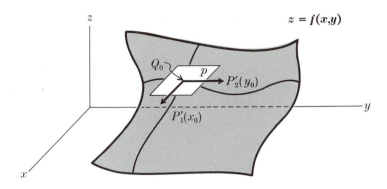

FIGURE 1

Here the coefficient vector (a, b, c) is perpendicular to both $(1, 0, f_x)$ and $(0, 1, f_y)$. We see by inspection that

(5) $$(f_x, f_y, -1)$$

is such a vector. The equation of the plane p is, then

(6) $$z - z_0 = f_x(x - x_0) + f_y(y - y_0)$$

$\big($where f_x and f_y are evaluated at $(x_0, y_0)\big)$.

We take equation (6) as the definition of the tangent plane.

Remark. Writing $\Delta x = x - x_0$, $\Delta y = y - y_0$, $\Delta z = z - z_0$, we have

(7) $$\Delta z = f_x\, \Delta x + f_y\, \Delta y.$$

But this is simply df, the total differential of f (§13C(2)). When Δx and Δy are small, df is the approximate change in f. Thus, the tangent plane is an approximation to the surface near the point of tangency.

Example 1. Find the tangent plane to the surface

$$z = f(x, y) = xy + y^2$$

at the point $(1, 2, 6)$.

Solution. Here $f_x = y$ and $f_y = x + 2y$. Hence $f_x(1, 2) = 2$ and $f_y(1, 2) = 5$, and the equation of the tangent plane is

$$z - 6 = 2(x - 1) + 5(y - 2).$$

Example 2. Find the tangent plane to the hemisphere

$$z = f(x, y) = \sqrt{1 - x^2 - y^2}$$

at a point (a, b, c) $(c > 0)$ on its surface.

Solution. Here

$$f_x = -\frac{x}{z} = -\frac{a}{c} \quad \text{and} \quad f_y = -\frac{y}{z} = -\frac{b}{c}.$$

Hence the equation is

$$z - c = -\frac{a}{c}(x - a) - \frac{b}{c}(y - b),$$

or, more symmetrically, $a(x - a) + b(y - b) + c(z - c) = 0$, or

$$ax + by + cz = 1.$$

Note that this plane is perpendicular to the radius vector (a, b, c) to the point, in agreement with the definition given in geometry.

13F2. Properties of the tangent plane. The tangent plane has the following property: if C is any curve at all that lies on the surface and passes through the point (x_0, y_0, z_0), and that has a tangent there, then that tangent lies in the plane. To see this, let

$$(8) \qquad\qquad C(t) = (X, Y, Z),$$

where X, Y, and Z are, of course, functions of t. Since C lies on the surface $z = f(x, y)$, we have

$$(9) \qquad\qquad Z = f(X, Y).$$

Then the tangent vector is

$$(10) \qquad C'(t) = (X', Y', Z') = (X', Y', f_x X' + f_y Y').$$

We note that this vector is perpendicular to

$$(11) \qquad\qquad (f_x, f_y, -1).$$

Therefore the tangent line to C lies in the tangent plane $\big(\text{see } (5)\big)$.

Problems (§13F)

1. Find the equation of the tangent plane to the given surface at the given point.

(a) $z = \dfrac{x}{\sqrt{x^2 + y^2}}$, $(4, -3, \frac{4}{5})$

(b) $z = x^2 - 2y^2$, $(2, 1, 2)$
(c) $z = xy$, $(-2, 3, -6)$
(d) $z = 3x^2 + 2y^2$, $(-1, 2, 11)$

(e) $z = \dfrac{x + y}{x - 2y}$, $(3, 1, 4)$

(f) $z = e^x \cos y$, $(0, 0, 1)$
(g) $z = 6/x$, $(2, 1, 3)$

(h) $z = \arctan \dfrac{y}{x}$, $(2, 2, \pi/4)$

2. The *normal line* to a surface at a point is the line through the point and perpendicular to the tangent plane there. Find a vector equation for the normal line at the given point on each surface in Problem 1.

3. Show that every plane tangent to the surface $z = \sqrt{x^2 + y^2}$ passes through the origin.

4. Show that every tangent plane to the surface $z = 1/xy$ meets the coordinate axes in points $(a, 0, 0)$, $(0, b, 0)$, and $(0, 0, c)$ such that $abc = 27$.

5. The *angle between two surfaces* at a common point is defined to be the angle between the tangent planes there. Find the angle between the given surfaces at the given point.

(a) $z = x^2$, $z = 100 - y^2$; $(-6, 8, 36)$
(b) $z = e^{xy}$, $z = x^2 + xy^2 + y^3$; $(1, 0, 1)$
(c) $z = x^2 + y^2$, $3x - 2y - z = 3$; $(2, -1, 5)$
(d) $z = \log\sqrt{x^2 + y^2}$, $z = e^{xy} - 1$; $(1, 0, 0)$

Answers to problems (§13F)

1. (a) $9x + 12y - 125z + 100 = 0$ (b) $4x - 4y - z = 2$
(c) $3x - 2y - z + 6 = 0$ (d) $6x - 8y + z + 11 = 0$ (e) $3x - 9y + z = 4$
(f) $x - z + 1 = 0$ (g) $3x + 2z = 12$ (h) $x - y + 4z = \pi$

2. (a) $P = (4, -3, \frac{4}{5}) + t(9, 12, -125)$
(b) $P = (2, 1, 2) + t(-4, 4, 1)$
(c) $P = (-2, 3, -6) + t(-3, 2, 1)$
(d) $P = (-1, 2, 11) + t(6, -8, 1)$
(e) $P = (3, 1, 4) + t(3, -9, 1)$
(f) $P = (0, 0, 1) + t(1, 0, -1)$
(g) $P = (2, 1, 3) + t(3, 0, 2)$
(h) $P = (2, 2, \pi/4) + t(1, -1, 4)$

5. (a) $\cos \theta = 1/\sqrt{145 \times 257} \approx 0.005, \quad \theta \approx 90°$
   (b) $\cos \theta = 1/\sqrt{10} \approx 0.32, \quad \theta \approx 70°$
   (c) $\cos \theta = 17/7\sqrt{6} \approx 0.99, \quad \theta \approx 5°$
   (d) $\cos \theta = \frac{1}{2}, \quad \theta = 60°$

§13G. Higher derivatives

13G1. *Derivatives of second order.* For a function of two variables, $z = f(x, y)$, the partials f_x and f_y are also functions of two variables and may themselves have partial derivatives. These are then second partials of f. As usual, several symbols are in common use. The partial of f_x with respect to x,

$$(1) \qquad \frac{\partial}{\partial x} f_x \quad \text{or} \quad (f_x)_x \quad \text{or} \quad \frac{\partial}{\partial x}\left(\frac{\partial z}{\partial x}\right),$$

is denoted more briefly by

$$(2) \qquad f_{xx} \quad \text{or} \quad \frac{\partial^2 z}{\partial x^2}.$$

Similarly, $(\partial/\partial y)f_y$ is written

$$(3) \qquad f_{yy} \quad \text{or} \quad \frac{\partial^2 z}{\partial y^2}.$$

The "mixed" partial $(\partial/\partial y)f_x$ is written

$$(4) \qquad f_{xy} \quad \text{or} \quad \frac{\partial^2 z}{\partial y \, \partial x};$$

and $(\partial/\partial x)f_y$ is denoted

$$(5) \qquad f_{yx} \quad \text{or} \quad \frac{\partial^2 z}{\partial x \, \partial y}.$$

The order of writing the variables here is logical, even if at first bewildering. It is of little importance, however, as f_{xy} and f_{yx} will usually turn out to be equal. In fact:

(6)   *If f_y and f_{xy} exist and if f_{xy} is continuous, then f_{yx} exists and $f_{yx} = f_{xy}$.*

Of course, this also holds with x and y interchanged.

This result is of obvious importance in the statement of theorems involving mixed partials. It also saves computing f_{yx} separately. In practice, however, it is good to compute both f_{xy} and f_{yx} anyhow, as a check.

One can prove (6) by a series of intricate maneuverings with the Mean-Value Theorem. We will give a proof that is easier to follow, based on the theory of the double integral. The proof appears in §14C4.

Example 1. Find the first and second partials of

$$f(x, y) = x^2 y + y^2 \sin x.$$

Solution. We have

$$f_x = 2xy + y^2 \cos x, \qquad f_{xy} = 2x + 2y \cos x.$$

At this stage we know that f_{yx} exists and equals f_{xy}, because f_y exists (by inspection of f) and f_{xy} is continuous (by inspection of f_{xy}). The other partials are

$$f_{xx} = 2y - y^2 \sin x, \qquad f_y = x^2 + 2y \sin x, \qquad f_{yy} = 2 \sin x.$$

13G2. Second directional derivative. Consider a surface $z = f(x, y)$. We will assume that z has continuous second partial derivatives. Consider the vertical plane through a point (x_0, y_0) and in the direction α from the x-axis (Figure 1). The directional derivative of f at (x_0, y_0), in the direction α, is the rate of change of f at (x_0, y_0) along the curve of intersection of the plane with the surface. As we saw in §13D(7), it is given by

$$(7) \qquad\qquad f_{(\alpha)} = f_x \cos \alpha + f_y \sin \alpha.$$

What about the *concavity* of the curve? For this we want the "second directional derivative in the direction α," i.e., the directional derivative of

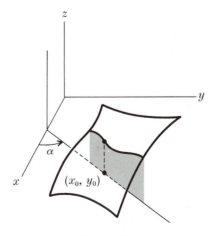

FIGURE 1

$f_{(\alpha)}$ in the direction α. To find it, we apply (7) to (7). The ordinary first partials of $f_{(\alpha)}$ are

(8)
$$\frac{\partial}{\partial x} f_{(\alpha)} = f_{xx} \cos \alpha + f_{yx} \sin \alpha$$

and

(9)
$$\frac{\partial}{\partial y} f_{(\alpha)} = f_{xy} \cos \alpha + f_{yy} \sin \alpha.$$

The second directional derivative is

(10)
$$\frac{\partial}{\partial x} f_{(\alpha)} \cos \alpha + \frac{\partial}{\partial y} f_{(\alpha)} \sin \alpha = f_{xx} \cos^2\alpha + 2f_{xy} \cos \alpha \sin \alpha + f_{yy} \sin^2\alpha.$$

(In computing this we have taken advantage of the equality $f_{xy} = f_{yx}$.) Where this is positive the curve is concave up; where negative, concave down.

Example 2. Compute and interpret the second directional derivative of

$$f(x, y) = x^3 - 2y^2 + x + 3y$$

in the direction $\pi/3$ at the point $(1, 2)$.

Solution. The vertical plane through $(1, 2)$ and in the direction $\pi/3$ intersects the surface $z = f(x, y)$ in a certain curve. The derivative in question tells us about the concavity of that curve at the point $(1, 2)$.
We have

$$f_x = 3x^2 + 1, \qquad f_y = -4y + 3,$$

and

$$f_{xx} = 6x, \qquad f_{xy} = 0, \qquad \text{and} \qquad f_{yy} = -4.$$

Hence

$$f_{xx}(1, 2) = 6, \qquad f_{xy}(1, 2) = 0, \qquad \text{and} \qquad f_{yy}(1, 2) = -4.$$

Since $\cos(\pi/3) = \tfrac{1}{2}$ and $\sin(\pi/3) = \sqrt{3}/2$, the required second derivative is

$$(6)(\tfrac{1}{2})^2 + (2)(0)(\tfrac{1}{2})(\tfrac{1}{2}\sqrt{3}) + (-4)(\tfrac{1}{2}\sqrt{3})^2 = -\tfrac{3}{2}.$$

As this is negative, the curve described is concave down at $(1, 2)$.
Note that we derived this information without computing the first directional derivative, $f_{(\pi/3)}$. If we wish to find it, we note that

$$f_x(1, 2) = 4 \qquad \text{and} \qquad f_y(1, 2) = -5.$$

Hence

$$f_{(\pi/3)} = (4)(\tfrac{1}{2}) - (5)(\tfrac{1}{2}\sqrt{3}) < 0.$$

Therefore the curve is decreasing at the point.

13G3. Derivatives of higher order. These are defined in the obvious way. For instance,

$$(11) \qquad\qquad f_{xyy} \qquad \text{means} \qquad \frac{\partial}{\partial y} f_{xy}.$$

Again it turns out that, under mild conditions of continuity, the order of differentiation is immaterial; for instance,

$$(12) \qquad\qquad f_{xyy} = f_{yxy} = f_{yyx}.$$

Example 3. In Example 1, we have:

$$f_{xxx} = -y^2 \cos x, \qquad f_{yxx} = 2 - 2y \sin x,$$
$$f_{xxy} = 2 - 2y \sin x, \qquad f_{yxy} = 2 \cos x,$$
$$f_{xyx} = 2 - 2y \sin x, \qquad f_{yyx} = 2 \cos x,$$
$$f_{xyy} = 2 \cos x, \qquad f_{yyy} = 0.$$

13G4. Functions of several variables. The corresponding results hold in more than two variables.

Example 4. If $f(x, y, z) = xyz + xy^2 + yz^2$, then

$$f_x = yz + y^2, \qquad f_y = xz + 2xy + z^2, \qquad f_z = xy + 2yz,$$
$$f_{xx} = 0, \qquad f_{yx} = z + 2y, \qquad f_{zx} = y,$$

etc.

Problems (§13G)

1. Compute f_{xx}, f_{yy}, and f_{xy}.
   (a) $f(x, y) = x^3 - 4x^2y + 3xy^2 + 2y^3$
   (b) $f(x, y) = e^x \cos xy + e^y \sin xy$

   (c) $f(x, y) = \dfrac{x}{x - y}$

   (d) $f(x, y) = \sin(x^2 + y^2)$

   (e) $f(x, y) = \dfrac{x^2}{y} + \dfrac{y^2}{x}$

   (f) $f(x, y) = (x^2y + xy^2)^5$

   (g) $f(x, y) = \arctan \dfrac{y}{x}$

   (h) $f(x, y) = x^y$

2. If $z = e^{-t}(\cos x + \sin y)$, show that

$$\frac{\partial^2 z}{\partial x^2} + \frac{\partial^2 z}{\partial y^2} = \frac{\partial z}{\partial t}.$$

3. Show that the following functions satisfy the *wave equation*,

$$a^2 \frac{\partial^2 z}{\partial x^2} = \frac{\partial^2 z}{\partial t^2}.$$

(a) $z = (x + at)^4$
(b) $z = (x - at)^4$
(c) $z = x^2 + a^2 t^2$
(d) $z = e^{x+at} \log(x + at)$
(e) $z = \sin x \cos at - \cos x \sin at$

4. Show that the following functions satisfy the *Laplace equation*,

$$\frac{\partial^2 z}{\partial x^2} + \frac{\partial^2 z}{\partial y^2} = 0.$$

(a) $z = x^2 - y^2$
(b) $z = 4x^3 - 6x^2 y - 12xy^2 + 2y^3 + 3xy - 2x + y - 4$
(c) $z = e^x \cos y$
(d) $z = e^{-2x} \cos 2y$

(e) $z = \arctan \dfrac{y}{x}$

(f) $z = \log \sqrt{x^2 + y^2}$
(g) $z = \log[(x + 1)^2 + (y + 2)^2]$

5. Let $z = f(x + 2y) + g(x - 2y)$, where f and g are twice differentiable. Show that

$$\frac{\partial^2 z}{\partial y^2} = 4 \frac{\partial^2 z}{\partial x^2}.$$

6. Find the second directional derivative of f in the direction α.
(a) $f(x, y) = \sin(x^2 + y)$, $\alpha = \pi/6$
(b) $f(x, y) = xe^y$, $\alpha = 5\pi/6$
(c) $f(x, y) = x^2 + xy + y^2$, $\alpha = \pi/4$
(d) $f(x, y) = x \cos xy$, $\alpha = \pi/3$

7. Compute $f_{xxx}, f_{xxy}, f_{xyy}$, and f_{yyy}.
(a) $f(x, y) = x^4 + 3x^3 y^3 + x^2 y - 2xy$
(b) $f(x, y = \sqrt{x^2 + y^2}$
(c) $f(x, y) = e^{xy} \cos y$
(d) $f(x, y) = y \sin xy$

8. Compute f_{xy}, f_{yz}, and f_{xyz}.
(a) $f(x, y, z) = x^3 + xy^2 - y^2 z^2 + x^2 z^3 + z^4$

(b) $f(x, y, z) = \dfrac{xyz}{x^2 + y^2 + z^2}$

(c) $f(x, y, z) = e^{xyz} + \dfrac{x}{yz}$

(d) $f(x, y, z) = \arcsin \dfrac{xy}{z}$ $(z > 0)$

Answers to problems (§13G)

1. (a) $f_{xx} = 6x - 8y$, $f_{yy} = 6x + 12y$, $f_{xy} = -8x + 6y$

   (b) with $C = \cos xy$, $S = \sin xy$:
   $$f_{xx} = (1 - y^2)e^xC - 2ye^xS - y^2e^yS,$$
   $$f_{yy} = (1 - x^2)e^yS + 2xe^yC - x^2e^xC,$$
   $$f_{xy} = e^y(C + yC - xyS) - e^x(S + xS + xyC)$$

   (c) with $a = (x - y)^{-3}$: $f_{xx} = 2ya$, $f_{yy} = 2xa$, $f_{xy} = -(x + y)a$

   (d) with $C = \cos(x^2 + y^2)$, $S = \sin(x^2 + y^2)$:
   $$f_{xx} = 2C - 4x^2S,\quad f_{yy} = 2C - 4y^2S,\quad f_{xy} = -4xyS$$

   (e) $f_{xx} = 2(1/y + y^2/x^3)$, $f_{yy} = 2(1/x + x^2/y^3)$, $f_{xy} = -2(x/y^2 + y/x^2)$

   (f) with $a = x^2y + xy^2$: $f_{xx} = 10a^3[ya + 2(2xy + y^2)^2]$,
   $$f_{yy} = 10a^3[xa + 2(x^2 + 2xy)^2],$$
   $$f_{xy} = 10a^3[(x + y)a + 2(x^2 + 2xy)(2xy + y^2)]$$

   (g) with $a = (x^2 + y^2)^{-2}$: $f_{xx} = 2xya$, $f_{yy} = -2xya$, $f_{xy} = (y^2 - x^2)a$

   (h) $f_{xx} = y(y - 1)x^{y-2}$, $f_{yy} = x^y(\log x)^2$, $f_{xy} = x^{y-1}(1 + y \log x)$

6. (a) $\frac{3}{2} \cos(x^2 + y) - (3x^2 + x\sqrt{3} + \frac{1}{4}) \sin(x^2 + y)$

   (b) $\frac{1}{4}e^y(x - 2\sqrt{3})$ (c) 3

   (d) $-\frac{1}{2}(y + 2\sqrt{3}) \sin xy - \frac{1}{4}(xy^2 + 2x^2y\sqrt{3} + 3x^3) \cos xy$

7. (a) $f_{xxx} = 24x + 18y^3$, $f_{xxy} = 54xy^2 + 2$, $f_{xyy} = 54x^2y$, $f_{yyy} = 18x^3$

   (b) with $a = (x^2 + y^2)^{-5/2}$: $f_{xxx} = -3xy^2a$, $f_{xxy} = (2x^2y - y^3)a$,
   $$f_{xyy} = (2xy^2 - x^3)a,\quad f_{yyy} = -3x^2ya$$

   (c) $f_{xxx} = y^3e^{xy} \cos y$, $f_{xxy} = ye^{xy}[(xy + 2) \cos y - y \sin y]$,
   $$f_{xyy} = (2x + x^2y - y)e^{xy} \cos y - 2(xy + 1)e^{xy} \sin y,$$
   $$f_{yyy} = (x^3 - 3x)e^{xy} \cos y + (1 - 3x^2)e^{xy} \sin y$$

   (d) $f_{xxx} = -y^4 \cos xy$, $f_{xxy} = -y^2(xy \cos xy + 3 \sin xy)$,
   $$f_{xyy} = -4xy \sin xy + (2 - x^2y^2) \cos xy,$$
   $$f_{yyy} = -x^2(3 \sin xy + xy \cos xy)$$

8. (a) $f_{xy} = 2y$, $f_{yz} = -4yz$, $f_{xyz} = 0$

   (b) with $a = (x^2 + y^2 + z^2)^{-1}$:
   $$f_{xy} = za - 2(y^2 + x^2)za^2 + 8x^2y^2za^3,$$
   $$f_{yz} = xa - 2(z^2 + y^2)xa^2 + 8xy^2z^2a^3,$$
   $$f_{xyz} = -a + 8(x^2y^2 + x^2z^2 + y^2z^2)a^3 - 48x^2y^2z^2a^4$$

   (c) with $a = xyz$: $f_{xy} = z(1 + a)e^a - 1/y^2z$,
   $$f_{yz} = x(1 + a)e^a + x/y^2z^2,\quad f_{xyz} = (1 + 3a + a^2)e^a + 1/y^2z^2$$

   (d) with $a = (z^2 - x^2y^2)^{-1/2}$: $f_{xy} = z^2a^3$,
   $$f_{yz} = -xza^3,\quad f_{xyz} = -(z^2 + 2x^2y^2)za^5$$

§13H. Maxima and minima

Problems concerning maxima and minima of functions of two variables quickly become quite complex. We confine ourselves here to an introduction to *local* maxima and minima, including a two-variable version of the second-derivative test (§4D5).

13H1. Local maximum and minimum. Let f be a smooth function on an open region R and let (x_0, y_0) be a point of R. We say that f has a *local maximum* at (x_0, y_0) if on some open box about (x_0, y_0), f attains a maximum value at (x_0, y_0). *Local minimum* is defined similarly. Corresponding to §4D(1), we have:

If f has a local maximum or local minimum at (x_0, y_0), then

$$(1) \qquad f_x(x_0, y_0) = 0 \qquad and \qquad f_y(x_0, y_0) = 0.$$

Proof. The function of one variable $f(x, y_0)$ has a local maximum or minimum at $x = x_0$. Hence its derivative is 0 there. That is, $f_x(x_0, y_0) = 0$. Similarly, $f_y(x_0, y_0) = 0$.

13H2. Saddle points. It can happen that f_x and f_y are both 0 at a point where f has neither a local maximum nor a local minimum.

One way is illustrated by the cylindrical surface $z = x^3$. This is generated by taking the curve $z = x^3$ in the xz-plane and sliding it parallel to the y-axis. The curve has an inflection point at the origin, and the resulting surface inherits this feature. See Figure 1.

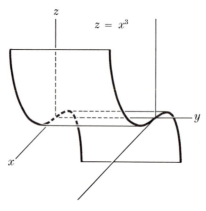

FIGURE 1

A more interesting illustration is a *saddle point*. This is a point at which f has a local *maximum* in one direction and a local *minimum* in a different direction.

Example 1. Consider the function

$$z = f(x, y) = y^2 - x^2.$$

Its graph is the hyperbolic paraboloid shown in §12B, Figure 10. Clearly, f has no local maximum or minimum at $(0, 0)$, yet $f_x(0, 0) = 0$ and $f_y(0, 0) = 0$.

In fact, $(0, 0, 0)$ is a saddle point. The curve $z = y^2$ in the yz-plane (which is the intersection of this plane with the surface) has a minimum at the origin, while the curve $z = -x^2$ in the xz-plane has a maximum.

Example 1 seems to suggest a simple second-derivative test for identifying local maxima and minima. Consider a point where f_x and f_y are 0. Perhaps if f_{xx} and f_{yy} are both positive we have a local minimum; and if both negative, a local maximum. Unfortunately, things are not so simple. The function may be increasing in both the x-direction and the y-direction, but decreasing in some other direction.

Example 2. Let

$$f(x, y) = x^2 - 3xy + y^2.$$

Then

$$f_x = 2x - 3y \qquad \text{and} \qquad f_y = 2y - 3x,$$

and

$$f_x(0, 0) = 0 \qquad \text{and} \qquad f_y(0, 0) = 0.$$

Furthermore, $f_{xx} = 2$ and $f_{yy} = 2$; thus both f_{xx} and f_{yy} are positive at $(0, 0)$. But f does not have a local minimum there. For, consider the intersection of the surface with the plane $x = y$. This is the curve

$$z = x^2 - 3x^2 + x^2 = -x^2$$

in that plane, and it goes in the wrong direction.

Clearly, a second-derivative test will have to take account of how f increases or decreases in *all* directions.

13H3. The second-derivative tests. Let f have continuous second partial derivatives on an open region. We consider a point (x_0, y_0) at which $f_x = 0$ and $f_y = 0$.

The results rest on the sign of the second directional derivative in the direction α. Let us call this derivative $f_{(\alpha\alpha)}$. From §13G(10),

(2) $$f_{(\alpha\alpha)} = f_{xx} \cos^2\alpha + 2f_{xy} \cos \alpha \sin \alpha + f_{yy} \sin^2\alpha.$$

Suppose there is an open box about (x_0, y_0) on which $f_{(\alpha\alpha)}$ is positive for every α. Then the curves *in every direction* are concave up. It follows that f has a local minimum at (x_0, y_0). Similarly, if $f_{(\alpha\alpha)}$ is negative for every α then f has a local maximum there. Finally, if $f_{(\alpha\alpha)}(x_0, y_0)$ is positive for some values of α and negative for others, then $\big(x_0, y_0, f(x_0, y_0)\big)$ is a saddle point.

The test itself is stated in terms of the sign of the quantity

(3)
$$\Delta = f_{xx}f_{yy} - f_{xy}^2.$$

Note that when $\Delta > 0$, f_{xx} and f_{yy} must have the same sign.

(a) *If $\Delta > 0$ and $f_{xx} > 0$ at (x_0, y_0) then f has a local minimum there.*

Proof. Since f_{xx} and Δ are continuous, they are positive throughout some open box about (x_0, y_0). Hence it suffices to show that $f_{(\alpha\alpha)} > 0$ wherever $\Delta > 0$ and $f_{xx} > 0$. If $\sin \alpha = 0$ we have $f_{(\alpha\alpha)} = f_{xx} > 0$, while if $\cos \alpha = 0$ then $f_{(\alpha\alpha)} = f_{yy} > 0$.

Henceforth assume $\sin \alpha \neq 0$ and $\cos \alpha \neq 0$. We make use of the following observation about numbers A, B, and C.

(4)
$$\text{If} \quad A + B > 0 \quad \text{and} \quad AB > C^2, \quad \text{then} \quad A + B > 2C.$$

To prove (4) note that $(A - B)^2 \geq 0$. Adding $4AB$ to both sides, we get $(A + B)^2 \geq 4AB$. Then $(A + B)^2 > 4C^2$, and hence $A + B > 2C$.

Now to show that $f_{(\alpha\alpha)} > 0$ we quote (4) with

(5)
$$A = f_{xx}\cos^2\alpha, \qquad B = f_{yy}\sin^2\alpha, \qquad C = -f_{xy}\cos\alpha\sin\alpha.$$

In similar fashion, we obtain:

(b) *If $\Delta > 0$ and $f_{xx} < 0$ at (x_0, y_0) then f has a local maximum there.*

Proof. To show that $f_{(\alpha\alpha)} < 0$ we quote (4) with

(6)
$$A = -f_{xx}\cos^2\alpha, \qquad B = -f_{yy}\sin^2\alpha, \qquad C = f_{xy}\cos\alpha\sin\alpha.$$

Finally, we prove:

(c) *If $\Delta < 0$ at (x_0, y_0) then $\left(x_0, y_0, f(x_0, y_0)\right)$ is a saddle point.*

Proof. We are to find two directions for which $f_{(\alpha\alpha)}$ has opposite signs.

If f_{xx} and f_{yy} are both 0 then $f_{xy} \neq 0$. By (2),
$$f_{(\alpha\alpha)} = 2f_{xy}\cos\alpha\sin\alpha,$$
and this has opposite signs at $\alpha = \pi/4$ and $\alpha = -\pi/4$.

If f_{xx} and f_{yy} are not both 0 then, say, $f_{xx} \neq 0$. Define

(7)
$$q = \sqrt{f_{xx}^2 + f_{xy}^2}.$$

Then $q \neq 0$. Hence we can define α_0 (between 0 and 2π) by

(8)
$$\cos\alpha_0 = -\frac{f_{xy}}{q}, \qquad \sin\alpha_0 = \frac{f_{xx}}{q}.$$

Then by (2),

(9)
$$f_{(\alpha_0\alpha_0)} = \frac{1}{q^2}(f_{xx}f_{xy}^2 - 2f_{xy}^2 f_{xx} + f_{yy}f_{xx}^2) = \frac{1}{q^2}f_{xx}\Delta.$$

By hypothesis, this has the opposite sign to f_{xx}. Since $f_{xx} = f_{(00)}$, we have two values of $f_{(\alpha\alpha)}$ with opposite signs (namely, $f_{(00)}$ and $f_{(\alpha_0\alpha_0)}$).

Remark. If $\Delta = 0$ the theorem gives no information, and, in fact, anything can happen (see Problem 2).

Example 3. Find the local maxima, local minima, and saddle points of

$$f(x, y) = x^3 - 6xy + y^3.$$

Solution. Here $f_x = 3x^2 - 6y$ and $f_y = 3y^2 - 6x$. Setting these equal to 0 and solving simultaneously, we find that f_x and f_y are both 0 at (0, 0) and at (2, 2), but not elsewhere.

Next, $f_{xx} = 6x$, $f_{xy} = -6$, and $f_{yy} = 6y$. Hence

$$\Delta = f_{xx}f_{yy} - f_{xy}^2 = 36xy - 36.$$

Then $\Delta(0, 0) = -36 < 0$, so (0, 0) is a saddle point. At (2, 2), we have $\Delta(2, 2) = 36 \cdot 4 - 36 > 0$ and $f_{xx}(2, 2) = 6 \cdot 2 > 0$, so f has a local minimum at (2, 2). This is no local maximum.

Problems (§13H)

1. Find the points at which $f_x = f_y = 0$, and classify them by means of the second derivative tests of §13H3: $f(x, y) =$

    (a) $x^3 + y^3 - 3xy$ (g) $\sin x + \sin y + \sin(x + y)$

    (b) $x^2 - y^2 + 6x - 2y + 4$ (h) $\dfrac{1}{x^2 - y^2 + 1}$

    (c) $x^2 - 2xy + 3y^2$ (i) $(x - 1)(xy - 1)$

    (d) $xy + \dfrac{2}{x} + \dfrac{4}{y}$ (j) $xy(4x + 2y + 1)$

    (e) $e^x \cos y$ (k) $x^3y^2(x + y - 24)$
    (f) $e^{-x} \sin^2 y$ (l) $xye^{-(3x+2y)}$

2. Show that for each of the following functions, f_x, f_y, and Δ are all 0 at the origin. Show further that in (a), there is a local maximum at the origin; in (b), a local minimum; and, in (c), neither.

    (a) $f(x, y) = -(x^4 + y^4)$ (c) $f(x, y) = x^3y^3$
    (b) $f(x, y) = x^4 + y^4$

Answers to problems (§13H)

1. (a) (0, 0), saddle point;   (1, 1), local minimum
    (b) $(-3, -1)$, saddle point   (c) (0, 0), local minimum
    (d) (1, 2), local minimum   (e) none
    (f) all points $(x, n\pi)$, n an integer;   the tests fail

(g) $\left(\dfrac{\pi}{3} + 2n\pi, \dfrac{\pi}{3} + 2n\pi\right)$, n an integer, local maximum;

$\left(-\dfrac{\pi}{3} + 2n\pi, -\dfrac{\pi}{3} + 2n\pi\right)$, n an integer, local minimum;

$(\pi + 2n\pi, \pi + 2n\pi)$, n an integer, the tests fail

(h) $(0, 0)$, saddle point

(i) $(0, -1)$ and $(1, 1)$, saddle points

(j) $(0, 0)$, $(-\frac{1}{4}, 0)$, and $(0, -\frac{1}{2})$, saddle points; $(-\frac{1}{12}, -\frac{1}{6})$, local maximum

(k) $x = 0$, y arbitrary, the tests fail; $y = 0$, x arbitrary, the tests fail;
$(12, 8)$, local minimum

(l) $(0, 0)$, saddle point; $(\frac{1}{3}, \frac{1}{2})$, local maximum

§13I. Implicit functions

13I1. Implicit functions. We begin with two examples.
Consider the equation

(1) $$x^2 + y^2 - 1 = 0.$$

It cannot be solved for y as a function of x. In other words, its graph is not the graph of a function $y = f(x)$. However, any point (x_0, y_0) of the graph lies on a *portion* of the graph that *is* the graph of a function—moreover, of a differentiable function (provided $y_0 \neq 0$) (Figure 1). Explicitly, if $y_0 > 0$, then (x_0, y_0) lies on the graph of the function

$$y = \sqrt{1 - x^2},$$

which is differentiable for $-1 < x < 1$. And if $y_0 < 0$, then (x_0, y_0) lies on the graph of the function

$$y = -\sqrt{1 - x^2},$$

which is differentiable for $-1 < y < 1$.

Knowing that y is definable as a differentiable function, we can find dy/dx from (1) by the chain rule, without solving for y. We have

(2) $$2x + 2yy' = 0,$$

(3) $$y' = -\frac{x}{y} \qquad (y \neq 0).$$

We really should know that y is definable as a differentiable function before we "find" dy/dx. Suppose, for example, that instead of (1) we had

(4) $$x^2 + y^2 + 1 = 0.$$

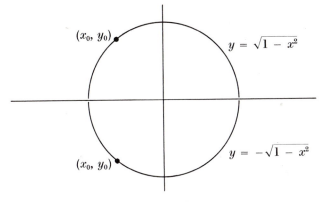

(x_0, y_0)

$y = \sqrt{1 - x^2}$

$y = -\sqrt{1 - x^2}$

(x_0, y_0)

FIGURE 1

Proceeding blindly, one could write down the same equation (2) "by the chain rule," and then (3). But all this would be meaningless, as (4) does not define anything at all.

Next, consider the equation

(5) $$e^{xy} + x^2 - y - 1 = 0.$$

In the first place, there *are* points that satisfy it; for example, $(0, 0)$. So the graph of the equation is not mythical. Is it the graph of a function $y = f(x)$? Perhaps. But we don't know how to solve for y.

If we *knew* that y were a differentiable function of x we could find y' from (5), by the chain rule. We would get

(6) $$e^{xy}(xy' + y) + 2x - y' = 0, \qquad y' = \frac{2x + ye^{xy}}{1 - xe^{xy}}$$

(provided $1 - xe^{xy} \neq 0$). But we don't know how we stand.

Most questions of this sort are settled by the following theorem.

1312. The Implicit-Function Theorem. *Let $F(x, y)$ be a smooth function and let (x_0, y_0) be a point such that $F(x_0, y_0) = 0$.*

Assume $F_y(x_0, y_0) \neq 0$. Then on a suitable interval I about x_0, there is exactly one differentiable function

(7) $$y = f(x)$$

such that $y_0 = f(x_0)$ and

(8) $$F\big(x, f(x)\big) = 0 \qquad\qquad (x \in I).$$

The derivative of y is given by

(9)
$$\frac{dy}{dx} = -\frac{F_x}{F_y}$$

(and hence is continuous).

Proof. Say $F_y(x_0, y_0) > 0$. (A similar proof holds if $F_y(x_0, y_0) < 0$.) Since F_y is continuous, there is an open box $I_0 \times J_0$ about (x_0, y_0) throughout which $F_y(x, y) > 0$. It follows that for each fixed $x \in I_0$, $F(x, y)$ is an increasing function of y.

Pick y_1 and y_2 in J_0 such that $y_1 < y_0 < y_2$. Since $F(x_0, y_0) = 0$ and F is increasing with y, we have $F(x_0, y_1) < 0$ and $F(x_0, y_2) > 0$. By continuity of F, there is an interval I about x_0 such that $F(x, y_1) < 0$ for every $x \in I$ and $F(x, y_2) > 0$ for every $x \in I$. See Figure 2.

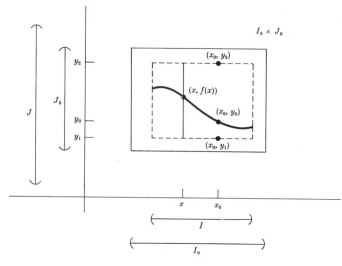

FIGURE 2

Consider any $x \in I$ and look at the vertical segment at x. By the Intermediate-Value Theorem (§2E5), there is a number, call it $f(x)$, between y_1 and y_2 such that $F(x, f(x)) = 0$. Since F is increasing with y, there is only one such number.

In this way, for each $x \in I$, we associate a number $f(x)$. We thus define a function f. This function is continuous at x_0. For, given any open interval J about y_0, we pick J_0 in the above so that $J_0 \subset J$. Then, as we have just shown, there is an interval I about x_0 that f takes into J.

Next, we show that f is differentiable at (x_0, y_0). Consider a change Δx and the corresponding change

(10)
$$\Delta y = f(x_0 + \Delta x) - f(x_0).$$

Since $F(x, f(x))$ is constant (namely, 0), $\Delta F = 0$. Thus, the Fundamental Lemma (§13B(2)) gives

(11) $$0 = F_x(x_0, y_0)\,\Delta x + F_y(x_0, y_0)\,\Delta y + A\,\Delta x + B\,\Delta y,$$

where A and B approach 0 as $(\Delta x, \Delta y) \to (0, 0)$. Now, since f is continuous at x_0, $\Delta y \to 0$ as $\Delta x \to 0$. Hence A and B approach 0 as $\Delta x \to 0$. Now, from (11),

(12) $$\frac{\Delta y}{\Delta x} = -\frac{F_x(x_0, y_0) + A}{F_y(x_0, y_0) + B}$$

(provided Δx is small so that B is small and the denominator is not 0). Now letting $\Delta x \to 0$ in (12), we get (9) at the point (x_0, y_0).

Finally, this proof of differentiability applies as well to any other point $(x, f(x))$ for $x \in I$.

As the proof shows, the graph of $y = f(x)$ is the only part of the graph of $F(x, y) = 0$ that intersects the box $I \times (y_1, y_2)$ (see Figure 2).

The corresponding theorem holds with x and y interchanged. If we assume $F_x(x_0, y_0) \neq 0$ (instead of $F_y(x_0, y_0) \neq 0$), then we conclude that on a suitable interval J about y_0 there is a unique differentiable function

(13) $$x = g(y)$$

such that $x_0 = g(y_0)$ and

(14) $$F(g(y), y) = 0.$$

Here,

(15) $$\frac{dx}{dy} = -\frac{F_y}{F_x}.$$

13I3. Implicit differentiation. The Implicit-Function Theorem justifies the method of "implicit differentiation." Let $F(x, y)$ be a smooth function given by a definite formula; for example,

$$F(x, y) = e^{xy} + x^2 - y - 1.$$

We ask whether the equation $F(x, y) = 0$ defines y as a differentiable function of x. The method of implicit differentiation is to differentiate both sides with respect to x, as though it does, using the chain rule for functions of one variable, and then solve for y'.

The fact is that if the procedure makes sense (at some point of the graph of F) then the assumption was correct. This is because the expression we get for $(d/dx)F(x, y)$ turns out to be $F_x + F_y y'$ (as can be seen by tracing the process back to the Fundamental Lemma). If we were able to solve the equation $F_x + F_y y' = 0$ for y' we must have had $F_y \neq 0$. And this is just the condition in the Implicit-Function Theorem that tells us that y *is* a differentiable function of x.

When the equation $F(x, y) = 0$ defines y as a differentiable function of x, we have, by the chain rule,

(16) $$F_x + F_y y' = 0.$$

Hence, if $F_y(x_0, y_0) = 0$ and $F_x(x_0, y_0) \neq 0$, then y is definitely *not* a differentiable function of x near x_0, since equation (16) is impossible. Similarly, if $F_x = 0$ while $F_y \neq 0$, then x is *not* a differentiable function of y.

When F_x and F_y are both 0, the issue is in doubt.

Example 1. Let

$$F(x, y) = x^2 + y^2 - 1.$$

Then $F_x = 2x$ and $F_y = 2y$. The equation

$$x^2 + y^2 - 1 = 0$$

determines y as a differentiable function of x about any point on the graph where $F_y = 2y \neq 0$, i.e., $y \neq 0$; and then

$$\frac{dy}{dx} = -\frac{F_x}{F_y} = -\frac{x}{y}.$$

It also determines x as a differentiable function of y about any point where $F_x = 2x \neq 0$, i.e., $x \neq 0$; and then

$$\frac{dx}{dy} = -\frac{F_y}{F_x} = -\frac{y}{x}.$$

At $(1, 0)$ we have $F_y = 0$ and $F_x \neq 0$, so y is not a differentiable function of x there. (The tangent line is vertical.) Observe that *every* open box about $(1, 0)$ intersects *both* $y = \sqrt{1 - x^2}$ *and* $y = -\sqrt{1 - x^2}$. The same is true at $(-1, 0)$. Similarly, x is not a differentiable function of y at $(0, 1)$ or at $(0, -1)$.

Example 2. Let

$$F(x, y) = e^{xy} + x^2 - y - 1.$$

Then $F_x = ye^{xy} + 2x$ and $F_y = xe^{xy} - 1$. The equation

$$e^{xy} + x^2 - y - 1 = 0$$

determines y as a differentiable function of x about any point of the graph where $F_y = xe^{xy} - 1 \neq 0$; for instance, at $(0, 0)$. And then

$$\frac{dy}{dx} = -\frac{F_x}{F_y} = \frac{2x + ye^{xy}}{1 - xe^{xy}}.$$

On the other hand, since $F_x(0, 0) = 0$ and $F_y(0, 0) \neq 0$, x is not a differentiable function of y there.

1314. Implicit functions of several variables. The theorem for three or more variables takes the following form.

Let $F(x, y, z)$ be a smooth function and let (x_0, y_0, z_0) be a point such that $F(x_0, y_0, z_0) = 0$.

Assume $F_z(x_0, y_0, z_0) \neq 0$. Then on a suitable box $I \times J$ about (x_0, y_0), there is exactly one smooth function

$$(17) \hspace{3cm} z = f(x, y)$$

such that $z_0 = f(x_0, y_0)$ and

$$(18) \hspace{2cm} F\big(x, y, f(x, y)\big) = 0 \hspace{2cm} \big((x, y) \in I \times J\big).$$

The partial derivatives of z are given by

$$(19) \hspace{2cm} \frac{\partial z}{\partial x} = -\frac{F_x}{F_z} \hspace{1cm} \text{and} \hspace{1cm} \frac{\partial z}{\partial y} = -\frac{F_y}{F_z}$$

(and hence are continuous).

The proof is analogous to the earlier one.

The following detail may be noted. When finding $\partial z/\partial x$ from the Fundamental Lemma, first set $\Delta y = 0$; when finding $\partial z/\partial y$, first set $\Delta x = 0$.

Similarly, if $F_y(x_0, y_0, z_0) \neq 0$ then y is defined as a smooth function of x and z; and if $F_x(x_0, y_0, z_0) \neq 0$ then x is defined as a smooth function of y and z.

Example 3. Let

$$F(x, y, z) = xy + xz + yz - 11.$$

Then $F(1, 2, 3) = 0$. We have $F_x = y + z$, $F_y = x + z$, and $F_z = x + y$. Since $F_z(1, 2, 3) = 3 \neq 0$, the equation

$$F(x, y, z) = 0$$

determines z as a function of x and y on a suitable box $I \times J$ about $(1, 2)$; and we have

$$\frac{\partial z}{\partial x} = -\frac{F_x}{F_z} = -\frac{y + z}{x + y} \quad \text{and} \quad \frac{\partial z}{\partial y} = -\frac{F_y}{F_z} = -\frac{x + z}{x + y}.$$

Problems (§13I)

1. Differentiate implicitly to find dy/dx in terms of x and y.

   (a) $y^2 - 3x^2 + 2 = 0$
   (b) $e^x \cos y + e^y \cos x = 1$
   (c) $\sin xy + x = 1$
   (d) $x^{2/3} + y^{2/3} = 4$
   (e) $\log \cos y = \sqrt{x}$
   (f) $x^3 + 2x^2y - xy^2 + y^3 = 2$
   (g) $\sin y + \tan y = x^2 + x$
   (h) $xe^x = y^2 + xy$

2. Evaluate dy/dx at the given point.

   (a) $x^2 - xy - y^2 + 5 = 0, \quad (1, 2)$
   (b) $e^{xy} + 2 = x + y, \quad (3, 0)$
   (c) $x^{2/3} + y^{2/3} = 5, \quad (1, 8)$
   (d) $x^2 + y^2 = 25, \quad (-4, 3)$

3. Differentiate implicitly to find $\partial u/\partial x$ and $\partial u/\partial y$.

   (a) $2x - 3y + u = \log u$
   (b) $2x^2 - 3y^2 + u^2 + x - 2y + 7 = 0$
   (c) $\sin xy + \sin xu + \sin yu = 1$
   (d) $u = xy \sin xu$
   (e) $x^2 + y \sin u + x^3u^2 - 6 = 0$
   (f) $\sqrt{xyu} + \sin xy \cos u = 5$
   (g) $u + e^u = xy + 1$
   (h) $\arctan x + \arctan y + \arctan u = 5$
   (i) $\sin x + \sin y + \sin u = 5$

4. (a) If $x^2 + xy^2 + yz^2 = 6$, what is $\dfrac{\partial z}{\partial x} \dfrac{\partial x}{\partial y} \dfrac{\partial y}{\partial z}$?

   (b) Let $f(x, y, z) = 0$, where f is any smooth function of three variables. If $f_x f_y f_z \neq 0$, what is $\dfrac{\partial z}{\partial x} \dfrac{\partial x}{\partial y} \dfrac{\partial y}{\partial z}$?

5. (a) If

   $$y \tan(x^2 + z^2) + \log(x^2 + z^2) - \exp \frac{x^2 + z^2}{y} = 0,$$

   what is $\partial z/\partial x$?

   (b) If $f(u, y)$ is smooth, $f_u \neq 0$, and $f(x^2 + z^2, y) = 0$, what is $\partial z/\partial x$?

6. Show that if the equation $F(x, y) = 0$ defines y as a twice-differentiable function of x, then

$$\frac{d^2 y}{dx^2} = -\frac{F_{xx}F_y^2 - 2F_{xy}F_xF_y + F_{yy}F_x^2}{F_y^3}.$$

Answers to problems (§13I)

1. (a) $3x/y$ (b) $(e^y \sin x - e^x \cos y)/(e^y \cos x - e^x \sin y)$
   (c) $-(y + \sec xy)/x$ (d) $-(y/x)^{1/3}$ (e) $-(\cot y)/2\sqrt{x}$
   (f) $-(3x^2 + 4xy - y^2)/(2x^2 - 2xy + 3y^2)$ (g) $(2x + 1)/(\cos y + \sec^2 y)$
   (h) $(e^x + xe^x - y)/(2y + x)$

2. (a) 0 (b) $\frac{1}{2}$ (c) -2 (d) $\frac{4}{3}$

3. (We give $\partial u/\partial x$, then $\partial u/\partial y$.)
   (a) $2u/(1 - u)$; $3u/(u - 1)$ (b) $-(4x + 1)/2u$; $(3y + 1)/u$

   (c) $-\dfrac{y \cos xy + u \cos xu}{x \cos xu + y \cos yu}$; $-\dfrac{x \cos xy + u \cos yu}{x \cos xu + y \cos yu}$

   (d) $(y \sin xu + xyu \cos xu)/(1 - x^2 y \cos xu)$; $(x \sin xu)/(1 - x^2 y \cos xu)$
   (e) $-(2x + 3x^2 u^2)/(2x^3 u + y \cos u)$; $-(\sin u)/(2x^3 u + y \cos u)$

   (f) with $a = 2\sqrt{xyu}$: $\dfrac{ay \cos xy \cos u + yu}{a \sin xy \sin u - xy}$; $\dfrac{ax \cos xy \cos u + xu}{a \sin xy \sin u - xy}$

   (g) $y/(e^u + 1)$; $x/(e^u + 1)$ (h) $-(1 + u^2)/(1 + x^2)$; $-(1 + u^2)/(1 + y^2)$
   (i) no function defined

4. (a) -1 (b) -1

5. (a) $-x/z$ (b) $-x/z$

§13J. Implicit functions: tangent lines and planes

13J1. Tangent lines. As we saw in §13I, if $F(x, y)$ is smooth, then the graph of $F(x, y) = 0$ can be broken up into graphs of differentiable functions $y = f(x)$ or $x = g(y)$ except about points at which F_x and F_y are both 0.

Thus, if $F(x_0, y_0) = 0$, and either $F_x(x_0, y_0) \neq 0$ or $F_y(x_0, y_0) \neq 0$, then there is a portion of the graph passing through (x_0, y_0) for which a tangent line at (x_0, y_0) is defined. We call this line the tangent line to the graph of $F(x, y) = 0$. If this equation is solvable for y as a function of x, then $dy/dx = -F_x/F_y$. If it is solvable for x in terms of y then $dx/dy = -F_y/F_x$. In either case, a direction vector for the tangent line is

(1)
$$\left(F_y(x_0, y_0), \quad -F_x(x_0, y_0)\right).$$

The vector equation of the line is

(2)
$$T(t) = (x_0, y_0) + t(F_y, -F_x),$$

where F_x and F_y are evaluated at (x_0, y_0).

Example 1. We return again to the circle

$$F(x, y) = x^2 + y^2 - 1 = 0.$$

Here $F_x = 2x$ and $F_y = 2y$, at least one of which is not 0. Hence a direction vector for the tangent line at any point (x_0, y_0) is $(2y_0, -2x_0)$, or, more simply, $(y_0, -x_0)$. The vector equation of the tangent line is

$$T(t) = (x_0, y_0) + t(y_0, -x_0).$$

The equation of the tangent line at $(1, 0)$, for instance, is

$$T(t) = (1, 0) + t(0, 1) = (1, -t).$$

Parametrically, $x = 1$, $y = -t$; in coordinate form, $x = 1$.

Example 2. Find a vector equation of the tangent line to the graph of

$$F(x, y) = x^3 + xy^5 + y^3 - 11 = 0$$

at the point $(2, 1)$.

Solution. Here

$$F_x = 3x^2 + y^5 \quad \text{and} \quad F_y = 5xy^4 + 3y^2.$$

Then $F_x(2, 1) = 13$ and $F_y(2, 1) = 13$, at least one of which is not 0. Hence $(13, -13)$, or more simply, $(1, -1)$, is a direction vector for the tangent line. The equation of the line is

$$T(t) = (2, 1) + t(1, -1).$$

Parametrically, $x = 2 + t$, $y = 1 - t$. In coordinate form, $x + y = 3$.

13J2. Tangent planes. Consider an equation

$$(3) \qquad F(x, y, z) = 0,$$

where F is a smooth function of three variables. If (x_0, y_0, z_0) is a point on the graph such that $F_z(x_0, y_0, z_0) \neq 0$, then we may use the results of the Implicit-Function Theorem (§13I4). Thus, (3) determines a smooth surface

$$(4) \qquad z = f(x, y)$$

containing the point (x_0, y_0, z_0), and

$$(5) \qquad f_x = -\frac{F_x}{F_z} \quad \text{and} \quad f_y = -\frac{F_y}{F_z}.$$

When we substitute these expressions for the derivatives into the equation of the tangent plane $\big(\text{§13F(6)}\big)$:

$$(6) \qquad z - z_0 = f_x(x - x_0) + f_y(y - y_0),$$

we get the following symmetric equation for the plane:

(7)
$$F_x(x - x_0) + F_y(y - y_0) + F_z(z - z_0) = 0$$

(where F_x, F_y, and F_z are evaluated at (x_0, y_0, z_0)).

Now, this is the equation of a plane provided only that *at least one of* the coefficients, F_x, F_y, F_z *be different from zero*. A discussion of surfaces given in the form $y = g(x, z)$ or $x = h(y, z)$ would lead as above to this same equation. Hence we take (7) as an extension of the definition of the tangent plane.

Example 3. Find the equation of the tangent plane to the *ellipsoid*

$$\frac{x^2}{a^2} + \frac{y^2}{b^2} + \frac{z^2}{c^2} - 1 = 0$$

at an arbitrary point (x_0, y_0, z_0).

Solution. If F denotes the function on the left side of the equation, then at (x_0, y_0, z_0),

$$F_x = \frac{2x_0}{a^2}, \qquad F_y = \frac{2y_0}{b^2}, \qquad F_z = \frac{2z_0}{c^2}.$$

Note that not all three derivatives can be zero simultaneously (since (x_0, y_0, z_0) satisfies the equation of the ellipsoid). Hence the equation of the tangent plane at the point is

$$\frac{x_0}{a^2}(x - x_0) + \frac{y_0}{b^2}(y - y_0) + \frac{z_0}{c^2}(z - z_0) = 0.$$

This may also be written as

$$\frac{x_0 x}{a^2} + \frac{y_0 y}{b^2} + \frac{z_0 z}{c^2} = 1.$$

Problems (§13J)

1. Find the equation of the line tangent to the given curve at the given point.
   (a) $xy = 2$, $(2, 1)$
   (b) $x^2 - 3xy + 3y^2 = 3$, $(3, 2)$
   (c) $e^{xy} + \sin y + y^2 = 1$, $(2, 0)$
   (d) $x^{2/3} + y^{2/3} = 5$, $(8, 1)$

2. Find the equation of the plane tangent to the given surface at the given point.
   (a) $x^2 + y^2 + z^2 = 14$, $(1, -2, 3)$
   (b) $x^2 + 3y^2 + 2z^2 = 9$, $(2, -1, 1)$
   (c) $x^2 + y^2 = 169$, $(5, -12, 3)$
   (d) $\sqrt{x} + \sqrt{y} = 3$, $(1, 4, 0)$

(e) $\dfrac{x^2}{4} + \dfrac{y^2}{9} - \dfrac{z^2}{25} = 1,$ $(2, 3, 5)$

(f) $xy + yz + zx = -1,$ $(1, 2, -1)$

(g) $3x + 2y - 3z = 0,$ $(2, 3, 4)$

(h) $\dfrac{x^2}{4} - y^2 - \dfrac{z^2}{9} = 1,$ $(4, \sqrt{2}, 3)$

3. Find the angle between the given surfaces (i.e., between their tangent planes) at the given point.

(a) $x^2 + y^2 + z^2 = 8,$ $(x - 1)^2 + (y - 2)^2 + (z - 3)^2 = 6;$ $(2, 0, 2)$

(b) $z = x^2 + y^2,$ $3x^2 + z^2 - 2y^2 = 20;$ $(1, 2, 5)$

(c) $z^2 + x^2 - y^2 = 4,$ $2x + y - z = 1;$ $(1, 1, 2)$

Answers to problems (§13J)

1. (a) $x + 2y = 4$ (b) $y = 2$ (c) $y = 0$ (d) $x + 2y = 10$

2. (a) $x - 2y + 3z = 14$ (b) $2x - 3y + 2z = 9$ (c) $5x - 12y = 169$

(d) $2x + y = 6$ (e) $15x + 10y - 6z = 30$ (f) $x + 3z + 2 = 0$

(g) $3x + 2y - 3z = 0$ (h) $3x - 3\sqrt{2}y - z = 3$

3. (a) $90°$ (b) $\cos \theta = 3/\sqrt{42} \approx 0.46,$ $\theta \approx 65°$

(c) $\cos \theta = \tfrac{1}{6} \approx 0.17,$ $\theta \approx 80°$

CHAPTER 14

Multiple Integrals

This chapter is about integration in several variables. The development is modeled after the treatment of the single integral in Chapters 5 and 6. The present theory is much more complicated, of course. Portions set in small type may be omitted without interrupting the main ideas.

§14A. The double integral

14A1. Volume. Let f be a continuous function defined on a closed box $K = [a, b] \times [\alpha, \beta]$. If f is nonnegative everywhere then its graph is a surface lying above the region K. We ask how to define and measure the volume of the solid lying under this surface and above K. This is analogous to the problem of area posed in Chapter 5.

The solution, too, will be analogous to the one found there. We consider the volume over sub-boxes of K. Let V_H denote the volume over the sub-box H. Obviously, we demand that volume be additive. That is,

(A) $$V_H + V_{H'} = V_{H \cup H'}$$

whenever H and H' are adjacent, i.e., share exactly one edge and nothing else, so that their union is also a box (Figure 1).

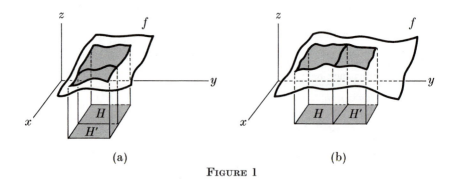

(a)   (b)

FIGURE 1

Next, we surely want V_H to be greater than the volume enclosed by the cylinder upon H of height $\min_H f$, and less than that of height $\max_H f$ (Figure 2). (Recall that f does assume both a maximum and a minimum on H (§12C2).) Thus, we require:

(B) $$\text{(area of } H) \times \min_H f \leq V_H \leq \text{(area of } H) \times \max_H f.$$

As we will see, there is exactly one way of defining V_H so that **(A)** and **(B)** are satisfied.

14A2. Definition of the integral of a continuous function. Continuing the analogy with Chapter 5, we now examine these ideas in a more general setting. Let f be a continuous function (not necessarily nonnegative) defined on a closed box $K = [a, b] \times [\alpha, \beta]$. With each sub-box, H, of K (including K itself), we wish to associate a number,

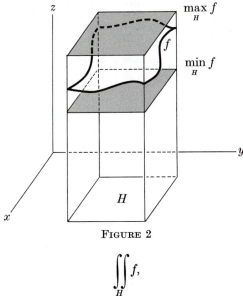

FIGURE 2

$$\iint_H f,$$

such that the following are true:

(A) (Additivity)
$$\iint_H f + \iint_{H'} f = \iint_{H \cup H'} f$$

whenever H and H' are adjacent sub-boxes of K; and

(B) (Betweenness)

$$(area\ of\ H) \times \min_H f \le \iint_H f \le (area\ of\ H) \times \max_H f.$$

We will see shortly that there is one and only one way of making this association. The number $\iint_H f$ is called the *double integral*, or, simply, *integral*, of f over H. Another notation is

$$\iint_H f(x,\ y)\ dA.$$

The two integral signs and the letter A (for "area") are used to emphasize the two-dimensional character of the concept.

Note that

$$\iint_H 1\ dA = area\ of\ H,$$

as is evident from **(B)**.

14A3. Existence and uniqueness of the integral. To establish the existence and uniqueness of the double integral, one may argue exactly as in §5D3 and §5D4 for the single integral.

We give the outline, omitting details. Let a sub-box H of K be given. Consider any partition of H (§12C3). On each segment, G, of the partition, we look at

(1)   (area of G) $\times \min\limits_{G} f$

and

(2)   (area of G) $\times \max\limits_{G} f$.

Adding all the terms (1), one for each segment of the partition, we get a *lower sum* on H; adding the terms (2) gives us an *upper* sum on H. It can be seen that every lower sum is $\leq$ every upper sum. Moreover, one can show that, given $\epsilon > 0$, there exist lower and upper sums L and U such that $U - L < \epsilon$.

Let S_H denote the *least upper bound* (§1H2) of the set of all lower sums on H. It turns out that S_H satisfies **(A)** and **(B)** and is the only function with these properties. Therefore it is the integral, $\iint\limits_H f$, and the integral is unique.

In the course of this discussion one finds that the integral over K is the *unique* number that is $\geq$ every lower sum L on K and $\leq$ every upper sum U.

We also obtain information on approximating by lower and upper sums. Let the *norm* of a partition mean the largest dimension among its segments. Then we have, as in §5G1:

Darboux's Theorem. *Given $\epsilon > 0$, there exists $\delta > 0$ such that, for any partition of norm $< \delta$, $U - L < \epsilon$.*

14A4. Volume as an integral. Let f be nonnegative and continuous on a closed box $K = [a, b] \times [\alpha, \beta]$. We wish to define the volume under the graph of f over K. We agreed in §14A1 that the volume function had to satisfy **(A)** and **(B)**. Hence there is just one way we can define the volume; namely,

(3) $$V = \iint\limits_K f.$$

Let us see how we might compute this volume. Consider any fixed number x_0 in $[a, b]$. The plane at x_0 parallel to the yz-plane intersects the solid in a plane region. This is the region under a curve: under the graph of the function $f(x_0, y)$, which is a function of y alone. (See Figure 3.) Hence its area is

$$A(x_0) = \int_\alpha^\beta f(x_0, y)\, dy.$$

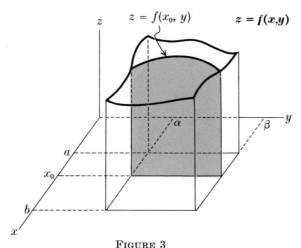

FIGURE 3

By analogy with the area under a curve as the integral of the height, one might now conjecture that the volume under the surface is the integral of $A(x)$. Indeed, this is certainly the case if $f(x, y)$ is a constant, so that the solid is a (three-dimensional) box; or, more generally, if $f(x, y)$ is constant for each fixed y, so that the surface is a cylinder (looked at sideways). (See Figure 4.)

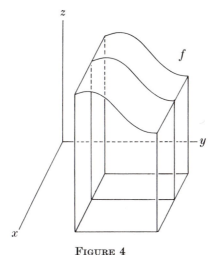

FIGURE 4

We will see that the conjecture is true. In the next section (§14B), we lay the groundwork for the proof. The main theorem and its proof are given in §14C.

§14B. Integration with respect to one variable

14B1. Integration with respect to one variable. Let f be continuous on a closed box $K = [a, b] \times [\alpha, \beta]$. For each fixed $x \in [a, b]$, $f(x, y)$ is a continuous function of y on $[\alpha, \beta]$. The integral

$$\int_\alpha^\beta f(x, y) \, dy$$

is therefore defined. This integral depends on x. We might call it $\varphi(x)$:

(1) $$\varphi(x) = \int_\alpha^\beta f(x, y) \, dy.$$

Example 1. If $f(x, y) = xy^2 + x^2 y$ on $K = [0, 1] \times [-1, 1]$, then

$$\varphi(\tfrac{1}{2}) = \int_{-1}^1 f(\tfrac{1}{2}, y) \, dy = \int_{-1}^1 (\tfrac{1}{2}y^2 + \tfrac{1}{4}y) \, dy = \tfrac{1}{3},$$

and

$$\varphi(\tfrac{1}{3}) = \int_{-1}^1 f(\tfrac{1}{3}, y) \, dy = \int_{-1}^1 (\tfrac{1}{3}y^2 + \tfrac{1}{9}y) \, dy = \tfrac{2}{9}.$$

In general,

$$\varphi(x) = \int_{-1}^1 f(x, y) \, dy = \int_{-1}^1 (xy^2 + x^2 y) \, dy$$

$$= (x)(\tfrac{1}{3}y^3) + (x^2)(\tfrac{1}{2}y^2) \Big|_{y=-1}^{y=1}$$

$$= \tfrac{2}{3}x + 0$$

$$= \tfrac{2}{3}x.$$

Likewise, for each fixed $y \in [\alpha, \beta]$, $f(x, y)$ is a continuous function of x in $[a, b]$ and we may consider the integral

$$\int_a^b f(x, y) \, dx.$$

This is a function of y, say $\psi(y)$:

(2) $$\psi(y) = \int_a^b f(x, y) \, dx.$$

Example 2. With f as in Example 1,

$$\psi(\tfrac{1}{2}) = \int_0^1 f(x, \tfrac{1}{2}) \, dx = \int_0^1 (\tfrac{1}{4}x + \tfrac{1}{2}x^2) \, dx = \tfrac{7}{24},$$

and

$$\psi(-\tfrac{1}{2}) = \int_0^1 f(x, -\tfrac{1}{2}) \, dx = \int_0^1 (\tfrac{1}{4}x - \tfrac{1}{2}x^2) \, dx = -\tfrac{1}{24}.$$

In general,

$$\psi(y) = \int_0^1 f(x, y) \, dx = \int_0^1 (xy^2 + x^2y) \, dx$$

$$= (y^2)(\tfrac{1}{2}x^2) + (y)(\tfrac{1}{3}x^3) \, \Big|_{x=0}^{x=1}$$

$$= \tfrac{1}{2}y^2 + \tfrac{1}{3}y.$$

We show next that the functions (1) and (2) are always continuous.

14B2. Continuity of the single integral. *Let f be continuous on a closed box $K = [a, b] \times [\alpha, \beta]$.*
Then the integral

$$\int_\alpha^\beta f(x, y) \, dy \qquad\qquad [= \varphi(x)]$$

is a continuous function of x on $[a, b]$.
Likewise,

$$\int_a^b f(x, y) \, dx \qquad\qquad [= \psi(y)]$$

is a continuous function of y on $[\alpha, \beta]$.

Proof. We prove the first statement, the other proof being similar. Consider any point x_0. Given any interval J about $\varphi(x_0)$, we are to find an interval I about x_0 that φ takes into J. We may assume that J is of the form $(\varphi(x_0) - \epsilon, \varphi(x_0) + \epsilon))$.

We work with the number $\epsilon' = \epsilon/(\beta - \alpha)$. Pick an ϵ'-partition of K (§12C4). The interval I will run from the point of subdivision immediately to the left of x_0 to the first one on its right (Figure 1). Then for each $x \in I$ and for every $y \in [\alpha, \beta]$, the points (x, y) and (x_0, y) belong to the same segment of the partition. Therefore

$$|f(x, y) - f(x_0, y)| < \epsilon'.$$

Thus,

(3) $$\qquad\qquad f(x_0, y) - \epsilon' < f(x, y) < f(x_0, y) + \epsilon' \qquad (x \in I, \, y \in [\alpha, \beta]).$$

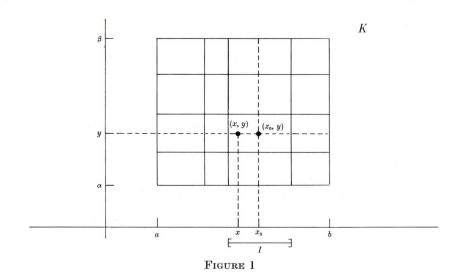

FIGURE 1

Now we integrate (with respect to y) from α to β. Since $\int_\alpha^\beta f(x, y)\, dy = \varphi(x)$, and $\int_\alpha^\beta \epsilon' = \epsilon'(\beta - \alpha) = \epsilon$, we get

(4) $\varphi(x_0) - \epsilon < \varphi(x) < \varphi(x_0) + \epsilon$ $(x \in I)$.

This is what we were to prove.

14B3. Integrating the single integral. Since the integral

(5) $$\int_\alpha^\beta f(x, y)\, dy$$ $[= \varphi(x)]$

is a continuous function of x in $[a, b]$, we can integrate *it*. Thus, we consider

(6) $$\int_a^b \left(\int_\alpha^\beta f(x, y)\, dy \right) dx$$ $\left[= \int_a^b \varphi(x)\, dx \right]$,

or, simply,

(7) $$\int_a^b \int_\alpha^\beta f(x, y)\, dy\, dx.$$

Similarly, the integral

(8) $$\int_a^b f(x, y)\, dx$$ $[= \psi(y)]$

is a continuous function of y and we may integrate *it*:

(9) $$\int_\alpha^\beta \left(\int_a^b f(x, y)\, dx \right) dy \qquad \left[= \int_\alpha^\beta \psi(y)\, dy \right]$$

(10) $$= \int_\alpha^\beta \int_a^b f(x, y)\, dx\, dy.$$

Example 3. Let $f(x, y) = x^2 + xy + 2y^2$ on the box $K = [0, 1] \times [0, 2]$. Then

$$\int_0^2 (x^2 + xy + 2y^2)\, dy = x^2 y + \tfrac{1}{2}xy^2 + \tfrac{2}{3}y^3 \Big|_{y=0}^{y=2}$$

$$= 2x^2 + 2x + \tfrac{16}{3} \qquad [= \varphi(x)].$$

And then

$$\int_0^1 (2x^2 + 2x + \tfrac{16}{3})\, dx = \tfrac{2}{3} + 1 + \tfrac{16}{3} = 7 \qquad \left[= \int_0^1 \varphi(x)\, dx \right].$$

When we integrate in the other order, we get

$$\int_0^1 (x^2 + xy + 2y^2)\, dx = \tfrac{1}{3}x^3 + \tfrac{1}{2}yx^2 + 2y^2 x \Big|_{x=0}^{x=1}$$

$$= \tfrac{1}{3} + \tfrac{1}{2}y + 2y^2 \qquad [= \psi(y)],$$

and then

$$\int_0^2 (\tfrac{1}{3} + \tfrac{1}{2}y + 2y^2)\, dy = \tfrac{2}{3} + 1 + \tfrac{16}{3} = 7 \qquad \left[= \int_0^2 \psi(y)\, dy \right].$$

Thus,

$$\int_0^1 \int_0^2 (x^2 + xy + 2y^2)\, dy\, dx = 7$$

and

$$\int_0^2 \int_0^1 (x^2 + xy + 2y^2)\, dx\, dy = 7.$$

Note that the answers are the same.

Another common notation for the iterated integral is:

(11) $$\int_a^b dx \int_\alpha^\beta f(x, y)\, dy = \int_a^b \int_\alpha^\beta f(x, y)\, dy\, dx$$

and

(12) $$\int_\alpha^\beta dy \int_a^b f(x, y)\, dx = \int_\alpha^\beta \int_a^b f(x, y)\, dx\, dy.$$

Problems (§14B)

1. Compute $\varphi(0)$ and $\varphi(1)$ (where defined) by substituting $x = 0$ and $x = 1$ into the integrand, then performing the resulting integrations with respect to y.

(a) $\varphi(x) = \int_0^1 (x + y)\, dy$

(b) $\varphi(x) = \int_1^2 3y^2\, dy$

(c) $\varphi(x) = \int_1^3 (x^2 + 2x)\, dy$

(d) $\varphi(x) = \int_{-1}^1 e^{xy}\, dy$

(e) $\varphi(x) = \int_1^4 \log(x^2 y)\, dy$

(f) $\varphi(x) = \int_0^1 \frac{1}{x^2 + y^2}\, dy$

(g) $\varphi(x) = \int_0^{\pi/4} \sin(x + y)\, dy$

(h) $\varphi(x) = \int_0^{\pi/2} \cos(xy)\, dy$

2. In Problem 1, find an expression for $\varphi(x)$ by performing the integration with respect to y.

3. Compute $\psi(0)$ and $\psi(1)$ (where defined) by substituting $y = 0$ and $y = 1$ into the integrand, then performing the resulting integrations with respect to x.

(a) $\psi(y) = \int_0^1 (x^2 + xy)\, dx$

(b) $\psi(y) = \int_0^2 \frac{x}{y^2}\, dx$

(c) $\psi(y) = \int_0^1 xe^{x^2 y}\, dx$

(d) $\psi(y) = \int_3^5 y^2\, dx$

(e) $\psi(y) = \int_3^4 x^3\, dx$

(f) $\psi(y) = \int_0^\pi \cos(x + y)\, dx$

(g) $\psi(y) = \int_{-1}^0 \sqrt{x^2 + y^2}\, dx$

[See §8G, Problem 1.]

(h) $\psi(y) = \int_0^1 \sqrt{y^2 - x^2}\, dx$

[See §8G, Problem 1.]

4. In Problem 3, find an expression for $\psi(y)$ by performing the integration with respect to x.

5. Evaluate the following iterated integrals.

(a) $\int_1^2 \int_2^3 (x^2 + 2y^2 - xy + 1)\, dy\, dx$

(b) $\int_0^\pi \int_0^{\pi/2} y \sin x\, dy\, dx$

(c) $\int_1^2 \int_1^3 x^2\, dy\, dx$

(d) $\displaystyle\int_0^1 \int_1^2 \left(x^2 y + \frac{x}{y} \right) dy\, dx$

(e) $\displaystyle\int_0^4 \int_1^4 (x + y)^{10}\, dy\, dx$

(f) $\displaystyle\int_0^1 \int_0^1 x e^{x+y}\, dy\, dx$

(g) $\displaystyle\int_1^2 \int_1^2 \frac{x}{x + y}\, dy\, dx$ [See §7C, Problem 2.]

(h) $\displaystyle\int_1^2 \int_0^1 x \log(x + y)\, dy\, dx$ [See §7C, Problem 2.]

6. Evaluate the following iterated integrals.

(a) $\displaystyle\int_2^3 \int_1^2 (x^2 + 2y^2 - xy + 1)\, dx\, dy$

(b) $\displaystyle\int_0^{\pi/2} \int_0^{\pi} y \sin x\, dx\, dy$

(c) $\displaystyle\int_1^3 \int_1^2 x^2\, dx\, dy$

(d) $\displaystyle\int_1^2 \int_0^1 \left(x^2 y + \frac{x}{y} \right) dx\, dy$

(e) $\displaystyle\int_1^4 \int_0^4 (x + y)^{10}\, dx\, dy$

(f) $\displaystyle\int_0^1 \int_0^1 x e^{x+y}\, dx\, dy$

(g) $\displaystyle\int_1^2 \int_1^2 \frac{x}{x + y}\, dx\, dy$ [See §7C, Problem 2.]

(h) $\displaystyle\int_0^1 \int_1^2 x \log(x + y)\, dx\, dy$ [See §7C, Problem 2.]

Answers to problems (§14B)

1. (a) $\varphi(0) = \displaystyle\int_0^1 y\, dy = \frac{1}{2}, \quad \varphi(1) = \int_0^1 (1 + y)\, dy = \frac{3}{2}$

   (b) $\varphi(0) = \varphi(1) = \displaystyle\int_1^2 3y^2\, dy = 7$

(c) $\varphi(0) = \int_1^3 0 \, dy = 0, \quad \varphi(1) = \int_1^3 3 \, dy = 6$

(d) $\varphi(0) = \int_{-1}^1 1 \, dy = 2, \quad \varphi(1) = \int_{-1}^1 e^y \, dy = e - \dfrac{1}{e}$

(e) $\varphi(0)$ not defined, $\quad \varphi(1) = \int_1^4 \log y \, dy = 8 \log 2 - 3$

(f) $\varphi(0)$ not defined, $\quad \varphi(1) = \int_0^1 \dfrac{1}{1 + y^2} \, dy = \pi/4$

(g) $\varphi(0) = \int_0^{\pi/4} \sin y \, dy = 1 - \tfrac{1}{2}\sqrt{2},$

$\quad \varphi(1) = \int_0^{\pi/4} \sin(1 + y) \, dy = \cos 1 - \cos(1 + \pi/4)$

(h) $\varphi(0) = \int_0^{\pi/2} 1 \, dy = \pi/2, \quad \varphi(1) = \int_0^{\pi/2} \cos y \, dy = 1$

2. (a) $\varphi(x) = x + \tfrac{1}{2}$ (b) $\varphi(x) = 7$ (c) $\varphi(x) = 2(x^2 + 2x)$

(d) $\varphi(x) = \dfrac{1}{x}(e^x - e^{-x}) \quad (x \neq 0), \quad \varphi(0) = 2$

(e) $\varphi(x) = 3 \log x^2 + 8 \log 2 - 3 \quad (x \neq 0)$

(f) $\varphi(x) = \dfrac{1}{x} \arctan \dfrac{1}{x} \quad (x \neq 0)$ (g) $\varphi(x) = \cos x - \cos(x + \pi/4)$

(h) $\varphi(x) = \dfrac{1}{x} \sin \dfrac{\pi}{2} x \quad (x \neq 0), \quad \varphi(0) = \pi/2$

3. (a) $\psi(0) = \int_0^1 x^2 \, dx = \tfrac{1}{3}, \quad \psi(1) = \int_0^1 (x^2 + x) \, dx = \tfrac{5}{6}$

(b) $\psi(0)$ not defined, $\quad \psi(1) = \int_0^2 x \, dx = 2$

(c) $\psi(0) = \int_0^1 x \, dx = \tfrac{1}{2}, \quad \psi(1) = \int_0^1 x e^{x^2} \, dx = \tfrac{1}{2}(e - 1)$

(d) $\psi(0) + \int_3^5 0 \, dx = 0, \quad \psi(1) = \int_3^5 1 \, dx = 2$

(e) $\psi(0) = \psi(1) = \int_3^4 x^3 \, dx = \tfrac{175}{4}$

(f) $\psi(0) = \int_0^\pi \cos x \, dx = 0, \quad \psi(1) = \int_0^\pi \cos(x + 1) \, dx = -2 \sin 1$

(g) $\psi(0) = \int_{-1}^0 |x| \, dx = \tfrac{1}{2}, \quad \psi(1) = \int_{-1}^0 \sqrt{x^2 + 1} \, dx = \tfrac{1}{2}[\sqrt{2} + \log(1 + \sqrt{2})]$

(h) $\psi(0)$ not defined, $\quad \psi(1) = \int_0^1 \sqrt{1 - x^2} \, dx = \pi/4$

4. (a) $\psi(y) = \frac{1}{3} + \frac{1}{2}y$ (b) $\psi(y) = 2/y^2$ $(y \neq 0)$

(c) $\psi(y) = \dfrac{1}{2y}(e^y - 1)$ $(y \neq 0)$, $\psi(0) = \frac{1}{2}$

(d) $\psi(y) = 2y^2$ (e) $\psi(y) = \frac{175}{4}$ (f) $\psi(y) = -2\sin y$

(g) $\psi(y) = \frac{1}{2}[\sqrt{1 + y^2} + y^2 \log|y| - y^2 \log(-1 + \sqrt{1 + y^2})]$ $(y \neq 0)$,

     $\psi(0) = \frac{1}{2}$

(h) $\psi(y) = \frac{1}{2}\left(\sqrt{y^2 - 1} + y^2 \arcsin\dfrac{1}{y}\right)$

5. (a) $\frac{49}{4}$ (b) $\pi^2/4$ (c) $\frac{14}{3}$ (d) $\frac{1}{2}(1 + \log 2)$

(e) $\frac{1}{132}(8^{12} - 5^{12} - 4^{12} + 1)$ (f) $e - 1$ (g) $\frac{1}{2}$

(h) $\frac{9}{2}\log 3 - \frac{10}{3}\log 2 - \frac{19}{12}$

6. (a) $\frac{49}{4}$ (b) $\pi^2/4$ (c) $\frac{14}{3}$ (d) $\frac{1}{2}(1 + \log 2)$

(e) $\frac{1}{132}(8^{12} - 5^{12} - 4^{12} + 1)$ (f) $e - 1$ (g) $\frac{1}{2}$

(h) $\frac{9}{2}\log 3 - \frac{10}{3}\log 2 - \frac{19}{12}$

§14C. Iterated integrals and double integrals

14C1. The double integral as an iterated integral. We are now ready to present the main result.

Let f be continuous on a closed box $K = [a, b] \times [\alpha, \beta]$.

Then the double integral of f over K and the two iterated integrals are all equal:

$$(1) \qquad \iint\limits_K f = \int_a^b \int_\alpha^\beta f(x, y)\, dy\, dx = \int_\alpha^\beta \int_a^b f(x, y)\, dx\, dy.$$

Proof. We show that the iterated integral

$$(2) \qquad \int_u^v \int_\sigma^\tau f(x, y)\, dy\, dx$$

satisfies **(A)** and **(B)** and hence is the double integral. A similar proof holds for the iterated integral in the other direction.

Additivity. Let u, v, w and σ, τ, ω be as in Figure 1. For convenience, write

$$(3) \qquad g(x) = \int_\sigma^\tau f(x, y)\, dy, \qquad h(x) = \int_\tau^\omega f(x, y)\, dy, \qquad k(x) = \int_\sigma^\omega f(x, y)\, dy.$$

Additivity of (2) in the x-direction is expressed by

$$(4) \qquad \int_u^v g + \int_v^w g = \int_u^w g;$$

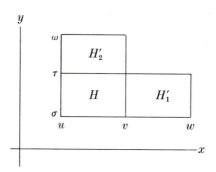

FIGURE 1

additivity in the y-direction is expressed by

$$(5) \qquad \int_u^v g + \int_u^v h = \int_u^v k.$$

To prove (4), we simply invoke additivity of the single integral. The same result yields $k = g + h$, and this establishes (5).

Betweenness. Let $H = [u, v] \times [\sigma, \tau]$ (Figure 1). We know that for the single integral, if $F \le G$ on an interval $[s, t]$, then $\int_s^t F \le \int_s^t G$ (§5E1(e)). So we have

$$(6) \qquad \int_\sigma^\tau f(x, y)\, dy \le \int_\sigma^\tau (\max_H f)\, dy = (\tau - \sigma) \max_H f.$$

Applying the principle again, we have, in turn,

$$(7) \quad \int_u^v \int_\sigma^\tau f(x, y)\, dy\, dx \le \int_u^v (\tau - \sigma)(\max_H f)\, dx = (v - u)(\tau - \sigma)(\max_H f)$$

$$= (\text{area of } H) \times \max_H f.$$

Similarly,

$$(8) \qquad \int_u^v \int_\sigma^\tau f(x, y)\, dy\, dx \ge (\text{area of } H) \times \min_H f.$$

Example 1. Find the volume over the box $K = [0, 1] \times [0, 2]$ under the plane $z = x + y$.

Solution.

$$V = \iint_K (x + y)\, dA = \int_0^1 \int_0^2 (x + y)\, dy\, dx = \int_0^1 (2x + 2)\, dx = 3.$$

14C2. Properties of the double integral. *Let f and g be continuous on a box K. Then*

(a) $\displaystyle\iint_K (f + g) = \iint_K f + \iint_K g.$

(b) $\displaystyle\iint_K (f - g) = \iint_K f - \iint_K g.$

(c) $\displaystyle\iint_K cf = c \iint_K f \quad (c \text{ a constant}).$

(d) *If* $f \geq 0$ *on* K, *then* $\displaystyle\iint_K f \geq 0.$

(e) *If* $f \geq g$ *on* K, *then* $\displaystyle\iint_K f \geq \iint_K g.$

These are all easy consequences of the fact that the double integral is expressible as an iterated integral and that the corresponding relations hold for single integrals (§5E1).

Example 2. Show that $\iint_K \sqrt{1 + x^3 + y^3}\, dA$, where $K = [0, 1] \times [0, 1]$, lies between 1 and $1\frac{1}{2}$.

Solution. Since

$$1 \leq \sqrt{1 + x^3 + y^3} \leq 1 + x^3 + y^3 \qquad ((x, y) \in K),$$

we have

$$\iint_K 1\, dA \leq \iint_K \sqrt{1 + x^3 + y^3}\, dA \leq \iint_K (1 + x^3 + y^3)\, dA,$$

by (e). Evidently, $\iint_K 1\, dA = 1$. Also,

$$\iint_K (1 + x^3 + y^3)\, dA = \int_0^1 \int_0^1 (1 + x^3 + y^3)\, dy\, dx$$

$$= \int_0^1 (1 + x^3 + \tfrac{1}{4})\, dx = 1\tfrac{1}{2}.$$

This gives the required estimate.

14C3. The double integral as a function of two variables. We now present a two-variable version of the Fundamental Theorem of Calculus. Then we apply it to establish the theorem on the equality of the mixed partial derivatives (§13G(6)).

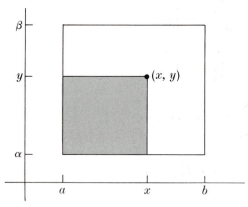

FIGURE 2

Let f be continuous on a box $K = [a, b] \times [\alpha, \beta]$. For $a \leq x \leq b$ and $\alpha \leq y \leq \beta$, consider the sub-box $[a, x] \times [\alpha, y]$ (Figure 2). The double integral of f over this sub-box is a function of x and y, say $G(x, y)$:

$$(9) \qquad G(x\ y) = \iint\limits_{[a,x] \times [\alpha,y]} f.$$

According to §14C1, we can express this as an iterated integral. Since x and y will appear as limits of integration, we use different symbols, say s and t, for the variables of integration. We have:

$$(10) \qquad G(x, y) = \int_a^x \int_\alpha^y f(s, t)\, dt\, ds = \int_\alpha^y \int_a^x f(s, t)\, ds\, dt.$$

By the Fundamental Theorem of Calculus,

$$(11) \qquad G_x(x, y) = \int_\alpha^y f(x, t)\, dt \qquad \text{and} \qquad G_y(x, y) = \int_a^x f(s, y)\, ds.$$

By the Fundamental Theorem of Calculus, once more,

$$(12) \qquad G_{xy}(x, y) = f(x, y) = G_{yx}(x, y).$$

So the mixed partials of G are equal. Using this fact, we shall now establish their equality in general. The proof is a good exercise in partial derivatives and the Fundamental Theorem of Calculus.

14C4. Equality of the mixed partials. *Let $F(x, y)$ be defined on an open region (§12A3). Assume that F_y and F_{xy} exist and that F_{xy} is continuous. Then F_{yx} exists and*

$$F_{yx} = F_{xy}.$$

Proof. The idea of the proof is very simple. Since F_{xy} is continuous, it has a double integral, G; and $G_{xy} = F_{xy} = G_{yx}$ (12). Since $F_{xy} = G_{xy}$, two integrations yield $F = G$ *plus* functions in x alone or y alone (corresponding to the "constants of integration" for the case of a single variable). Differentiating twice we then get $F_{yx} = G_{yx}$, which yields the result.

The details are as follows. To examine F at any point in the region and at points near it, we choose a closed box $K = [a, b] \times [\alpha, \beta]$ about the point and contained in the region, and work within it.

We apply §14C3 to the function $f = F_{xy}$. Thus, from (12),

(13) $$G_{xy} = F_{xy} = G_{yx}.$$

For brevity put $E = F - G$. Then

(14) $$E_{xy} = 0.$$

We now integrate twice, first with respect to y, then with respect to x.
The first integration gives, by the Fundamental Theorem of Calculus,

(15) $$0 = \int_{\alpha}^{y} E_{xt}(x, t)\, dt = E_x(x, y) - E_x(x, \alpha).$$

The second integration gives, also by the Fundamental Theorem of Calculus,

(16) $$0 = \int_{a}^{x} [E_s(s, y) - E_s(s, \alpha)]\, ds$$

(17) $$= [E(x, y) - E(a, y)] - [E(x, \alpha) - E(a, \alpha)].$$

Now we differentiate twice, first with respect to y, then with respect to x.
In (17), the second bracketed expression is independent of y; so its partial with respect to y is 0. As for the first bracket, we know that $E_y = F_y - G_y$ exists, since, F_y exists by hypothesis and G_y exists in any case (11). So we have

(18) $$0 = E_y(x, y) - E_y(a, y).$$

The last term here is independent of x, so its partial with respect to x is 0. To examine $E_y(x, y)$, we break it up, rewriting (18) as:

(19) $$F_y(x, y) = G_y(x, y) + E_y(a, y).$$

Since the partial with respect to x of the right side exists, so does that of the left. Hence we have

(20) $$F_{yx}(x, y) = G_{yx}(x, y).$$

With (13), this gives us $F_{yx} = F_{xy}$, which was to be proved.

Problems (§14C)

1. Evaluate the following.

(a) $\displaystyle\iint_{K} x^2 y^2\, dA,$ where $K = [-2, 2] \times [2, 4]$

(b) $\displaystyle\iint_K \frac{x}{y^2}\, dA$, where $K = [1, 2] \times [3, 4]$

(c) $\displaystyle\iint_K \log y^x\, dA$, where $K = [1, 2] \times [1, 3]$

(d) $\displaystyle\iint_K \sin(x + y)\, dA$, where $K = [0, \pi/4] \times [0, \pi/4]$

(e) $\displaystyle\iint_K (x + y^2)\, dA$, where $K = [0, 1] \times [0, 2]$

(f) $\displaystyle\iint_K xy \sin x\, dA$, where $K = [-\pi/2, \pi/2] \times [0, \pi/4]$

(g) $\displaystyle\iint_K \frac{xy}{x^2 + y^2}\, dA$, where $K = [1, 2] \times [1, 2]$

(h) $\displaystyle\iint_K \frac{x}{x^2 + y^2}\, dA$, where $K = [1, 2] \times [1, 2]$ [See §8F, Problem 4.]

2. Find the volume of the solid lying under the graph of f over the box K.
   (a) $f(x, y) = x^2 y + y$, $\quad K = [0, 1] \times [0, 2]$

   (b) $f(x, y) = x + \dfrac{1}{y}$, $\quad K = [0, 1] \times [1, 2]$

   (c) $f(x, y) = xy + \dfrac{1}{x}$, $\quad K = [1, 2] \times [0, 1]$

   (d) $f(x, y) = \dfrac{y}{(x + y)^2}$, $\quad K = [1, 2] \times [1, 2]$

   (e) $f(x, y) = \log(x^2 + y^2)$, $\quad K = [1, 2] \times [1, 2]$ [See §8F, Problems 4 and 3.]

3. Evaluate the following.
   (a) $\displaystyle\int_0^1 \int_{-1}^1 xe^{xy}\, dx\, dy$

   (b) $\displaystyle\int_1^2 \int_1^2 \frac{x^2}{x^2 + y^2}\, dx\, dy$ [See §8F, Problem 3.]

   (c) $\displaystyle\int_1^2 \int_1^2 \frac{y}{x^2(x^2 + y^2)}\, dx\, dy$ [See §8F, Problem 4.]

4. Let

$$I = \iint_K x \tan\left(x + \frac{y}{2}\right) dA,$$

where $K = [0, \pi/6] \times [0, \pi/3]$.

  (a) Decide, by comparing integrands, which of the following integrals are lower bounds for I and which are upper bounds. [*Hint.* See §8C, Problem 8.]

$$J = \iint_K x \sin\left(x + \frac{y}{2}\right) dA \qquad\qquad Q = \iint_K x \sec\left(x + \frac{y}{2}\right) dA$$

$$L = \frac{\pi^2}{72} \int_0^{\pi/3} \tan\left(\frac{\pi}{6} + \frac{y}{2}\right) dy \qquad R = \iint_K \tan^2\left(x + \frac{y}{2}\right) dA$$

$$M = \frac{\pi}{3} \int_0^{\pi/6} x \tan x\, dx \qquad\qquad S = \frac{\pi}{3} \int_0^{\pi/6} x \tan^2 x\, dx$$

$$N = \frac{\pi}{3} \int_0^{\pi/6} x \tan\left(x + \frac{\pi}{6}\right) dx \qquad T = \iint_K x \tan\frac{y}{2}\, dA$$

$$P = \iint_K x \left(x + \frac{y}{2}\right) dA \qquad\qquad U = \iint_K x \tan\left(\frac{\pi}{6} + \frac{y}{2}\right) dA$$

  (b) Estimate I by computing P and U.

5. Let

$$F(x, y) = \int_2^x \int_3^y s\sqrt{1 + s^3 t^3}\, ds\, dt.$$

Compute F_x, F_y, and F_{xy}.

  6. Show in detail how the properties of the double integral, listed as (a), (b), (c), (d), and (e) in §14C2, follow from corresponding properties of single integrals.

Answers to problems (§14C)

1. (a) $\frac{896}{9}$ (b) $\frac{1}{8}$ (c) $\frac{3}{2}(3 \log 3 - 2)$ (d) $\sqrt{2} - 1$
   (e) $\frac{11}{3}$ (f) $\pi^2/16$
   (g) $\frac{1}{2}(13 \log 2 - 5 \log 5)$ (h) $\frac{1}{2}(2 \arctan 2 + 7 \log 2 - 3 \log 5 - \frac{1}{2}\pi)$

2. (a) $\frac{8}{3}$ (b) $\frac{1}{2} + \log 2$ (c) $\frac{3}{4} + \log 2$
   (d) $5 \log 2 - 3 \log 3$ (e) $6 \arctan 2 + 13 \log 2 - 4 \log 5 - 3 - \frac{3}{2}\pi$

3. (a) $e - \dfrac{1}{e} - 2$ (b) $\frac{1}{2}$ (c) $\frac{1}{4}(\frac{1}{2}\pi + 3 \log 5 - 5 \log 2 - 2 \arctan 2)$

4. (a) lower bounds: J, M, P, S, T;   upper bounds: L, N, Q, R, U
   (b) $P = 7\pi^4/6^5 \approx 0.09$, $U = (\pi^2/72) \log 3 \approx 0.15$, $0.09 \le I \le 0.15$

5. $F_x = \displaystyle\int_3^y s\sqrt{1 + s^3 x^3}\, ds$, $F_y = \displaystyle\int_2^x y\sqrt{1 + y^3 t^3}\, dt$, $F_{xy} = y\sqrt{1 + y^3 x^3}$

§14D. Integration over standard regions

14D1. Standard regions. Most of the significant applications of double integrals are to functions defined over regions more general than closed boxes. Explicitly, the regions we will now consider are of one of the following types: (i) α and β are *any* continuous functions of x (not necessarily constant), with $\alpha(x) \leq \beta(x)$ for all $x \in [a, b]$; or (ii) a and b are any continuous functions of y, with $a(y) \leq b(y)$ for all $y \in [\alpha, \beta]$. See Figure 1. In each case, all bounding lines or curves are included as part of the region (i.e., the region is "closed"). There is no standard name for these regions. We will call them "standard" regions.

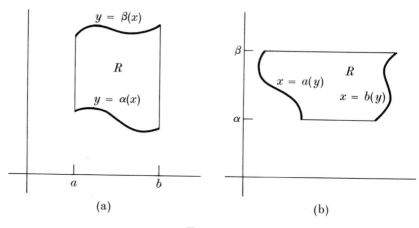

(a)   (b)

Figure 1

The definition of the integral of a continuous function f over a standard region R is the same as over a closed box (§14A2), with one modification: the subregion H one considers is now not necessarily a box but rather the result of intersecting R with a box. See Figure 2. Clearly, H is also a standard region. Hence its area, the area between two graphs, is a meaningful concept, and references to this area in **(A)** and **(B)** make sense.

In §14D3 below we indicate how to extend the theory of the integral to standard regions. It turns out that

$$\iint_R f$$

exists and is unique, and can be evaluated as an iterated integral.

First, however, we describe the results about iterated integrals.

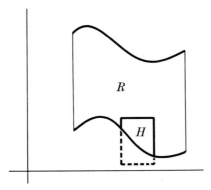

FIGURE 2

14D2. Iterated integrals over standard regions. The order of integration over a standard region must be chosen so that the limits in the last integral are constants.

Suppose f is continuous on the region R of Figure 1(a).

For each fixed $x \in [a, b]$, the integral

$$(1) \qquad \int_{\alpha(x)}^{\beta(x)} f(x, y)\, dy \qquad\qquad [= \varphi(x),\ \text{say}]$$

is a function of x, not only because of the x in the integrand but also because x appears in the limits of integration. This offers no difficulty in integrating $\varphi(x)$ (which, as it turns out, is continuous), and we have

$$(2) \qquad \iint_R f = \int_a^b \int_{\alpha(x)}^{\beta(x)} f(x, y)\, dy\, dx \qquad \left[= \int_a^b \varphi(x)\, dx \right].$$

However, complications arise if we try to iterate the integrals in the other order, and we shall not attempt it.

Similarly, if f is continuous on the region R of Figure 1(b), then for each fixed $y \in [\alpha, \beta]$ we may consider the integral

$$(3) \qquad \int_{a(y)}^{b(y)} f(x, y)\, dx \qquad\qquad [= \psi(y),\ \text{say}].$$

We have

$$(4) \qquad \iint_R f = \int_\alpha^\beta \int_{a(y)}^{b(y)} f(x, y)\, dx\, dy \qquad \left[= \int_\alpha^\beta \psi(y)\, dy \right];$$

but we do not attempt to integrate in the other order.

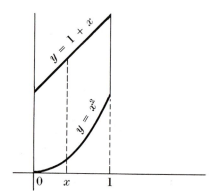

FIGURE 3

Example 1. Find $\iint_R xy \, dA$, where R is the standard region bounded by the lines $x = a = 0$ and $x = b = 1$ and the curves $y = \alpha(x) = x^2$ and $y = \beta(x) = 1 + x$ (Figure 3).

Solution.

$$\iint_R xy \, dA = \int_0^1 \int_{x^2}^{1+x} xy \, dy \, dx = \int_0^1 x \left(\int_{x^2}^{1+x} y \, dy \right) dx$$

$$= \tfrac{1}{2} \int_0^1 x[(1 + x)^2 - (x^2)^2] \, dx = \tfrac{5}{8}.$$

Example 2. Find $\iint_R ye^x \, dA$, where R is the standard region bounded by the lines $y = \alpha = 1$ and $y = \beta = 2$ and the curves $x = a(y) = \log y$ and $x = b(y) = y + 1$ (Figure 4).

Solution.

$$\iint_R ye^x \, dA = \int_1^2 \int_{\log y}^{y+1} ye^x \, dx \, dy = \int_1^2 y \left(\int_{\log y}^{y+1} e^x \, dx \right) dy$$

$$= \int_1^2 y(e^{y+1} - y) \, dy = e^3 - \tfrac{7}{3}.$$

Example 3. Find the volume under the surface $z = x + y$ over the standard region R bounded by the curves $y = 2x^3$ and $y = 2\sqrt{x}$.

Solution. This region falls under both types. We evaluate $\iint_R z \, dA$ both ways.

To determine the limits of integration, we solve to find where the bounding curves intersect: $(0, 0)$ and $(1, 2)$. Hence the limiting lines are $x = 0$ and $x = 1$, and $y = 0$ and $y = 2$.

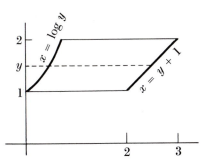

FIGURE 4

Method 1. See Figure 5(a).

$$V = \iint_R z \, dA = \int_0^1 \int_{2x^3}^{2\sqrt{x}} (x + y) \, dy \, dx$$

$$= \int_0^1 \left(x(2\sqrt{x} - 2x^3) + \tfrac{1}{2}[(2\sqrt{x})^2 - (2x^3)^2] \right) dx$$

$$= \tfrac{39}{35}.$$

Method 2. See Figure 5(b).

$$V = \iint_R z \, dA = \int_0^2 \int_{(y/2)^2}^{(y/2)^{1/3}} (x + y) \, dx \, dy$$

$$= \int_0^2 \left(\tfrac{1}{2}[(\tfrac{1}{2}y)^{2/3} - (\tfrac{1}{2}y)^4] + y[(\tfrac{1}{2}y)^{1/3} - (\tfrac{1}{2}y)^2] \right) dy$$

$$= \tfrac{39}{35}.$$

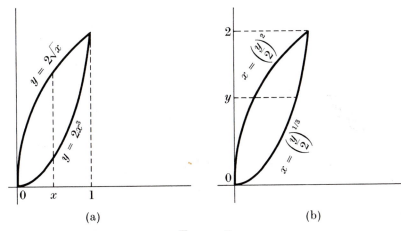

(a)   (b)

FIGURE 5

14D3. Continuous functions on standard regions. Basic results about continuous functions, including the development of the integral, extend to standard regions. Here we describe some of the results and outline their proofs.

Let f be continuous on a standard region R. We consider the case in which R is bounded by lines $x = a$ and $x = b$ and continuous curves $y = \alpha(x)$ and $y = \beta(x)$. See Figure 6. The arguments with the other form of standard region are analogous.

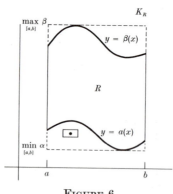

FIGURE 6

Maximum-Value Theorem. *f is bounded on R and assumes a maximum and a minimum there.*

Proof. By the Maximum-Value Theorem for functions of one variable (§2E7), $\alpha(x)$ and $\beta(x)$ both assume a minimum and a maximum on the interval $[a, b]$. It follows that R is contained in a closed box—for example, the box

$$K_R = [a, b] \times [\min_{[a,b]} \alpha, \max_{[a,b]} \beta].$$

Now, in reading the proof of §12C2, let us agree that f is bounded on any set where it is not defined. It is easy to see from continuity that each point of K_R that lies outside of R lies inside some box that does not meet R (cf. §12A3, *Remark*). By our agreement, f is bounded on that box. We now see that the proof of §12C2 carries through verbatim.

Existence of ϵ-partitions. A partition of R is understood to mean the decomposition obtained by intersecting R with a partition of K_R. Given any number $\epsilon > 0$, we agree that f varies by less than ϵ on any box where it is not defined. The proof of §12C4 then carries through to show that R admits an ϵ-partition.

The integral. The remaining modifications in the development of the integral are carried through in the same spirit and with the help of some technical skill.

Intermediate-Value Theorem. *If (x_1, y_1) and (x_2, y_2) are any two points of R, then f assumes all values between $f(x_1, y_1)$ and $f(x_2, y_2)$.*

Proof. First of all, $f(x_1, y)$ is a continuous function of y. Hence, by the Intermediate-Value Theorem for functions of a single variable (§2E5), $f(x_1, y)$ assumes all values between $f(x_1, y_1)$ and $f(x_1, \beta(x_1))$. See Figure 7. Likewise, $f(x_2, y)$ assumes all values between $f(x_2, y_2)$ and $f(x_2, \beta(x_2))$. Finally, it is easy to see that

$$F(x) = f(x, \beta(x))$$

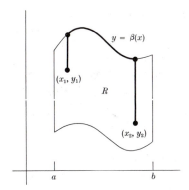

$$y = \beta(x)$$

(x_1, y_1)

R

(x_2, y_2)

a b

FIGURE 7

is a continuous function of x; therefore it assumes all values between $F(x_1) = f(x_1, \beta(x_1))$ and $F(x_2) = f(x_2, \beta(x_2))$. Putting all these facts together, we get our conclusion.

Image of a standard region. *The image of R under f is a closed interval.*

Proof. As we have just seen, f assumes a minimum, a maximum, and all values in between.

Problems (§14D)

1. Sketch the standard regions bounded by the given curves.
   (a) $x = 0$, $\quad x = 5$, $\quad y = e^x$, $\quad y = 0$
   (b) $x = 0$, $\quad x = 5$, $\quad y = e^x$, $\quad y = e^{-x}$
   (c) $x = 0$, $\quad x = \pi$, $\quad y = \sin x$, $\quad y = \sin^2 x$
   (d) $x = 0$, $\quad x = \pi$, $\quad y = \sin x$, $\quad y = x^2 + 1$
   (e) $y = 1$, $\quad y = 2$, $\quad x = 0$, $\quad x = \sqrt{4 - y^2}$
   (f) $x = 0$, $\quad x = 1$, $\quad y = x$, $\quad y = x^2$
   (g) $y = -1$, $\quad x = y^3$, $\quad x + y = 2$
   (h) $y = 0$, $\quad y = 2$, $\quad x = y^2$, $\quad (x - 5)^2 + (y - 1)^2 = 1$

2. Evaluate the following integrals.

   (a) $\displaystyle\int_0^1 \int_{x^2}^x x^2 y \, dy \, dx$

   (b) $\displaystyle\int_{-1}^1 \int_{-\sqrt{1-x^2}}^{\sqrt{1-x^2}} y \, dy \, dx$

   (c) $\displaystyle\int_0^{\pi/2} \int_0^{\cos y} x \sin y \, dx \, dy$

   (d) $\displaystyle\int_0^2 \int_0^y (4 - y^2)^{3/2} \, dx \, dy$

   (e) $\displaystyle\int_0^{\pi/4} \int_0^{\sec x} 2y \, dy \, dx$

   (f) $\displaystyle\int_0^2 \int_0^x \sqrt{4 - x^2} \, dy \, dx$

   (g) $\displaystyle\int_0^1 \int_0^{y^2} \sqrt{y^3 + 1} \, dx \, dy$

   (h) $\displaystyle\int_{-2}^2 \int_0^{4-x^2} (x + y) \, dy \, dx$

   (i) $\displaystyle\int_0^4 \int_{\sqrt{x}}^2 (y^2 + 2xy) \, dy \, dx$

   (j) $\displaystyle\int_0^{1/\sqrt{2}} \int_x^{\sqrt{1-x^2}} (2x^2 + y) \, dy \, dx$

3. Evaluate $\iint_R f(x, y)\, dA$ in the following cases.
   (a) $f(x, y) = xy - y^2$, R the region bounded by $x = 3, y = x, y = x/2$
   (b) $f(x, y) = 2x - 3y + 1$, R bounded by $x^2 + y^2 = 4$
   (c) $f(x, y) = xe^y$, R bounded by $y = 0, x = 1, y = x^2$
   (d) $f(x, y) = xy$, R bounded by $y = 0, x = 2y + 2, x = 3 - 3y$
   (e) $f(x, y) = \sqrt{1 + x^2}$, R bounded by $x = 0, y = 1, y = x$

   (f) $f(x, y) = \dfrac{x}{x^2 + y^2}$, R bounded by $y = -1$, $y = 1$, $x = 1$, $x = \sqrt{2 - y^2}$ [See §8F, Problem 4.]
   (g) $f(x, y) = x$, R bounded by $x = 0, x^2 + y^2 = 4, x^2 + (y - 2)^2 = 4$

4. Find the volume of the solid described.
   (a) Under the plane $z = 1 - y$, above the region in the xy-plane bounded by the lines $y = 0, x = 1$, and $y = x$
   (b) Under the graph of $z = xy/2$, above the region bounded by $x = 2$, $y = 0, y = x$
   (c) Under the plane $z = x + 2y$, above the region bounded by $x = 2$, $y = 0, y = x^3$
   (d) Under the graph of $z = xy$, over the region bounded by $y = 0, y = x^2$, $y = (x - 2)^2$
   (e) In the first octant, below the plane $z = x + y$ and above the region bounded by $x^2 + 4y^2 = 4$
   (f) In the first octant, below the plane $z = 3x + y$ and above the disc $x^2 + y^2 \le 16$

5. Write the given expression the form

$$\int_\alpha^\beta \int_{a(y)}^{b(y)} z\, dx\, dy$$

(or as a sum of such expressions).

(a) $\displaystyle\int_0^1 \int_{2x^2}^{2x} z\, dy\, dx$
                 (d) $\displaystyle\int_1^e \int_0^{\log x} z\, dy\, dx$

(b) $\displaystyle\int_{-1}^1 \int_{x^3}^{\sqrt{2 - x^2}} z\, dy\, dx$
       (e) $\displaystyle\int_0^{1/2} \int_0^x z\, dy\, dx + \int_{1/2}^1 \int_0^{1-x} z\, dy\, dx$

(c) $\displaystyle\int_{-2}^0 \int_{2x+4}^{4 - x^2} z\, dy\, dx$
          (f) $\displaystyle\int_{-3}^2 \int_{|x|}^{\sqrt{6-x}} z\, dy\, dx$

6. Write the given expression in the form

$$\int_a^b \int_{\alpha(x)}^{\beta(x)} z\, dy\, dx$$

(or as a sum of such expressions).

(a) $\displaystyle\int_1^2 \int_0^{\sqrt{4-y^2}} z\,dx\,dy$

(b) $\displaystyle\int_{-1}^0 \int_{y+2}^{1+\sqrt{1-y^2}} z\,dx\,dy$

(c) $\displaystyle\int_0^{\pi/2} \int_0^{\sin y} z\,dx\,dy$

(d) $\displaystyle\int_{-2}^{-2/\sqrt{3}} \int_{-\sqrt{4-y^2}/\sqrt{2}}^{\sqrt{4-y^2}/\sqrt{2}} z\,dx\,dy \;+\; \int_{-2/\sqrt{3}}^{2/\sqrt{3}} \int_y^{\sqrt{4-y^2}/\sqrt{2}} z\,dx\,dy$

(e) $\displaystyle\int_{-4}^{2\sqrt{2}} \int_{(y+4)/(1+\sqrt{2})}^{\sqrt{16-y^2}} z\,dx\,dy$

(f) $\displaystyle\int_{-1}^2 \int_{(y-1)^2}^{5-y^2} z\,dx\,dy$

7. Evaluate the following integrals.

(a) $\displaystyle\int_0^1 \int_{\sqrt[3]{y}}^1 \sqrt{x^4+1}\,dx\,dy$

(c) $\displaystyle\int_{-2}^2 \int_{-\sqrt{4-y^2}}^{\sqrt{4-y^2}} y\sqrt{x^3+1}\,dx\,dy$

(b) $\displaystyle\int_0^{\sqrt{2}} \int_x^{\sqrt{2}} e^{y^2}\,dy\,dx$

(d) $\displaystyle\int_0^{\sqrt{2}} \int_{x^2}^2 e^{x/\sqrt{y}}\,dy\,dx$

8. (a) Find the volume of the solid in the first octant bounded by the cylinder $x^2 + y^2 = 16$ and the plane $z = x$.
   (b) Find the volume inside both the cylinders $x^2 + y^2 = 9$ and $x^2 + z^2 = 9$.
   (c) The cylinder $y = 2 - e^{x^2}$ is cut by the cylinder $z = 2 - e^{x^2}$ and by the plane $z = y$. Show that the volume in the first octant enclosed by the first intersection is twice that of the second.

9. Let $\varphi(x)$ be a nonnegative continuous function on $[a, b]$ and let R be the region under its graph. Show that the area of R as given by a double integral agrees with the familiar formula from the calculus of one variable. What about the area between two graphs?

10. Show that if R is a standard region in the upper xy-plane, then the volume obtained by revolving R about the x-axis is

(5)
$$V = 2\pi \iint_R y\,dA.$$

[*Hint.* There are two problems, depending on the form of the standard region. In each case, replace the double integral by an iterated integral and perform one integration.]

Answers to problems (§14D)

2. (a) $\frac{1}{35}$ (b) 0   (c) $\frac{1}{6}$ (d) $\frac{32}{5}$ (e) 1   (f) $\frac{8}{3}$ (g) $\frac{2}{9}(2\sqrt{2}-1)$

   (h) $\frac{256}{15}$ (i) $\frac{256}{15}$ (j) $\frac{1}{2}(\frac{1}{8}\pi + \frac{1}{3}\sqrt{2} - \frac{1}{4})$

3. (a) $\frac{27}{16}$ (b) 4π (c) $\dfrac{e}{2} - 1$ (d) $\frac{49}{3000}$

   (e) $\frac{1}{3}(1 - \frac{1}{2}\sqrt{2}) + \frac{1}{2}\log(1 + \sqrt{2})$ (f) $2 - \dfrac{\pi}{2}$ (g) $\frac{5}{3}$

4. (a) $\frac{1}{3}$ (b) 1   (c) $\frac{864}{35}$ (d) $\frac{1}{5}$ (e) 2   (f) $\frac{256}{3}$

5. (a) $\displaystyle\int_0^2\int_{y/2}^{\sqrt{y/2}} z\,dx\,dy$ (b) $\displaystyle\int_{-1}^1\int_{-1}^{\sqrt[3]{y}} z\,dx\,dy + \int_1^{\sqrt2}\int_{-\sqrt{2-y^2}}^{\sqrt{2-y^2}} z\,dx\,dy$

   (c) $\displaystyle\int_0^4\int_{-\sqrt{4-y}}^{(y-4)/2} z\,dx\,dy$ (d) $\displaystyle\int_0^1\int_{e^y}^{e} z\,dx\,dy$

   (e) $\displaystyle\int_0^{1/2}\int_y^{1-y} z\,dx\,dy$ (f) $\displaystyle\int_0^2\int_{-y}^{y} z\,dx\,dy + \int_2^3\int_{-y}^{6-y^2} z\,dx\,dy$

6. (a) $\displaystyle\int_0^{\sqrt3}\int_1^{\sqrt{4-x^2}} z\,dy\,dx$ (b) $\displaystyle\int_1^2\int_{-\sqrt{1-(x-1)^2}}^{x-2} z\,dy\,dx$

   (c) $\displaystyle\int_0^1\int_{\arcsin x}^{\pi/2} z\,dy\,dx$ (d) $\displaystyle\int_{-2/\sqrt3}^{2/\sqrt3}\int_{-\sqrt{4-2x^2}}^{x} z\,dy\,dx + \int_{2/\sqrt3}^2\int_{-\sqrt{4-2x^2}}^{\sqrt{4-2x^2}} z\,dy\,dx$

   (e) $\displaystyle\int_0^{2\sqrt2}\int_{-\sqrt{16-x^2}}^{(1+\sqrt2)x-4} z\,dy\,dx + \int_{2\sqrt2}^4\int_{-\sqrt{16-x^2}}^{\sqrt{16-x^2}} z\,dy\,dx$

   (f) $\displaystyle\int_0^1\int_{1-\sqrt{x}}^{1+\sqrt{x}} z\,dy\,dx + \int_1^4\int_{1-\sqrt{x}}^{\sqrt{5-x}} z\,dy\,dx + \int_4^5\int_{-\sqrt{5-x}}^{\sqrt{5-x}} z\,dy\,dx$

7. (a) $\frac{1}{6}(2\sqrt2 - 1)$ (b) $\frac{1}{2}(e^2 - 1)$ (c) 0   (d) $\frac{4}{3}(e-1)\sqrt2$

8. (a) $\frac{64}{3}$ (b) 144

§14E. A general theorem for the double integral

There is a general theorem for the double integral like that for the single integral (§6A2). The ingredients are:

(i) One or more functions $f, g, \cdots$ continuous on a standard region R. For simplicity, we state the theorem for the case of two functions, f and g.

(ii) A continuous function of these functions, which we call $C(f, g)$. For instance, $C(f, g)$ might be fg, or $f - g$, or $\sqrt{f^2 + g^2}$, etc. We also have to consider

$$C\big(f(x_1, y_1), g(x_2, y_2)\big),$$

in which f and g are evaluated at different points.

(iii) A function that associates a number to each standard subregion of R. The number associated to the subregion H is denoted S_H. In particular, the number associated to R itself is S_R.

The theorem is:
Let R, f, g, C, and S_H be as above.
Suppose that

(A) $$S_H + S_{H'} = S_{H \cup H'}$$

whenever H and H' are adjacent subregions; and

(B) $(area\ of\ H) \times C(\min_H f, \max_H g) \le S_H$

$$\le (area\ of\ H) \times C(\max_H f, \min_H g).$$

Then

$$S_R = \iint_R C(f, g).$$

Remark. Hypothesis **(B)** may also be given with $\min_H f$ and $\min_H g$ on the left and $\max_H f$ and $\max_H g$ on the right; etc.

Proof of theorem. We know that the integral is the unique number lying between every lower sum L and every upper sum U for the function $C(f, g)$ (§14A3). So we will prove that

(1) $$L \le S_R \le U$$

for all L and U. We do this by showing that if ϵ is any positive number, then

(2) $$L - \epsilon < S_R < U + \epsilon.$$

(Then (1) follows. For, if $S_R > U$, say, we pick $\epsilon = S_R - U$ and find from (2) that $S_R < S_R$, an absurdity.)

Let L, U, and ϵ be given. In outline, what we do is to choose a partition of R so that, on each segment, H,

(3) $$C(\max_H f, \min_H g) < \max_H C(f, g) + \epsilon',$$

where

(4) $$\epsilon' = \frac{\epsilon}{area\ of\ R}.$$

This can be done because f, g, and C are continuous (details below). With **(B)**, we get

(5) $$S_H < (area\ of\ H) \max_H C(f, g) + \epsilon'(area\ of\ H).$$

We now sum over all segments H. By **(A)** (applied repeatedly), the sum of the numbers S_H is S_R. So we get (by definition of U)

(6) $$S_R < U + \epsilon.$$

This gives us half of (2). We get the other half similarly.

The details are as follows.

The images of f and g are closed intervals (§14D3). Therefore the function C is defined on the closed box

$$(\text{image of } f) \times (\text{image of } g).$$

Let P be an $(\epsilon'/2)$-partition of this box with respect to C (§12C4): on each segment of the partition, C varies by less that $\epsilon'/2$. Let δ denote the smallest dimension of a segment of P.

We now construct a special partition of R. Take a δ-partition of R with respect to the function f and a δ-partition with respect to g. Then take the partition of R that yields the given upper sum, U. Now pool all the lines of subdivision so as to form a partition that refines all three. Its upper sum is, if anything, less than U; hence in proving (2) we may assume it is the U we started with.

On any segment, H, f and g each vary by less than δ. Because of the choice of δ, the values of f on H cannot cut across more than one vertical line of subdivision of P; see Figure 1. Likewise, the values of g cannot cut across more than one horizontal line of subdivision of P. Consequently, the pairs of values

(7) $$(f(x_1, y_1), g(x_2, y_2))$$

meet at most four segments of P, related as shown in the figure. Since P is an $(\epsilon'/2)$-partition for C, C varies on this set by less than ϵ'. In particular, we have (3). The proof now continues as already described.

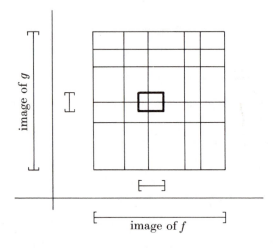

FIGURE 1

Problems (§14E)

1. In each case, S_H is defined for each standard subregion H of a standard region R and satisfies **(A)** and the additional condition stated, where f and g are continuous on R. Express S_H as a double integral.

Note. $|H|$ denotes the area of H.

(a) $|H|(\min_H f^2 - \max_H g^2) \le S_H \le |H|(\max_H f^2 - \min_H g^2)$

(b) $|H| \min_H(f)/\max_H g \le S_H \le |H| \max_H(f)/\min_H g$

(c) $|H| \min_H e^{xy} \max_H(x + y) \le S_H \le |H| \max_H e^{xy} \min_H(x + y)$

(d) $|H| \max_H \sin(xy) \min_H \cos^2 y \le S_H \le |H| \min_H \sin(xy) \max_H \cos^2 y$

(e) $|H| \exp(\min_H f + \min_H g) \le S_H \le |H| \exp(\max_H f + \max_H g)$

(f) $|H|[\min_H(x^2 y) + \max_H \sqrt{xy}] \le S_H \le |H|[\max_H(x^2 y) + \min_H \sqrt{xy}]$

(g) $|H|\sqrt{(\min_H f)^2 + (\min_H g)^2} \le S_H \le |H|\sqrt{(\max_H f)^2 + (\max_H g)^2}$

(h) $|H| \, x \sin(\min_H f) \le S_H \le |H| \, x \sin(\max_H f)$ for all $(x, y) \in H$

Answers to problems (§14E)

1. (a) $\displaystyle\iint_H (f^2 - g^2)$ (b) $\displaystyle\iint_H \frac{f}{g}$ (c) $\displaystyle\iint_H e^{xy}(x + y) \, dA$

(d) $\displaystyle\iint_H \sin(xy) \cos^2 y \, dA$ (e) $\displaystyle\iint_H \exp(f(x, y) + g(x, y)) \, dA$

(f) $\displaystyle\iint_H (x^2 y + \sqrt{xy}) \, dA$ (g) $\displaystyle\iint_H \sqrt{f(x, y)^2 + g(x, y)^2} \, dA$

(h) $\displaystyle\iint_H x \sin(f(x, y)) \, dA$

§14F. Moments

14F1. First moments: finite case. Moments in the one-dimensional case were defined in §6F. In two dimensions, the reasoning is very much the same. Consider n masses,

$$m_1, \cdots, m_n$$

situated at points

$$(x_1, y_1), \cdots, (x_n, y_n).$$

This time we define *two* first moments, M_y, the first moment about the y-axis:

(1) $$M_y = m_1 x_1 + \cdots + m_n x_n,$$

and M_x, the first moment about the x-axis:

(2) $$M_x = m_1 y_1 + \cdots + m_n y_n.$$

The total mass, M, is equal to $m_1 + \cdots + m_n$. The center of mass, $(\bar{x}, \bar{y})$, is defined by the equations

(3) $$\bar{x} = \frac{M_y}{M}, \qquad \bar{y} = \frac{M_x}{M}.$$

Example 1. What are the first moments and center of mass of the system of three particles distributed in the plane as follows (Figure 1): mass 3 at (4, 2), mass 2 at (4, 3), mass 3 at (1, 1)?

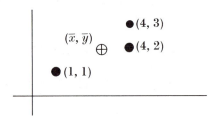

FIGURE 1

Solution. The first moment about the y-axis is

$$M_y = (3)(4) + (2)(4) + (3)(1) = 23.$$

The first moment about the x-axis is

$$M_x = (3)(2) + (2)(3) + (3)(1) = 15.$$

The total mass is $M = 3 + 2 + 3 = 8$. The center of mass is $(\bar{x}, \bar{y})$, where

$$\bar{x} = \frac{M_y}{M} = \frac{23}{8}, \qquad \bar{y} = \frac{M_x}{M} = \frac{15}{8}.$$

14F2. First moments: continuous case. We now consider a *continuous* distribution of mass on a standard region R. That is, we assume that a continuous density function $\rho(x, y)$ is defined on R so that, for each standard subregion H, the mass over H is $\iint_H \rho$. We wish to define the first moments about the axes and the center of mass of this distribution.

Consider the moment about the y-axis. Let $M_y(H)$ denote the contribution to this moment that comes from the subregion H. For adjacent subregions, H and H', we will assume

(A) $M_y(H) + M_y(H') = M_y(H \cup H').$

Moreover, on any subregion H,

(B)   (area of H)$(\min_H x)(\min_H \rho) \le M_y(H) \le$ (area of H)$(\max_H x)(\max_H \rho).$

These assumptions about moments are imposed by the physicists. By the General Theorem for the Double Integral (§14E), we are forced to the definition

(4)
$$M_y = \iint_R x\rho(x, y)\, dA$$

Similarly, we are forced to define

(5)
$$M_x = \iint_R y\rho(x, y)\, dA.$$

The total mass, M, is given by

(6)
$$M = \iint_R \rho.$$

Finally the center of mass, $(\bar{x}, \bar{y})$, is defined by the equations

(7) $$\bar{x} = \frac{M_y}{M} = \frac{\iint_R x\rho(x, y)\, dA}{\iint_R \rho(x, y)\, dA}, \qquad \bar{y} = \frac{M_x}{M} = \frac{\iint_R y\rho(x, y)\, dA}{\iint_R \rho(x, y)\, dA}.$$

The word *centroid* is used instead of center of mass when the plane region is being thought of purely geometrically, i.e., for a plane region of constant density 1. Thus, the centroid $(\bar{x}, \bar{y})$ is given by:

(8) $$\bar{x} = \frac{\iint_R x\, dA}{\text{area of } R}, \qquad \bar{y} = \frac{\iint_R y\, dA}{\text{area of } R}.$$

Example 2. Let $\rho(x, y) = x + y$ on the standard region R bounded by $x = 0$, $x = 2$, $y = 0$, and $y = x^2$ (Figure 2). Find M, M_y, M_x, and the center of mass.

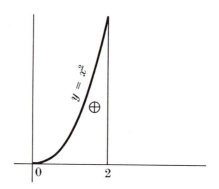

FIGURE 2

Solution. The mass is

$$M = \iint_R (x + y) \, dA = \int_0^2 \int_0^{x^2} (x + y) \, dy \, dx = \tfrac{36}{5}.$$

The moments are

$$M_y = \iint_R x(x + y) \, dA = \int_0^2 x \int_0^{x^2} (x + y) \, dy \, dx = \tfrac{176}{15};$$

$$M_x = \iint_R y(x + y) \, dA = \int_0^2 \int_0^{x^2} (xy + y^2) \, dy \, dx = \tfrac{80}{7}.$$

The center of mass is given by

$$\bar{x} = \frac{M_y}{M} = \frac{44}{27}, \qquad \bar{y} = \frac{M_x}{M} = \frac{100}{63}.$$

Example 3. Find the centroid of the plane region R of Example 2.

Solution. $M = \text{area} = \int_0^2 x^2 \, dx = \tfrac{8}{3}.$

$$M_y = \int_0^2 x \int_0^{x^2} dy \, dx = 4; \; M_x = \int_0^2 \int_0^{x^2} y \, dy \, dx = \tfrac{16}{5}.$$

$$\bar{x} = \frac{M_y}{M} = \frac{3}{2}, \qquad \bar{y} = \frac{M_x}{M} = \frac{6}{5}.$$

14F3. *Higher moments.* Higher moments are defined analogously. In the finite case the second moments are defined as

(9) $\qquad I_y = m_1 x_1^2 + \cdots + m_n x_n^2, \qquad I_x = m_1 y_1^2 + \cdots + m_n y_n^2.$

In the continuous case they are defined by

(10) $\qquad I_y = \iint\limits_R x^2 \rho(x, y) \, dA, \qquad I_x = \iint\limits_R y^2 \rho(x, y) \, dA.$

Here, I_y is the second moment or *moment of inertia* about the y-axis, I_x the second moment or moment of inertia about the x-axis.

Closely related to these quantities is the second moment about the origin, or *polar moment*, I_0. It is defined by

(11) $\qquad I_0 = m_1(x_1^2 + y_1^2) + \cdots + m_n(x_n^2 + y_n^2)$

in the finite case, and

(12) $\qquad I_0 = \iint (x^2 + y^2)\rho(x, y) \, dA$

in the continuous case. Clearly, in both cases,

(13) $\qquad I_0 = I_x + I_y.$

Example 4. Find the second moments of the distribution of Example 2.

Solution.

$$I_y = \int_0^2 x^2 \int_0^{x^2} (x + y) \, dy \, dx = \tfrac{13}{21} \times 2^5;$$

$$I_x = \int_0^2 \int_0^{x^2} (xy^2 + y^3) \, dy \, dx = \tfrac{7}{9} \times 2^5;$$

$$I_0 = I_y + I_x = \tfrac{11}{63} \times 2^8.$$

Problems (§14F)

1. In each case, ρ is the density function over the region R. Find the mass of the distribution, the first moments about the coordinate axes, and the center of mass.

(a) $\rho(x, y) = x^2 + y^2$, R the region bounded by $x = 0$, $x = 1$, $y = 0$, $y = 1$

(b) $\rho(x, y) = 2x$, R bounded by $y = \sqrt{x}$, $y = x^3$

(c) $\rho(x, y) = 2y$, R bounded by $y = 0$, $y = \sin x$ $(0 \le x \le \pi)$
(d) $\rho(x, y) = x$, R bounded by $x = 2$, $y^2 = 8x$
(e) $\rho(x, y) = 1$, R bounded by $y = x^2$, $y = 4x - x^2$
(f) $\rho(x, y) = x + y^2$, R bounded by $(x - 1)^2 + y^2 = 1$

2. Find the centroids of the following regions.
   (a) The region bounded by $x^2 + (y - 1)^2 = 4$
   (b) The region lying above the x-axis and below the graph of $y = \sin x$ $(0 \le x \le \pi)$
   (c) The region bounded by the triangle with vertices $(0, 0)$, $(a, 0)$, and (b, c)
   (d) The set of all points (x, y) such that $x \ge 0$, $y \ge 0$, and $x^2 + y^2 \le 1$
   (e) The region lying above the x-axis and below the graph of $y = 4 - x^2$
   (f) The region bounded by the graphs of $y = x$ and $y = x^2$

3. In each case, ρ is the density function over the region R. Find the second moments of the distribution.
   (a) $\rho(x, y) = 6$, R the region bounded by $x = 4$, $x = -4$, $y = 4$, $y = -4$
   (b) $\rho(x, y) = 1$, R bounded by $\dfrac{x^2}{4} + \dfrac{y^2}{9} = 1$

   (c) $\rho(x, y) = 2x$, R bounded by $y = x^3$, $y = \sqrt{x}$
   (d) $\rho(x, y) = 2$, R bounded by $y = 0$, $y = \sin x$ $(0 \le x \le \pi)$

4. We have defined the force on a vertical surface submerged in a fluid of constant density k (§6G). Show that the force on the region of Figure 3 is $k\bar{y}A$, where A is the area of the region.

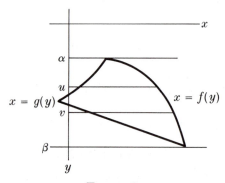

$x = g(y)$ $x = f(y)$

FIGURE 3

5. Let R be a standard region in the xy-plane, having centroid $(\bar{x}, \bar{y})$. Choose a new coordinate system with axes parallel to the original axes, and with the new origin at the point originally (h, k). Show that the centroid of R, expressed in the new system, is $(\bar{x} - h, \bar{y} - k)$.

Answers to problems (§14F)

1. (a) $M = \frac{2}{3}$, $M_y = \frac{5}{12}$, $M_x = \frac{5}{12}$, $(\bar{x}, \bar{y}) = (\frac{5}{8}, \frac{5}{8})$
   (b) $M = \frac{2}{5}$, $M_y = \frac{5}{21}$, $M_x = \frac{5}{24}$, $(\bar{x}, \bar{y}) = (\frac{25}{42}, \frac{25}{48})$
   (c) $M = \pi/2$, $M_y = \pi^2/4$, $M_x = \frac{8}{9}$, $(\bar{x}, \bar{y}) = (\pi/2, 16/9\pi)$
   (d) $M = \frac{64}{5}$, $M_y = \frac{128}{7}$, $M_x = 0$, $(\bar{x}, \bar{y}) = (\frac{10}{7}, 0)$
   (e) $M = \frac{8}{3}$, $M_y = \frac{8}{3}$, $M_x = \frac{16}{3}$, $(\bar{x}, \bar{y}) = (1, 2)$
   (f) $M = 5\pi/4$, $M_y = 3\pi/2$, $M_x = 0$, $(\bar{x}, \bar{y}) = (\frac{6}{5}, 0)$
2. (a) $(0, 1)$ (b) $(\pi/2, \pi/8)$ (c) $((a + b)/3, c/3)$
   (d) $(4/3\pi, 4/3\pi)$ (e) $(0, \frac{8}{5})$ (f) $(\frac{1}{2}, \frac{2}{5})$
3. (a) $I_y = I_x = 2048$, $I_0 = 4096$ (b) $I_y = 6\pi$, $I_x = 27\pi/2$, $I_0 = 39\pi/2$
   (c) $I_y = \frac{10}{63}$, $I_x = \frac{10}{77}$, $I_0 = \frac{200}{693}$
   (d) $I_y = 2\pi^2 - 8$, $I_x = \frac{8}{9}$, $I_0 = 2\pi^2 - \frac{64}{9}$

§14G. The double integral in polar coordinates

14G1. Polar boxes. Let $F(r, \theta)$ be a continuous function of the two variables r and θ, defined on a box K. Then the double integral $\iint\limits_{K} F$ is defined. For the case $F(r, \theta) = 1 = $ constant, this integral, $\iint\limits_{K} 1$, is equal to the area of K. These facts were developed in §14A. There the variables were called x and y rather than r and θ; but in both cases they are simply two real variables.

Things change when we come to expressing the double integral as an iterated integral, as this involves actually computing areas. Assume that $K = [a, b] \times [\alpha, \beta]$. Let us now interpret this as a "polar box," i.e., $a \leq r \leq b$ and $\alpha \leq \theta \leq \beta$, where (r, θ) are polar coordinates. We assume $0 \leq a$ and $\beta - \alpha \leq 2\pi$. Consider any polar sub-box $H = [u, v] \times [\sigma, \tau]$ (see Figure 1). The area of H is the area of the sector of radius v minus that of radius u. Now, the area of a sector of radius r and angle θ is $\frac{1}{2}r^2\theta$ (§8G4). Therefore,

(1) $$\text{area of } H = \tfrac{1}{2}(v^2 - u^2)(\tau - \sigma).$$

Note that this may be written:

(2) $$\text{area of } H = \int_u^v r \, dr \int_\sigma^\tau d\theta.$$

14G2. The double integral in polar coordinates as an iterated integral. *Let $F(r, \theta)$ be continuous on a closed box $K = [a, b] \times [\alpha, \beta]$. If (r, θ) are polar coordinates, then*

(3) $$\iint\limits_{K} F = \int_a^b \int_\alpha^\beta F(r, \theta) \, d\theta \, r \, dr = \int_\alpha^\beta \int_a^b F(r, \theta) \, r \, dr \, d\theta.$$

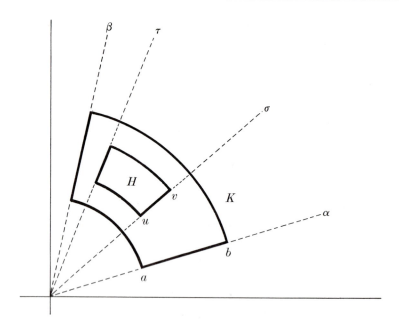

Figure 1

Proof. The proof is modeled after that of §14C1. We show that the iterated integral

$$(4) \qquad\qquad \int_u^v \int_\sigma^\tau F(r,\,\theta)\,d\theta\,r\,dr$$

satisfies **(A)** and **(B)** and hence is the double integral. A similar proof holds for the iterated integral in the other direction.

Additivity. This was done in §14C1. (Replace y by θ, x by r, and let $f(x,\,y) = rF(r,\,\theta)$.)

Betweenness. Let H be as in Figure 1. We know that for the single integral, if $G_1 \le G_2$ on an interval $[s,\,t]$, then $\int_s^t G_1 \le \int_s^t G_2$ (§5E1(e)). So we have

$$(5) \qquad \int_\sigma^\tau F(r,\,\theta)\,d\theta \le \int_\sigma^\tau (\max_H F)\,d\theta = (\tau - \sigma)\max_H F.$$

Applying the principle again, we have, in turn,

$$\int_u^v \int_\sigma^\tau F(r, \theta) \, d\theta \, r \, dr \leq \int_u^v (\tau - \sigma)(\max_H F) \, r \, dr$$

(6)
$$= \tfrac{1}{2}(v^2 - u^2)(\tau - \sigma) \max_H F$$

$$= (\text{area of } H) \times \max_H F.$$

Similarly, this integral is $\geq$ (area of H) $\times \min_H F$.

Example 1. Find the volume enclosed by a sphere of radius a.

Solution. With the center at the origin, the equation of the sphere is $x^2 + y^2 + z^2 = a^2$, or, in terms of polar coordinates in the plane,

$$z^2 = a^2 - r^2.$$

The volume is twice that of the upper half. Hence

$$V = \iint_R z \, dA = 2 \int_0^{2\pi} d\theta \int_0^a \sqrt{a^2 - r^2} \, r \, dr$$

$$= 4\pi \tfrac{1}{3}(a^2 - r^2)^{3/2} \Big|_a^0 = \tfrac{4}{3}\pi a^3.$$

Example 2. Find the polar moment of the distribution over the upper half-disc

$$x^2 + y^2 \leq a^2$$

(where $a > 0$) given by the density function

$$\sigma(r, \theta) = r \sin \theta$$

(where (r, θ) are polar coordinates).

Solution. We have (see §14F3):

$$I_0 = \iint_R r^2 \sigma(r, \theta) \, dA = \int_0^\pi \sin \theta \, d\theta \int_0^a r^4 \, dr = \tfrac{2}{5}a^5.$$

14G3. Standard regions in polar coordinates. Standard regions in polar coordinates are principally of the form shown in Figure 2. The bounding curves are (i) rays $\theta = \alpha$ and $\theta = \beta$, where $\alpha \leq \beta$ and $\beta - \alpha \leq 2\pi$; and (ii) curves $r = a(\theta)$ and $r = b(\theta)$, where a and b are continuous functions of θ, and $0 \leq a(\theta) \leq b(\theta)$ ($\alpha \leq \theta \leq \beta$).

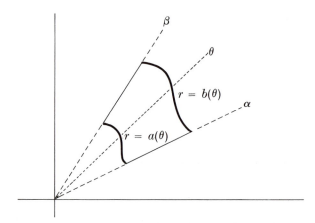

FIGURE 2

The theory of the double integral can be shown, as before, to extend to these regions. For the iterated integral, we have

(7)
$$\iint_R F = \int_\alpha^\beta \int_{a(\theta)}^{b(\theta)} F(r,\,\theta)\, r\, dr\, d\theta.$$

The other form of standard region, in which a and b are constant while α and β may be variable, is rarely encountered and we will not consider it.

Example 3. Evaluate $\iint_R \sin\theta\, dA$, where R is the region in the upper half-plane bounded by the cardioid $r = 1 - \cos\theta$ and the circle $r = 2$ (Figure 3).

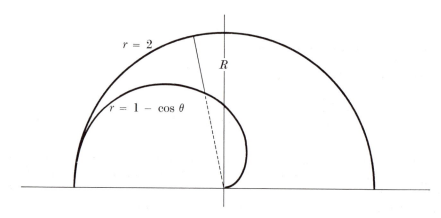

FIGURE 3

Solution.

$$\iint\limits_{R} \sin\theta \, dA = \int_{0}^{\pi} \int_{1-\cos\theta}^{2} r \, dr \sin\theta \, d\theta$$

$$= \tfrac{1}{2} \int_{0}^{\pi} [4 - (1 - \cos\theta)^2] \sin\theta \, d\theta = \tfrac{8}{3}.$$

Example 4. Let $r = \varphi(\theta)$ be a nonnegative continuous function on $[\alpha, \beta]$, where $\beta - \alpha \leq 2\pi$, and let R be the region bounded by the graph of φ (in polar coordinates) and the rays $\theta = \alpha$ and $\theta = \beta$. See Figure 4. Since

$$\text{area of } R = \iint\limits_{R} dA,$$

we have

$$\text{area of } R = \int_{\alpha}^{\beta} \int_{0}^{\varphi(\theta)} r \, dr \, d\theta = \tfrac{1}{2} \int_{\alpha}^{\beta} \varphi(\theta)^2 \, d\theta,$$

the formula obtained in §10D1.

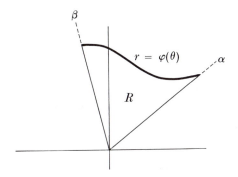

FIGURE 4

Problems (§14G)

1. Sketch the polar standard regions bounded by the given curves.
   (a) $r = 1, r = 2, \theta = 0, \theta = \pi/2$
   (b) $r = 1, r = 2 - \sin\theta, \theta = 0, \theta = \pi/2$
   (c) $r = \sin\theta, \theta = 0, \theta = \pi/4$
   (d) $r = 2\cos\theta, \theta = \pi/4, \theta = \pi/2$
   (e) $r = \cos 3\theta, \theta = -\pi/6, \theta = \pi/6$
   (f) $r = 1 + \cos\theta, r = 1, \theta = 0, \theta = \pi/2$
   (g) $r = 1 + \cos\theta, r = 2, \theta = 0, \theta = \pi/2$
   (h) $r = 3\cos\theta, r = \cos\theta, \theta = 0, \theta = \pi/2$
   (i) $r^2 = \sin\theta, r = \sin\theta, \theta = 0, \theta = \pi/2$

2. Evaluate the following integrals (in which (r, θ) are polar coordinates).

(a) $\displaystyle\iint\limits_{R} \sqrt{1 - r^2}\, dA,$ where R is the disc $x^2 + y^2 \leq 1$

(b) $\displaystyle\iint\limits_{R} e^{r^2}\, dA,$ where R is the disc $x^2 + y^2 \leq 4$

(c) $\displaystyle\iint\limits_{R} r\, dA,$ where R is the disc bounded by the circle $x^2 + y^2 = 2x.$

[*Caution*. Choose the limits of integration for θ so that $r \geq 0.$]

(d) $\displaystyle\iint\limits_{R} \sin\theta\, dA,$ where R is the interior of the triangle bounded by the

lines $x = 2,\, y = 0,$ and $y = \sqrt{3}\, x$

(e) $\displaystyle\iint\limits_{R} \sqrt{r}\,\sin\theta\, dA,$ where R is the interior of the triangle bounded by

the lines $x = 2,\, y = 0,$ and $y = \sqrt{3}\, x$

3. Find the volume of the solid described.
   (a) Under the surface $z = 4 - x^2 - y^2$, above the disc enclosed by the circle $r = 2\cos\theta$
   (b) Under the plane $z = 4 - y$, above the disc of radius 2 centered at the origin
   (c) Under the sphere $z^2 + r^2 = 4$, above the cone $z = \sqrt{x^2 + y^2}$
   (d) Under the sphere $x^2 + y^2 + z^2 = 16$, above the sector bounded by $\theta = 0,\, \theta = \pi/4,$ and $r = 4$
   (e) Under the paraboloid $z = x^2 + y^2$, above the disc enclosed by the circle $r = 2\sin\theta$
   (f) Under the paraboloid $z = x^2 + y^2$, above the triangular region bounded by the lines $\theta = 0,\, \theta = \pi/4,\, r = 3\sec\theta$
   (g) Bounded by the sphere $x^2 + y^2 + z^2 = 4$ and the cylinder $x^2 + y^2 = 1$
4. Write the given expression in the form

$$\int_{\alpha}^{\beta} \int_{a(\theta)}^{b(\theta)} zr\, dr\, d\theta.$$

(a) $\displaystyle\int_{0}^{x} \int_{0}^{\sqrt{1-x^2}} z\, dy\, dx$

(b) $\displaystyle\int_{0}^{2} \int_{-\sqrt{4-x^2}}^{\sqrt{4-x^2}} z\, dy\, dx$

(c) $\displaystyle\int_{-3}^{0}\int_{-\sqrt{9-y^2}}^{\sqrt{9-y^2}} z\ dx\ dy$

(d) $\displaystyle\int_{0}^{1/\sqrt{2}}\int_{y}^{\sqrt{1-y^2}} z\ dx\ dy$

(e) $\displaystyle\int_{0}^{2}\int_{0}^{\sqrt{1-(x-1)^2}} z\ dy\ dx$

(f) $\displaystyle\int_{0}^{1}\int_{x^2}^{x} z\ dy\ dx$

(g) $\displaystyle\int_{0}^{1}\int_{1-\sqrt{1-y^2}}^{1+\sqrt{1-y^2}} z\ dx\ dy$

(h) $\displaystyle\int_{0}^{1/\sqrt{2}}\int_{0}^{x} z\ dy\ dx\ +\ \int_{1/\sqrt{2}}^{1}\int_{}^{\sqrt{1-x^2}} z\ dy\ dx$

(i) $\displaystyle\int_{0}^{1}\int_{y}^{\sqrt{y}} z\ dx\ dy$

(j) $\displaystyle\int_{0}^{1}\int_{0}^{\sqrt{y^2/3-y^2}} z\ dx\ dy$

5. Evaluate the following integrals.

(a) $\displaystyle\int_{0}^{1}\int_{0}^{\sqrt{1-x^2}} \frac{1}{1+\sqrt{x^2+y^2}}\ dy\ dx$

(b) $\displaystyle\int_{0}^{2}\int_{-\sqrt{4-x^2}}^{\sqrt{4-x^2}} (x^2+y^2)^{3/2}\ dy\ dx$

(c) $\displaystyle\int_{-3}^{0}\int_{-\sqrt{9-y^2}}^{\sqrt{9-y^2}} (x^2+y^2)^{1/6}\ dx\ dy$

(d) $\displaystyle\int_{0}^{1/\sqrt{2}}\int_{y}^{\sqrt{1-y^2}} x^6 y\ dx\ dy$

(e) $\displaystyle\int_{0}^{2}\int_{0}^{\sqrt{1-(x-1)^2}} y(x^2+y^2)^2\ dy\ dx$

(f) $\displaystyle\int_{-1}^{1}\int_{1-\sqrt{1-y^2}}^{1+\sqrt{1-y^2}} y^3\ dx\ dy$

(g) $\displaystyle\int_{0}^{1/\sqrt{2}}\int_{0}^{x} e^{x^2+y^2}\ dy\ dx\ +\ \int_{1/\sqrt{2}}^{1}\int_{0}^{\sqrt{1-x^2}} e^{x^2+y^2}\ dy\ dx$

(h) $\displaystyle\int_{0}^{1}\int_{0}^{\sqrt{y^2/3-y^2}} dx\ dy$

6. Find the polar moment of the distribution over R given by the density function σ.

   (a) $\sigma(r, \theta) = r \cos \theta$, R bounded by $\theta = 0$, $\theta = \pi/3$, $r = \sin \theta$

   (b) $\sigma(r, \theta) = 1/(1 + \cos \theta)$, R bounded by $\theta = \pi/2$, $\theta = -\pi/2$, $r = 1 + \cos \theta$

   (c) $\sigma(r, \theta) = r|\sin \theta|$, R the disc $x^2 + y^2 \le 4$

Answers to problems (§14G)

2. (a) $2\pi/3$ (b) $\pi(e^4 - 1)$ (c) $\frac{32}{9}$ (d) 2   (e) $\frac{16}{15}(4 - \sqrt{2})$

3. (a) $5\pi/2$ (b) 16π (c) $\frac{8}{3}\pi(2 - \sqrt{2})$ (d) $16\pi/3$ (e) $3\pi/2$

   (f) 27   (g) $\frac{4}{3}\pi(8 - 3\sqrt{3})$

4. (a) $\displaystyle\int_0^{\pi/2} \int_0^1 zr\, dr\, d\theta$ (b) $\displaystyle\int_{-\pi/2}^{\pi/2} \int_0^2 zr\, dr\, d\theta$

   (c) $\displaystyle\int_\pi^{2\pi} \int_0^3 zr\, dr\, d\theta$ (d) $\displaystyle\int_0^{\pi/4} \int_0^1 zr\, dr\, d\theta$

   (e) $\displaystyle\int_0^{\pi/2} \int_0^{2\cos\theta} zr\, dr\, d\theta$ (f) $\displaystyle\int_0^{\pi/4} \int_0^{\tan\theta\sec\theta} zr\, dr\, d\theta$

   (g) $\displaystyle\int_0^{\pi/2} \int_0^{2\cos\theta} zr\, dr\, d\theta$ (h) $\displaystyle\int_0^{\pi/4} \int_0^1 zr\, dr\, d\theta$

   (i) $\displaystyle\int_0^{\pi/4} \int_0^{\tan\theta\sec\theta} zr\, dr\, d\theta$ (j) $\displaystyle\int_0^{\pi/2} \int_0^{\sqrt{\sin\theta}} zr\, dr\, d\theta$

5. (a) $\frac{1}{2}\pi(1 - \log 2)$ (b) $32\pi/5$ (c) $\frac{27}{7}\pi\sqrt[3]{3}$ (d) $\frac{1}{1008}(16 - \sqrt{2})$

   (e) $\frac{16}{7}$ (f) 0   (g) $\frac{1}{8}\pi(e - 1)$ (h) $\frac{1}{2}$

6. (a) $\frac{9}{640}$ (b) $\frac{1}{2}\pi + \frac{4}{3}$ (c) $\frac{128}{5}$

§14H. Area of a surface

14H1. Properties of surface area. We have previously developed a formula for the area of a surface of revolution $(\S6E(5))$. Now we are in a position to consider more general surfaces.

Let $z = f(x, y)$ be continuous on a box $K = [a, b] \times [\alpha, \beta]$. Consider the portion of the surface over any sub-box H. Let S_H denote its area; this is what we wish to define. Naturally, we assume

(A) $$S_H + S_{H'} = S_{H \cup H'}$$

whenever H and H' are adjacent boxes.

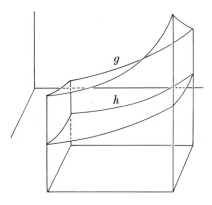

FIGURE 1

To get bounds for S_H we have to be able to *compare* areas. We will reason as in §6D (for arc length), stepping up the argument to two dimensions. Consider two functions, g and h, for which g_x, g_y, h_x, and h_y are all positive at every point, so that both surfaces are rising as we move toward increasing x or increasing y (Figure 1). If *at each point*, the surface g is steeper *in both directions* than h, then surely we feel that g has the larger area. (If g and h are both *plane* surfaces, this can be verified as a fact.) By considering absolute values we include decreasing as well as increasing functions. Our formulation is:

(1)   *If* $|g_x(x, y)| \geq |h_x(x, y)|$ *and* $|g_y(x, y)| \geq |h_y(x, y)|$ *for all* (x, y), *then the area of the surface* g *is greater than the area of* h.

We shall adopt this as a guiding principle. Actually, we need it only when one or the other surface is planar. So we need to know the area over H in a plane, p. We may assume that p passes through the origin, so that its equation takes the simple form

(2) $z = ax + by.$

A normal line to p has direction numbers $(a, b, -1)$. Hence if γ is its acute angle with the z-axis, then

(3) $\cos \gamma = \dfrac{1}{\sqrt{a^2 + b^2 + 1}}.$

We now show that the area over H in p is equal to

(4)
$$\frac{\text{area of } H}{\cos \gamma} \; ;$$

i.e., to

(5)
$$(\text{area of } H) \times \sqrt{a^2 + b^2 + 1}.$$

To see this, rotate the xy-axes (temporarily) about the z-axis so that the intersection of p with the xy-plane is the (new) x-axis (Figure 2). This does

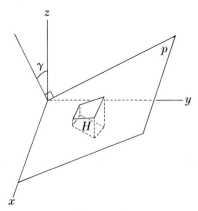

FIGURE 2

not affect γ. Now consider a line segment lying in p. If it is parallel to the x-axis, then it is the same length as its projection in the xy-plane. If it is perpendicular to the x-axis, its projection is foreshortened by the factor $\cos \gamma$. See Figure 3. Since areas of boxes can be found from measurements in these directions, the area in p is as stated in (4) and (5).

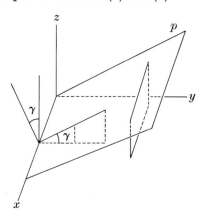

FIGURE 3

14H2. Definition of the area of a surface. Our assumption (1) now leads easily to a definition and formula for surface area. Let $z = f(x, y)$ be defined on a box $K = [a, b] \times [\alpha, \beta]$ and have continuous first partials there (i.e., f is smooth). Then $|f_x|$ and $|f_y|$ assume maxima and minima on any sub-box H. Let S_H denote the area over H in the surface f. By (1), S_H is less than the area over H in the plane

$$(6) \qquad z = (\max_H |f_x|)x + (\max_H |f_y|)y.$$

In view of (5), we have

$$(7) \qquad S_H \leq (\text{area of } H) \times \sqrt{(\max_H |f_x|)^2 + (\max_H |f_y|)^2 + 1}.$$

Similarly, S_H is bounded from below by the corresponding expression in $\min_H |f_x|$ and $\min_H |f_y|$. The two together assert condition **(B)** of §14E with respect to the function

$$(8) \qquad \sqrt{f_x^2 + f_y^2 + 1}.$$

By the General Theorem for the Double Integral (§14E), we are forced to the following definition of the surface area over K:

$$(9) \qquad S = \iint_K \sqrt{f_x^2 + f_y^2 + 1}\, dA.$$

This discussion applies as well to the area of a surface over any standard region, not just a box. The proof of (4) in that case is fussy, however. We omit the details.

Example 1. Find the area of the portion of the cone

$$z = f(x, y) = \sqrt{x^2 + y^2}$$

lying above the disc $R: (x - 1)^2 + y^2 \leq 1$.

Solution. We have

$$f_x = \frac{x}{\sqrt{x^2 + y^2}}, \qquad f_y = \frac{y}{\sqrt{x^2 + y^2}}.$$

Hence

$$S = \iint\limits_{R} \sqrt{1 + \frac{x^2}{x^2 + y^2} + \frac{y^2}{x^2 + y^2}} \, dA$$

$$= \iint\limits_{R} \sqrt{2} \, dA = \pi\sqrt{2},$$

since the area of the disc is π.

Example 2. Find the area of the portion of the hyperbolic paraboloid

$$z = f(x, y) = y^2 - x^2$$

over the disc $R: x^2 + y^2 \le 1$. (See §12B, Figure 10.)

Solution. We have

$$f_x = -2x, \qquad f_y = 2y,$$

and

$$S = \iint\limits_{R} \sqrt{4x^2 + 4y^2 + 1} \, dA.$$

We iterate in polar coordinates, obtaining

$$S = \int_0^{2\pi} \int_0^1 \sqrt{4r^2 + 1} \, r \, dr \, d\theta = \frac{\pi}{6}(5\sqrt{5} - 1).$$

Problems (§14H)

1. Find the area of the surface described.
   (a) the portion of the plane $2x + 3y + 4z = 12$ lying in the first octant
   (b) the part of the plane $z = 2x$ lying inside the cylinder $x^2 + y^2 = 9$
   (c) the portion of the cylinder $x^2 + z^2 = 4$ lying above the triangle bounded by the lines $x = 0$, $y = 0$, and $x + y = 2$ (in the xy-plane)
   (d) the part of the cylinder $y^2 + z^2 = 4$ lying inside the cylinder $x^2 + y^2 = 4$
   (e) the "cap" of height h cut off from a sphere of radius a by a plane
   (f) the portion of the sphere $x^2 + y^2 + z^2 = 25$ lying between the planes $z = 3$ and $z = 4$ [See (e).]
   (g) the portion of the surface $z = xy$ lying above the circle $x^2 + y^2 = 1$
   (h) the part of the cone $z^2 = x^2 + y^2$ inside the sphere $x^2 + y^2 + z^2 = 2x$
   (i) the part of the cone $z = \sqrt{x^2 + y^2}$ lying under the plane $x + y + z = 6$ in the first octant
   (j) the part of the plane $x + y + z = 6$ lying inside the cone $z = \sqrt{x^2 + y^2}$ in the first octant [See (i).]

§14I. Triple integrals

Triple integrals are defined in terms of additivity and betweenness in a manner analogous to single and double integrals. The representation of a triple integral as an iterated integral also follows the same pattern as before. A typical standard region is now a three-dimensional region Q bounded by surfaces

$$z = A(x, y) \qquad \text{and} \qquad z = B(x, y),$$

where A and B are continuous and satisfy $A(x, y) \le B(x, y)$ on a two-dimensional standard region bounded by continuous curves

$$y = \alpha(x) \qquad \text{and} \qquad y = \beta(x)$$

for $a \le x \le b$. If $f(x, y, z)$ is a continuous function of three variables, then the triple integral

(1)
$$\iiint_Q f, \qquad \text{or} \qquad \iiint_Q f(x, y, z)\, dV,$$

is defined, and we have

(2)
$$\iiint_Q f = \int_a^b \int_{\alpha(x)}^{\beta(x)} \int_{A(x,y)}^{B(x,y)} f(x, y, z)\, dz\, dy\, dx.$$

As before, different arrangements of the variables give other forms of standard regions. When "cylindrical" coordinates, (r, θ, z), are used— i.e., (r, θ) are polar coordinates in the plane—the iteration takes the form

(3)
$$\iiint_Q F = \int_\alpha^\beta \int_{a(\theta)}^{b(\theta)} \int_{A(r,\theta)}^{B(r,\theta)} F(r, \theta, z)\, dz\, r\, dr\, d\theta,$$

etc. For the constant function 1, we have

(4)
$$\iiint_Q 1\, dV = \text{volume of } Q.$$

Example 1. Consider the solid bounded by the xy-plane and the surface $z = xy$, over the region bounded by the x-axis and the curve $y = \sqrt{x}$, for $0 \leq x \leq 1$. If the mass is distributed according to the density function $\rho(x, y, z) = z$, what is the total mass?

Solution.

$$M = \iiint_Q \rho(x, y, z) \, dV = \int_0^1 \int_0^{\sqrt{x}} \int_0^{xy} z \, dz \, dy \, dx$$

$$= \tfrac{1}{2} \int_0^1 \int_0^{\sqrt{x}} x^2 y^2 \, dy \, dx = \tfrac{1}{6} \int_0^1 x^{7/2} \, dx = \tfrac{1}{27}.$$

Problems (§14I)

1. Evaluate the following iterated integrals.

(a) $\displaystyle \int_0^1 \int_1^2 \int_2^4 xyz \, dx \, dy \, dz$

(b) $\displaystyle \int_0^2 \int_0^y \int_0^{x+y} (x + y + z) \, dz \, dx \, dy$

(c) $\displaystyle \int_0^6 \int_0^{x^2} \int_0^{6-x-y} 2x \, dz \, dy \, dx$

(d) $\displaystyle \int_0^2 \int_0^{\sqrt{4-y^2}} \int_0^{4-y^2-x^2} y \, dz \, dx \, dy$

(e) $\displaystyle \int_0^{\pi/2} \int_0^1 \int_0^2 zr^2 \cos \theta \, dz \, dr \, d\theta$

(f) $\displaystyle \int_{-\pi/2}^{\pi/2} \int_0^{\sqrt{\cos \theta}} \int_{-r}^r r^2 \, dz \, dr \, d\theta$

2. Evaluate $\displaystyle \iiint_Q f(x, y, z) \, dV$ for the given function f and region Q.

(a) $f(x, y, z) = xy + 2z$; Q is the region lying above the xy-plane, inside the cylinder $x^2 + y^2 = 16$, and below the plane $z = 4$

(b) $f(x, y, z) = x^2 + y^2$; Q is the region in the first octant cut off by the plane $6x + 3y + 2z = 6$

(c) $f(x, y, z) = xy$; Q is the region in the first octant lying under the cone $z = \sqrt{x^2 + y^2}$ and inside the cylinder $x^2 + y^2 = 4$

(d) $f(x, y, z) = x$; Q is the region in the first octant under the surface $z = 4 - x^2 - y^2$ and above the xy-plane

3. Express the volume of the given region as a three-fold iterated integral, and evaluate.

(a) interior to the ellipsoid $\dfrac{x^2}{a^2} + \dfrac{y^2}{b^2} + \dfrac{z^2}{c^2} = 1$ $(a > 0,\, b > 0,\, c > 0)$

(b) inside the cylinder $x^2 + y^2 = 4x$, above the xy-plane under the plane $z = 1 + x$

(c) above the region in the xy-plane bounded by the graph of $x^2 + y^2 = 4x$, and below the surface $x^2 + y^2 = 4z$

(d) bounded above by the surface $z = 9 - x^2 - y^2$ and below by the plane $z = 6 - 2y$

Answers to problems (§14I)

1. (a) $\frac{9}{2}$ (b) 14   (c) $-6998\frac{2}{5}$ (d) $\frac{64}{15}$ (e) $\frac{2}{3}$ (f) $\pi/4$

2. (a) 256π (b) $\frac{1}{2}$ (c) $\frac{16}{5}$ (d) $\frac{64}{15}$

3. (a) $V = 8 \displaystyle\int_0^a \int_0^{b\sqrt{1-x^2/a^2}} \int_0^{c\sqrt{1-x^2/a^2-y^2/b^2}} dz\, dy\, dx = \frac{4}{3}\pi abc$

(b) $V = \displaystyle\int_{-\pi/2}^{\pi/2} \int_0^{4\cos\theta} \int_0^{1+r\cos\theta} r\, dz\, dr\, d\theta = 12\pi$

(c) $V = \displaystyle\int_{-\pi/2}^{\pi/2} \int_0^{4\cos\theta} \int_0^{r^2/4} r\, dz\, dr\, d\theta = 6\pi$

(d) $V = \displaystyle\int_{-2}^{2} \int_{1-\sqrt{4-x^2}}^{1+\sqrt{4-x^2}} \int_{6-2y}^{9-x^2-y^2} dz\, dy\, dx = 8\pi$

CHAPTER 15

Infinite Series

This chapter returns to functions of one variable. It may be taken up at any time after Chapter 8. In addition, §15D on L'Hôpital's Rule is independent of the rest of the chapter.

§15A covers preliminary material on infinite sequences. §§15B, C and §§15E, F discuss series in general and the principal tests for convergence. The last four sections, §§15G, H, I, J, treat power series, including differentiation and integration, examples of standard series, and Taylor's Theorem.

§15A. Sequences

15A1. Definition of a sequence. We now take up a special sort of function called a "sequence." A *sequence* is a function whose domain is the set of natural numbers: 0, 1, 2, $\cdots$, or one of its subsets.

Functional values in a sequence are indicated by subscripts: the nth term is denoted a_n, for example. Technically, the letter a itself then represents the sequence—i.e., the function. However, we will not need to use it that way.

As with other functions, we may define a sequence by means of a formula. Often, the first few terms and a general term are listed in a row: $a_0, a_1, \cdots,$ $a_n, \cdots$. Other notations for a sequence are (a_n) (formal), or just a_n (informal, like speaking of "the function x^2.")

Example 1. The formula $a_n = n^2$ defines the sequence (n^2), that is, the sequence

$$0, 1, 4, \cdots, n^2, \cdots.$$

Example 2. The formula $a_n = n/(n^2 + 1)$ defines the sequence

$$0, \frac{1}{2}, \frac{2}{5}, \cdots, \frac{n}{n^2 + 1}, \cdots.$$

Various self-explanatory conventions are in use. For example, when we write $a_n = 1/n$, we understand that the domain is the set of *positive* integers.

15A2. Limit of a sequence; the symbol ∞. A sequence (a_n) is said to have the *limit s* and to *converge* to s:

$$(1) \qquad\qquad \lim a_n = s, \qquad a_n \to s,$$

if, given any (open) interval J about s, there is an integer N such that $a_n \in J$ whenever $n > N$. In other words, all terms of the sequence after the Nth are in J. A handy way to express this is that $a_n \in J$ *for all sufficiently large n.*

A sequence that has a limit is said to be *convergent*; otherwise, *divergent*.

Other notations for $\lim a_n = s$ are

(2) $$\lim_n a_n = s \quad \text{and} \quad \lim_{n \to \infty} a_n = s.$$

The expression $\lim_{n \to \infty} a_n$ is read, "the limit of a_n as n approaches infinity." The symbol ∞, read "infinity," has no meaning by itself but only as part of the larger expression—just as the middle bar in the letter "E" has no meaning by itself but only as part of the larger expression.

Example 3. Verify that

$$\lim \left(2 + \frac{1}{n} \right) = 2.$$

Solution. If J is any interval about 2, then J contains an interval J_0 of the form $J_0 = (2 - 1/N, 2 + 1/N)$. Then for $n > N$, $2 + 1/n \in J_0 \subset J$.

Convergence of a sequence is not affected by the addition or deletion of any (finite) number of terms at the beginning. The sequence $1, \frac{1}{2}, \frac{1}{3}, \cdots$ approaches 0 just as well as $\frac{1}{3}, \frac{1}{4}, \frac{1}{5}, \cdots$ does. This is essentially just a change of labeling. If $a_n \in J$ for all sufficiently large n then $a_{n-2} \in J$ for all sufficiently large $n - 2$.

15A3. Properties of limits. The properties of limits discussed in §2D and §2E for functions defined on intervals carry over directly to limits of sequences. The following results are particularly useful. There is no need to include the proofs.

If $a_n = s$ for every n, then $\lim a_n = s$.

If $\lim a_n = s$ and $\lim b_n = t$, then $\lim(a_n + b_n) = s + t$, $\lim(a_n - b_n) = s - t$, and $\lim(a_n b_n) = st$. If, in addition, $b_n \neq 0$ and $b \neq 0$, then $\lim a_n/b_n = a/b$. Moreover, if $a_n > 0$ and $s > 0$, then $\lim a_n^{b_n} = s^t$.

If $\lim |a_n| = 0$ then $\lim a_n = 0$. If $\lim a_n = s$, then $\lim |a_n| = |s|$.

If $\lim a_n = s$, and if f is a function continuous at s, then $\lim f(a_n) = f(s)$.

Example 4.

$$\lim \frac{2n^2 + 3n - 7}{3n^2 - n + 1} = \lim \frac{2 + 3/n - 7/n^2}{3 - 1/n + 1/n^2}$$

$$= \frac{\lim 2 + \lim 3/n - \lim 7/n^2}{\lim 3 - \lim 1/n + \lim 1/n^2}$$

$$= \frac{2 + 0 - 0}{3 - 0 + 0} = \frac{2}{3} ;$$

therefore

$$\lim \exp \left(\frac{2n^2 + 3n - 7}{3n^2 - n + 1} \right)^{2 - 1/n} = \exp \left(\frac{2}{3} \right)^2 .$$

15A4. Increasing, decreasing, and bounded sequences. The notions of increasing, decreasing, and bounded function apply in particular to sequences. Thus, (a_n) is increasing if $a_m < a_n$ whenever $m < n$, decreasing if $a_m > a_n$ whenever $m < n$, and bounded if there is a number M such that all $|a_n| \le M$.

The important results about increasing sequences hold as well for nondecreasing sequences—i.e., sequences in which $a_m \le a_n$ whenever $m < n$. Because "nondecreasing" is such a clumsy word, we will state results for increasing sequences, leaving the reader to observe for himself that they also hold for nondecreasing sequences.

(3)   *Every convergent sequence is bounded.*

Proof. If $a_n \to s$, there is a number N such that $a_n \in J = (s - 1, s + 1)$ for all $n > N$. For these n, $|a_n| < |s| + 1$. Hence if M is the largest of the $N + 2$ numbers $|a_0|, |a_1|, \cdots , |a_N|, |s| + 1$, then $|a_n| \le M$ for *every* n.

The following partial converse is basic.

15A5. Boundedness and convergence. *Every bounded, increasing sequence is convergent.*

Proof. Let (a_n) be bounded and increasing. The set of numbers a_n has an upper bound; by the Axiom of Completeness (§1H2), it has a least upper bound, say s. We show that $a_n \to s$. Let $J = (u, v)$ be an interval about s: $u < s < v$. Pick N so that $a_N > u$; such a term exists, since otherwise u would be an upper bound smaller than s, the smallest. Since the sequence is increasing and is bounded by s, $u < a_N < a_n \le s$ for all $n > N$. Thus, $a_n \in J$ for all $n > N$.

Similarly, *every bounded, decreasing sequence is convergent.*

Example 5. The sequence $(1 + 1/n)^n$ is increasing and is bounded by the number e. (See §4D, Problem 14, and §7D, Problem 17(b).) Therefore this sequence has a limit.

The latter reference shows that the number e is contained in the interval $((1 + 1/n)^n, (1 + 1/n)^{n+1})$, whose length is $(1 + 1/n)^n/n$ and hence $< e/n$. It follows easily that:

(4) $$\lim \left(1 + \frac{1}{n} \right)^n = e.$$

The same conclusion follows from the formula $\lim_{x \to 0} (1 + x)^{1/x} = e$ (§7E(11)).

Problems (§15A)

1. Write the first four terms of the sequence.

(a) $(n^2 - 1)$

(d) $(1 + (-1)^n)$

(b) $\left(1 - \dfrac{1}{2^n}\right)$

(e) $(\cos n\pi)$

(c) $\left(\dfrac{(-1)^{n+1}}{n+1}\right)$

(f) $(e^{\sin n\pi/2})$

2. Find the limit of the sequence defined by the given formula.

(a) $a_n = \dfrac{n^3 - n + 1}{n^3 + n - 1}$

(c) $a_n = \dfrac{n+1}{n^2 + 1}$

(b) $a_n = \dfrac{n^2 - 21n + 2}{2n^2}$

(d) $a_n = n - \dfrac{n^3 - 2n^2 + n - 1}{n^2 - 1}$

3. Find the limit of the sequence having the given nth term.

(a) $\cos \dfrac{1}{n}$

(d) $\dfrac{\sin(1/n)}{e^{1/n}}$

(b) $(2 - 1/n^2)^2$

(e) $\log \left(1 + \dfrac{1}{1+n}\right)$

(c) $10^{n/(2n+1)}$

(f) $\sqrt{(2n^2 + 1)/(n^2 - 5)}$

4. Determine whether the sequence with the given nth term converges or diverges.

(a) $\dfrac{n^2 + 1}{n}$

(f) $\dfrac{\sin n}{n}$

(b) $1 + \cos n\pi$

(g) $2^{1/n}$

(c) $1 + \sin n\pi$

(h) e^n

(d) $\dfrac{2^n}{1 \cdot 2 \cdots n}$

(i) $(3^{-n})^{1/n}$

(e) $1 - \dfrac{\log n}{n}$ [*Hint.* §15A5.]

(j) $\sqrt{n+2} - \sqrt{n-1}$
[*Hint.* Rationalize the numerator.]

5. Let $a_n > 0$, $b_n > 0$, $\lim a_n = s > 0$, $\lim b_n = t > 0$, and $\lim c_n = 0$. Evaluate the following.

(a) $\lim(2a_n + 3b_n)^{1/n}$

(b) $\lim(2a_n^{b_n} + 3^{c_n})$

(c) $\lim \left[\left(a_n + \dfrac{n}{n^2 + 1}\right)^2 + \left(b_n - \dfrac{n^2}{n^2 + 1}\right)^3 \right]$

(d) $\lim(e^{a_n} + \log b_n)$

6. Using the fact that $(1 + 1/n)^n \to e$ (4), find the limits of the following sequences.

(a) $\left(1 + \dfrac{1}{n}\right)^{2n}$

(c) $\left(1 + \dfrac{2}{n}\right)^{n}$

(b) $\left(1 + \dfrac{1}{2n}\right)^{n}$

(d) $\left(1 + \dfrac{1}{n^2}\right)^{n}$

7. Let (a_n) be a sequence such that $a_{2n} \to s$ and $a_{2n+1} \to s$. Show that $a_n \to s$.

8. (a) Prove that if $s > 1$, then s^n increases without limit. [*Hint.* If $s = 1 + h$, then $s^2 > 1 + 2h$.]

(b) Conclude that if $0 < r < 1$, then $r^n \to 0$.

Answers to problems (§15A)

1. (a) $-1, 0, 3, 8$ (b) $0, \frac{1}{2}, \frac{3}{4}, \frac{7}{8}$ (c) $-1, \frac{1}{2}, -\frac{1}{3}, \frac{1}{4}$
 (d) $2, 0, 2, 0$ (e) $1, -1, 1, -1$ (f) $1, e, 1, 1/e$

2. (a) 1 (b) $\frac{1}{2}$ (c) 0 (d) 2

3. (a) 1 (b) 4 (c) $\sqrt{10}$ (d) 0 (e) 0 (f) $\sqrt{2}$

4. (a) diverges   (b) diverges   (c) converges   (d) converges
 (e) converges   (f) converges   (g) converges   (h) diverges
 (i) converges   (j) converges

5. (a) 1 (b) $2s^t + 1$ (c) $s^2 + (t-1)^3$ (d) $e^s + \log t$

6. (a) e^2 (b) $\sqrt{e}$ (c) e^2 (d) 1

§15B. Series

15B1. The $\sum$-notation. An infinite sequence often arises in the form of an *infinite series*, representing "infinite addition." Everybody remembers the famous geometric series

(1)
$$1 + \frac{1}{2} + \frac{1}{4} + \cdots + \frac{1}{2^n} + \cdots,$$

whose "sum" is universally acknowledged to be 2—though not everyone may remember just why. The why is that we start with 1 (total so far: 1), then add $\frac{1}{2}$ (total so far: $1\frac{1}{2}$), then add $\frac{1}{4}$ (total so far: $1\frac{3}{4}$), and so on, creating the *sequence of partial sums*,

(2)
$$1, 1\frac{1}{2}, 1\frac{3}{4}, \cdots;$$

since this sequence converges to the limit 2 we say that the sum of the series is 2. This series will be discussed carefully in §15B3.

In general, an *infinite series* is an expression of the form

(3) $$u_0 + u_1 + \cdots + u_n + \cdots.$$

To write this compactly we use the symbol $\sum$ (sigma, for *sum*), as follows:

(4) $$\sum_0^\infty u_n = u_0 + u_1 + \cdots + u_n + \cdots$$

("the sum of u_n from 0 to ∞"). The subscript n is the "index of summation," corresponding to the variable of integration in an integral. For finite sums we write

(5) $$\sum_p^q u_n = u_p + u_{p+1} + \cdots + u_q \qquad \text{(where } p \leq q\text{)}.$$

Other self-explanatory variations are also used. In particular, the index of summation may be i, j, k, or any other appropriate letter.

Example 1.

(a) $\sum_0^\infty \dfrac{1}{2^n} = 1 + \dfrac{1}{2} + \dfrac{1}{4} + \cdots + \dfrac{1}{2^n} + \cdots.$

This is the series (1).

(b) $\sum_0^\infty \dfrac{1}{2^k} = 1 + \dfrac{1}{2} + \dfrac{1}{4} + \cdots + \dfrac{1}{2^k} + \cdots.$

This is the same as (a).

(c) $\sum_5^7 \sqrt{n} = \sqrt{5} + \sqrt{6} + \sqrt{7}.$

Example 2.

(a) $\sum_1^\infty \dfrac{1}{n^2} = 1 + \dfrac{1}{4} + \dfrac{1}{9} + \cdots + \dfrac{1}{n^2} + \cdots.$

(b) $\sum_0^\infty \dfrac{1}{(n+1)^2} = 1 + \dfrac{1}{4} + \dfrac{1}{9} + \cdots + \dfrac{1}{(n+1)^2} + \cdots.$

This is the same series as (a).

When limits are understood from context, they are often omitted. In particular,

(6) $$\sum u_n \quad \text{will usually mean} \quad \sum_0^\infty u_n.$$

Let

$$(7) \qquad \sum u_n = u_0 + u_1 + \cdots + u_n + \cdots$$

be an infinite series. For each index n, we define the *partial sum*:

$$(8) \qquad S_n = \sum_0^n u_k = u_0 + \cdots + u_n.$$

Example 3. In the geometric series $\sum 1/2^n$,

$$S_0 = 1, \qquad S_1 = 1 + \tfrac{1}{2}, \qquad S_2 = 1 + \tfrac{1}{2} + \tfrac{1}{4},$$

and so on.

15B2. Convergence. Consider an infinite series $\sum u_n$. If the sequence of partial sums, (S_n), converges to a limit, then we say that the series $\sum u_n$ *converges*, and we define its *sum* to be that limit. That is, by definition,

$$(9) \qquad \sum u_n = s \quad means \quad S_n \to s$$

(where S_n is given by (8)). In other words, $\sum u_n = s$ provided that, given any interval J about s, $S_n \in J$ for all sufficiently large n. Note that we are using $\sum u_n$ to stand both for the series (an expression) and its sum (a number).

If the sequence of partial sums diverges we say that the series diverges.

Evidently, convergence of a series is not affected by the addition or deletion of any (finite) number of terms at the beginning. Cf. §15A2. (The value of the sum is affected, but not the fact of convergence.) This leads to the following *necessary* condition for the convergence of a series.

$$(10) \qquad If \ \sum u_k \ \ converges \ \ then \ \ u_n \to 0.$$

Proof. If $\sum u_k = s$ then $S_n \to s$ and $S_{n-1} \to s$, so $u_n = S_n - S_{n-1} \to 0$. This result has the following corollary.

$$(11) \qquad If \ \sum u_n \ \ converges \ \ then \ its \ terms \ are \ bounded.$$

Proof. By (10), the sequence (u_n) converges (to 0); by 15A(3), it is bounded.

Example 4. The series

$$\sum \frac{n}{n + 10} = \frac{1}{11} + \frac{2}{12} + \frac{3}{13} + \cdots$$

diverges, since $\lim(n/(n + 10)) = 1 \neq 0$.

The converse to (10) is not true. It is possible that $u_n \to 0$ and yet $\sum u_n$ diverges.

Example 5. The series

$$1 + \tfrac{1}{2} + \tfrac{1}{2} + \tfrac{1}{3} + \tfrac{1}{3} + \tfrac{1}{3} + \cdots,$$

where for each k, there are k consecutive terms equal to $1/k$, obviously diverges, yet the nth term approaches 0.

If $\sum u_n$ and $\sum v_n$ are convergent series and c is a constant, then

$$(12) \qquad \sum (u_n + v_n) = \sum u_n + \sum v_n,$$

$$\sum (u_n - v_n) = \sum u_n - \sum v_n,$$

and

$$(13) \qquad \sum cu_n = c \sum u_n.$$

That is, in each case, the series on the left converges, and its sum is as stated. The proofs are elementary and are omitted. Cf. §15A3.

Finally, we make the following observation. Assume that $\sum u_k$ converges. For each n, we may consider the series

$$(14) \qquad \sum_{n+1}^{\infty} u_k = u_{n+1} + u_{n+2} + \cdots ;$$

this converges too, as it differs from the original series by a finite number of terms—in fact, by the partial sum S_n.

$$(15) \qquad \textit{If } \sum_{0}^{\infty} u_k \textit{ converges then, as } n \to \infty, \sum_{n+1}^{\infty} u_k \to 0.$$

For, if $\sum u_k = s$, then $\sum_{n+1}^{\infty} u_k = s - S_n \to 0$.

15B3. Geometric series. It is only rarely possible to obtain a useful formula for the nth partial sum of a series. The chief exceptions are *geometric series*; as a consequence, they play a significant role in the theory. By definition, a geometric series is a series of the form

$$(16) \qquad \sum ar^n = a + ar + \cdots + ar^n + \cdots.$$

The number r is called the *ratio*. If $r = 1$ the nth partial sum is $a + \cdots + a = (n + 1)a$. For $r \neq 1$ we use the identity

$$(17) \qquad (1 - r)(1 + r + \cdots + r^n) = 1 - r^{n+1}.$$

to obtain

(18)
$$a + ar + \cdots + ar^n = \frac{a}{1 - r}(1 - r^{n+1}).$$

We now see that *the series* (16) *converges for* $|r| < 1$ *and diverges for* $|r| \geq 1$ (*if* $a \neq 0$). For, if $|r| < 1$, then $r^{n+1} \to 0$. (If you are not sure of this, see §15A, Problem 8). Then $1 - r^{n+1} \to 1$ and hence, by (18),

(19)
$$\sum_0^\infty ar^n = \frac{a}{1 - r} \qquad (|r| < 1).$$

On the other hand, if $a \neq 0$ and $|r| \geq 1$, then $|ar^n| \geq |a| > 0$; then ar^n does not approach 0, so the series diverges.

Example 6.

(a) $\displaystyle\sum (2)\left(\frac{1}{3}\right)^n = 2 + \frac{2}{3} + \cdots + \frac{2}{3^n} + \cdots = \frac{2}{1 - 1/3} = 3.$

(b) $\displaystyle\sum (2)\left(-\frac{1}{3}\right)^n = 2 - \frac{2}{3} + \cdots (-1)^n \frac{2}{3^n} + \cdots = \frac{2}{1 + 1/3} = \frac{3}{2}.$

Problems (§15B)

1. Write out the first four terms of the following series.

Example: $\sum 1/n = 1 + \frac{1}{2} + \frac{1}{3} + \frac{1}{4} + \cdots.$

(a) $\displaystyle\sum \left(1 + \frac{n}{n^2 + 1}\right)$

(e) $\displaystyle\sum (2^{i+1} + 2^i)$

(b) $\displaystyle\sum \left[1 - \left(\frac{m}{m^2 + 1}\right)\right]$

(f) $\displaystyle\sum \frac{n + 1}{n + 2}$

(c) $\displaystyle\sum (-1)^n \frac{1}{n}$

(g) $\displaystyle\sum \frac{k^2 + 1}{k^2 + k + 1}$

(d) $\displaystyle\sum (-1)^k \frac{1}{k^2}$

(h) $\displaystyle\sum \sin(n\pi/2)$

2. In Problem 1, in each case, write the first four terms of the sequence of partial sums.

3. Evaluate:

(a) $\displaystyle\sum_{2}^{4} 2 \log n$ (c) $\displaystyle\sum_{2}^{6} (n + 2)$

(b) $\displaystyle\sum_{5}^{6} \frac{n}{n^2 + 1}$ (d) $\displaystyle\sum_{1}^{5} 2$

4. Show that the following series diverge.

(a) $\displaystyle\sum 2^{1/n}$

(b) $\displaystyle\sum (-1)^n$

(c) $\displaystyle\sum \frac{2^n}{n^3}$ [*Hint.* Show that the terms are eventually increasing.]

5. In the series $\frac{1}{3} - \frac{2}{5} + \frac{3}{7} - \frac{4}{9} + \cdots$, the signs alternate, the numerators are the positive integers $1, 2, \cdots$, and the denominators are the successive odd integers, starting with 3.

(a) Write the series in the $\sum$-notation.

(b) Show that the series diverges.

6. Determine whether or not $\sum_{2}^{\infty} \log(1 - 1/n)$ converges by first finding an explicit formula for the nth partial sum.

7. Find the sum of the following geometric series.

(a) $\displaystyle\sum 3(\tfrac{1}{5})^n$ (c) $\frac{1}{3} + \frac{2}{9} + \frac{4}{27} + \cdots$

(b) $\displaystyle\sum (-1)^{n+1}(\tfrac{2}{3})^n$

8. Express each repeating decimal as a quotient of two integers.

(a) $0.\overline{12}$ (i.e., $0.121212\cdots$)   (c) $12.1\overline{02}$

(b) $0.\overline{243}$ (d) $0.002\overline{001}$

9. The first term of a geometric series is 3, and the fourth is $\frac{81}{64}$. Find the sum of the series.

10. The second term of a geometric series is 9, and the fourth term is 1. The sum is not $\frac{81}{2}$. What is it?

11. A rubber ball rebounds to $\frac{2}{5}$ the height from which it falls. If it is dropped from 10 feet, and allowed to continue bouncing, how far does it travel before coming to rest?

12. Two cyclists, 30 miles apart, approach each other, each pedaling at 15 mph. A deer fly starts on the nose of one and flies back and forth between their noses at 50 mph. When the noses (and the fly) come together, how far has the fly flown?

The hard way to solve this problem is to express the total distance as a geometric series, and sum. There is also an easy way. Do it both ways.

13. Determine for what values of x the given series converges, and express the sum as a simple function of x.

(a) $\sum 5x^n$ (c) $\sum (\log x)^n$

(b) $\sum (\sin x)^n$

14. Let sgn $x = 1$ for $x > 0$, -1 for $x < 0$, and 0 for $x = 0$. For what values of x does $\sum (\text{sgn } x)^n$ converge, and to what?

15. Show that every infinite sequence is the sequence of partial sums of some series.

Answers to problems (§15B)

1. (a) $1 + \frac{3}{2} + \frac{7}{5} + \frac{13}{10} + \cdots$ (b) $1 + \frac{1}{2} + \frac{3}{5} + \frac{7}{10} + \cdots$
   (c) $-1 + \frac{1}{2} - \frac{1}{3} + \frac{1}{4} - \cdots$ (d) $-1 + \frac{1}{4} - \frac{1}{9} + \frac{1}{16} - \cdots$
   (e) $3 + 6 + 12 + 24 + \cdots$ (f) $\frac{1}{2} + \frac{2}{3} + \frac{3}{4} + \frac{4}{5} + \cdots$
   (g) $1 + \frac{2}{3} + \frac{5}{7} + \frac{10}{13} + \cdots$ (h) $0 + 1 + 0 - 1 + \cdots$

2. (a) $1, \frac{5}{2}, \frac{39}{10}, \frac{26}{5}$ (b) $1, \frac{3}{2}, \frac{21}{10}, \frac{14}{5}$ (c) $-1, -\frac{1}{2}, -\frac{5}{6}, -\frac{7}{12}$
   (d) $-1, -\frac{3}{4}, -\frac{31}{36}, -\frac{115}{144}$ (e) $3, 9, 21, 45$ (f) $\frac{1}{2}, \frac{7}{6}, \frac{23}{12}, \frac{163}{60}$
   (g) $1, \frac{5}{3}, \frac{50}{21}, \frac{860}{273}$ (h) $0, 1, 1, 0$

3. (a) $2 \log 24$ (b) $\frac{341}{962}$ (c) 30 (d) 10

5. (a) $\sum (-1)^{n+1} \dfrac{n}{2n + 1}$

6. $S_n = -\log n$; the series diverges

7. (a) $\frac{15}{4}$ (b) $\frac{3}{5}$ (c) 1

8. (a) $\frac{4}{33}$ (b) $\frac{9}{37}$ (c) $\dfrac{11,981}{990}$ (d) $\dfrac{1,999}{999,000}$

9. 12

10. $-\frac{81}{4}$

11. $\frac{70}{3}$

12. 50 miles

13. (a) converges for $|x| < 1$, sum $5/(1 - x)$
    (b) converges for $x \neq (k + \frac{1}{2})\pi$, sum $1/(1 - \sin x)$
    (c) converges for $1/e < x < e$, sum $1/(1 - \log x)$

14. converges for $x = 0$ only, sum 0

§15C. Positive series

To obtain results about series in general we first study a special type of series: those in which every term is positive. For brevity, we call them *positive series*.

Usually it will be enough to assume that a series $\sum u_n$ is *eventually* positive, i.e., that $u_n > 0$ for sufficiently large n. In many cases, "positive" can be replaced by "nonnegative." We leave it to the reader to make these observations for himself when they come up.

15C1. Boundedness and convergence.
(a) *If a series converges, its sequence of partial sums is bounded.*
(b) *Conversely, if the sequence of partial sums of a* positive *series is bounded, then the series converges.*

This is a rephrasing of §15A4 and §15A5.

The basic test for convergence of a positive series is the following comparison test.

15C2. The comparison test. Let $\sum u_n$ and $\sum v_n$ be positive series such that $u_n \leq v_n$ for sufficiently large n. Then if $\sum v_n$ converges, so does $\sum u_n$. (Equivalently, if $\sum u_n$ diverges, so does $\sum v_n$.)

Proof. If $\sum v_n$ converges, its sequence of partial sums is bounded $\big(\S15C1(a)\big)$; since $u_n \leq v_n$, the sequence of partial sums of $\sum u_n$ is also bounded; hence $\sum u_n$ converges $\big(\S15C1(b)\big)$.

Example 1. Show that $\sum 1/n^n$ converges.

Solution. For $n \geq 2$,

$$\frac{1}{n^n} \leq \frac{1}{2^n}.$$

Hence the series converges by comparison with the convergent geometric series $\sum 1/2^n$.

15C3. Factorial notation. The number $n!$ ("n factorial") is defined for $n \geq 0$ as follows:

$$0! = 1,$$
$$n! = 1 \times 2 \times \cdots \times n \qquad (n \geq 1).$$

Evidently, for all $n \geq 0$, $(n+1)! = (n+1)n!$.

The number $n!$ increases very rapidly with n:

$0! = 1,$	$1! = 1,$	$6! =$	$720,$
	$2! = 2,$	$7! =$	$5,040,$
	$3! = 6,$	$8! =$	$40,320,$
	$4! = 24,$	$9! =$	$362,880,$
	$5! = 120,$	$10! =$	$3,628,800.$

Example 2. Show that $\sum 1/n!$ converges.

Solution. For $n \geq 2$,

$$n! = 2 \times \cdots \times n \geq 2 \times \cdots \times 2 = 2^{n-1}.$$

Hence $1/n! \leq 1/2^{n-1}$, and the series converges by comparison with a geometric series.

15C4. *The harmonic series.* This is the series

(1)
$$\sum \frac{1}{n} = 1 + \frac{1}{2} + \cdots + \frac{1}{n} + \cdots$$

(so named because the harmonics of a musical tone have wavelengths 1/2, 1/3, $\cdots$ of the fundamental tone). It is a perhaps surprising but easily established fact that the harmonic series *diverges*. To see this we compare the harmonic series:

(2)
$$\sum \frac{1}{n} = 1 + \frac{1}{2} + \left(\frac{1}{3} + \frac{1}{4}\right) + \left(\frac{1}{5} + \frac{1}{6} + \frac{1}{7} + \frac{1}{8}\right) + \left(\frac{1}{9} + \cdots + \frac{1}{16}\right) + \cdots,$$

with the series

(3)
$$1 + \frac{1}{2} + \left(\frac{1}{4} + \frac{1}{4}\right) + \left(\frac{1}{8} + \frac{1}{8} + \frac{1}{8} + \frac{1}{8}\right) + \left(\frac{1}{16} + \cdots + \frac{1}{16}\right) + \cdots,$$

which it dominates. The parentheses are used here to help the eye. Now, the series (3) diverges, as clearly there exist partial sums as large as we please. (The sum of the first 2^{2n-2} terms is n). Hence the harmonic series diverges.

Suppose we modify the harmonic series by squaring each term: $\sum 1/n^2$. This series *converges*. To see this, we compare it:

(4)
$$\sum \frac{1}{n^2} = 1 + \left(\frac{1}{2^2} + \frac{1}{3^2}\right) + \left(\frac{1}{4^2} + \frac{1}{5^2} + \frac{1}{6^2} + \frac{1}{7^2}\right) + \cdots$$

with the series

(5)
$$1 + \left(\frac{1}{2^2} + \frac{1}{2^2}\right) + \left(\frac{1}{4^2} + \frac{1}{4^2} + \frac{1}{4^2} + \frac{1}{4^2}\right) + \cdots$$

In (5), the expressions in parentheses are

$$2 \times \frac{1}{2^2} = \frac{1}{2}, \qquad 4 \times \frac{1}{4^2} = \frac{1}{4}, \qquad \cdots.$$

Hence the partial sums in (5) are all less than $1 + \frac{1}{2} + \frac{1}{4} + \cdots = 2$. (In fact, the series (5) converges to 2.) Since (5) dominates (4), (4) also converges.

Example 3. Decide whether $\sum 1/\sqrt{n}$ converges.

Solution. $1/\sqrt{n} \geq 1/n$, so the series diverges by comparison with the harmonic series.

Example 4. Decide whether $\sum (n + 1)/n^3$ converges.

Solution.

$$\frac{n + 1}{n^3} \leq \frac{n + n}{n^3} = \frac{2}{n^2},$$

and $\sum 2/n^2$ converges, since $\sum 1/n^2$ does. Hence the given series converges, by the comparison test.

These and related series will be analyzed by other methods in §15E3 below.

15C5. Infinite limits. When a sequence (a_n) of positive terms is unbounded, we often find it convenient to say that $a_n \to \infty$, or $\lim a_n = \infty$. This is an extension of the notion of limit, originally required to be a real number. When using this terminology, we will attempt to state results in such a way that misinterpretation is either not possible or not harmful.

Probably the most commonly used test for convergence is the so-called "ratio test," which we now discuss.

15C6. The ratio test. *Let $\sum u_n$ be a positive series, and let*

(6)
$$l = \lim \frac{u_{n+1}}{u_n}$$

(assuming this limit exists—∞ permitted).

> *If $l < 1$, the series converges.*
>
> *If $l > 1$ (or if $l = \infty$), the series diverges.*

Proof. If $l < 1$, pick t satisfying $l < t < 1$. There is an integer N such that $u_{n+1}/u_n < t$ for $n \geq N$. Then $u_{N+1} < tu_N$, $u_{N+2} < tu_{N+1} < t^2u_N$, and, in general, $u_{N+n} < t^n u_N$. Hence from the Nth term on, $\sum u_n$ is dominated by the convergent geometric series $u_N \sum t^n$. By the comparison test, $\sum u_n$ converges.

If $l > 1$ then $u_{n+1}/u_n > 1$ for sufficiently large n; then u_n increases with n, hence does not approach 0, and so the series diverges.

Example 5. Decide whether the series

$$\sum \frac{n^2 + 7n}{n^2 \cdot n!}$$

converges.

Solution. The test ratio is

$$\frac{u_{n+1}}{u_n} = \frac{(n + 1)^2 + 7(n + 1)}{(n + 1)^2 \cdot (n + 1)!} \frac{n^2 \cdot n!}{n^2 + 7n}.$$

Since

$$\frac{n!}{(n + 1)!} = \frac{1}{n + 1},$$

we have

$$\frac{u_{n+1}}{u_n} = \frac{1 + 7/(n + 1)}{1 + 7/n} \frac{1}{n + 1} \to 0 < 1.$$

Therefore the series converges.

Example 6. Decide whether the series $\sum 2^n/n^2$ converges.

Solution. The test ratio is

$$\frac{2^{n+1}}{(n + 1)^2} \frac{n^2}{2^n} = 2 \left(\frac{n}{n + 1} \right)^2 \to 2 > 1.$$

Hence the series diverges.

Remark. If $l = 1$, the test gives no information.

Example 7. For the harmonic series, $\sum 1/n$, the test ratio is $n/(n + 1) \to 1$. For the series $\sum 1/n^2$, the test ratio is $n^2/(n + 1)^2 \to 1$. Hence for both series $l = 1$. As we saw in §15C4, the first diverges, the second converges.

Remark. In order for the ratio test to apply, the ratio u_{n+1}/u_n must have a limit. In the series $1 + 2 + 6 + 12 + \cdots$, in which the ratios alternate between 2 and 3, u_{n+1}/u_n has no limit; this series diverges. In the series $1 + \frac{1}{2} + \frac{1}{6} + \frac{1}{12} + \cdots$, in which the ratios alternate between $\frac{1}{2}$ and $\frac{1}{3}$, u_{n+1}/u_n again has no limit; but this series clearly converges. (For example, it is dominated by the geometric series $\sum 1/2^n$.)

Problems (§15C)

1. Express in terms of factorials and powers of 2:
   (a) $2 \cdot 4 \cdots (2n)$
   (b) $1 \cdot 3 \cdots (2n - 1)$
   (c) $n(n - 1) \cdots (n - r + 1)$

2. Decide whether the following series converge or diverge.

(a) $\sum \dfrac{n+2}{n^2+1}$

(b) $\sum \dfrac{n^3}{3^n}$

(c) $\sum \dfrac{\log n}{n^3}$

(d) $\sum \left(\dfrac{5^n}{n!} - \dfrac{1}{n^3}\right)$

(e) $\sum \dfrac{n}{2^n}$

(f) $\sum n(\tfrac{3}{4})^n$

(g) $\sum \dfrac{n^3}{(\log 2)^n}$

(h) $\sum \dfrac{n^3}{(\log 3)^n}$

(i) $\sum \dfrac{2^n}{10^n}$

(j) $\sum \dfrac{\log n}{n}$

(k) $\sum \dfrac{n^2+n^3}{2^n}$

(l) $\sum n^{-1/3}$

(m) $\sum \dfrac{n^n}{n!}$ [*Hint.* §15A(4).]

(n) $\sum \dfrac{3^n}{(n^2+5n+6)2^n}$

(o) $\sum \left(\dfrac{1}{\sqrt{n}} - \dfrac{1}{n}\right)$

(p) $\sum \dfrac{1}{\sqrt{n^2+1}}$

(q) $\sum \dfrac{1}{1+\log n}$

(r) $\sum \dfrac{1}{1+e^n}$

3. For what values of x does the given series converge:

(a) $\sum \dfrac{x^n}{n^n}$?

(b) $\sum \dfrac{x^n}{n!}$?

4. Prove that if $\sum a_n$ is a convergent positive series, then so is $\sum a_n^2$.

5. Prove that if $a_n \geq 0$ for all n, and $\sum a_n^2$ converges, then $\sum a_n/n$ converges. [*Hint.* Compare with $\sum (a_n^2 + 1/n^2)$.]

6. Establish the **ratio-comparison test**:

Let $\sum u_n$ and $\sum v_n$ be positive series. If the sequence u_n/v_n is bounded (in particular, if it has a limit), and if $\sum v_n$ converges, then $\sum u_n$ converges.

Hence if both u_n/v_n and v_n/u_n are bounded, then $\sum u_n$ and $\sum v_n$ either both converge or both diverge. In particular, this is the case if u_n/v_n has a limit $\neq 0$.

Answers to problems (§15C)

1. (a) $2^n n!$ (b) $(2n)!/2^n n!$ (c) $n!/(n-r)!$

2. (C denotes converge, D diverge.)

(a) D   (b) C   (c) C   (d) C   (e) C   (f) C   (g) D   (h) C

(i) C   (j) D   (k) C   (l) D   (m) D   (n) D   (o) D   (p) D

(q) D   (r) C

3. (a) all x (b) all x

§15D. L'Hôpital's Rule

In the ratio test and other tests for convergence of a series, we are often called upon to evaluate the limit of a quotient where both numerator and denominator approach zero or both become infinite. There is a particularly handy procedure for doing this, known as *L'Hôpital's Rule*. Before stating it, we make some remarks about limits and absolute values.

15D1. Absolute value. We will need to feel comfortable with absolute values, so we begin by reviewing their properties. By definition, $|a|$ is a or $-a$, whichever is ≥ 0. Therefore $|a| = |-a| \geq 0$. Other fundamental inequalities are

(1) $$a \leq |a|, \qquad -a \leq |a|,$$

and

(2) $$|a + b| \leq |a| + |b|.$$

Statement (1) should be obvious: if $a \geq 0$ then $-a \leq 0 \leq a = |a|$, while if $a < 0$ then $a < 0 < -a = |a|$. To obtain (2), we apply (1) to conclude that $a \leq |a|$ and $b \leq |b|$, from which $a + b \leq |a| + |b|$. Similarly, $(-a) + (-b) \leq |a| + |b|$. Since $|a + b|$ is one or the other: $a + b$ or $-(a + b)$, (2) follows.

Of course, (2) extends to any finite number of summands.

Next, the following statements all say the same thing:

(3) $\qquad |y - b| < \epsilon, \qquad b - \epsilon < y < b + \epsilon, \qquad y \in (b - \epsilon, b + \epsilon).$

15D2. Limits; the symbol ∞. Recall that $\lim\limits_{x \to a} f(x) = b$ provided that, given any interval J about b, there is an interval I cut at a that f takes into J (§2B3). We may assume that J is centered at b; so we have $\lim\limits_{x \to a} f(x) = b$ provided that, given any number $\epsilon > 0$, there is an interval I cut at a such that $|f(x) - b| < \epsilon$ for all $x \in I$. Otherwise expressed: $\lim\limits_{x \to a} f(x) = b$ provided that, given any number $\epsilon > 0$, $|f(x) - b| < \epsilon$ *for all x sufficiently near a.*

In the work to follow we need an extension of the definition to limits at infinity and to infinite limits. The first context is the analogue of the limit of a sequence. We define

(4) $$\lim_{x \to \infty} f(x) = b$$

to mean that given any interval J about b, there is a number N such that $f(x) \in J$ whenever $x > N$. In terms of absolute values: given $\epsilon > 0$, there is a number N such that $|f(x) - b| < \epsilon$ for all $x > N$. Otherwise expressed: given $\epsilon > 0$, $|f(x) - b| < \epsilon$ *for all sufficiently large x.*

Example 1. Show that $\lim\limits_{x \to \infty} (\sin x)/x = 0$.

Solution. Let $\epsilon > 0$ be given. Since $|\sin x| \leq 1$, we simply choose $N > 1/\epsilon$. Then for $x > N$, $|(\sin x)/x| \leq 1/x < 1/N < \epsilon$.

Secondly, we define

$$(5) \qquad\qquad \lim_{x \to a} f(x) = \infty$$

to mean that given any number $M > 0$, there is an interval I cut at a such that $f(x) > M$ for $x \in I$.

Example 2. Show that $\lim\limits_{x \to 0} 1/x^2 = \infty$.

Solution. Given $M > 0$, let I be the interval $(-\sqrt{M}, \sqrt{M})$ minus the point 0. Then if $x \in I$ we have $x^2 < 1/M$, so $1/x^2 > M$.

Next, the two definitions can be combined. We say that

$$(6) \qquad\qquad \lim_{x \to \infty} f(x) = \infty$$

provided that, given any number $M > 0$, there is a number N such that $f(x) > M$ whenever $x > N$.

Finally, we use the symbol $-\infty$ as follows: $u \to -\infty$ means that $-u \to \infty$. The reader should have no trouble formulating the variants of (4), (5), and (6) that make use of this symbol (there are five in all).

Keep in mind that in each case, the symbol ∞ has no meaning by itself, but only as part of the larger expression.

Observe that if $f(x) \to \infty$ or $f(x) \to -\infty$ then $1/f(x) \to 0$. Conversely, if $1/f(x) \to 0$ *and if f does not change sign*, then either $f(x) \to \infty$ or $f(x) \to -\infty$.

One-sided limits are denoted, when emphasis is required, by

$$\lim_{x \to a+} f(x) \quad \text{and} \quad \lim_{x \to a-} f(x) ,$$

the first denoting approach from the right (i.e., through values $> a$), the second denoting approach from the left. When working with infinite limits, be on the watch for different limits from opposite sides.

Example 3.

$$\lim_{x \to 0+} \frac{1}{x} = \infty, \qquad \lim_{x \to 0-} \frac{1}{x} = -\infty.$$

15D3. *L'Hôpital's Rule.* *Let f and g be differentiable on (a, b), with g' never zero. If $\lim_{x \to a} f(x) = \lim_{x \to a} g(x) = 0$ and if $\lim_{x \to a} f'(x)/g'(x)$ exists, then*

(7)
$$\lim_{x \to a} \frac{f(x)}{g(x)} = \lim_{x \to a} \frac{f'(x)}{g'(x)}.$$

Proof. It is sufficient to show that for each $x \in (a, b)$ there is a number $z \in (a, x)$ for which

(8)
$$\frac{f(x)}{g(x)} = \frac{f'(z)}{g'(z)}.$$

For then when $x \to a$, $z \to a$ and $f'(z)/g'(z)$ approaches the given limit. So, then, does $f(x)/g(x)$.

We extend f and g by defining $f(a) = g(a) = 0$. Then f and g are continuous on $[a, b)$. Notice that $g(x) \neq 0$ for $x > a$; otherwise, by Rolle's Theorem (§4E2), g' would have a zero somewhere in (a, x), contrary to hypothesis.

Now, given $x \in (a, b)$, define

(9)
$$h(s) = f(s) - \frac{f(x)}{g(x)} g(s) \qquad (s \in [a, x]).$$

Then $h(a) = h(x) = 0$. By Rolle's Theorem, $h'(z) = 0$ for some $z \in (a, x)$; that is,

$$0 = f'(z) - \frac{f(x)}{g(x)} g'(z).$$

This gives (8.)

In the above statement, $x \to a$ from the right. The corresponding result holds for approach from the left. Hence the result holds in addition for approach from both sides.

The theorem is also true for $x \to \infty$ (or $x \to -\infty$). For then $t = 1/x \to 0$ and we have, by the original rule,

(10)
$$\lim_{x \to \infty} \frac{f(x)}{g(x)} = \lim_{t \to 0} \frac{f(1/t)}{g(1/t)} = \lim_{t \to 0} \frac{f'(x)\, dx/dt}{g'(x)\, dx/dt}.$$

In other words,

$$(11) \qquad \lim_{x \to \infty} \frac{f(x)}{g(x)} = \lim_{x \to \infty} \frac{f'(x)}{g'(x)}.$$

Finally, in all these statements, $\lim f'(x)/g'(x)$ may be ∞ or $-\infty$.

Example 4. Find $\lim\limits_{x \to 0} (\sin x - \tan x)/x$.

Solution. Here, numerator and denominator both approach 0. Hence L'Hôpital's Rule applies, and we have

$$\lim_{x \to 0} \frac{\sin x - \tan x}{x} = \lim_{x \to 0} \frac{\cos x - \sec^2 x}{1} = 1 - 1 = 0.$$

Example 5. Find $\lim\limits_{x \to 0} (e^x - 1)/x$.

Solution. Since numerator and denominator both approach 0, L'Hôpital's Rule applies. We have

$$\lim_{x \to 0} \frac{e^x - 1}{x} = \lim_{x \to 0} \frac{e^x}{1} = 1.$$

We might have noticed that the original limit $= \exp'(0) = 1$.

As these examples suggest, the advantage of L'Hôpital's Rule is to provide a routine that requires little thought.

Example 6. Find $\lim\limits_{x \to 0} (\sin x)/(x + 1)$.

Solution. If we are careless we may write

$$\lim_{x \to 0} \frac{\sin x}{x + 1} = \lim_{x \to 0} \frac{\cos x}{1} = 1. \qquad \textbf{X}$$

The correct solution is

$$\lim_{x \to 0} \frac{\sin x}{x + 1} = \frac{0}{1} = 0.$$

In this example the numerator and denominator do not both approach 0, and L'Hôpital's Rule is inapplicable.

Sometimes we apply the rule successively.

Example 7.

$$\lim_{x \to 1} \frac{x^3 + 3e^{1-x} - 4}{x - \log x - 1} = \lim_{x \to 1} \frac{3(x^2 - e^{1-x})}{1 - 1/x} = \lim_{x \to 1} \frac{3(2x + e^{1-x})}{1/x^2} = 9.$$

Functions not presented as fractions can often be transformed into fractions so that the rule will apply.

Example 8. Find $\lim\limits_{x \to 0} (\csc x - 1/x)$.

Solution. Here each term becomes infinite, so that the limit of their difference is in doubt. But we have

$$\lim_{x \to 0} \left(\csc x - \frac{1}{x} \right) = \lim_{x \to 0} \frac{x - \sin x}{x \sin x}$$

$$= \lim_{x \to 0} \frac{1 - \cos x}{x \cos x + \sin x} \qquad \text{(by L'Hôpital's Rule)}$$

$$= \lim_{x \to 0} \frac{\sin x}{-x \sin x + 2 \cos x} \qquad \text{(by L'Hôpital's Rule)}$$

$$= 0.$$

Example 9. Find $\lim\limits_{x \to 0} x/(2e^x - e^{2x} - 1)$.

Solution. By L'Hôpital's Rule,

$$\lim_{x \to 0} \frac{x}{2e^x - e^{2x} - 1} = \lim_{x \to 0} \frac{1}{2e^x(1 - e^x)}$$

—provided the latter limit exists. In fact, it does not. But, by L'Hôpital's Rule,

$$\lim_{x \to 0+} \frac{x}{2e^x - e^{2x} - 1} = \lim_{x \to 0+} \frac{1}{2e^x(1 - e^x)} = -\infty$$

and

$$\lim_{x \to 0-} \frac{x}{2e^x - e^{2x} - 1} = \lim_{x \to 0-} \frac{1}{2e^x(1 - e^x)} = +\infty.$$

15D4. L'Hôpital's Rule (continued). There is a corresponding theorem for the case in which f and g both become infinite.

Let f and g be differentiable on (a, b), with g' never zero. If $\lim\limits_{x \to a} f(x) = \infty$ or $-\infty$ and $\lim\limits_{x \to a} g(x) = \infty$ or $-\infty$, and if $\lim\limits_{x \to a} f'(x)/g'(x)$ exists, then

(12)
$$\lim_{x \to a} \frac{f(x)}{g(x)} = \lim_{x \to a} \frac{f'(x)}{g'(x)}.$$

Proof. This proof is considerably more difficult than the other.

Since g' is never 0, g is monotonic on (a, b) (§4A4). Hence g' does not change sign. For definiteness, say $g' > 0$.

Case 1. $f'(x)/g'(x) \to 0$. Note. This part of the proof does not use the hypothesis that the numerator becomes infinite.

Given any positive number ϵ, we are to show that there is an interval (a, b_0) throughout which $|f/g| < \epsilon$. We begin with the following simple observation.

(a)   *If $G - F$ and $G + F$ are both increasing on an interval $(\alpha, \beta]$, then there is a constant C such that*

$$ |F(x)| < |G(x)| + C \qquad\qquad (\alpha < x < \beta). \tag{13}$$

Proof of (a). Since $G - F$ is increasing,

$$ G(x) - F(x) < G(\beta) - F(\beta) \qquad\qquad (\alpha < x < \beta). \tag{14}$$

Hence if $C = |G(\beta)| + |F(\beta)|$ then $-F(x) < |G(x)| + C$. Similarly, $F(x) < |G(x)| + C$. This gives us (13), as required.

Now, since $f'/g' \to 0$, there is an interval $(a, \beta]$ on which $|f'|/g' < \epsilon/2$. Then $2|f'| < \epsilon g'$ on $(a, \beta]$; that is, $2f' < \epsilon g'$ and $-2f' < \epsilon g'$. Consequently, $\epsilon g - 2f$ and $\epsilon g + 2f$ are both increasing on $(a, \beta]$. By (a), there is a constant C such that

$$ 2|f(x)| < \epsilon|g(x)| + C \qquad\qquad (a < x < \beta). \tag{15}$$

Since $|g(x)| \to \infty$, we can pick $b_0 < \beta$ so that $\epsilon|g(x)| > C$ for all $x \in (a, b_0)$. Then

$$ 2|f(x)| < 2\epsilon|g(x)| \qquad\qquad (a < x < b_0), \tag{16}$$

and therefore $|f(x)/g(x)| < \epsilon$.

Case 2. $f'(x)/g'(x) \to L$, any real number. Define

$$ h(x) = f(x) - Lg(x). \tag{17}$$

Then $h'/g' = f'/g' - L \to 0$. By Case 1 (see *Note*), $h/g \to 0$; that is, $f/g - L \to 0$, as required.

Case 3. $\lim f'(x)/g'(x) = \infty$ *or* $-\infty$. Here $g'/f' \to 0$, so by Case 1, $g/f \to 0$.

Say $f'/g' \to +\infty$. Then f' and g' have the same sign near a. Then f and g both approach ∞ or both approach $-\infty$. So f and g have the same sign near a. Since $g/f \to 0$, we get $f/g \to +\infty$.

Again, the results hold for approach from the left or from both sides, as well as for $x \to \infty$ or $x \to -\infty$.

Our principal application of L'Hôpital's Rule will be to $\lim f(x)/g(x)$ as $x \to \infty$.

Example 10.

(a) $\lim\limits_{x \to \infty} e^x/x = \lim\limits_{x \to \infty} e^x/1 = \infty$:

$$ \lim_{x \to \infty} \frac{e^x}{x} = \infty. \tag{18}$$

(b) $\lim_{x \to \infty} e^x/x^n = \lim_{x \to \infty} e^x/nx^{n-1} = \cdots = \lim_{x \to \infty} e^x/n! = \infty$

(where n is a positive integer).

Example 11. $\lim_{x \to \infty} (\log x)/x = \lim_{x \to \infty} (1/x)/1 = 0$:

(19)
$$\lim_{x \to 0} \frac{\log x}{x} = 0.$$

To handle an expression of the form $\lim f(x)^{g(x)}$, we begin by taking logarithms.

Example 12. Show that $\lim_{x \to \infty} (\log x)^{1/x} = 1$.

Solution. Write $y = (\log x)^{1/x}$ $(x \geq 1)$. Then $\log y = (1/x) \log \log x$ and

$$\lim_{x \to \infty} \log y = \lim_{x \to \infty} \frac{\log \log x}{x} = \lim_{x \to \infty} \frac{(1/(\log x))(1/x)}{1} = 0.$$

Since the exponential is continuous,

$$\lim_{x \to \infty} y = \lim_{x \to \infty} \exp \log y = \exp \lim_{x \to \infty} \log y = e^0 = 1.$$

Problems (§15D)

1. Find the following limits.

(a) $\lim_{x \to 0} \dfrac{e^x - 1}{x^2}$

(b) $\lim_{x \to \infty} \dfrac{1}{x} \sqrt{1 + x^2}$

(c) $\lim_{x \to 0+} \dfrac{\log \sin^2 x}{\log x}$

(d) $\lim_{x \to \pi/2} \dfrac{\sin x}{x}$

(e) $\lim_{x \to 0} (x \log x)$

(f) $\lim_{x \to 0} x^x$

(g) $\lim_{x \to 0+} x^{1/\log x}$

(h) $\lim_{x \to (\pi/2)-} \sec x \, e^{-\tan x}$

(i) $\lim_{x \to (\pi/2)+} \sec x \, e^{-\tan x}$

(j) $\lim_{x \to 0+} x^{(x^2)}$

(k) $\lim_{x \to 0} \dfrac{x^3 - x^2 + 1}{3x^3 + x + 2}$

(l) $\lim_{x \to \infty} \dfrac{x^3 - x^2 + 1}{3x^3 + x + 2}$

(m) $\lim_{x \to 0} \left(\dfrac{1}{x} + \dfrac{1}{1 - e^{-x}} \right)$

(n) $\lim_{x \to 1} x^{1/(1-x)}$

(o) $\lim_{x \to 0} (\cos x)^{1/x}$

(p) $\lim_{x \to \infty} \dfrac{\log x}{\sqrt{x}}$

(q) $\lim_{x \to 0} \dfrac{(\sec^2 x - \cos x)^2}{x \cos x - \sin x}$

(r) $\lim_{x \to \infty} (x + e^{\alpha x})^{\beta/x}$

2. Prove that

(20)
$$\lim_{x\to 0} (1 + \alpha x)^{1/x} = e^{\alpha}.$$

Answers to problems (§15D)

1. (a) ∞ (b) 1   (c) 2   (d) $2/\pi$ (e) 0   (f) 1   (g) e (h) 0

(i) $-\infty$ (j) 1   (k) $\frac{1}{2}$ (l) $\frac{1}{3}$ (m) $\lim_{x\to 0+} = +\infty$, $\lim_{x\to 0-} = -\infty$

(n) $1/e$ (o) 1   (p) 0   (q) 0   (r) $e^{\alpha\beta}$

§15E. Improper integrals and the integral test

In this section we develop a test for convergence of a series by comparing it with the limit of an integral. First we look at the integral.

15E1. Improper integrals. We have defined $\int_a^b f(x)\,dx$ in case f is continuous on the *bounded* interval $[a, b]$. In case f is continuous on the *unbounded* interval $x \geq a$, then $\int_a^t f(x)\,dx$ is defined for every $t \geq a$ and we are invited to inquire whether this integral has a finite limit as $t \to \infty$. If so, we abbreviate the notation to $\int_a^\infty f(x)\,dx$:

(1)
$$\int_a^\infty f(x)\,dx = \lim_{t\to\infty} \int_a^t f(x)\,dx;$$

and we say that $\int_a^\infty f(x)\,dx$ *converges*. Otherwise, we say it diverges. In either case, $\int_a^\infty f$ is called an *improper* integral.

In case f is a positive function and the set of numbers $\int_a^t f$ for $t \geq a$ is bounded, then $\int_a^\infty f$ does converge: its value is simply the least upper bound of that set. The proof is the analogue of the one for series (§15A5).

Example 1.
$$\int_1^\infty \frac{1}{x^2}\,dx = \lim_{t\to\infty} \int_1^t \frac{1}{x^2}\,dx = \lim_{t\to\infty} \left(1 - \frac{1}{t}\right) = 1.$$

Example 2.
$$\int_0^\infty \frac{1}{1 + x^2}\,dx = \lim_{t\to\infty} \int_0^t \frac{1}{1 + x^2}\,dx = \lim_{t\to\infty} \arctan t = \pi/2.$$

Example 3.
$$\int_0^\infty xe^{-x^2}\,dx = \lim_{t\to\infty} \int_0^t xe^{-x^2}\,dx = \lim_{t\to\infty} \tfrac{1}{2}(1 - e^{-t^2}) = \tfrac{1}{2}.$$

Example 4.

$$\int_0^\infty x \, dx = \lim_{t \to \infty} \int_0^t x \, dx = \lim_{t \to \infty} \tfrac{1}{2}t^2 = \infty.$$

This integral diverges, because the limit is infinite.

Example 5.

$$\int_0^\infty \sin x \, dx = \lim_{t \to \infty} \int_0^t \sin x \, dx = \lim_{t \to \infty} (1 - \cos t)$$

—which does not exist. This integral diverges, because the limit does not exist.

15E2. The integral test. *If f is a positive, decreasing, continuous function on $x \geq 1$, then the improper integral $\int_1^\infty f$ and the infinite series $\sum f(n)$ either both converge or both diverge.*

Proof. Since we are dealing with positive numbers only, convergence is just a matter of whether the partial sums and integrals are bounded. Comparing areas in Figure 1 shows that

$$f(1) \geq \int_1^2 f(x) \, dx \geq f(2),$$

(2)

$$f(2) \geq \int_2^3 f(x) \, dx \geq f(3),$$

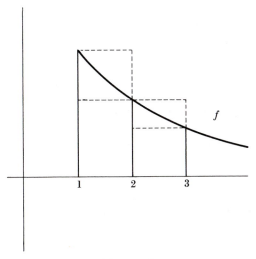

FIGURE 1

and so on. Therefore, for each m,

(3)
$$\sum_1^m f(n) \geq \int_1^{m+1} f(x)\, dx \geq \sum_2^{m+1} f(n).$$

It is clear from this that if the series is bounded then so is the integral, and conversely. Hence if the series converges then so does the integral, and conversely.

Of course the result holds for any interval $x \geq N$, not just $x \geq 1$.

In practice, we are given a positive series $\sum u_n$, and we look for an f to match: $f(n) = u_n$. Often when u_n is given by a formula, we can define $f(x)$ by the same formula.

Example 6. Let us use the integral test to show that the harmonic series diverges. We have $u_n = 1/n$. Define $f(x) = 1/x$. Then f is positive, decreasing, and continuous on $x \geq 1$, and $f(n) = u_n$. The integral test applies and we have

$$\int_1^\infty \frac{1}{x}\, dx = \lim_{t \to \infty} \log t = \infty.$$

Since this diverges, so does the series.

Example 7. Test the series $\sum_2^\infty 1/n(\log n)^2$ for convergence.

Solution. The function $f(x) = 1/x(\log x)^2$ is positive, decreasing, and continuous on $x \geq 2$, and $f(n) = 1/n(\log n)^2$. The integral test applies and we have

$$\int_2^\infty \frac{1}{x(\log x)^2}\, dx = \lim_{t \to \infty} \left(\frac{1}{\log 2} - \frac{1}{\log t} \right) = \frac{1}{\log 2}.$$

Since the integral converges, so does the series.

15E3. The *p*-series. The harmonic series is a particular instance ($p = 1$) of the so-called *p-series*, namely, those series of the form

$$\sum \frac{1}{n^p} = 1 + \frac{1}{2^p} + \frac{1}{3^p} + \cdots,$$

where p is a positive real number. The ratio test fails to give information about convergence; in each case, $u_{n+1}/u_n \to 1$. With the help of the integral test, however, we obtain the following result.

(4)   *The p-series converges if* $p > 1$ *and diverges if* $p \leq 1$.

Proof. We define $f(x) = 1/x^p$. Then f is positive, decreasing, and continuous on $x \geq 1$, and $f(n) = 1/n^p$. The integral test applies, and we have, for $p \neq 1$,

$$\int_1^\infty \frac{1}{x^p}\, dx = \lim_{t \to \infty} \frac{1}{1-p}(t^{1-p} - 1),$$

which is equal to $1/(p-1)$ if $p > 1$, and ∞ if $p < 1$. We already know the result for $p = 1$. Hence the series converges as stated.

The p-series provide us with a new stock of series to use in comparison tests.

Problems (§15E)

1. Evaluate the following improper integrals.

(a) $\displaystyle\int_2^\infty \frac{1}{4+x^2}\, dx$

(b) $\displaystyle\int_1^\infty \frac{1}{x^2}\, dx$

(c) $\displaystyle\int_1^\infty \frac{1}{x\sqrt{x}}\, dx$

(d) $\displaystyle\int_1^\infty \frac{1}{x}\, dx$

(e) $\displaystyle\int_1^\infty \frac{1}{\sqrt{x}}\, dx$

(f) $\displaystyle\int_0^\infty \frac{1}{(1+x)^3}\, dx$

(g) $\displaystyle\int_0^\infty \frac{x}{1+x^2}\, dx$

(h) $\displaystyle\int_3^\infty \frac{x}{(1+x^2)^2}\, dx$

(i) $\displaystyle\int_0^\infty \frac{1}{\sqrt{2x+1}}\, dx$

(j) $\displaystyle\int_0^\infty \frac{1}{2x+1}\, dx$

(k) $\displaystyle\int_0^\infty \frac{1}{(2x+1)^2}\, dx$

(l) $\displaystyle\int_2^\infty \frac{1}{x^2-1}\, dx$

(m) $\displaystyle\int_1^\infty \left(\frac{1}{\sqrt{x}} - \frac{1}{\sqrt{x+1}}\right) dx$

(n) $\displaystyle\int_1^\infty \frac{1}{x^2}\sin\frac{1}{x}\, dx$

(o) $\displaystyle\int_0^\infty e^{-x}\, dx$

(p) $\displaystyle\int_0^\infty e^{-2x}\, dx$

(q) $\displaystyle\int_0^\infty \frac{1}{\sqrt{e^x}}\, dx$

(r) $\displaystyle\int_0^\infty xe^{-x}\, dx$

(s) $\displaystyle\int_0^\infty \frac{1}{e^x+e^{-x}}\, dx$

(t) $\displaystyle\int_0^\infty e^{-(x+e^{-x})}\, dx$

2. Test for convergence or divergence.

(a) $\sum \dfrac{1}{n \log n}$ (e) $\sum \dfrac{1}{n(\log n)^2}$

(b) $\sum \dfrac{\log n}{n}$ (f) $\sum \dfrac{\sqrt{n}}{n^2 - 2}$

(c) $\sum \dfrac{\log n}{n^2}$ (g) $\sum \dfrac{1}{(\log n)^3}$

(d) $\sum \dfrac{1}{(n - 2)^2}$ (h) $\sum \dfrac{1}{(\log n)^{\log n}}$

[*Hint.* $(\log n)^{\log n} = n^{\log (\log n)}$.]

3. Using the ratio-comparison test (§15C, Problem 6), or otherwise, show that $\sum n^{-(1 + 1/n)}$ diverges. Notice that the terms of this series eventually lie between those of the harmonic series and any convergent p-series ($p > 1$).

4. To each series $\sum u_n$ we associate the "condensed" series

(5) $$\sum 2^n u_{2^n} = u_0 + 2u_2 + 4u_4 + 8u_8 + \cdots.$$

For each series in Problem 2, write the first three terms and the general term of the condensed series.

5. Establish the **condensation test**:

A positive decreasing series $\sum u_n$ converges or diverges accordingly as the condensed series

$$\sum 2^n u_{2^n}$$

converges or diverges. [*Hint.* Write the condensed series as $u_0 + u_2 + u_2 + u_4 + u_4 + u_4 + u_4 + \cdots$, and show that this lies between $\sum_1^\infty u_n$ and $u_0 + 2\sum_1^\infty u_n$.]

Example 8. Use the condensation test to show that $\sum 1/n^3$ converges.

Solution. The condensed series is

$$\sum \dfrac{2^n}{(2^n)^3} = \sum \dfrac{1}{(2^n)^2} = \sum \dfrac{1}{4^n},$$

a convergent geometric series.

6. In Problem 2, in each case, test for convergence or divergence by using the condensation test (Problem 5).

Answers to problems (§15E)

1. (a) $\pi/8$ (b) 1   (c) 2   (d) ∞ (e) ∞ (f) $\frac{1}{2}$ (g) ∞ (h) $\frac{1}{20}$
(i) ∞ (j) ∞ (k) $\frac{1}{2}$ (l) $\frac{1}{2}\log 3$ (m) $2(\sqrt{2} - 1)$ (n) $1 - \cos 1$
(o) 1   (p) $\frac{1}{2}$ (q) 2   (r) 1   (s) $\pi/4$ (t) $1 - 1/e$

2. (C denotes converges; D, diverges)
(a) D   (b) D   (c) C   (d) C   (e) C   (f) C   (g) D   (h) C

§15F. Absolute convergence; alternating series

We now take up series with mixed signs. We gain information from the corresponding series of absolute values, which is a positive series and therefore easier to analyze.

15F1. Absolute convergence. A series $\sum u_n$ is said to *converge absolutely* if the series of absolute values,

$$(1) \qquad \sum |u_n| = |u_0| + |u_1| + \cdots,$$

converges. In this definition it is not assumed that $\sum u_n$ converges. But we expect that it does; convergence seems easier in a series with mixed signs, with their cancelling effects, than when all terms are pulling in the same direction.

15F2. Convergence. *If a series converges absolutely, it converges.*

Proof. Let $\sum u_n$ converge absolutely, i.e., $\sum |u_n|$ converges. Now, for each n, $0 \le u_n + |u_n| \le 2|u_n|$ (§15D(1)). Therefore $\sum (u_n + |u_n|)$ converges, by the comparison test. Since $\sum (u_n + |u_n|)$ and $\sum |u_n|$ both converge, so does their difference, $\sum u_n$.

Example 1.

$$\sum_{1}^{\infty} (-1)^{n+1} \frac{1}{n^2} = 1 - \frac{1}{4} + \frac{1}{9} - \cdots$$

converges, because

$$\sum_{1}^{\infty} \frac{1}{n^2} = 1 + \frac{1}{4} + \frac{1}{9} + \cdots$$

converges (§15C4, §15E3).

15F3. The ratio test. If we combine the result just proved with the ratio test of §15C6, we get the following rule.

Let $\sum u_n$ be an infinite series, and let

$$(2) \qquad l = \lim \left| \frac{u_{n+1}}{u_n} \right|$$

(assuming this limit exists—∞ permitted).

If $l < 1$, the series converges (absolutely).
If $l > 1$ (or if $l = \infty$), the series diverges.

Proof. If $l < 1$ then $\sum |u_n|$ converges (§15C6), i.e., $\sum u_n$ converges absolutely. If $l > 1$ then $|u_{n+1}/u_n| > 1$ for sufficiently large n, so that $|u_n|$ increases with n; then u_n does not approach 0, and so $\sum u_n$ diverges.

Example 2. The series

$$\sum (-1)^n \frac{n+1}{2^n} = 1 - \frac{2}{2} + \frac{3}{4} - \frac{4}{8} + \frac{5}{16} - \cdots$$

converges, as we have

$$\frac{(n+2)/2^{n+1}}{(n+1)/2^n} = \frac{1}{2}\frac{n+2}{n+1} \to \frac{1}{2} < 1.$$

15F4. Alternating series. We now turn to series with mixed signs. The most important of these are the *alternating* series, in which the terms are alternately positive and negative.

For an alternating series in which the terms decrease in absolute value, there is a very simple and convenient test for convergence.

Let

(3) $a_0 - a_1 + a_2 - \cdots$ $(a_n > 0)$

be a given alternating series. If (a_n) is a decreasing sequence, and $a_n \to 0$, then the series (3) converges. Moreover,

(4) $S_1 < S_3 < S_5 < \cdots < s < \cdots < S_4 < S_2 < S_0,$

where s is the sum and S_n is the nth partial sum. Hence

(5) $|S_n - s| < a_{n+1}.$

Proof. Since $S_1 + (a_2 - a_3) = S_3$ we have $S_1 < S_3$. Similarly, $S_3 < S_5$, etc. Likewise, $S_0 > S_2 > \cdots$. Next, $S_1 = S_0 - a_1$; hence $S_1 < S_0$, and, similarly, $S_3 < S_2$, etc. Finally, $S_p < S_q$ for *any* odd p and even q, as the following examples make clear: $S_7 < S_2$ because $S_7 < S_6 < S_2$; and $S_3 < S_8$ because $S_3 < S_9 < S_8$.

Since the sequence (S_{2n+1}) is bounded (e.g., by S_0) and increasing, it converges, say to s (§15A5). Of course, each $S_{2n+1} < s$. Since $S_{2n} - S_{2n+1} = a_{2n+1} \to 0$, the decreasing sequence (S_{2n}) also converges to s; and each $S_{2n} > s$.

Example 3. The alternating series

$$\sum (-1)^n \frac{1}{(n+1)!} = 1 - \frac{1}{2!} + \frac{1}{3!} - \cdots$$

converges, since $1/(n+1)!$ decreases to 0; and the sum lies between

$$S_3 = 1 - \tfrac{1}{2} + \tfrac{1}{6} - \tfrac{1}{24} = 0.625 \quad \text{and} \quad S_4 = S_3 + \tfrac{1}{120} \approx 0.633;$$

hence $s = 0.63$, correct to two places.

In applying the theorem, both parts of the hypothesis must be verified; a common error is to check only that $a_n \to 0$.

Example 4. In the alternating series

$$2 - 1 + \tfrac{2}{2} - \tfrac{1}{2} + \tfrac{2}{3} - \tfrac{1}{3} + \cdots,$$

the nth term approaches 0—but the terms are not decreasing. In fact, S_{2n} is equal to the nth partial sum of the harmonic series (§15C4); hence (S_{2n}) is unbounded, and the series diverges.

15F5. Conditional convergence. A convergent series that is not absolutely convergent is said to be *conditionally* convergent.

Example 5. The alternating series

$$\sum (-1)^n \frac{1}{n} = 1 - \frac{1}{2} + \frac{1}{3} - \frac{1}{4} + \cdots$$

converges, since obviously $1/n$ is decreasing to 0. But this series is not absolutely convergent, as the corresponding series of absolute values is the harmonic series (§15C4). Here, then, is an example of a conditionally convergent series.

Problems (§15F)

1. Classify the following series as absolutely convergent, conditionally convergent, or divergent.

(a) $\sum (-1)^{n+1} \dfrac{1}{\sqrt{n+1}}$

(b) $\sum (-2)^{n+1} \dfrac{1}{(2n+1)!}$

(c) $\sum (-1)^{n+1} \dfrac{n}{n+1}$

(d) $\sum (-1)^{n+1} \dfrac{1}{2^{1/n}}$

(e) $\sum (-1)^{n+1} \dfrac{1}{\log n}$

(f) $\sum (-1)^{n+1} \dfrac{2^n}{n!}$

(g) $\sum (-1)^{n+1} \dfrac{n^2}{n^4 - 1}$

(h) $\sum (-1)^{n+1} \dfrac{\log n}{n}$

(i) $\sum (-1)^{n+1} \dfrac{\log n}{\sqrt{n}}$

(j) $\sum (-1)^{n+1} n^2 (0.9)^n$

(k) $\sum (-1)^{n+1} \dfrac{\sqrt{n}}{n+1}$

(l) $1 - \dfrac{1}{2} + \dfrac{1}{3!} - \dfrac{1}{4} + \dfrac{1}{5!} - \dfrac{1}{6} + \cdots$

(m) $1 - \tfrac{1}{2} + \tfrac{1}{2} - \tfrac{1}{3} + \tfrac{1}{3} - \tfrac{1}{4} +$

(n) $1 - \dfrac{1}{2^2} - \dfrac{1}{3^2} + \dfrac{1}{4^2} + \dfrac{1}{5^2} + \dfrac{1}{6^2} - \cdots$

(1 plus, 2 minus, 3 plus, 4 minus, $\cdots$)

2. Find the values of x for which the series converges.

(a) $\sum (-1)^{n+1} \dfrac{x^{4n}}{n}$

(c) $\sum \dfrac{\cos nx}{n^2}$

(b) $\sum \dfrac{x^{2n+1}}{2n+1}$

3. Find the largest open interval I such that the series

$$\sum \frac{n!}{n^n} x^n$$

converges for all $x \in I$. [*Hint.* §15A(4).]

4. Exhibit a convergent series $\sum a_n$ for which $\sum a_n^2$ diverges. [*Hint.* Compare §15C, Problem 4.]

Answers to problems (§15F)

1. (abs = absolutely convergent, cond = conditionally convergent, div = divergent)

(a) cond   (b) abs   (c) div   (d) div   (e) cond   (f) abs
(g) abs   (h) cond   (i) cond   (j) abs   (k) cond   (l) div
(m) cond   (n) abs

2. (a) $-1 \le x \le 1$ (b) $-1 < x < 1$ (c) all x

3. $-e < x < e$

§15G. Power series

15G1. Power series. We have been discussing series whose terms are constants. Now we turn to series whose terms are variables. The most important series of this type are the *power series*, i.e., the series of the form

(1) $$\sum a_n x^n = a_0 + a_1 x + a_2 x^2 + \cdots + a_n x^n + \cdots.$$

These are the ones we will study.

Consider, for example, the particular power series

(2) $$\sum x^n = 1 + x + x^2 + \cdots + x^n + \cdots.$$

This is a geometric series with ratio x. When $|x| < 1$ it converges; when $|x| \ge 1$ it diverges. Thus, we will expect a power series to converge for some values of x and to diverge for others. Note that every power series converges for $x = 0$.

Clearly, we will be interested in knowing for which x a given power series converges. At any such x, the sum of the series is a number—whose value depends on x. We might call the value $f(x)$. We thus define a function: $f(x) = \sum a_n x^n$ (wherever the series converges). For instance, for $|x| < 1$, the series (2) defines the function $1/(1 - x)$.

Then we can ask questions about $f(x)$. Is it continuous? Is it differentiable?

Moreover, we can look at the whole thing the other way around. *Given a function, f, does there exist a power series that converges to it?* If so, then the power series may furnish new information about the original function.

Every *polynomial* is a power series (for which only a finite number of co-efficients are different from zero). What about, say, the exponential function —can we represent *it* as a power series?

These are questions we will look into.

We begin with the following fundamental result about power series.

15G2. Radius of convergence. *With each power series $\sum a_n x^n$ is associ-ated its radius of convergence, r, with the following property. The series:*

(a) *converges absolutely for $|x| < r$, and*

(b) *diverges for $|x| > r$.*

Comments. Either r is a number ≥ 0, or $r = \infty$. The extreme cases are (i) $r = 0$ (the series converges at $x = 0$ only, (a) being true by default), and (ii) $r = \infty$ (the series converges absolutely for every x, (b) being true by default).

Proof of theorem. Since we are dealing with absolute convergence we may simplify the notation by considering $x \geq 0$ only.

First we prove:

(3)   *Let $y > 0$. If $\sum a_n y^n$ converges, then $\sum a_n x^n$ converges absolutely for $0 \leq x < y$.*

Proof of (3). Since $\sum a_n y^n$ converges, its terms are bounded $\big(\S15\text{B}(11)\big)$; say

$$|a_n y^n| \leq M \qquad\qquad (n = 1, 2, \cdots).$$

Then, given $0 \leq x < y$, we have

$$|a_n x^n| = |a_n y^n| \left(\frac{x}{y}\right)^n \leq M \left(\frac{x}{y}\right)^n.$$

Hence $\sum |a_n x^n|$ converges, by comparison with the geometric series $\sum M(x/y)^n$.

For emphasis, we also state (3) the other way around:

(4)   *Let $x \geq 0$. If $\sum a_n x^n$ does not converge absolutely, then $\sum a_n y^n$ diverges for all $y > x$.*

We now define r, and prove that it satisfies (a) and (b).

Let S denote the set of all $x \geq 0$ for which $\sum a_n x^n$ converges absolutely. Notice that $0 \in S$. In case S is the set of *all* $x \geq 0$ we define $r = \infty$ and we are done—that is, (a) and (b) are true.

Assume, then, that there is a point x at which $\sum a_n x^n$ does not converge absolutely. By (4), $\sum a_n y^n$ diverges for all $y > x$. Thus, x is an upper bound of S. By the Axiom of Completeness (§1H2), S has a *least* upper bound, which we call r. Evidently, $r \geq 0$.

(a) *If $0 \leq x < r$ then $\sum a_n x^n$ converges absolutely.* For otherwise, by (4), $\sum a_n y^n$ diverges for all $y > x$, so that x is an upper bound of S; but this contradicts the fact that r is the *least* upper bound.

(b) *If $y > r$ then $\sum a_n y^n$ diverges.* For otherwise, by (3), $\sum a_n x^n$ converges absolutely (i.e., $x \in S$) for $0 \leq x < y$, and hence for $r < x < y$; but this contradicts the fact that r is an upper bound of S.

15G3. Interval of convergence. The *interval of convergence* of a power series is the set of all x for which the series converges. If the radius of convergence is a number $r > 0$, then, as we have just seen, the interval of convergence includes the interval $(-r, r)$. It may include one or both endpoints as well; to find out, we test each separately. If $r = \infty$, the interval of convergence is $\mathscr{R}$; if $r = 0$, it is $\{0\}$.

15G4. The ratio test. Given a power series, $\sum a_n x^n$, how do we find its interval of convergence?

The first step is to find r, the radius of convergence. A simple method that works in most standard situations is the ratio test:

(5)
$$r = \lim \left| \frac{a_n}{a_{n+1}} \right|$$

(assuming this limit exists—∞ permitted). For, by the ratio test of §15C6, the series converges or diverges according as

(6) $$\lim \left| \frac{a_{n+1} x^{n+1}}{a_n x^n} \right| \quad \text{is} \quad < 1 \quad \text{or} \quad > 1,$$

that is to say, according as

(7) $$|x| < \lim |a_n/a_{n+1}| \quad \text{or} \quad |x| > \lim |a_n/a_{n+1}|.$$

Hence (5) holds (§15G2).

The second step is to test the endpoints.

Example 1. Find the interval of convergence of the series

$$\sum \frac{n+1}{2^n} x^n = 1 + \frac{2}{2} x + \frac{3}{2^2} x^2 + \cdots .$$

Solution.

$$\left| \frac{a_n}{a_{n+1}} \right| = \frac{(n+1)/2^n}{(n+2)/2^{n+1}} = 2 \frac{n+1}{n+2} \to 2;$$

so $r = 2$. In this example, the series diverges at both ends of the interval: at $x = 2$ it is the series $1 + 2 + 3 + \cdots$, and at $x = -2$ it is the series $1 - 2 + 3 - \cdots$. Therefore the interval of convergence is $(-2, 2)$.

Example 2. Find the interval of convergence of the series

$$\sum \frac{x^n}{n} = x + \frac{x^2}{2} + \frac{x^3}{3} + \cdots .$$

Solution. $|a_n/a_{n+1}| = (n+1)/n \to 1$, so $r = 1$. For $x = r = 1$ the series is $1 + \frac{1}{2} + \frac{1}{3} + \cdots$, the harmonic series, which diverges (§15C4, §15E3). For $x = -r = -1$ we get the alternating series $1 - \frac{1}{2} + \frac{1}{3} + \cdots$, which converges (§15F4). Therefore the interval of convergence is $[-1, 1)$.

Example 3. Find the interval of convergence of the series

$$\sum \frac{x^n}{n^2} = x + \frac{x^2}{4} + \frac{x^3}{9} + \cdots .$$

Solution. $|a_n/a_{n+1}| = (n+1)^2/n^2 \to 1$, so $r = 1$. At $x = 1$ the series is $1 + \frac{1}{4} + \frac{1}{9} + \cdots$, which converges (§15C4, §15E3). For $x = -1$ we get the alternating series $-1 + \frac{1}{4} - \frac{1}{9} + \cdots$, which also converges. Therefore the interval of convergence is $[-1, 1]$.

Example 4. The series

$$\sum \frac{x^n}{n!} = 1 + x + \frac{x^2}{2!} + \cdots$$

converges everywhere; its interval of convergence is $\mathscr{R}$. The series

$$\sum n! \, x^n = 1 + x + 2! \, x^2 + \cdots$$

converges at $x = 0$ only; its interval of convergence is $\{0\}$.

Two power series $\sum a_n x^n$ and $\sum b_n x^n$ both converge within the smaller of their two intervals of convergence; hence their term-by-term sum and difference also converge there:

(8)

$$\sum a_n x^n + \sum b_n x^n = \sum (a_n + b_n) x^n,$$

$$\sum a_n x^n - \sum b_n x^n = \sum (a_n - b_n) x^n$$

(see §15B2). Furthermore, if c is a constant and k is a positive integer then $\sum a_n x^n$ and $\sum c a_n x^{n+k}$ have the same interval of convergence, and

(9)
$$\sum c a_n x^{n+k} = c x^k \sum a_n x^n.$$

15G5. Uniform convergence. Let $\sum a_n x^n$ be a power series, and let x_0 be a point at which it converges. For each n we may consider the series

(10)
$$\sum_{n+1}^{\infty} a_k x_0^k = a_{n+1} x_0^{n+1} + a_{n+2} x_0^{n+2} + \cdots.$$

As we know, $\sum_{n+1}^{\infty} a_k x_0^k \to 0$ as $n \to \infty$ (§15B(15)). Thus, given $\epsilon > 0$,

(11)
$$\left| \sum_{n+1}^{\infty} a_k x_0^k \right| < \epsilon \quad \text{for sufficiently large } n.$$

According to the next theorem, we can say much more—namely, that on any closed interval within the interval of convergence, (11) holds *for all x simultaneously.*

Let $\sum a_n x^n$ have radius of convergence r. Consider any number s, $0 \le s < r$, and let $\epsilon > 0$ be given. Then for sufficiently large n,

(12)
$$\left| \sum_{n+1}^{\infty} a_k x^k \right| < \epsilon \quad \text{for all} \quad x \in [-s, s].$$

Proof. For $x \in [-s, s]$,

(13)
$$\left| \sum_{n+1}^{\infty} a_k x^k \right| \le \sum_{n+1}^{\infty} |a_k x^k| \le \sum_{n+1}^{\infty} |a_k s^k|.$$

Since $\sum a_k s^k$ converges absolutely (§15G2), $\sum_{n+1}^{\infty} |a_k s^k| < \epsilon$ for sufficiently large n (§15B(15)). Therefore $|\sum_{n+1}^{\infty} a_k x^k| < \epsilon$ for such n.

This result may be expressed by saying that $\sum a_n x^n$ converges *uniformly* on $[-s, s]$.

Problems (§15G)

1. Find the interval of convergence of the following power series.

(a) $\sum x^n$

(j) $\sum \dfrac{x^n}{(2n)!}$

(b) $\sum \left(\dfrac{x}{2}\right)^n$

(k) $\sum \dfrac{2^n x^n}{n!}$

(c) $\sum (ax)^n$

(l) $\sum \dfrac{(2n)!}{n!} x^n$

(d) $\sum (-1)^n x^{2n}$

(m) $\sum \dfrac{2^n}{n^3} x^n$

(e) $\sum \dfrac{x^n}{n^2 + 1}$

(n) $\sum \dfrac{3^n}{n \cdot 5^n} x^n$

(f) $\sum \dfrac{(-1)^n}{2n + 1} x^{2n+1}$

(o) $\sum \dfrac{\log n}{n} x^n$

(g) $\sum \dfrac{(2x)^n}{n^2}$

(p) $\sum \dfrac{(-1)^{n+1} x^n}{n \log n}$

(h) $\sum \dfrac{x^n}{n \cdot 2^n}$

(q) $\sum \dfrac{x^n}{\log n}$

(i) $\sum n^2 x^n$

(r) $\sum \dfrac{(-1)^n x^n}{n(\log n)^2}$

2. (a) Establish the **root test**:
   Let $\sum u_n$ be a positive series, and let $l = \lim u_n^{1/n}$. If $l < 1$ the series converges; if $l > 1$ or $l = \infty$ the series diverges.
   (b) Give examples to show that when $l = 1$ the series may converge or diverge.

3. Use the root test (Problem 2) to establish the root test for power series: If $\lim |a_n|^{1/n} = L$, or ∞, or 0, then the radius of convergence of $\sum a_n x^n$ is $1/L$, or 0, or ∞ (resp.).

4. Using the root test (Problems 2 and 3), or otherwise, find the radius of convergence of the following series.

(a) $\sum 2^n x^n$

(d) $\sum \dfrac{x^n}{(\log n)^n}$

(b) $\sum \left(\dfrac{x}{n}\right)^n$

(e) $\sum (1 + \cos n\pi)^n \dfrac{x^n}{2^n}$

(c) $\sum n^n x^n$

(f) $\sum \dfrac{n^{n^2} x^n}{(n + 1)^{n^2}}$ [*Hint.* §15A(4).]

1. (a) $(-1, 1)$ (b) $(-2, 2)$ (c) $(-1/a, 1/a)$ (d) $(-1, 1)$
   (e) $[-1, 1]$ (f) $[-1, 1]$ (g) $[-\frac{1}{2}, \frac{1}{2}]$ (h) $[-2, 2)$ (i) $(-1, 1)$
   (j) $\mathcal{R}$ (k) $\mathcal{R}$ (l) $\{0\}$ (m) $[-\frac{1}{2}, \frac{1}{2}]$ (n) $[-\frac{5}{3}, \frac{5}{3})$ (o) $[-1, 1)$
   (p) $(-1, 1]$ (q) $[-1, 1)$ (r) $[-1, 1]$
4. (a) $\frac{1}{2}$ (b) ∞ (c) 0 (d) ∞ (e) 1 (f) e

§15H. Power series as functions

In this section we look at functions defined by power series. Let $f(x) = \sum a_n x^n$. By this we mean that the series $\sum a_n x^n$ converges to the function $f(x)$ (within some interval). For each index n, we define

$$(1) \qquad S_n(x) = \sum_0^n a_k x^k, \qquad R_n(x) = f(x) - S_n(x);$$

$S_n(x)$ is the partial sum, $R_n(x)$ the *remainder*. Evidently,

$$(2) \qquad R_n(x) = \sum_{n+1}^{\infty} a_k x^k.$$

15H1. Continuity of power series. *A power series $f(x) = \sum a_n x^n$ is continuous within its interval of convergence.*

Proof. Let $x_0 \in (-r, r)$, where r is the radius of convergence. Given $\epsilon > 0$, we are to find an interval I about x_0 such that

$$(3) \qquad |f(x) - f(x_0)| < \epsilon \quad \text{for} \quad x \in I.$$

Pick s such that $|x_0| < s < r$. By uniform convergence (§15G5), there is an integer n such that

$$(4) \qquad |R_n(x)| < \epsilon/3 \quad \text{for all} \quad x \in [-s, s].$$

Next, the polynomial $S_n(x)$ is continuous, and so there is an open interval I, $x_0 \in I \subset [-s, s]$, such that

$$(5) \qquad |S_n(x) - S_n(x_0)| < \epsilon/3 \qquad (x \in I).$$

Then for $x \in I$,

$$(6) \qquad \begin{aligned} |f(x) - f(x_0)| &= |S_n(x) + R_n(x) - S_n(x_0) - R_n(x_0)| \\ &\le |S_n(x) - S_n(x_0)| + |R_n(x)| + |R_n(x_0)| \\ &< \epsilon/3 + \epsilon/3 + \epsilon/3 = \epsilon. \end{aligned}$$

Next we will prove that f is not just continuous but differentiable. First we need the following lemma.

15H2. Series with the same radius of convergence. *If (c_n) is a sequence of positive numbers for which $c_{n+1}/c_n \to 1$, then the two series*

$$\text{(7)} \qquad \sum a_n x^n \quad \text{and} \quad \sum c_n a_n x^n$$

have the same radius of convergence.

Proof. Note that in case a_{n+1}/a_n has a limit, the proof is immediate: $c_{n+1}a_{n+1}/c_n a_n$ has the same limit, and hence the radii are equal. In those series we will encounter, this argument will apply.

The general proof goes as follows. Let r and s be the radii of convergence of $\sum a_n x^n$ and $\sum c_n a_n x^n$. First we show that $r \le s$.

To do this, we show that whenever x satisfies $0 \le x < r$, then $x \le s$. (For then if $s < r$ we could pick a number x, $s < x < r$, a contradiction.) Let $0 \le x < r$, then, and pick y so that $x < y < r$. Since $\sum a_n y^n$ converges, its terms are bounded $\big(\S15B(11)\big)$; say

$$\text{(8)} \qquad\qquad |a_n y^n| \le M \qquad\qquad (n = 1, 2, \cdots).$$

Then

$$\text{(9)} \qquad |c_n a_n x^n| = c_n |a_n y^n| \left(\frac{x}{y}\right)^n \le c_n M \left(\frac{x}{y}\right)^n.$$

Now, $\sum c_n M(x/y)^n$ converges, by the ratio test:

$$\text{(10)} \qquad\qquad \frac{c_{n+1}}{c_n}\frac{x}{y} \to \frac{x}{y} < 1.$$

Therefore $\sum c_n a_n x^n$ converges. Consequently, $x \le s$.

As noted, we conclude that $r \le s$.

Now, since $c_{n+1}/c_n \to 1$, so also $(1/c_{n+1})/(1/c_n) \to 1$. Hence we can apply the above argument to show that $s \le r$.

Hence, finally, $s = r$.

15H3. Differentiation and integration of power series. We learned a long time ago that the derivative or integral of the sum of two functions is the sum of the derivatives or integrals. The rule holds as well for three, four, or any *finite* number of functions. This fact tells us nothing about the case of *infinitely* many functions.

We shall prove that if the infinitely many functions happen to be the terms in a power series, then the rules hold, within the interval of convergence.

Observe first that the two series

$$(11) \qquad \sum a_n x^n \quad \text{and} \quad \sum a_n x^{n+1}$$

converge for the same values of x. For if $\sum a_n x^n$ converges for a particular x, then $\sum a_n x^{n+1} = \sum x a_n x^n$ converges for that same x (§15B(13)). Conversely, if $\sum a_n x^{n+1}$ converges for any particular $x \neq 0$, then $\sum a_n x^n = \sum (1/x) a_n x^{n+1}$ converges for that same x (and for $x = 0$ it converges in any case).

We now prove the following theorem.

Let $f(x) = \sum_0^\infty a_n x^n$, with radius of convergence r. Then

$$(12) \qquad f'(x) = \sum_1^\infty n a_n x^{n-1} \qquad (|x| < r)$$

and

$$(13) \qquad \int_0^x f = \sum_0^\infty \frac{1}{n+1} a_n x^{n+1} \qquad (|x| < r).$$

Moreover, the radii of convergence of the series (12) *and* (13) *are also equal to r.*

Briefly: A power series may be differentiated and integrated term by term within its interval of convergence.

Proof. First we show that all three series have radius r. By the remark preceding the theorem, the series (12) and (13) have the same radii of convergence as the series

$$(14) \qquad \sum_1^\infty n a_n x^n \quad \text{and} \quad \sum_0^\infty \frac{1}{n+1} a_n x^n,$$

respectively. And by §15H2, each of these latter has the same radius of convergence as $\sum a_n x^n$ (namely, r).

As we know, $f(x)$ is continuous on $(-r, r)$ (§15H1). Since $S_n(x)$, as a polynomial, is also continuous, so is $R_n(x) = f(x) - S_n(x)$. Hence all these functions have integrals, and, for $x \in (-r, r)$,

$$(15) \qquad \int_0^x f = \int_0^x S_n + \int_0^x R_n.$$

We know how to integrate the polynomial $S_n(x) = \sum_0^n a_k x^k$, and we get

(16)
$$\int_0^x f = \sum_0^n \frac{1}{k+1} a_k x^{k+1} + \int_0^x R_n.$$

Hence we will prove (13) if we can show that for fixed $x \in (-r, r)$, $\int_0^x R_n \to 0$ as $n \to \infty$. To emphasize that x is fixed, we denote it by x_0. As the result is true (trivially) for $x_0 = 0$, let us assume, say, $x_0 > 0$. Let $\epsilon > 0$ be given. By uniform convergence (§15G5), for sufficiently large n,

(17)
$$|R_n(t)| < \frac{\epsilon}{x_0} \quad \text{for all} \quad t \in [-x_0, x_0].$$

Hence

(18)
$$\left| \int_0^{x_0} R_n \right| \le \int_0^{x_0} |R_n| \le \int_0^{x_0} \frac{\epsilon}{x_0} = \epsilon.$$

As noted, this proves (13).

Applying (13) to the function $g(x) = \sum_1^\infty n a_n x^{n-1}$ we have

(19)
$$\int_0^x g = \sum_1^\infty a_n x^n = f(x) - a_0.$$

On now differentiating, we get $f' = g$, which is (12).

§15I. Some special series

15I1. The geometric series. We know one series that defines a function we can express in closed form—namely, the geometric series. For $|x| < 1$ we have

(1)
$$1 + x + x^2 + \cdots = \frac{1}{1-x} \qquad (|x| < 1),$$

or, alternatively,

(2)
$$1 - x + x^2 - \cdots = \frac{1}{1+x} \qquad (|x| < 1).$$

By differentiating or integrating, as explained in §15H3, we can use these formulas to obtain others.

15I2. The logarithm. Integrating term by term in (2), we get

$$(3) \qquad x - \frac{x^2}{2} + \frac{x^3}{3} - \cdots = \log(1 + x) \qquad (|x| < 1).$$

(Using (1) instead would give an equivalent expression.) We can now compute values of $\log t$ for $0 < t < 2$—i.e., values of $\log(1 + x)$ for $-1 < x < 1$. In practice, one does not compute from this series, but from others that converge more rapidly.

The computation is also valid for $x = 1$, but to see this we have to argue more carefully. There is a subtle point here. Of course the series *converges* at $x = 1$ (§15F4). The issue is whether it converges *to the function* $\log(1 + x)$ when $x = 1$. To see that it does, we begin with the formula for the *finite* geometric series. For $x \neq -1$,

$$(4) \qquad \frac{1}{1 + x} = 1 - x + x^2 - \cdots + (-x)^n + \frac{(-x)^{n+1}}{1 + x} \qquad (x \neq -1).$$

Integrating, we get

$$(5) \qquad \log(1 + x) = x - \frac{x^2}{2} + \frac{x^3}{3} - \cdots + (-1)^n \frac{x^{n+1}}{n + 1} + R_n$$

$$(x \neq -1),$$

where

$$(6) \qquad R_n = (-1)^{n+1} \int_0^x \frac{t^{n+1}}{1 + t} \, dt \qquad (x \neq -1).$$

For $x = 1$,

$$(7) \qquad |R_n| = \int_0^1 \frac{t^{n+1}}{1 + t} \, dt \leq \int_0^1 t^{n+1} \, dt = \frac{1}{n + 2} \to 0.$$

Therefore the series (3) converges at $x = 1$. This gives us the following remarkable formula for the alternating harmonic series:

$$(8) \qquad 1 - \tfrac{1}{2} + \tfrac{1}{3} - \cdots = \log 2.$$

Of course, this series converges too slowly for a practical calculation of $\log 2$. To be sure of an approximation to within $1/100$ (roughly speaking, 2-place accuracy), we need 100 terms (§15F4).

15I3. The arctangent. Next, let us replace x by x^2 in (2). Note that $x^2 < 1$ is equivalent to $|x| < 1$. We get

$$(9) \qquad\qquad 1 - x^2 + x^4 - \cdots = \frac{1}{1 + x^2} \qquad\qquad (|x| < 1).$$

Integrating, we obtain

$$(10) \qquad\qquad x - \frac{x^3}{3} + \frac{x^5}{5} - \cdots = \arctan x \qquad (|x| < 1).$$

This series also converges too slowly for practical computation.

Again, by examining the remainder, let us show that the equation (10) holds at $x = 1$. We start with the equation

$$(11) \qquad \frac{1}{1 + x^2} = 1 - x^2 + x^4 - \cdots + (-x^2)^n + \frac{(-x^2)^{n+1}}{1 + x^2}.$$

Integrating, we get

$$(12) \qquad \arctan x = x - \frac{x^3}{3} + \frac{x^5}{5} - \cdots + (-1)^n \frac{x^{2n+1}}{2n+1} + R_n,$$

where

$$(13) \qquad\qquad R_n = (-1)^{n+1} \int_0^x \frac{t^{2n+2}}{1 + t^2} \, dt.$$

At $x = 1$,

$$(14) \qquad |R_n| = \int_0^1 \frac{t^{2n+2}}{1 + t^2} \, dt \leq \int_0^1 t^{2n+2} \, dt = \frac{1}{2n+3} \to 0.$$

Therefore (10) converges at $x = 1$. Since $\arctan 1 = \pi/4$, we find that

$$(15) \qquad\qquad 1 - \tfrac{1}{3} + \tfrac{1}{5} - \cdots = \frac{\pi}{4}.$$

Again, this series converges too slowly for a practical calculation of $\pi/4$.

By considering other variations on the geometric series, and by doing tricks with derivatives and integrals, one can come up with additional expansions. A certain amount of facility in these manipulations is worth while.

Example 1. What function does the series

$$1 + 2x + 3x^2 + \cdots$$

represent?

Solution. We notice that the series is the derivative of the series (1). Since the latter converges to $1/(1 - x)$, for $|x| < 1$, we have

$$(16) \qquad 1 + 2x + 3x^2 + \cdots = \frac{d}{dx}\frac{1}{1 - x} = \frac{1}{(1 - x)^2} \qquad (|x| < 1).$$

15I4. The exponential. Can we construct a series equal to its own derivative? If so, we will have the exponential; for, the only function equal to its own derivative everywhere, and equal to 1 at $x = 0$, is e^x. (We found this out in §7D4).

When we differentiate a series, the degree of each term goes down by 1; hence we want each new term to be the antiderivative of the preceding one. Choosing the constant term as 1, we then take the next as x, then $x^2/2, \cdots$, obtaining the series

$$(17) \qquad 1 + x + \frac{x^2}{2!} + \frac{x^3}{3!} + \cdots.$$

This series converges for all values of x. (The ratio test gives $|u_n/u_{n+1}| = (n + 1)!/n! = n + 1 \to \infty$.) When we differentiate term by term we get back the same series; and the value at $x = 1$ is 0. So we have:

$$(18) \qquad 1 + x + \frac{x^2}{2!} + \frac{x^3}{3!} + \cdots = e^x.$$

Example 2. Let us compute e. Putting $x = 1$ in (18), we get

$$(19) \qquad e = 1 + 1 + \frac{1}{2!} + \frac{1}{3!} + \cdots.$$

This converges pretty well. If we stop with the term $1/6!$, then what we leave off is

$$\frac{1}{7!} + \frac{1}{8!} + \frac{1}{9!} + \cdots = \frac{1}{7!}\left(1 + \frac{1}{8} + \frac{1}{8 \cdot 9} + \cdots\right)$$

$$< \frac{1}{7!}\left(1 + \frac{1}{8} + \frac{1}{8^2} + \cdots\right)$$

$$= \left(\frac{1}{5040}\right)\left(\frac{1}{1 - 1/8}\right) < 0.00023.$$

Carrying 4 places, we get

$$1 \quad = 1.0000$$
$$1 \quad = 1.0000$$
$$1/2! = 0.5000$$
$$1/3! = 0.1667$$
$$1/4! = 0.0417$$
$$1/5! = 0.0083$$
$$1/6! = 0.0014$$
$$\overline{}$$
$$e \approx 2.7181,$$

which is correct to 3 places.

1515. Hyperbolic functions. Exponential functions are often expressed conveniently in terms of the so-called "hyperbolic" functions, which are certain analogues of the trigonometric functions. The most important are the *hyperbolic sine*, denoted *sinh* (universally pronounced "cinch"), and the *hyperbolic cosine*, denoted *cosh* (and pronounced "cosh"). They are defined as follows:

$$e^x = \frac{x^n}{n!}$$

(20) $$\sinh x = \tfrac{1}{2}(e^x - e^{-x}), \qquad \cosh x = \tfrac{1}{2}(e^x + e^{-x}).$$

The following identities are immediate consequences of the definitions:

(21) $$e^x = \sinh x + \cosh x, \qquad e^{-x} = \cosh x - \sinh x,$$

(22) $$\cosh^2 x = 1 + \sinh^2 x,$$

(23) $$\frac{d}{dx} \sinh x = \cosh x, \qquad \frac{d}{dx} \cosh x = \sinh x.$$

Note also that $\sinh 0 = 0$, $\cosh 0 = 1$.

Example 3.

$$\frac{d}{dx} (\cosh 3x - \sinh 4x) = 3 \sinh 3x - 4 \cosh 4x.$$

To obtain power series expansions for cosh x and sinh x, we write down

$$e^x = 1 + x + \frac{x^2}{2!} + \frac{x^3}{3!} + \cdots,$$

(24)
$$e^{-x} = 1 - x + \frac{x^2}{2!} - \frac{x^3}{3!} + \cdots,$$

which converge (absolutely) for all values of x. Hence from (20) we have, at once:

(25)
$$\sinh x = x + \frac{x^3}{3!} + \frac{x^5}{5!} + \cdots,$$

(26)
$$\cosh x = 1 + \frac{x^2}{2!} + \frac{x^4}{4!} + \cdots,$$

which also converge (absolutely) for all values of x.

Problems (§15I)

1. Find series expansions for the following functions, and determine for what values of x the expansions are valid.

(a) $\log(1 - x)$

(c) $\dfrac{1}{(1 + x)^2}$

(b) $\dfrac{1}{1 - x^2}$

(d) $\displaystyle\int_0^x e^{-t^2}\, dt$

2. What functions do the following series represent?

(a) $1 + \dfrac{x}{2!} + \dfrac{x^2}{3!} + \cdots + \dfrac{x^{n-1}}{n!} + \cdots$

(b) $x + 2x^2 + 3x^3 + \cdots + nx^n + \cdots$

3. (a) Show that

$$\log \frac{1 + x}{1 - x} = 2\left(x + \frac{x^3}{3} + \frac{x^5}{5} + \cdots + \frac{x^{2n+1}}{2n + 1} + \cdots\right)$$

$$(-1 < x < 1.)$$

(b) Taking two terms, obtain the estimate

$$\log 2 \approx \tfrac{56}{81} \approx 0.69.$$

4. Verify the following identities.
   (a) $\sinh(x + y) = \sinh x \cosh y + \cosh x \sinh y$
   (b) $\cosh(x + y) = \cosh x \cosh y + \sinh x \sinh y$
   (c) $\sinh 2x = 2 \sinh x \cosh x$
   (d) $\cosh 2x = \cosh^2 x + \sinh^2 x$

5. Problem 4 illustrates some striking similarities (and differences) between hyperbolic functions and trigonometric functions. In this problem, we carry the parallel further. Define $\tanh x = (\sinh x)/(\cosh x)$, $\coth x = (\cosh x)/(\sinh x)$, $\operatorname{sech} x = 1/(\cosh x)$, $\operatorname{csch} x = 1/(\sinh x)$. Verify the following.

(a) $\tanh(x + y) = \dfrac{\tanh x + \tanh y}{1 + \tanh x \tanh y}$

(b) $\tanh 2x = \dfrac{2 \tanh x}{1 + \tanh^2 x}$

(c) $\tanh^2 x + \operatorname{sech}^2 x = 1$

(d) $\dfrac{d}{dx} \tanh x = \operatorname{sech}^2 x$

(e) $\dfrac{d}{dx} \coth x = -\operatorname{csch}^2 x$

(f) $\dfrac{d}{dx} \operatorname{sech} x = -\operatorname{sech} x \tanh x$

(g) $\dfrac{d}{dx} \operatorname{csch} x = -\operatorname{csch} x \coth x$

6. If $\sinh x = -\tfrac{4}{3}$, find $\cosh x$, $\tanh x$, $\coth x$, $\operatorname{sech} x$, and $\operatorname{csch} x$.

Answers to problems (§15I)

1. (a) $-\sum \dfrac{1}{n} x^n$, $-1 \le x < 1$ (b) $\sum x^{2n}$, $-1 < x < 1$

   (c) $\sum (-1)^{n+1} n x^{n-1}$, $-1 < x < 1$ (d) $\sum (-1)^n \dfrac{1}{(2n+1)n!} x^{2n+1}$, all x

2. (a) $f(x) = (e^x - 1)/x$ for all $x \ne 0$, $f(0) = 1$
   (b) $f(x) = x/(1-x)^2$, $|x| < 1$

6. (a) $\tfrac{5}{3}$, $-\tfrac{4}{5}$, $-\tfrac{5}{4}$, $\tfrac{3}{5}$, $-\tfrac{3}{4}$

§15J. Taylor series

15J1. Taylor coefficients. Let $f(x) = \sum a_n x^n$ have radius of convergence r. For $|x| < r$, we may differentiate $f(x) = \sum a_n x^n$ term by term, to get $f'(x)$ (§15H3). Having done this, we may do it again. We thus obtain series for $f, f', f'', \cdots$.

What is the nth derivative of $a_k x^k$? If $n > k$, it is 0. For $n = k$, we have

$$(1) \qquad \frac{d^n}{dx^n} a_n x^n = n!\, a_n.$$

Finally, if $n < k$ then $(d^n/dx^n)a_k x^k$ contains x as a factor. Consequently,

$$(2) \qquad f^{(n)}(x) = \sum_{k=0}^{\infty} \frac{d^n}{dx^n} a_k x^k$$

$$= n!\, a_n + \text{terms containing } x \text{ as a factor.}$$

In particular,

$$(3) \qquad f^{(n)}(0) = n!\, a_n.$$

This is a remarkable formula. Solved for a_n, it reads:

$$(4) \qquad a_n = \frac{1}{n!} f^{(n)}(0).$$

Thus for $|x| < r$, our power series becomes

$$(5)$$
$$f(x) = \sum_{0}^{\infty} \frac{1}{n!} f^{(n)}(0)x^n$$
$$= f(0) + f'(0)x + \frac{1}{2!} f''(0)x^2 + \cdots + \frac{1}{n!} f^{(n)}(0)x^n + \cdots.$$

This is called the *Taylor* series for f (about $x = 0$). As we see, it is valid throughout the interior of the interval of convergence. As (4) shows, it is the *only* power series for f.

Example 1. As we saw in §15I4,

$$e^x = 1 + x + \frac{x^2}{2!} + \cdots$$

for all values of x. Hence this is the Taylor series for e^x. It is easy to check formula (4) for the coefficients: $\exp^{(n)}(x) = e^x$, so $\exp^{(n)}(0) = 1$.

15J2. Taylor expansions. Given a function, f, infinitely differentiable on an interval about $x = 0$, we ask about its possible expansion as a power series in x. If f is infinitely differentiable on an interval about a point a, we may ask about its possible expansion in powers of $x - a$:

$$(6) \quad f(x) = \sum_0^\infty a_n(x - a)^n = a_0 + a_1(x - a) + a_2(x - a)^2 + \cdots.$$

This question is equivalent to the first, for (6) is the same as the series $g(t) = a_0 + a_1 t + a_2 t^2 + \cdots$, where $t = x - a$ and $g(t) = f(x)$. Since from (4), $a_n = (1/n!)g^{(n)}(0)$, we have

$$(7) \qquad\qquad a_n = \frac{1}{n!} f^{(n)}(a).$$

The series (6), which we may accordingly write as

$$(8) \qquad \begin{aligned} f(x) &= \sum_0^\infty \frac{1}{n!} f^{(n)}(a)(x - a)^n \\ &= f(a) + f'(a)(x - a) + \frac{1}{2!} f''(a)(x - a)^2 + \cdots \end{aligned}$$

is the Taylor series for f about the point $x = a$.

The partial sums in a Taylor series are, of course, polynomials:

$$(9) \qquad\qquad \sum_0^n \frac{1}{k!} f^{(k)}(a)(x - a)^k.$$

To say that the series converges to $f(x)$ is to say that the sequence of polynomials approaches $f(x)$.

The remainder is defined by

$$(10) \qquad R_n(x) = f(x) - \sum_0^n \frac{1}{k!} f^{(k)}(a)(x - a)^k.$$

To say that the series converges to $f(x)$ is to say that $R_n(x) \to 0$.

The polynomials are approximations to $f(x)$. To find out how good an approximation, we try to discover how close $R_n(x)$ is to 0. Of course, we have to find out whether $R_n(x)$ approaches 0 in the first place. We investigate this with the help of the following theorem.

15J3. The Mean-Value Theorem for integrals. *Let f and g be continuous on $[a, b]$, and with g of one sign. Then*

$$
(11) \qquad \int_a^b fg = f(z) \int_a^b g
$$

for suitable $z \in (a, b)$.

Proof. For definiteness, say $g \geq 0$. Then for each x,

$$
(12) \qquad (\min_{[a,b]} f)\, g(x) \leq f(x)g(x) \leq (\max_{[a,b]} f)\, g(x)
$$

(§2E7). Therefore

$$
(13) \qquad (\min_{[a,b]} f) \int_a^b g \leq \int_a^b fg \leq (\max_{[a,b]} f) \int_a^b g.
$$

In case $\int_a^b g = 0$, then by (13), $\int_a^b fg = 0$; then (11) holds because both sides are zero. Otherwise, $\int_a^b g > 0$ and we have

$$
(14) \qquad \min_{[a,b]} f \leq \frac{\int_a^b fg}{\int_a^b g} \leq \max_{[a,b]} f.
$$

By the Intermediate-Value Theorem (§2E5), there is a number $z \in (a, b)$ for which

$$
(15) \qquad \frac{\int_a^b fg}{\int_a^b g} = f(z).
$$

Conclusion (11) now follows.

We can now present the main result.

15J4. Taylor's Theorem. *Let f be infinitely differentiable on an interval about a point a. Then in this interval we have, for each $n = 0, 1, \cdots$,*

$$
(16) \qquad
\begin{aligned}
f(x) = f(a) &+ f'(a)(x - a) + \frac{1}{2!} f''(a)(x - a)^2 + \cdots \\
&+ \frac{1}{n!} f^{(n)}(a)(x - a)^n + R_n,
\end{aligned}
$$

where

$$(17) \qquad R_n = \frac{1}{(n+1)!} f^{(n+1)}(z)(x-a)^{n+1}$$

for some z between a and x.

Proof. Equation (16) is just the definition of R_n (10).

To obtain the expression (17) efficiently, we hold x fixed in (16) and differentiate with respect to a (!). On the left side we get 0, since x is constant. On the right we start with $f'(a)$, then immediately cancel it with $-f'(a)$ as we begin differentiating the product $f'(a)(x-a)$. Continuing, we find that everything cancels except at the end, the final result being

$$(18) \qquad 0 = \frac{1}{n!} f^{(n+1)}(a)(x-a)^n + R_n'.$$

Next, since R_n is 0 when $a = x$ (from (16)), we have $R_n = \int_x^a R_n'$. Hence if we transpose and then integrate in (18), we get

$$(19) \qquad R_n = \frac{1}{n!} \int_a^x f^{(n+1)}(s)(x-s)^n \, ds.$$

Since $(x-s)^n$ is of one sign for s between a and x, the Mean-Value Theorem for integrals applies (§15J3), and we obtain

$$(20) \qquad R^n = \frac{1}{n!} f^{(n+1)}(z) \int_a^x (x-s)^n \, ds$$

for suitable z between a and x. Finally, on performing this last integration, we get (17).

For the expansion about $a = 0$ we have the simpler formula:

$$(21) \qquad f(x) = f(0) + f'(0)x + \frac{1}{2!}f''(0)x^2 + \cdots + \frac{1}{n!}f^{(n)}(0)x^n + R_n,$$

where

$$(22) \qquad R_n = \frac{1}{(n+1)!} f^{(n+1)}(z)x^{n+1}$$

for suitable z between 0 and x.

Remark. When $n = 0$, Taylor's formula becomes

$$f(x) = f(a) + f'(z)(x - a)$$

—i.e., the Mean-Value Theorem. Thus, Taylor's Theorem is a generalization of that important theorem.

15J5. Sine and cosine. It is easy to compute the Taylor coefficients for the sine and the cosine. First, $\sin 0 = 0$, $\cos 0 = 1$, and $\sin'(0) = \cos 0 = 1$, $\cos'(0) = -\sin 0 = 0$. Next, $\sin'' = -\sin$ and $\cos'' = -\cos$. Hence the derivatives at 0 repeat in cycles of four: for the sine, $0, 1, 0, -1, 0, 1, 0, -1, \cdots$; and, for the cosine, $1, 0, -1, 0, 1, 0, -1, 0, \cdots$. The series are

(23)
$$\sin x = x - \frac{x^3}{3!} + \frac{x^5}{5!} - \cdots,$$

(24)
$$\cos x = 1 - \frac{x^2}{2!} + \frac{x^4}{4!} - \cdots.$$

Next, all derivatives of the sine or cosine are plus or minus sine or cosine, so their absolute value is ≤ 1. Hence for the remainder (22) we have, for the sine,

(25)
$$|R_n| = \frac{1}{(n+1)!} |\sin^{(n+1)}(z)| |x|^{n+1} \leq \frac{1}{(n+1)!} |x|^{n+1},$$

and this approaches 0 as $n \to \infty$, as we know (e.g., because the exponential series converges (§15I4)). The same reasoning applies to the cosine. Therefore the Taylor expansions of these functions converge to the function.

Example 2. The series (23) and (24) converge rapidly. At $x = \pi/6$, we get, carrying 4 places:

		$1 =$	1.0000
$\pi/6 =$	0.5236	$-(\pi/6)^2/2! =$	-0.1371
$-(\pi/6)^3/3! =$	-0.0239	$(\pi/6)^4/4! =$	0.0031
$\sin(\pi/6) \approx$	0.4997	$\cos(\pi/6) \approx$	0.8660

The first is correct to 3 places (exact value, $\frac{1}{2}$) and the second is correct to 4 places (exact value, $\frac{1}{2}\sqrt{2}$).

15J6. Applications of Taylor's Theorem. Taylor's Theorem gives us an expression for the remainder that we can work with. We can analyze the remainder to decide whether the Taylor series for a given function converges to that function, as we just did for the sine and cosine. We can study the remainder to get a bound on the numerical error in approximating a function by the corresponding Taylor polynomial; this tells us how to calculate functional values within stated limits of accuracy. In a word, the explicit expression for R_n tells whether, and how fast, the series is approaching the function.

We now apply the formula in calculating functional values.

Example 3. Compute $e^{0.1}$ to within 10^{-5}.

Solution. The expansion of e^x about $x = 0$ is

$$e^x = 1 + x + \cdots + \frac{x^n}{n!} + R_n, \qquad R_n = \frac{1}{(n+1)!} e^z x^{n+1},$$

where z lies between 0 and x. For $x = 0.1$ we have $e^z < e^{0.1} < e^{0.5} < 2$ (since $e < 4$). Therefore

$$R_n < \frac{2}{(n+1)!} \frac{1}{10^{n+1}}.$$

For $n = 3$ this is

$$\frac{2}{24} \frac{1}{10^4}, \quad \text{which is} \quad < \frac{1}{10^5};$$

so $n = 3$ will do. We have

$$e^{0.1} \approx 1 + 0.1 + \tfrac{1}{2}(0.01) + \tfrac{1}{6}(0.001) = 1.00517$$

(to five places). Since the error is $< 10^{-5}$, we see in any case that $e^{0.1} = 1.0052$ correct to 4 places.

Example 4. Compute $\log 0.98$ to within 10^{-5}.

Solution. We use the Taylor expansion about $x = 1$. This is

$$\log x = \log 1 + \log'(1)(x - 1) + \cdots + \frac{1}{n!} \log^{(n)}(1)(x - 1)^n + R_n,$$

$$R_n = \frac{1}{(n+1)!} \log^{(n+1)}(z)(x - 1)^{n+1}.$$

The coefficients are as follows. $\log 1 = 0$. Next, $\log'(x) = 1/x$, $\log''(x) = -1/x^2$, $\log'''(x) = 2/x^3, \cdots$, and we see that the derivatives at $x = 1$ are 1, -1, $2!$, $-3!, \cdots$. Hence the series is

$$\log x = (x - 1) - \tfrac{1}{2}(x - 1)^2 + \tfrac{1}{3}(x - 1)^3 - \cdots + (-1)^{n-1} \frac{1}{n}(x - 1)^n + R_n,$$

$$R_n = (-1)^n \frac{1}{n+1} \frac{1}{z^{n+1}}(x - 1)^{n+1}.$$

When $x = 0.98$, $x - 1 = -0.02$; the series is

$$\log(0.98) = -\left[0.02 + \tfrac{1}{2}(0.02)^2 + \cdots + \frac{1}{n}(0.02)^n\right] + R_n,$$

$$R_n = -\frac{1}{n+1}\frac{1}{z^{n+1}}(0.02)^{n+1},$$

where z lies between 0.98 and 1. Then

$$|R_n| < \frac{1}{n+1}\frac{1}{(0.98)^{n+1}}(0.02)^{n+1} = \frac{1}{n+1}\left(\frac{1}{49}\right)^{n+1}.$$

We compute:

m	$(1/49)^m$	$(1/m)(1/49)^m$
1	0.02041	0.02041
2	0.00042	0.00021
3	0.000008	0.000003

Hence $R_n < 10^{-5}$ when $n + 1 = 3$, i.e., $n = 2$; and we have

$$\log(0.98) \approx -[0.02 + 0.0002] = -0.02020$$

(to five places). Since the error is $< 10^{-5}$, we see in any case that $\log(0.98) = -0.0202$ correct to 4 places. (Actually, from our data, we can see that $\log(0.98) = -0.02020$ correct to 5 places.)

Problems (§15J)

1. Write the Taylor expansion about $x = 0$, with remainder R_3, for the following functions.

(a) $\sin^2 x$

(b) $\tan x$

(c) e^{-x^2}

(d) $\sec x$

(e) $\log \sec x$

(f) $\log \cos x$

2. Find the Taylor series about $x = 0$ for each of the following functions, and show that the series converges to the function. [*Hint.* Any power series converging to the function in an interval about 0 must be the Taylor series.]

(a) xe^x

(b) e^{-x^2}

(c) $\cos \sqrt{x}$

(d) $\sin x \cos x$

(e) $1 - 3\sin^2 x$

(f) $s(x) = \dfrac{\sin x}{x}$ $(x \neq 0)$, $s(0) = 1$

3. Estimate the following integrals to within 10^{-2}.

(a) $\displaystyle\int_0^1 e^{-x^2}\,dx$ (Use Problem 2(b).)

(b) $\displaystyle\int_0^{1/4} \cos \sqrt{x}\, dx$ (Use Problem 2(c).)

(c) $\displaystyle\int_0^1 s(x)\, dx,$ where s is the function defined in Problem 2(f).

4. Compute the following numbers.
   (a) $\sin 0.1$, to within 10^{-4}
   (b) $\cos 0.5$, to within 10^{-2}
   (c) $1/e$, to within 10^{-2}
   (d) $\log 1.2$, to within 10^{-2}
   (e) $\sin 1°$, to within 10^{-5} $[\pi/180 \approx 0.01745.]$
   (f) $\cos 5°$, to within 10^{-5} $[(\pi/36)^2 = 0.00762.]$

5. For what values of x is x an estimate for $\sin x$ to within 10^{-3}?

6. (a) Find the Taylor series for e^x about $x = a$ (and show that it converges to e^x) by substituting $x - a$ for x in the Taylor series for e^x about $x = 0$.
   (b) Find the Taylor series for $\sin x - \cos x$ about $x = \pi/4$ (and show that it converges to $\sin x - \cos x$) by substituting $x - \pi/4$ for x in the Taylor series for $\sin x$ about $x = 0$.

7. Find the Taylor series about $x = 1$ for:
   (a) $\sqrt{x}$ (b) $\log(1 + x)$

8. Estimate $\sin 62°$ to within 10^{-4}. $[\sqrt{3} \approx 1.73205,\ \pi/90 \approx 0.03491,$ $(\pi/90)^2 \approx 0.00122.]$

Answers to problems (§15J)

1. In each case, z denotes a suitable number between 0 and x.
   (a) $x^2 - \frac{1}{3}(\cos 2z)x^4$
   (b) $x + \frac{1}{3}x^3 + \frac{1}{3}(\sec^2 z \tan^3 z + 2 \sec^4 z \tan z)x^4$
   (c) $1 - x^2 + \frac{1}{6}e^{-z^2}(4z^4 - 12z^2 + 3)x^4$
   (d) $1 + \frac{1}{2}x^2 + \frac{1}{24}(5 \sec^5 z + 18 \sec^3 z \tan^2 z + \sec z \tan^4 z)x^4$
   (e) $\frac{1}{2}x^2 + \frac{1}{12}(\sec^4 z + 2 \sec^2 z \tan^2 z)x^4$
   (f) $-\frac{1}{2}x^2 - \frac{1}{12}(\sec^4 z + 2 \sec^2 z \tan^2 z)x^4$

2. (a) $\displaystyle\sum \frac{x^{n+1}}{n!}$ (b) $\displaystyle\sum (-1)^n \frac{x^{2n}}{n!}$ (c) $\displaystyle\sum (-1)^n \frac{x^n}{(2n)!}$

   (d) $\displaystyle\frac{1}{2}\sum (-1)^n \frac{(2x)^{2n+1}}{(2n+1)!}$ (e) $\displaystyle-\frac{1}{2} + \frac{3}{2}\sum (-1)^n \frac{(2x)^{2n}}{(2n)!}$

   (f) $\displaystyle\sum (-1)^n \frac{x^{2n}}{(2n+1)!}$

3. (a) 0.74   (b) 0.23   (c) 0.94

4. (a) 0.0998   (b) 0.88   (c) 0.37   (d) 0.18
   (e) 0.01745   (f) 0.99619

5. $|x| < \frac{1}{10}6^{1/3} \approx 0.18$

6. (a) $e^a \sum \frac{(x-a)^n}{n!}$ (b) $\sqrt{2} \sum (-1)^n \frac{(x-\pi/4)^{2n+1}}{(2n+1)!}$

7. (a) $1 + \sum_1^\infty (-1)^n \frac{(-1)(1)(3)\cdots(2n-3)}{2^n \cdot n!} (x-1)^n$

   (b) $\sum (-1)^{n+1} \frac{1}{n \cdot 2^n} (x-1)^n$

8. 0.8830

CHAPTER 16

Differential Equations

This chapter is a brief introduction to ordinary differential equations. The discussion is limited to a few special, but important, types of equations of the first order or second order.

The material can be taken up at any time after Chapter 8, with two exceptions. One is §16B3, on exact equations, which uses partial derivatives (Chapter 13), and the accompanying example in §16B2 and §16B4, which uses polar coordinates (Chapter 10); however, this material is independent of the rest of the chapter. The other is the uniqueness proof in §16D3, based on Taylor series (Chapter 15); to the extent that uniqueness may be accepted without proof, this subsection, too, is independent of the rest of the chapter.

§16A. Introduction

16A1. Differential equations. Differential equations are equations involving derivatives. They arise in many branches of applied mathematics.

A *differential equation* is an equation in x (the independent variable), y (the dependent variable), and some of the derivatives of y. To solve the differential equation is to find a function φ such that $y = \varphi(x)$ satisfies the equation.

Examples of differential equations are $y' = x$ and $y' = y$. The first is trivial: its solution is $y = \int x\, dx$. Likewise, any equation of the form $y' = f(x)$

has a solution that we already know, at least in principle: $y = \int f(x)\, dx$. In contrast, the equation $y' = y$, which we studied in §7D4, required a special attack.

Differential equations may also involve more than one independent variable, so that the derivatives in question are partial derivatives; these are called *partial* differential equations. For emphasis, equations involving only one independent variable—the ones we will discuss—are known as *ordinary* differential equations.

Differential equations are classified according to the order of the highest derivative that appears. We are going to consider only equations of the *first* order (highest derivative, y') or *second* order (highest derivative, y'').

The *general first-order differential equation* is of the form

$$(1) \qquad y' = f(x, y).$$

We will study a few special cases. As for the second-order equation, we will limit ourselves to equations with constant coefficients, that is to say, equations of the form

$$(2) \qquad ay'' + by' + cy = f(x),$$

where a, b, and c are constants. We will usually assume without mention that needed hypotheses are satisfied—for example, that $f(x, y)$ in (1) has continuous first partials.

Example 1. Show that $y = 3e^{2x} - 4x$ satisfies the differential equation

$$(\tfrac{1}{2}x - \tfrac{1}{4})y'' + y = xy'.$$

Solution. $y' = 6e^{2x} - 4$ and $y'' = 12e^{2x}$, and the verification is routine.

16A2. Differential notation. First-order differential equations are conveniently presented in terms of differentials. The function f in equation (1) can be written (in many ways) as a quotient, $f(x, y) = -M(x, y)/N(x, y)$, so that (1) assumes the form

$$(3) \qquad M(x, y) + N(x, y)y' = 0.$$

We then have

$$(4) \qquad \int [M(x, y) + N(x, y)y']\, dx = 0.$$

Now, by the formula for change of variable (§5F(17)), $\int Ny' \, dx = \int N \, dy$; hence (4) is the same as

$$(5) \qquad \int M(x, y) \, dx + \int N(x, y) \, dy = 0.$$

With this in mind, we see that the differential equation (3) may be written in the alternative form

$$(6) \qquad M(x, y) \, dx + N(x, y) \, dy = 0.$$

Example 2. The equation $y' = ky$ may be written as $dy = ky \, dx$.

Problems (§16A)

1. Write the following equations using differential notation.
 (a) $y' = -y$ (c) $y' \sin x = y \cos y$
 (b) $y' = x^2 - xy$ (d) $(1/x)y' = x/(x + y)$
2. Write the following equations using derivative notation.
 (a) $\cos(x + y) \, dx + \sin(x + y) \, dy = 0$
 (b) $x \log x \, dx - \log(x + y) \, dy = 0$
 (c) $(1 + x) \, dx + (1 - x) \, dy = 0$
 (d) $e^y \, dx - e^x \, dy = 0$
3. Verify that the given function is a solution of the given differential equation.
 (a) $3x - 4; \; yy'' + y'^2 = 9$
 (b) $3e^x - 2e^{-x}; \; y'' = y$
 (c) $4e^{2x} + 5e^{-2x}; \; y'' = 4y$
 (d) $5 \sinh 3x - 4 \cosh 3x; \; y'' = 9y$
 (e) $5 \sin 3x - 4 \cos 3x; \; y'' = -9y$
 (f) $3e^{2x} + 2xe^{2x}; \; y'' - 4y' + 4y = 0$
 (g) $-2e^{-x/12} + xe^{-x/12}; \; 6y'' + y' + \frac{1}{24}y = 0$
 (h) $\frac{3}{2}x^2 + 2x - 1; \; y'' - (1/x)y' + 2/x = 0$
 (i) $3/x - 1; \; xy'' + 2y' = 0$
 (j) $2e^x - 1; \; yy'' - y'^2 + y' = 0$

Answers to problems (§16A)

1. (a) $y \, dx + dy = 0$ (b) $(x^2 - xy) \, dx - dy = 0$
 (c) $y \cos y \, dx - \sin x \, dy = 0$ (d) $x^2 \, dx - (x + y) \, dy = 0$
2. (a) $y' = -\cot(x + y)$ (b) $\log(x + y)y' = x \log x$
 (c) $y' = (x - 1)/(x + 1)$ (d) $y' = e^{y-x}$

§16B. Exact equations

16B1. Separable equations. The simplest type of differential equation is one in which M is a function of x alone and N is a function of y alone. Then equation (6) above (§16A) assumes the form

$$(1) \qquad M(x)\,dx + N(y)\,dy = 0,$$

in which x and y appear in different terms; such an equation is said to be *separable*. When an equation can be reduced to this form we say that the variables are separable; for example, in the equation $y' = F(x)G(y)$ the variables are separable.

To solve (1) we simply integrate:

$$(2) \qquad \int M(x)\,dx + \int N(y)\,dy = C,$$

where C is an arbitrary constant. This defines y, implicitly at least, as a function of x—or, rather, as a family of functions, one for each value of C. Each of these functions is a solution of (1), as we see by differentiating in (2). The family of solutions defined by (2) is called the *general* solution of (1).

Example 1. The differential equation $y' = ky$ $(y > 0)$, which we solved in §7D4, is separable: $(1/y)\,dy = k\,dx$, with general solution $\log y = kx + C$, i.e., $y = ce^{kx}$.

16B2. The dog and the rabbit. We now present an example to show how one can set up a differential equation to describe a physical situation and then derive information from it.

Consider a rabbit running in a straight line at constant speed 1, while a dog runs at constant speed λ, headed at each instant directly toward the rabbit. What is the equation of the dog's path of pursuit?

We will derive the equation of greatest interest to the rabbit—that of the dog's *relative* path. It is convenient to use polar coordinates, with the rabbit at the origin, velocity directed downward.

In the relative picture, the dog has an upward velocity component of 1. See Figure 1. The resultant relative velocity is

$$(3) \qquad \frac{dr}{dt} = \sin\theta - \lambda.$$

Next, we want $d\theta/dt$. The dog alone, running along the radius, makes no contribution. But there is a component $\cos\theta$ along the perpendicular at the

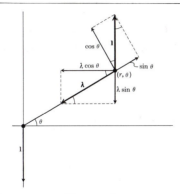

FIGURE 1

distance r; since the rate of change of angle is inversely proportional to the radius,

$$(4) \qquad \frac{d\theta}{dt} = \frac{1}{r} \cos \theta.$$

From (3) and (4) we obtain the differential equation

$$(5) \qquad \frac{dr}{d\theta} = r(\tan \theta - \lambda \sec \theta),$$

in which the variables are separable. We get

$$(6) \qquad \frac{1}{r} \, dr = (\tan \theta - \lambda \sec \theta) \, d\theta,$$

from which

$$(7) \qquad \log r = \log \sec \theta - \lambda \log(\sec \theta + \tan \theta) + C.$$

Taking exponentials and doing some rearranging, we obtain, finally,

$$(8) \qquad r = r_0 \sec \theta \, (\sec \theta - \tan \theta)^\lambda,$$

in which r_0 is a positive constant, the value of r at $\theta = 0$.

A great deal of information can be derived from these data. Let us assume $\lambda \geq 1$: the dog runs at least as fast as the rabbit. By (3), $dr/dt < 0$, as we expect. Writing

$$(9) \qquad r = r_0 \frac{(\cos \theta)^{\lambda - 1}}{(1 + \sin \theta)^\lambda},$$

we see that if $\lambda > 1$ then $r \to 0$ as $\theta \to \pi/2$. On the other hand, if $\lambda = 1$ then $r \to \frac{1}{2}r_0$ as $\theta \to \pi/2$: wherever the dog was at $\theta = 0$, he never gets closer than half that distance.

Next, we have, from Figure 1,

$$(10) \qquad \frac{dx}{dt} = -\lambda \cos \theta \quad \text{and} \quad \frac{dy}{dt} = 1 - \lambda \sin \theta;$$

therefore

$$(11) \qquad \frac{dy}{dx} = \frac{1}{\lambda} \frac{\lambda \sin \theta - 1}{\cos \theta}.$$

Thus, y increases with t until $\sin \theta = 1/\lambda$, then decreases. If $\lambda = 2$, for example, the dog's maximum distance behind the rabbit occurs at $\theta = 30°$; for $\lambda = 1$, the dog falls continually farther and farther behind.

Next, for $\lambda > 1$, $dy/dx \to \infty$ as $\theta \to \pi/2$; thus, the dog's (relative) path closes in tangent to the rabbit's path. For $\lambda = 1$ we have

$$(12) \qquad \frac{dy}{dx} = \frac{\sin \theta - 1}{\cos \theta} = -\frac{\cos \theta}{\sin \theta + 1} \to 0 \qquad\qquad (\lambda = 1)$$

as $\theta \to \pi/2$, so that the path approaches the horizontal.

Finally, we find from (11) that

$$(13) \qquad \frac{d}{d\theta}\left(\frac{dy}{dx}\right) = \frac{1}{\lambda} \sec^2\theta \, (\lambda - \sin \theta) > 0.$$

Hence

$$(14) \qquad \frac{d^2y}{dx^2} = \frac{d}{d\theta}\left(\frac{dy}{dx}\right) \frac{d\theta}{dt} \frac{dt}{dx} < 0.$$

Therefore the entire path is concave down.

Figures 2 and 3 show paths for $\lambda = 2$ and $\lambda = 1$.

This discussion is continued in §16B4 below.

16B3. Exact equations. The so-called *exact* equations are those first-order equations

$$(15) \qquad M(x, y) \, dx + N(x, y) \, dy = 0$$

in which

$$(16) \qquad \frac{\partial M}{\partial y} = \frac{\partial N}{\partial x}.$$

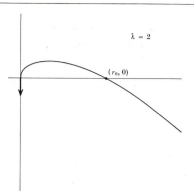

FIGURE 2

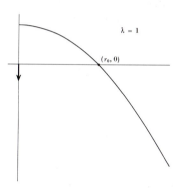

FIGURE 3

(Separable equations are the special case $M_y = N_x = 0$.) This is the same as saying that there is a function u for which

(17) $$\frac{\partial u}{\partial x} = M \qquad \text{and} \qquad \frac{\partial u}{\partial y} = N,$$

as we proceed to show. If (17) is true, then, assuming the mixed partials of u to be equal $\big(\S 13\mathrm{G}(6)\big)$, we immediately get (16).

The argument in the other direction is a little more involved. Assume (16) to hold, and consider the functions

(18) $$u_1 = \int^x M \, dx \qquad \text{and} \qquad u_2 = \int^y N \, dy,$$

in which the symbol $\int^x$ indicates integration with respect to x, holding y fixed, and $\int^y$ has the corresponding meaning. Then, of course,

$$(19) \qquad \frac{\partial u_1}{\partial x} = M \quad \text{and} \quad \frac{\partial u_2}{\partial y} = N.$$

The mixed partials are

$$(20) \qquad \frac{\partial}{\partial y}\frac{\partial u_1}{\partial x} = \frac{\partial M}{\partial y}, \quad \frac{\partial}{\partial x}\frac{\partial u_2}{\partial y} = \frac{\partial N}{\partial x};$$

and these are equal, by hypothesis (16). Thus the mixed partial of $u_1 - u_2$ is 0; it follows (from two integrations) that $u_1 - u_2$ is a function of x alone plus a function of y alone. For notational symmetry, we write this as a difference (rather than a sum):

$$(21) \qquad u_1 - u_2 = F(x) - G(y).$$

Finally, we define

$$(22) \qquad u = u_1 + G(y) = u_2 + F(x).$$

Then $\partial u/\partial x = M$ and $\partial u/\partial y = N$, as required.

To solve the differential equation (15), we construct u as described. Then (15) is the assertion that the total differential of u is 0;

$$(23) \qquad du = \frac{\partial u}{\partial x}\,dx + \frac{\partial u}{\partial y}\,dy = 0$$

(§13C). This implies that u is a constant (see remark following); hence the general solution of (15) is $u = C$.

Remark. The differential notation is set up so that things work out as they should. But as we have not developed that notation systematically, we should check. Equation (15) arose in the first place from the equation $M + Ny' = 0$. Since $M = u_x$ and $N = u_y$, we have $u_x + u_y y' = 0$. But this says that $du/dx = 0$, by the chain rule (§13E(7)). So $u = $ constant.

Example 2. Show that the equation

$$(x + y)\,dx + (x - y)\,dy = 0$$

is exact, and solve.

Solution. Here $M = x + y$ and $N = x - y$. Then $M_y = 1 = N_x$, so the equation is exact. We define

$$u_1 = \int^x M \, dx = \int^x (x + y) \, dx = \tfrac{1}{2}x^2 + xy,$$

$$u_2 = \int^y N \, dy = \int^y (x - y) \, dy = xy - \tfrac{1}{2}y^2.$$

Then $u_1 - u_2 = \tfrac{1}{2}x^2 + \tfrac{1}{2}y^2 \, [= F(x) - G(y)]$, so

$$u = \tfrac{1}{2}x^2 - \tfrac{1}{2}y^2 + xy,$$

and the solution of the equation is $\tfrac{1}{2}x^2 - \tfrac{1}{2}y^2 + xy = C$.

Example 3. Solve the equation

$$y \, dx - x \, dy = 0.$$

Solution. Here the variables are separable. But one should also notice that

(24)
$$\frac{y \, dx - x \, dy}{y^2}$$

is the total differential of x/y. So we have $d(x/y) = 0$, $x/y = C$, $x = Cy$.

16B4. The dog and the rabbit (continued). How long does the chase last ? (See §16B2.) To find out, we have to solve another differential equation. From (3) and (4) we see that, for arbitrary M and N,

(25) $M \, dr = M(\sin \theta - \lambda) \, dt$ and $N \, d\theta = (N/r) \cos \theta \, dt.$

Therefore

(26)
$$dt = \frac{M \, dr + N \, d\theta}{M(\sin \theta - \lambda) + (N/r) \cos \theta \, dt}.$$

A little study of the denominator suggests choosing

(27) $M = \sin \theta + \lambda$ and $N = r \cos \theta.$

Moreover, we then have $\partial M/\partial \theta = \partial N/\partial r \, [= \cos \theta]$, so that the numerator becomes a total differential. We get

(28)
$$dt = \frac{1}{1 - \lambda^2} [(\sin \theta \, dr + r \cos \theta \, d\theta) + \lambda \, dr]$$

$$= \frac{1}{1 - \lambda^2} (dy + \lambda \, dr)$$

—provided $\lambda \neq 1$. Hence the time from a given point to the end of the pursuit (at which time $y = r = 0$) is given by

$$(29) \qquad\qquad t = \frac{1}{\lambda^2 - 1}\,(y + \lambda r) \qquad\qquad (\lambda \neq 1).$$

(The signs get reversed because t refers to the end of the chase while y and r refer to the beginning.)

The distance the dog travels is

$$(30) \qquad\qquad s = \lambda t = \frac{\lambda}{\lambda^2 - 1}\,(y + \lambda r).$$

In particular, the distance the dog runs from the point $r = r_0$, $\theta = 0$ to the end is

$$(31) \qquad\qquad \frac{\lambda^2}{\lambda^2 - 1}\,r_0.$$

When $\lambda = 2$, for example, this is $\frac{4}{3}r_0$.

When $\lambda = 1$ we expect the chase to go on for ever. From (4) and (8), we have

$$(32) \qquad dt = r \sec\theta\, d\theta = r_0(\sec^3\theta - \sec^2\theta \tan\theta)\, d\theta \qquad (\lambda = 1).$$

Recalling the formula for $\int \sec^3\theta\, d\theta$ ($\S 8\mathrm{E}(17)$), we get

$$ t = \tfrac{1}{2}r_0[\sec\theta \tan\theta + \log(\sec\theta + \tan\theta) - \tan^2\theta]$$

$$(33) \qquad\quad = \tfrac{1}{2}r_0\left[\frac{\sin\theta}{1 + \sin\theta} + \log(\sec\theta + \tan\theta)\right] \qquad (\lambda = 1).$$

Consequently, $t \to \infty$ as $\theta \to \pi/2$, confirming our expectation.

Finally, let us compute the radius of curvature of the dog's actual path on the ground. Since the dog heads toward the rabbit at each instant, the slope of his ground path is simply $\tan\theta$. Therefore, by definition $\left(\S 9\mathrm{I}(6)\right)$,

$$(34) \qquad\qquad \rho = \left|\frac{ds}{d\theta}\right|.$$

Since $ds/dt = \lambda$ and $dt/d\theta = r \sec\theta$ (from (4)), we have

$$(35) \qquad\qquad \rho = \lambda r \sec\theta.$$

In particular, at $r = r_0$, $\theta = 0$, the radius of curvature is equal to λr_0.

Problems (§16B)

1. Solve the following differential equations.

(a) $x \, dx = y \, dy$

(b) $y' + \dfrac{y}{x} = 0$

(c) $(1 - x)y^2 = x^3 y'$

(d) $x^2 y' + y^2 = 0$

(e) $y' = \dfrac{1 - y^2}{1 + x^2}$

(f) $\tan x \csc 2x \sec y + y' = 0$

2. Find the particular solution of the differential equation that goes through the given point.

(a) $y' = \cos^2 y \csc^2 x$, $(\pi/2, \pi/4)$

(b) $xy' + 2y = xyy'$, $(e, 1)$

(c) $y' = \dfrac{xy}{x - 1}$, $(0, 1)$

(d) $(x + 1)y = x^2 y'$, $(1, 1)$

3. Find the value of k that makes the equation exact.

(a) $(2y + x) \, dx + (kx - y^2) \, dy = 0$

(b) $(xy^2 + 1) \, dx + (x^k y - 1) \, dy = 0$

(c) $(ky + y \cos xy) \, dx + (1 + x \cos xy) \, dy = 0$

(d) $\dfrac{y}{x^2 + y^2} \, dx + \dfrac{kx}{x^2 + y^2} \, dy = 0$

(e) $(1 + k)xy \, dx + (1 + kx^2) \, dy = 0$

(f) $\dfrac{x + y}{y} \, dx + k\left(\dfrac{x}{y}\right)^2 \, dy = 0$

4. Solve the following differential equations.

(a) $(x + y) \, dx + x \, dy = 0$

(b) $(3x - 2y) \, dx + (5y - 2x) \, dy = 0$

(c) $(x^2 + y^2) \, dx + 2xy \, dy = 0$

(d) $(x^2 - y^2) \, dx - 2xy \, dy = 0$

(e) $(x^2 + x^2 y^2) \, dx + 2(x^2 + 1)y \, dy = 0$

(f) $\log xy \, dx + \dfrac{x}{y} \, dy = 0$

(g) $\tan y \, dx + x \sec^2 y \, dy = 0$

(h) $(1 - r \sin \theta) \, d\theta + \cos \theta \, dr = 0$

(i) $(x^2 y - x + 1) \, dx + (\tfrac{1}{3}x^3 + y) \, dy = 0$

(j) $ye^{xy} \, dx + xe^{xy} \, dy = 0$

Answers to problems (§16B)

1. (a) $y^2 = x^2 + C$ (b) $xy = c$ (c) $y = 2x^2/(1 - 2x + Ax^2)$

(d) $y = x/(Ax - 1)$ (e) $y = (ce^{2 \arctan x} - 1)/(ce^{2 \arctan x} + 1)$

(f) $\sin y + \tfrac{1}{2} \tan x = C$

2. (a) $\tan y = 1 - \cot x$ (b) $x^2 = e^{1+y}/y$
   (c) $y = (1 - x)e^x$ (d) $y = xe^{1-1/x}$

3. (a) 2   (b) 2   (c) 0   (d) -1 (e) 1   (f) $-\frac{1}{2}$

4. (a) $x^2 + 2xy = C$ (b) $3x^2 - 4xy + 5y^2 = C$ (c) $x^3 + 3xy^2 = C$
   (d) $x^2 - 3xy^2 = C$ (e) $1 + y^2 = ce^{\arctan x - x}$

   (f) $y = \dfrac{1}{x} e^{1+c/x}$ (g) $x \tan y = C$ (h) $r = (A - \theta)\sec\theta$

   (i) $\frac{1}{3}x^3 y - \frac{1}{2}x^2 + x + \frac{1}{2}y^2 = C$ (j) $xy = c$

§16C. Other first-order equations

16C1. Homogeneous equations. A function $f(x, y)$ is said to be *homogeneous of degree n* if it satisfies the formula

$$(1) \qquad f(tx, ty) = t^n f(x, y).$$

(Cf. §13E, Problems 10–12.) For instance, $f(x, y) = x^2 + y^2$ is homogeneous of degree 2, since $f(tx, ty) = (tx)^2 + (ty)^2 = t^2 f(x, y)$.

To say that f is homogeneous *of degree 0* $\big(f(tx, ty) = f(x, y)\big)$ is to say that f is a function of y/x alone (or of x/y). For if $f(x, y) = f(tx, ty)$ for all t, we may take $t = 1/x$ to get $f(x, y) = f(1, y/x)$, a function of y/x alone. Conversely, if $f(x, y) = F(y/x)$, then clearly $f(tx, ty) = f(x, y)$.

The differential equation

$$(2) \qquad M(x, y)\, dx + N(x, y)\, dy = 0$$

is said to be *homogeneous* if M and N are homogeneous functions of the same degree. This is the same as saying that the function f, in the equivalent equation $y' = f(x, y)$, is homogeneous of degree 0. For, $f = -M/N$, so that if M and N are homogeneous of degree n, then $f(tx, ty) = f(x, y)$, as is easily seen. Conversely, given that $f(tx, ty) = f(x, y)$, we may construct a homogeneous equation (2) by choosing, say, $M(x, y) = f(x, y)$ and $N(x, y) = -1$.

We solve a homogeneous differential equation by means of the substitutions

$$(3) \qquad y = vx, \qquad dy = v\, dx + x\, dv$$

(or by means of $x = vy$ etc.). This transforms the equation $dy/dx = F(y/x) = F(v)$ to the equation $v\, dx + x\, dv = F(v)\, dx$, in which the variables are separable.

Example 1. Show that the differential equation

$$(x - y)\,dx + (x + y)\,dy = 0$$

is homogeneous, and solve.

Solution. Here $M = x - y$ and $N = x + y$, which are homogeneous of degree 1. With the substitutions (3), we get

$$x(1 - v)\,dx + x(1 + v)(v\,dx + x\,dv) = 0,$$

which reduces to

$$\frac{1}{x}\,dx + \frac{1}{1 + v^2}\,dv + \frac{v}{1 + v^2}\,dv = 0.$$

Multiplying through by 2 and integrating, we get

$$2\log x + 2\arctan v + \log(1 + v^2) = C,$$

that is,

$$\log(x^2 + y^2) + 2\arctan(y/x) = C.$$

We leave the answer in this form.

Check to see that the function y defined implicitly by this equation satisfies the original differential equation.

16C2. Linear equations. By a *linear differential equation of the first order* is meant an equation of the form

(4)
$$y' + P(x)y = Q(x).$$

Note that P and Q are functions of x alone.

There is a trick that enables us to solve these equations handily. It comes from noticing that the expression $y' + Py$ in (4) looks like part of a product formula. For a closer resemblance, introduce

(5)
$$z = \int P(x)\,dx, \qquad \text{so that} \qquad z' = P.$$

Equation (4) becomes

(6)
$$y' + z'y = Q.$$

It is not hard to recognize the expression $y' + z'y$ as arising when one differentiates ye^z:

(7)
$$(ye^z)' = (yz' + y')e^z.$$

At any rate, we do not have to think this up every time, but just remember it.

We rewrite (6) in the equivalent form

(8) $$(y' + z'y)e^z = Qe^z.$$

Then $(ye^z)' = Qe^z$. Therefore $ye^z = \int Qe^z\,dx + C$. Hence, finally,

(9) $$y = e^{-z}\left(\int Qe^z\,dx + C\right), \qquad \text{where} \qquad z' = P.$$

Example 2. Solve the differential equation

$$y' + \frac{y}{x} = x.$$

Solution. This is linear, with $P(x) = 1/x$ and hence $z = \int P(x)\,dx = \log x$, and with $Q(x) = x$. Then $e^z = x$ and the solution is

$$y = \frac{1}{x}\left(\int x^2\,dx + C\right) = \frac{1}{3}x^2 + \frac{C}{x}.$$

Check that this solves the original differential equation.

Problems (§16C)

1. Solve the following homogeneous equations.

    (a) $x\cos\dfrac{y}{x}\,dy = \left(y\cos\dfrac{y}{x} + x\right)dx$

    (b) $(3x - 2y)\,dx + (5y - 2x)\,dy = 0$

    (c) $(2x - y)\,dx + (3x - 4y)\,dy = 0$

    (d) $\sqrt{x^2 + y^2}\,dx = x\,dy - y\,dx$

    (e) $(x + \sqrt{x^2 + y^2})y\,dx - x^2\,dy = 0$

2. Solve the following linear equations.

    (a) $xy' - 2y = x^3$ (f) $y' - y = e^x$

    (b) $y' - 2y = 2x$ (g) $y' - 2y = e^{2x}$

    (c) $y' + \dfrac{y}{x} = 1$ (h) $y\,dx + x\,dy = \sec^2x\,dx$

    (d) $xy' - 2y = 2x$ (i) $y' + y = \cos x + \sin x$

    (e) $y' - y = 4e^{-x}$ (j) $y' + z'y = e^{-z}, \qquad z = \tan\log x$

Answers to problems (§16C)

1. (a) $y = x\arcsin\log(cx)$ (b) $5y^2 - 4xy + 3x^2 = c$
   (c) $x^5 = c(2y^2 - 3xy - x^2)^2$ (d) $y = \frac{1}{2}(1/c - cx^2)$
   (e) $y = cx(\sqrt{x^2 + y^2} + x)$

2. (a) $y = x^3 + cx^2$ (b) $y = ce^{2x} - x - \frac{1}{2}$ (c) $y = \frac{1}{2}x + C/x$
    (d) $y = Cx^2 - 2x$ (e) $y = Ce^x - 2e^{-x}$ (f) $y = e^x(x + C)$
    (g) $y = e^{2x}(x + C)$ (h) $y = (\tan x + C)/x$ (i) $y = \sin x + Ce^{-x}$
    (j) $y = e^{-z}(x + C)$

§16D. Linear second-order equations with constant coefficients

Fundamental physical laws about the universe are expressed in terms of acceleration, which is the *second* derivative of distance with respect to time. Clearly, then, it is important to know something about differential equations of the *second* order.

16D1. The homogeneous equation. The *general linear differential equation of the second order with constant coefficients* is

$$(1) \qquad\qquad ay'' + by' + cy = f(x),$$

where a, b, and c are constants. When f is the zero function, the equation is said to be *homogeneous*:

$$(2) \qquad\qquad ay'' + by' + cy = 0.$$

We begin by solving the homogeneous equation. Later on (§16E) we treat the nonhomogeneous equation.

Observe that if y_1 and y_2 are solutions of the homogeneous equation, (2), then so are all "linear combinations" of y_1 and y_2—i.e., all functions of the form

$$(3) \qquad\qquad Ay_1 + By_2,$$

where A and B are constants.

16D2. The special case. Before tackling the general homogeneous equation (2), we consider the equation

$$(4) \qquad\qquad y'' = ky.$$

We can write down solutions of this by inspection. If $k = 0$, (4) reduces to $y'' = 0$. Then y' is a constant, and y is a linear function; say

$$(5) \qquad\qquad y = A + Bx \qquad\qquad (k = 0).$$

This gives us a two-parameter family of solutions; and there are no others.

If $k > 0$ let us write $k = \lambda^2$. Equation (4) becomes $y'' = \lambda^2 y$, for which there are two obvious solutions: $y = e^{\lambda x}$ and $y = e^{-\lambda x}$. Then

$$y = Ae^{\lambda x} + Be^{-\lambda x} \qquad (6) \qquad (k = \lambda^2)$$

is also a solution, where A and B are arbitrary constants. Other obvious solutions of $y'' = \lambda^2 y$ are $y = \cosh \lambda x$ and $y = \sinh \lambda x$ $\left(\S15\mathrm{I}(23)\right)$; hence their linear combinations,

$$(7) \qquad y = A_1 \cosh \lambda x + B_1 \sinh \lambda x$$

are also solutions. But $\cosh \lambda x$ and $\sinh \lambda x$ are themselves linear combinations of $e^{\lambda x}$ and $e^{-\lambda x}$, and vice versa (see formulas (20) and (21) of §15I6), and so (6) and (7) produce the same set of functions.

Finally, if $k < 0$ we write $k = -\omega^2$. Then equation (4) becomes $y'' = -\omega^2 y$, for which we recognize the solutions $y = \cos \omega x$ and $y = \sin \omega x$. Then

$$y = A \cos \omega x + B \sin \omega x \qquad (8) \qquad (k = -\omega^2)$$

is also a solution, where A and B are arbitrary constants.

It is easy to check that in all cases $\left((5), (6), \text{ and } (8)\right)$, any pair of prescribed values for y and y' at any given point x_0 will determine the constants A and B. Such a prescription is called a set of *initial conditions*.

Example 1. Find the solution of $y'' = 4y$ satisfying the initial conditions $y = 2$ and $y' = 1$ when $x = 0$.

Solution. We have $k = 4$, $\lambda = 2$, and therefore

$$y = Ae^{2x} + Be^{-2x}.$$

Then

$$y' = 2Ae^{2x} - 2Be^{-2x}.$$

The initial conditions take the form

$$2 = A + B,$$

$$1 = 2A - 2B.$$

Solving, we get $A = \frac{5}{4}$, $B = \frac{3}{4}$. Hence, finally,

$$y = \tfrac{5}{4}e^{2x} + \tfrac{3}{4}e^{-2x}.$$

16D3. Uniqueness of the solutions. We now show that the solutions of $y'' = ky$ ($k \neq 0$) expressed in (6) $\big($or (7)$\big)$ and (8) are the only possible ones. Suppose $y = f(x)$ is a solution of $y'' = ky$. To simplify the notation we assume that f is defined at $x = 0$; we will see that this is always true in any case. Since $f'' = kf$ we have $f''' = kf'$. Then $f^{(4)} = kf'' = k^2 f$ and $f^{(5)} = kf''' = k^2 f'$; and so on. In general,

$$(9) \qquad\qquad f^{(2n)} = k^n f \qquad \text{and} \qquad f^{(2n+1)} = k^n f'.$$

Thus, $f(x)$ has a Taylor series, whose even and odd terms divide naturally as follows:

$$(10) \qquad f(0)\left(1 + \frac{kx^2}{2!} + \frac{k^2 x^4}{4!} + \cdots\right) + f'(0)\left(x + \frac{kx^3}{3!} + \frac{k^2 x^5}{5!} + \cdots\right).$$

It is not hard to show that this series converges to $f(x)$ (see below). Then for $k = \lambda^2 > 0$ we have $\big($from §15I (25 and 26)$\big)$

$$(11) \qquad\qquad f(x) = f(0) \cosh \lambda x + \frac{1}{\lambda} f'(0) \sinh \lambda x;$$

and for $k = -\omega^2 < 0$ we have $\big($from §15J (23 and 24)$\big)$

$$(12) \qquad\qquad f(x) = f(0) \cos \omega x + \frac{1}{\lambda} f'(0) \sin \omega x.$$

Finally, if $F'' = kF$ and F is defined at $x = a$, set $f(x) = F(x + a)$. Then $f'' = kf$, and f is defined at 0. Then, as we have just seen, f is defined everywhere. Since $F(x) = f(x - a)$, F, too, is defined everywhere.

To verify convergence to $f(x)$ we look at the remainder term $\big(§15J(22)\big)$. From (9), we get

$$(13) \qquad\qquad R_{2n} = \frac{1}{(2n+1)!} k^n f(z) x^{2n+1}.$$

We may assume (to avoid absolute value signs) that $k > 0$ and $x > 0$. The continuous function f is bounded on $[0, x]$, say by M; so $|f(z)| \leq M$. Since $k^n < (k+1)^n < (k+1)^{2n+1}$, we see that

$$(14) \qquad\qquad |R_{2n}| \leq \frac{1}{(2n+1)!} M[(k+1)x]^{2n+1}.$$

As we know, this approaches 0 as $n \to \infty$ $\big($e.g., because the exponential series converges (§15I4)$\big)$. A similar proof shows that $|R_{2n+1}| \to 0$.

16D4. The general case. To solve the general homogeneous equation

$$(15) \qquad\qquad ay'' + by' + cy = 0,$$

we refer it to the special case by means of the substitution $y = ze^{-(b/2a)x}$. Equation (15) becomes

$$(16) \qquad\qquad z'' = \frac{1}{4a^2}(b^2 - 4ac)z.$$

We know how to solve (16) (§16D2). And if z is a solution of (16), then $y = ze^{-(b/2a)x}$ is a solution of the original equation, (15).

The substitution is arrived at in the first place by educated guessing. Someone once thought of trying $y = ze^{hx}$, where h is a constant to be determined. Then

$$(17) \qquad y' = (hz + z')e^{hx}, \qquad \text{and} \qquad y'' = (h^2z + 2hz' + z'')e^{hx},$$

and (15) becomes

$$(18) \qquad\qquad az'' + (2ah + b)z' + (ah^2 + bh + c)z = 0$$

(since e^{hx} is never 0). Now we choose $h = -b/2a$ so as to make the term in z' disappear. We then get (16).

16D5. Summary. The equation

$$(19) \qquad\qquad ay'' + by' + cy = 0$$

has the general solution

$$(20) \qquad\qquad y = Au + Bv,$$

where u and v are as follows.

(i) If $b^2 - 4ac = 0$, then

$$u = e^{-(b/2a)x}, \qquad v = xe^{-(b/2a)x}.$$

(ii) If $b^2 - 4ac > 0$, set $\lambda = \frac{1}{2a}\sqrt{b^2 - 4ac}$; then

$$u = e^{(\lambda - b/2a)x}, \qquad v = e^{(-\lambda - b/2a)x}.$$

(iii) If $b^2 - 4ac < 0$, set $\omega = \dfrac{1}{2a} \sqrt{4ac - b^2}$; then

$$u = e^{-(b/2a)x} \cos \omega x, \qquad v = e^{-(b/2a)x} \sin \omega x.$$

In each case, a particular solution is determined by prescribing the values of y and y' at some point x_0. To find the particular solution, impose these initial conditions and solve for the constants A and B.

To see that this is possible, we write

(21)
$$y_0 = Au_0 + Bv_0,$$
$$y_0' = Au_0' + Bv_0',$$

where the subscript indicates values at x_0. These are the equations to be solved for A and B. The condition for solvability is $u_0v_0' - v_0u_0' \neq 0$. (Start solving and you will see why.) We proceed to verify that $uv' - vu'$ is never 0. If $uv' - vu' = 0$ at any point, then $(v/u)' = 0$ there (provided $u \neq 0$). Now, in (i), $v/u = x$ and we have $(v/u)' = 1 \neq 0$. In (ii), $v/u = e^{-2\lambda x}$ and so $(v/u)' = -2\lambda e^{-2\lambda x} \neq 0$, since $\lambda \neq 0$. In (iii), where $u \neq 0$, we have $v/u = \tan \omega x$ and therefore $(v/u)' = \omega \sec^2 \omega x \neq 0$, since $\omega \neq 0$. If $u_0 = 0$ then there is an interval about x_0 in which $v \neq 0$, and we may consider $u/v = \cot \omega x$; then $(u/v)' = -\omega \csc^2 \omega x \neq 0$.

Example 2. Find the solution of $y'' - 2y' + y = 0$ satisfying the initial conditions $y = 1$ and $y' = 0$ when $x = 0$.

Solution. Here $b^2 - 4ac = 0$, $b/2a = -1$. Therefore $u = e^x$, $v = xe^x$, and

$$y = e^x(A + Bx).$$

Then

$$y' = e^x(A + B + Bx)$$

and the initial conditions take the form

$$1 = A,$$

$$0 = A + B.$$

Hence $A = 1$, $B = -1$, and the solution is

$$y = e^x(1 - x).$$

Example 3. Find the solution of $y'' + 4y' + 3y = 0$ satisfying the initial conditions $y = 1$ and $y' = 2$ when $x = 1$.

Solution. Here $b^2 - 4ac = 4$, $1/2a = \frac{1}{2}$, $\lambda = \frac{1}{2}\sqrt{4} = 1$, $b/2a = 2$. Therefore $u = e^{-x}$, $v = e^{-3x}$, and

$$y = Ae^{-x} + Be^{-3x}.$$

Then

$$y' = -Ae^{-x} - 3Be^{-3x},$$

and the initial conditions take the form

$$1 = Ae^{-1} + Be^{-3},$$

$$2 = -Ae^{-1} - 3Be^{-3}.$$

Solving, we get $A = \frac{5}{2}e$, $B = -\frac{3}{2}e^3$. Hence, finally,

$$y = \frac{5}{2}e^{1-x} - \frac{3}{2}e^{3(1-x)}.$$

Example 4. Find the solution of $\frac{1}{2}y'' + y' + 2y = 0$ satisfying the initial conditions $y = \sqrt{3}$ and $y' = \sqrt{3}$ when $x = 0$.

Solution. Here $b^2 - 4ac = -3$, $1/2a = 1$, $\omega = (1)(\sqrt{3}) = \sqrt{3}$, $b/2a = 1$. Therefore $u = e^{-x} \cos \sqrt{3}x$, $v = e^{-x} \sin \sqrt{3}x$, and

$$y = e^{-x}(A \cos \sqrt{3}x + B \sin \sqrt{3}x).$$

Then

$$y' = e^{-x}[-(\sqrt{3}A + B) \sin \sqrt{3}x + (-A + \sqrt{3}B) \cos \sqrt{3}x].$$

The initial conditions take the form

$$\sqrt{3} = A,$$

$$\sqrt{3} = -A + B\sqrt{3}.$$

Hence $A = \sqrt{3}$, $B = 2$, and the solution is

$$y = e^{-x}(\sqrt{3} \cos \sqrt{3}x + 2 \sin \sqrt{3}x).$$

Problems (§16D)

1. Find the general solution of the following equations. (The solutions will be used in Problem 2 following and in §16E, Problem 1.)

(a) $y'' - 4y = 0$ (d) $4y'' + y = 0$

(b) $4y'' - y = 0$ (e) $y'' - 4y' = 0$

(c) $y'' + 4y = 0$ (f) $y'' + 4y' = 0$

2. In Problem 1, in each case, find the particular solution satisfying the following initial conditions.

(a) $y = 2$ and $y' = 0$ when $x = \log 2$

(b) $y = e + 1$ and $y' = \frac{1}{2}(e - 1)$ when $x = 2$

(c) $y = 2$ and $y' = 2$ when $x = \pi$

(d) $y = 2$ and $y' = 2$ when $x = \pi$

(e) $y = 4$ and $y' = 4$ when $x = 0$

(f) $y = 4$ and $y' = 4$ when $x = 0$

3. Find the general solution of the following equations. (The solutions will be used in Problem 4 following and in §16E, Problem 3.)

(a) $y'' - y' - 2y = 0$ (d) $y'' + 2y' + 2y = 0$
(b) $y'' - y' + 2y = 0$ (e) $y'' + 3y' + 2y = 0$
(c) $y'' + 2y' + y = 0$ (f) $2y'' + 3y' + y = 0$

4. In Problem 3, in each case, find the particular solution satisfying the following initial conditions.

(a) $y = 3$ and $y' = 3$ when $x = 0$
(b) $y = \sqrt{7}$ and $y' = \sqrt{7}$ when $x = 0$
(c) $y = 2$ and $y' = -1$ when $x = 1$
(d) $y = 2$ and $y' = 1$ when $x = 0$
(e) $y = 2$ and $y' = -3$ when $x = \log \frac{1}{3}$
(f) $y = 0$ and $y' = \frac{1}{2}$ when $x = \log \frac{1}{3}$

Answers to problems (§16D)

1. (a) $y = Ae^{2x} + Be^{-2x}$ (b) $y = Ae^{x/2} + Be^{-x/2}$
   (c) $y = A \cos 2x + B \sin 2x$ (d) $y = A \cos \frac{1}{2}x + B \sin \frac{1}{2}x$
   (e) $y = Ae^{4x} + B$ (f) $y = A + Be^{-4x}$
2. (a) $y = \frac{1}{4}e^{2x} + 4e^{-2x}$ (b) $y = e^{x/2} + e^{1-x/2}$
   (c) $y = 2 \cos 2x + \sin 2x$ (d) $y = -4 \cos \frac{1}{2}x + 2 \sin \frac{1}{2}x$
   (e) $y = e^{4x} + 3$ (f) $y = 5 - e^{-4x}$
3. (a) $y = Ae^{2x} + Be^{-x}$ (b) $y = e^{x/2}(A \cos \frac{1}{2}\sqrt{7}x + B \sin \frac{1}{2}\sqrt{7}x)$
   (c) $y = e^{-x}(A + Bx)$ (d) $y = e^{-x}(A \cos x + B \sin x)$
   (e) $y = Ae^{-x} + Be^{-2x}$ (f) $y = Ae^{-x/2} + Be^{-x}$
4. (a) $y = 2e^{2x} + e^{-x}$ (b) $y = e^{x/2}(\sqrt{7} \cos \frac{1}{2}\sqrt{7}x + \sin \frac{1}{2}\sqrt{7}x)$
   (c) $y = e^{1-x}(1 + x)$ (d) $y = e^{-x}(2 \cos x + 3 \sin x)$
   (e) $y = \frac{1}{3}e^{-x} + \frac{1}{9}e^{-2x}$ (f) $y = \frac{1}{3}\sqrt{3}e^{-x/2} - \frac{1}{3}e^{-x}$

§16E. The nonhomogeneous equation

16E1. Analysis. We solve the equation

$$(1) \qquad\qquad ay'' + by' + cy = f(x)$$

by a method known as "variation of parameters." The solution is made to depend on the solution of the corresponding homogeneous equation

$$(2) \qquad\qquad ay'' + by' + cy = 0.$$

Observe that if u is a solution of (2) and y is a solution of (1), then $u + y$ is a solution of (1). Conversely, the difference of two solutions of (1) is a solution of (2). It follows that the general solution of (1) is

$$(3) \qquad\qquad y = Au + Bv + w,$$

where $Au + Bv$ is the general solution of (2) $\big(\S16\mathrm{D}(20)\big)$ and w is a particular solution of (1). Our problem, then, is to find a particular solution of (1).

The idea of the method is to replace the general solution $Au + Bv$ of (2) by

$$(4) \qquad\qquad w = \alpha u + \beta v,$$

where α and β are now functions, not constants, in the hope of producing a solution of (1). Differentiating in (4), we get

$$(5) \qquad\qquad w' = (\alpha u' + \beta v') + (\alpha' u + \beta' v).$$

We now impose the condition

$$(6) \qquad\qquad \alpha' u + \beta' v = 0,$$

in the hope of simplifying the computations without sacrificing the solution. Then (5) becomes

$$(7) \qquad\qquad w' = \alpha u' + \beta v'.$$

Differentiating in (7), we get

$$(8) \qquad\qquad w'' = (\alpha u'' + \beta v'') + (\alpha' u' + \beta' v').$$

The condition that w satisfy the differential equation (1) is $aw'' + bw' + cw = f$; that is, from (8), (7), and (4),

$$(9) \qquad \alpha(au'' + bu' + cu) + \beta(av'' + bv' + cv) + a(\alpha' u' + \beta' v') = f.$$

Since u and v satisfy (2), the first two parentheses in (9) are 0. So (9) reduces to

$$(10) \qquad\qquad a(\alpha' u' + \beta' v') = f.$$

With (6), this gives us a pair of linear equations in the unknowns α' and β':

$$(11) \qquad\qquad \begin{aligned} \alpha' u + \beta' v &= 0, \\[2mm] \alpha' u' + \beta' v' &= \frac{f}{a}. \end{aligned}$$

These can be solved in the usual way, to give

$$(12) \qquad \alpha' = -\frac{1}{a}\frac{vf}{uv' - vu'}, \qquad \beta' = \frac{1}{a}\frac{uf}{uv' - vu'}.$$

(Recall from §16D5 that $uv' - vu' \neq 0$.) Finally, we find α and β from

(13) $$\alpha = \int \alpha', \qquad \beta = \int \beta'.$$

16E2. Summary. The general solution of

(14) $$ay'' + by' + cy = f(x)$$

is

(15) $$y = Au + Bv + w, \qquad \text{where} \qquad w = \alpha u + \beta v,$$

in which u and v are solutions of the homogeneous equation (2) as described in §16D5, and α and β are given by

(16) $$\alpha = -\frac{1}{a}\int \frac{vf}{uv' - vu'}, \qquad \beta = \frac{1}{a}\int \frac{uf}{uv' - vu'}.$$

A particular solution is determined by prescribing the values of y and y' at any given point x_0 and solving for A and B. §16D5 shows that this is always possible. (To adapt the discussion there to the present situation, replace y in §16D(21) by $y - w$.)

Remark. The denominators in (16) may often be computed handily by taking advantage of the identity

(17) $$uv' - vu' = u^2 \left(\frac{v}{u}\right)'.$$

Example 1. Find the solution of $y'' - 2y' + y = e^x$ satisfying the initial conditions $y = 1$ and $y' = 0$ when $x = 0$.

Solution. We solved the homogeneous equation in §16D, Example 2, getting $u = e^x$, $v = xe^x$. Then $uv' - vu' = e^{2x}$. We are given $f(x) = e^x$; hence $vf = xe^{2x}$, $uf = e^{2x}$. We have

$$\alpha = (-1)\int x \, dx = -\tfrac{1}{2}x^2, \qquad \beta = (1)\int 1 \, dx = x.$$

Therefore $w = (-\tfrac{1}{2}x^2)(e^x) + (x)(xe^x) = \tfrac{1}{2}x^2 e^x$, and

$$y = e^x(A + Bx + \tfrac{1}{2}x^2).$$

Then

$$y' = e^x[A + B + (B + 1)x + \tfrac{1}{2}x^2]$$

and the initial conditions take the form

$$1 = A,$$

$$0 = A + B.$$

Hence $A = 1$, $B = -1$, and the solution is

$$y = e^x(1 - x + \tfrac{1}{2}x^2).$$

Example 2. Find the solution of $y'' + 4y' + 3y = x$ satisfying the initial conditions $y = 1$ and $y' = -2$ when $x = 0$.

Solution. We solved the homogeneous equation in §16D, Example 3, getting $u = e^{-x}$, $v = e^{-3x}$.

We may now continue with the formal procedure to find w, a particular solution of the given (nonhomogeneous) equation. But it is obvious that $w = Cx + D$ will work, for suitable constants C and D. We have $w' = C$, $w'' = 0$, and we easily find that $C = \tfrac{1}{3}$, $D = -\tfrac{4}{9}$, so that $w = \tfrac{1}{3}x - \tfrac{4}{9}$. Then

$$y = Ae^{-x} + Be^{-3x} + \tfrac{1}{3}x - \tfrac{4}{9}$$

and

$$y' = -Ae^{-x} - 3Be^{-3x} + \tfrac{1}{3}.$$

The initial conditions take the form

$$1 = A + B - \tfrac{4}{9},$$

$$-2 = -A - 3B + \tfrac{1}{3}.$$

Then $A = 1$, $B = \tfrac{4}{9}$, and we have, finally,

$$y = e^{-x} + \tfrac{4}{9}e^{-3x} + \tfrac{1}{3}x - \tfrac{4}{9}.$$

Example 3. Find the solution of $\tfrac{1}{2}y'' + y' + 2y = e^{-x}$ satisfying the initial conditions $y = 0$ and $y' = 3$ when $x = 0$.

Solution. We solved the homogeneous equation in §16D, Example 4, getting $u = e^{-x} \cos \sqrt{3}x$, $v = e^{-x} \sin \sqrt{3}x$.

It is obvious that $w = Ce^{-x}$, for a suitable constant C, is a solution of the given (nonhomogeneous) equation; but this time let us illustrate the general procedure. We have $v/u = \tan \sqrt{3}x$ and $uv' - vu' = u^2(v/u)' = \sqrt{3}e^{-2x}$. We are given $f(x) = e^{-x}$; hence $vf = e^{-2x} \sin \sqrt{3}x$, $uf = e^{-2x} \cos \sqrt{3}x$. Also, $a = \tfrac{1}{2}$. We have

$$\alpha = -\frac{2}{\sqrt{3}} \int \sin \sqrt{3}x \, dx = \frac{2}{3} \cos \sqrt{3}x, \quad \beta = \frac{2}{\sqrt{3}} \int \cos \sqrt{3}x \, dx = \frac{2}{3} \sin \sqrt{3}x.$$

Therefore $w = \alpha u + \beta v = \tfrac{2}{3}e^{-x}$, and

$$y = e^{-x}(A \cos \sqrt{3}x + B \sin \sqrt{3}x + \tfrac{2}{3}).$$

Then
$$y' = e^{-x}[-(\sqrt{3}A + B)\sin\sqrt{3}x + (-A + \sqrt{3}B)\cos\sqrt{3}x) - \tfrac{2}{3}]$$
and the initial conditions take the form
$$0 = A + \tfrac{2}{3},$$
$$3 = -A + \sqrt{3}B - \tfrac{2}{3}.$$
Then $A = -\tfrac{2}{3}$, $B = \sqrt{3}$, and we have, finally,
$$y = e^{-x}(-\tfrac{2}{3}\cos\sqrt{3}x + \sqrt{3}\sin\sqrt{3}x + \tfrac{2}{3}).$$

Problems (§16E)

1. Find the general solution of the following equations. (The corresponding homogeneous equations appeared in §16D, Problem 1. The solutions will be used in Problem 2 following.)

In each case, check that w is a solution of the given (nonhomogeneous) equation.

(a) $y'' - 4y = e^{2x}$

(b) $4y'' - y = x + e^x$

(c) $y'' + 4y = \sec 2x$

(d) $4y'' + y = \tan \tfrac{1}{2}x$

(e) $y'' - 4y' = e^x$

(f) $y'' + 4y' = e^x + e^{-x}$

2. In Problem 1, in each case, find the particular solution satisfying the following initial conditions.

(a) $y = \tfrac{1}{8}$ and $y' = 0$ when $x = 0$

(b) $y = 3$ and $y' = 0$ when $x = 0$

(c) $y = 2$ and $y' = 2$ when $x = 0$

(d) $y = 1$ and $y' = 0$ when $x = 0$

(e) $y = 2$ and $y' = 1$ when $x = 0$

(f) $y = \tfrac{1}{6}$ and $y' = \tfrac{2}{3}$ when $x = \log 2$

3. Find the general solution of the following equations. (The corresponding homogeneous equations appeared in §16D, Problem 3. The solutions will be used in Problem 4 following.)

In each case, check that w is a solution of the given (nonhomogeneous) equation.

(a) $y'' - y' - 2y = e^x \sin x$

(b) $y'' - y' + 2y = e^{x/2}$

(c) $y'' + 2y' + y = e^{-x}$

(d) $y'' + 2y' + 2y = e^{-x}$

(e) $y'' + 3y' + 2y = 1 - e^{-x}$

(f) $2y'' + 3y' + y = x + e^x$

4. In Problem 3, in each case, find the particular solution satisfying the following initial conditions.

(a) $y = -1$ and $y' = \tfrac{4}{5}$ when $x = 0$

(b) $y = 0$ and $y' = 0$ when $x = 0$

(c) $y = 0$ and $y' = 0$ when $x = 1$

(d) $y = 0$ and $y' = -e^{-\pi/2}$ when $x = \pi/2$

(e) $y = \tfrac{1}{2}$ and $y' = 0$ when $x = \log\tfrac{1}{2}$

(f) $y = -2$ and $y' = 0$ when $x = 0$

Answers to problems (§16E)

1. (a) $y = e^{2x}(A - \frac{1}{16} + \frac{1}{4}x) + Be^{-2x}$
   (b) $y = Ae^{x/2} + Be^{-x/2} - x + \frac{1}{3}e^x$
   (c) $y = A \cos 2x + B \sin 2x + \frac{1}{2}x \sin 2x + \frac{1}{4} \cos 2x \log \cos 2x$
   (d) $y = A \cos \frac{1}{2}x + B \sin \frac{1}{2}x - \cos \frac{1}{2}x \log(\sec \frac{1}{2}x + \tan \frac{1}{2}x)$
   (e) $y = Ae^{4x} + B - \frac{1}{5}e^x$ (f) $y = A + Be^{-4x} + \frac{1}{5}e^x - \frac{1}{3}e^{-x}$

2. (a) $y = \frac{1}{4}xe^{2x} + \frac{1}{8}e^{-2x}$ (b) $y = 2e^{x/2} + \frac{2}{3}e^{-x/2} - x + \frac{1}{3}e^x$
   (c) $y = 2 \cos 2x + \sin 2x + \frac{1}{2}x \sin 2x + \frac{1}{4} \cos 2x \log \cos 2x$
   (d) $y = \cos \frac{1}{2}x + \sin \frac{1}{2}x - \cos \frac{1}{2}x \log(\sec \frac{1}{2}x + \tan \frac{1}{2}x)$
   (e) $y = \frac{1}{3}e^{4x} + 2 - \frac{1}{3}e^x$ (f) $y = -\frac{1}{24} - \frac{2}{5}e^{-4x} + \frac{1}{5}e^x - \frac{1}{3}e^{-x}$

3. (a) $y = Ae^{2x} + Be^{-x} - \frac{1}{10}e^x(3 \sin x + \cos x)$
   (b) $y = e^{x/2}(A \cos \frac{1}{2}\sqrt{7}x + B \sin \frac{1}{2}\sqrt{7}x + \frac{4}{7})$ (c) $y = e^{-x}(A + Bx + \frac{1}{2}x^2)$
   (d) $y = e^{-x}(A \cos x + B \sin x + 1)$
   (e) $y = e^{-x}(A - 1 + x) + Be^{-2x} + \frac{1}{2}$
   (f) $y = Ae^{-x/2} + Be^{-x} + \frac{1}{6}e^x + x - 3$

4. (a) $y = \frac{1}{10}e^{2x} - e^{-x} - \frac{1}{10}e^x(3 \sin x + \cos x)$
   (b) $y = \frac{4}{7}e^{x/2}(1 - \cos \frac{1}{2}\sqrt{7}x)$ (c) $y = e^{-x}(\frac{1}{2} - x + \frac{1}{2}x^2)$
   (d) $y = e^{-x}(\cos x - \sin x + 1)$
   (e) $y = e^{-x}(x - 1 + \log 2) + \frac{1}{2}e^{-2x} + \frac{1}{2}$
   (f) $y = \frac{1}{6}e^{-x/2} + \frac{2}{3}e^{-x} + \frac{1}{6}e^x + x - 3$

Index

For additional information, see the individual chapter tables of contents.

669

convergence of series, 593, 598
  absolute, 615
  conditional, 617
  interval of, 620
  radius of, 619, 620, 625
  tests for: *see* convergence tests
  uniform, 622
convergence tests:
  alternating series test, 616
  comparison test, 598
  condensation test, 614
  integral test, 611
  ratio-comparison test, 602
  ratio test, 600, 615, 620
  root test, 623
coordinates, 3, 10, 428
  cylindrical, 583
  polar, 404, 571, 583
cos, cosine, 299f, 638
cosecant: *see* csc
cosh, 631, 632, 633
cot, cotangent, 311
coth, 633
critical point, 140
  just one, 143
cross, 468
csc, 311
csch, 633
curvature, 389f, 402, 460f
curve, 364, 453
cycloid, 366, 367
cylinder, 224, 474, 475
  area and volume of, 130, 224, 244

Δ, 522
Δ-notation, 88, 489
Darboux integral, 215
Darboux's Theorem, 212, 538
decreasing function, 40, 98, 156
decreasing sequence, 589
density, 254, 566
derivative, 95, 101; *see also* slope, tangent, velocity
  directional, 497f, 515
  of function defined parametrically, 378
  of implicit function, 126, 526f
  of inverse function, 173
  mixed partial, 514, 550
  partial, 485
  second and higher, 134, 514f
  sign of, 98, 156
  of vector-valued function, 371, 454
difference quotient, 90
differentiability, 104

differentiable function, 95, 104
  behavior of, 139f
  inverse of, 173
  vector-valued, 373, 454
differential, 491f, 644
differential calculus, 95
differential equation, 288, 643
  exact, 648
  of first order, 644
  general solution of, 646, 650, 660, 665
  homogeneous, 654, 657
  linear, 655, 657
  nonhomogeneous, 663
  order of, 644
  ordinary, 644
  partial, 643
  of second order, 644, 657
  separable, 646
  uniqueness of solution of, 659
differentiation, 95
  implicit, 126
  logarithmic, 296
  of power series, 625–626
differentiation formulas:
  algebraic, 105f, 111f, 124
  the chain rule, 119, 374, 454, 502f
  exponential, 287, 294
  inverse trigonometric, 334f
  logarithmic, 274, 293
  trigonometric, 301, 309f
  vector, 373f, 454
direction angles, 434
direction cosines, 434
direction numbers, 437
direction vectors, 357, 437, 442
directrix, 19
disc: *see* circle
disc method, 226
distance, 10, 408, 429
  directed, 3
  to a line, 29, 31
divergence: *see* convergence
dog and rabbit, 646, 651
domain of a function, 37f, 468
dot product: *see* scalar product
double integral, 537
  in polar coordinates, 571

e, 280, 294, 589, 630
e^x, 285, 608, 630, 634
ellipse, ellipsoid, 229, 368, 477
equality of the mixed partials, 514, 550
equation of a curve, 10
Euler's formula, 509